Rawat

Handbook of Environmental Isotope Geochemistry

VOLUME 2

The Terrestrial Environment, B

HANDBOOK OF ENVIRONMENTAL ISOTOPE GEOCHEMISTRY

P. FRITZ and J.Ch. FONTES (Editors)

1. THE TERRESTRIAL ENVIRONMENT, A
2. THE TERRESTRIAL ENVIRONMENT, B
3. THE MARINE ENVIRONMENT, A
4. THE MARINE ENVIRONMENT, B
5. THE HIGH TEMPERATURE ENVIRONMENT

Handbook of Environmental Isotope Geochemistry

Edited by

P. FRITZ

*Department of Earth Science, University of Waterloo,
Waterloo, Ontario, Canada*

and

J.Ch. FONTES

*Laboratoire d'Hydrologie et de Géochimie Isotopique, Université
de Paris-Sud, Orsay, France*

VOLUME **2**

The Terrestrial Environment, B

ELSEVIER
Amsterdam — Oxford — New York — Tokyo 1986

ELSEVIER SCIENCE PUBLISHERS B.V.
Sara Burgerhartstraat 25
P.O. Box 211, 1000 AE Amsterdam, The Netherlands

Distributors for the United States and Canada:

ELSEVIER SCIENCE PUBLISHING COMPANY INC.
52, Vanderbilt Avenue
New York, NY 10017, U.S.A.

Library of Congress Cataloging in Publication Data
(Revised for volume 2B)
Main entry under title:

Handbook of environmental isotope geochemistry.

   Includes bibliographies and indexes.
   Contents: v. 1. The terrestrial environment, A. --
v. 2. The terrestrial environment, B.
   1. Isotope geology. 2. Environmental chemistry.
3. Geochemistry. I. Fritz, P. (Peter), 1937-   .
II. Fontes, J. Ch. (Jean-Charles), 1936-   .
QE501.4.N9H36   551.9   79-21332
ISBN 0-444-41781-8 (set)

ISBN 0-444-42225-0 (Vol. 2)
ISBN 0-444-41781-8 (Series)

© Elsevier Science Publishers B.V., 1986

All rights reserved. No part of this publication may be reproduced, stored in a retrieval system or transmitted in any form or by any means, electronic, mechanical, photocopying, recording or otherwise, without the prior written permission of the publisher, Elsevier Science Publishers B.V./Science & Technology Division, P.O. Box 330, 1000 AH Amsterdam, The Netherlands.

Special regulations for readers in the USA — This publication has been registered with the Copyright Clearance Center Inc. (CCC), Salem, Massachussetts. Information can be obtained from the CCC about conditions under which photocopies of parts of this publication may be made in the USA. All other copyright questions, including photocopying outside of the USA, should be referred to the publisher.

Printed in The Netherlands

PREFACE

This volume of our series deals again with "The Terrestrial Environment". It includes topics not covered in Volume 1 (The Terrestrial Environment A) and expands on selected areas where a more complete coverage was felt to be desirable. One will still find fields which are not treated, as for example an extensive discussion on radiocarbon analyses on Upper Quaternary materials and, furthermore, it may be justified to dedicate in future specific volumes to the review of specialized subject areas in which environmental isotope analyses revealed themselves as particularly useful tools. This may be the case for topics such as the terrestrial water cycle, paleoenvironmental studies, the geochemistry of contaminants and others.

For this volume we obtained again the collaboration of well known scientists whose availability to write review papers tends to be inverse proportional to their degree of expertise. This is special reason to acknowledge their efforts, contributions which, we believe, will also be appreciated by the readers.

Despite repeated editorial checks and the attention of the publisher, errors may still persist. We would like to point out that, in Volume 1, page 13, the correct values are:

$$\delta^2 H_{SLAP/V\text{-}SMOW} = -428\text{\textperthousand} \quad \text{and} \quad \delta^2 H_{NBS\text{-}1/SMOW} = -47.6\text{\textperthousand}.$$

J.Ch. Fontes and P. Fritz
Orsay and Waterloo, December 1985

LIST OF CONTRIBUTORS

R.S. BARNES	*Laboratory of Radiation Ecology, College of Fisheries, University of Washington, Seattle, Washington, U.S.A.*
H.W. BENTLEY	*Department of Hydrology and Water Resources, College of Earth Sciences, University of Arizona, Tucson, Arizona, U.S.A.*
W.W. CAROTHERS	*Water Resources Department, U.S. Geological Survey, Menlo Park, California, U.S.A.*
S.N. DAVIS	*Department of Hydrology and Water Resources, University of Arizona, Tucson, Arizona, U.S.A.*
T. FLORKOWSKI	*International Atomic Energy Agency, Vienna, Austria*
R. GONFIANTINI	*Isotope Hydrology Section, International Atomic Energy Agency, Vienna, Austria*
G. HILLAIRE-MARCEL	*Department of Geology, University of Quebec at Montreal, Montreal, Quebec, Canada*
H. HÜBNER	*Zentral Institut für Isotopenforschung, Akademie der Wissenschaften der DDR, Leipzig, G.D.R.*
J. JOUZEL	*Département de Physico-Chimie, CEN/SACLAY, Gif/Yvette, France*
Y.K. KHARAKA	*Water Resources Department, U.S. Geological Survey, Menlo Park, California, U.S.A.*
W.G. MOOK	*Isotope Physics Laboratory, University of Groningen, Groningen, The Netherlands*
F.M. PHILLIPS	*Geoscience Department and Geophysical Research Center, New Mexico Institute of Mining and Technology, Socorro, New Mexico, U.S.A.*
K. RÓŻANSKI	*Institute of Physics and Nuclear Techniques, University of Mining and Metallurgy, Cracow, Poland*
W. SALOMONS	*Delft Hydraulics Laboratory, Haren Branch, Haren, The Netherlands*

VIII

W.R. SCHELL   *Department of Radiation Health, University of Pittsburgh, Pittsburgh, Pennsylvania, U.S.A.*

H.P. SCHWARCZ   *Department of Geology, McMaster University, Hamilton, Ontario, Canada*

B. TURI   *Institute of Geochemistry, University of Rome, Rome, Italy*

A. ZUBER   *Institute of Physics and Nuclear Techniques, University of Mining and Metallurgy, Cracow, Poland*

CONTENTS

Preface V
List of Contributors... VII

CHAPTER 1. MATHEMATICAL MODELS FOR THE INTERPRETATION OF ENVIRONMENTAL RADIOISOTOPES IN GROUNDWATER SYSTEMS.. 1
    A. Zuber
Introduction... 1
Principles of the tracer method 3
Some basic concepts and the mathematical tool 8
Models and their hydrological significance 13
Dispersivity and the dispersion parameter........................... 22
Infiltration studies.. 24
Stagnant water in the saturated zone................................ 27
The movement of delayed tracers..................................... 29
Cases of constant tracer input 31
Variable tracer input 33
Combined interpretation... 48
Concluding remarks.. 54
References.. 55

CHAPTER 2. ISOTOPES IN CLOUD PHYSICS: MULTIPHASE AND MULTISTAGE CONDENSATION PROCESSES 61
    J. Jouzel
Introduction.. 61
Basic principles.. 63
Isotopic effects during growth of individual elements............... 66
Isotopic cloud models 70
Sampling of water vapour and condensed phases in the atmosphere..... 85
A brief survey of isotopic studies in rain and snow................. 87
The study of hailstone growth mechanisms by means of isotopic analyses 89
Conclusion.. 104
References.. 105

CHAPTER 3. ENVIRONMENTAL ISOTOPES IN LAKE STUDIES........... 113
    R. Gonfiantini
Introduction.. 113
Isotopic fractionations during evaporation of water 114
Water balance studies of lakes 130
Water mixing studies 146
Miscellaneous studies... 161
References.. 163

CHAPTER 4. ENVIRONMENTAL ISOTOPE AND ANTHROPOGENIC TRACERS OF RECENT LAKE SEDIMENTATION 169
W.R. Schell and R.S. Barnes

Introduction... 169
Biogeochemical cycling in lakes 169
Environmental isotope and element tracers............................. 171
Applications... 185
Summary and conclusions.. 201
References.. 202

CHAPTER 5. STABLE ISOTOPE GEOCHEMISTRY OF TRAVERTINES 207
B. Turi

Introduction... 207
Mineralogy, petrology and physical properties of travertines 208
Geochemical conditions controlling travertine deposition 209
$^{13}C/^{12}C$ and $^{18}O/^{16}O$ variations in travertines........................... 211
Isotopic fractionations in travertine formation 215
Diagnetic effects.. 222
Radiocarbon dating of travertines....................................... 225
Sulphur, strontium and lead isotopes.................................... 225
Summary and conclusions.. 233
References.. 235

CHAPTER 6. ISOTOPE GEOCHEMISTRY OF CARBONATES IN THE WEATHERING ZONE 239
W. Salomons and W.G. Mook

Introduction... 239
Isotopic composition of carbon and oxygen sources in the weathering zone 241
The formation of dissolved inorganic carbon............................ 246
Reprecipitation processes in the weathering zone....................... 251
Isotopic composition of carbonates in the weathering zone 256
Summary and conclusions.. 265
References.. 265

CHAPTER 7. GEOCHRONOLOGY AND ISOTOPIC GEOCHEMISTRY OF SPELEOTHEMS 271
H.P. Schwarcz

The cave environment 271
Geochronoly of speleothems: methods 275
Applications of geochronology of speleothems 282
Stable isotope geochemistry of speleothems 286
Conclusions 299
References.. 300

CHAPTER 8. OXYGEN AND HYDROGEN ISOTOPE GEOCHEMISTRY OF DEEP BASIN BRINES.. 305
Y.K. Kharaka and W.W. Carothers

Introduction... 305
Terms applied to subsurface waters..................................... 306
Isotopic composition of surface waters 307
Origin of water in deep basin brines 314
Modification of stable isotopes in deep basin waters..................... 332

Other applications of stable isotopes of water.................................. 349
Summary.. 352
References.. 353

CHAPTER 9. ISOTOPE EFFECTS OF NITROGEN IN THE SOIL AND BIOSPHERE 361
H. Hübner

Introduction.. 361
^{15}N in the pedosphere 365
Isotope input in the pedosphere.. 369
Interpretation of isotopic variations of NO_3-N in the pedosphere, groundwater, and surface waters.. 405
Outlook... 418
References.. 419

CHAPTER 10. CHLORINE-36 IN THE TERRESTRIAL ENVIRONMENT 427
H.W. Bentley, F.M. Phillips and S.N. Davis

Introduction.. 427
Natural production of ^{36}Cl 428
Studies of ^{36}Cl in the lithosphere.. 434
Dating saline sediments with ^{36}Cl.. 440
Dating old groundwater with ^{36}Cl.. 441
Atmospheric nuclear weapons testing ^{36}Cl as a hydrologic tracer 460
References.. 475

CHAPTER 11. RADIOACTIVE NOBLE GASES IN THE TERRESTRIAL ENVIRONMENT.. 481
T. Florkowski and K. Różanski

Introduction.. 481
^{85}Kr... 482
^{81}Kr... 493
^{39}Ar and ^{37}Ar 494
^{133}Xe... 501
References.. 502

CHAPTER 12. ISOTOPES AND FOOD 507
C. Hillaire-Marcel

Introduction.. 507
Carbon and plants... 507
^{18}O and ^{2}H in plants 525
The radioactive isotopes ^{3}H and ^{14}C 537
Other possible applications 539
Conclusions.. 543
References.. 544

Subject Index.. 549
Index of geographical names 555

Chapter 1

MATHEMATICAL MODELS FOR THE INTERPRETATION OF ENVIRONMENTAL RADIOISOTOPES IN GROUNDWATER SYSTEMS

A. ZUBER

INTRODUCTION

The concepts of mathematical models discussed in this chapter are well founded on a number of original studies. However for a consistent presentation it appeared necessary to reexamine some basic concepts related to the tracer method.

Special attention has been paid to the problems encountered in hydrogeology and other related fields. An effort to achieve a consistent and practical presentation of these basic concepts is given in introductory sections. The next sections contain some ideas which cannot be treated as generally recognized. In the section "Dispersivity and the dispersion parameters" the author tried to explain in a consistent manner the values of the dispersion constant which appear in experiments performed in different scales and conditions. Some basic problems encountered in the application of mathematical models to the interpretation of tracer experiments in the unsaturated zone are discussed in the section "Infiltration studies". The movement of delayed tracers in the saturated zone is treated in the sections "Stagnant water in the saturated zone" and "The movement of delayed tracers".

The environmental tracer methods provoked in the past discussions on their applicability in hydrology. Increasing number of publications on both the development of principles of the methods and examples of practical applications prove that these methods found a permanent place in hydrology as a complementary tool to other well established methods, and in special cases even as a basic tool supplying information unobtainable by other methods. Nir and Lewis (1975) gave the following classification of the conceptual contribution of tracer data: (a) determination of specific hydrologic parameters, (b) confirmation or rejection of concurrent hypotheses or models, (c) extension or reconstruction of time series of data for use as input in system analysis, (d) strengthening the link between physical hydrology and system analysis, and (e) supply of preliminary description of regional hydrological systems lacking basic hydrologic data. The contribution of the tracer methods is usually in categories (a), (b), and (e). It seems that to this classification we should add: (f) supply of information on the origin of water. This last point appears to be particularly useful in mining hydrology, in the management of wells exploiting mineral waters, and for some problems related to the prediction of pollution hazards. In categories (b), (e), and (f) the interpretation is very often based on a qualitative approach. This approach means that it is sufficient to determine if a given radioisotope is present or not at a given sampling site, or if its concentration differs from that found at other sampling points, or if this concentration is variable or not in time. However, the experimenter is seldom satisfied with such an approach and usually tries to develop a mathematical model leading to a quantitative interpretation. The quantitative approach consists in solving the inverse problem, i.e., from the set of experimental data the parameters of the model are determined. Even when the existing experimental data are not sufficient to solve the inverse problem, the use of models still gives a better understanding of the investigated system than the purely qualitative approach.

It would be difficult to discuss here all the models developed so far. Therefore we shall concentrate our attention on the most common and most applicable approach which is based on the so-called lumped-parameter, or black-box models. Such models are well known in hydrology; however, their development in relation to the tracer method took place mainly in chemical engineering.

In a lumped-parameter model or a black-box model, spatial variations are ignored and the various properties and the state of the system can be considered to be homogeneous throughout the entire system (Himmelblau and Bischoff, 1968). If the spatial variations of parameters in a given system were known there would probably be no need for the radiotracer methods. In other words, the tracer methods may be particularly useful in investigating systems with unknown distribution of parameters, i.e., systems with the lack of detailed conventional hydrological observations or systems where the conventional methods do not yield satisfactory results (e.g., in karst formations or fractured rocks).

It has to be stressed once more that the discussion is limited to models which allow to solve the inverse problem at the present stage of the tracer method. There are many other models developed for a better understanding of the involved phenomena. For instance, Geyh and Backhaus (1979) tried to estimate the influence of diffusion and leaking into a confined aquifer from an unconfined one through a semipermeable aquitard on the age determination by the ^{14}C method. An opposite effect may be caused by diffusive losses of the tracer into fine-grained aquitard (Sudicky and Frind, 1981) or from a fractured aquifer into a porous rock matrix (Glueckauf, 1980, 1981; Grisak and Pickens, 1980, 1981; Grisak et al., 1980; Neretnieks, 1980; Tang et al., 1981).

The paper does not deal either with the multi-box models, which have been applied recently by many authors. These models should rather be treated as the distributed-parameter models because by variation of the number of boxes, their volumes, inflow ratios, and the interconnections between the boxes, it is possible to model, to a certain degree, the geometry of the aquifer and the spatial variations of the parameters. Such models may be useful in some cases. However, in general, they should not be recommended for solving the inverse problem, because a larger number of fitting parameters makes the obtained solutions doubtful. More detailed critical discussion of the multi-box models is given in this respect by Małoszewski and Zuber (1982).

PRINCIPLES OF THE TRACER METHOD

Before going into detailed discussion on the models it is necessary to consider the principles of the tracer method and some basic concepts. The considerations of this section are not only applicable to our main subject but they are also of importance to any tracer experiments or solute transport studies in hydrology, soil physics, and in some other fields. It will not be possible to give here a full description of the involved phenomena because it is beyond the scope of our work. However, we shall try to outline the problems which are involved in the tracer method. The interested reader, who would like to deepen or extend his knowledge, will be referred to original works.

The author's experience is that during all the scientific meetings on the tracer methods in hydrology the most lively discussions are on the tracer behaviour and they take place in private. In written form they do not appear as the discussed problems are either ignored or considered self-evident. These difficulties of the tracer method are not only inherent in hydrology. Gardner and Ely (1967) stated that: "Despite the extensive use of the tracer method, there have been few efforts to define it or clearly enunciate its principles. Perhaps the two main reasons for this lack are that the definition and principles are thought to be self-evident and that it is difficult to arrive at a general definition and a statement of the principles involved that are both meaningful and exact". The definition accepted by these authors is also acceptable for our purposes. It reads: The tracer method is a technique for obtaining information about a system or some part of a system by observing the behaviour of a specific substance, the tracer, that has been added to the system. In the case of environmental tracers they are added (injected) to the system by natural processes, whereas their production is either natural or results from the global activity of man.

The above definition needs no special comments or discussion. However, we shall see below that several "self-evident" terms adherent to the tracer method have to be carefully defined or redefined if we want to clarify some controversial ideas which appear in literature concerning both the tracer method in general and the tracer methods in hydrology in particular.

There have been several attempts to define an ideal tracer. The concept of an ideal tracer is useful in selecting artificial tracers, in searching for environmental tracers, and first of all in the development of mathematical models. The most commonly applied definition states that an ideal tracer must behave exactly like the traced material and must have one property that distinguishes it from the traced material, so that it can be easily detected (Gardner and Ely, 1967). This definition is not quite satisfactory as in some cases it leads to misunderstandings. Imagine, for instance, tritiated water ($^1H^3HO$) as a tracer in dynamic water systems where the transport of water is of interest. It is well known (e.g., see Kaufman and Orlob, 1956a) that in systems with bound water in the solid matrix (e.g., in organic soils or clay minerals), the movement of this tracer is delayed with respect to the movement of the water flux. This delay is caused by the exchange of the traced molecules with the molecules of the bound water. The tracer behaves exactly as the traced material because H_2O particles also undergo exchange. However, each tracer particle which temporarily disappears from the water flux is replaced by another untraced particle, thus the water flux remains unchanged.

Two other instructive examples concerning the behaviour of "an ideal tracer" as defined above were given by Kaufman and Orlob (1956b) and Małoszewski et al. (1980). Both examples are related to the movement of calcium ions traced with the radioactive $^{45}Ca^{2+}$, although they were performed under different experimental conditions. In both cases, in spite of the ionic exchange with the solid phase, the front of the increased Ca^{2+} concentration moved more or less with the same velocity as the water moved through the experimental columns. This was caused by the fact that Ca^{2+} cations exchanged mainly with Ca in the solid matrix. Thus each Ca^{2+} cation disappearing from the solute was at the same time replaced by another Ca^{2+} cation entering the solute. In such a case, the observer measuring the movement of Ca^{2+} cations in the flowing water is not able to recognize if the exchange process takes place or not. However, the movement of $^{45}Ca^{2+}$ cations was considerably delayed because they exchanged with inactive particles in the matrix. Thus, the active Ca^{2+} cations were ideal for tracing the behaviour of the calcium in the system, but not for determining the mass transport of the traced material.

Similarly, in systems with bound water, many other substances which do not behave exactly as the traced material (water molecules) will represent the water flux (the volumetric flow rate through the system) much better than the tritiated water. On the other hand, the tritiated water which is ideal for groundwater systems without bound water, or ^{85}Kr which as a chemically inert element may be ideal for confined aquifers, in surface waters will have a tendency to be in equilibrium with the atmospheric concentrations. A further discussion of the gaseous tracers is given on p. 52, where examples of the influence of their diffusion through unsaturated zone are considered.

At this stage we start to feel that some of the difficulties result from the fact that the hydrologist is usually interested in the flux of water, i.e., in the volumetric flow rate, or velocity, or in other related parameters. However, tracers exist in atomic form (e.g., ^{85}Kr), in ionic form (e.g., Cl$^-$, H^{14}CO$_3^-$), in molecular form (e.g., tritiated water or ^{14}CO$_2$), and in grain form (e.g., sand grains or pebbles artificially traced with a radioisotope or dye in sediment movement studies), but they do not exist in the required flux form.

In artificial groundwater tracing the required properties of an ideal tracer are usually listed (e.g., see Kaufman and Orlob, 1956a, or Davis and De Wiest, 1967). When summarized these properties lead to the following definition: an ideal tracer is an easily detected substance that will correctly describe the velocity variations of the traced liquid without in any way modifying the transmission characteristics of the system. Nir and Lewis (1975) in their search for a general definition defined and ideal tracer as a substance which has the same response function (the transit time distribution) as the traced material. In later sections it will be shown that neither of these definitions is sufficiently precise because the transit time distribution of the tracer depends on the measuring conditions. A modified definition of Nir and Lewis is given in the next section.

In order to obtain a general definition of an ideal tracer it is proposed here to modify somewhat the first definition. The modified definition reads: an ideal tracer is a substance that behaves in the system exactly as the traced material as far as the sought parameters are concerned, and which has one property that distinguishes it from the traced material. This definition means that for an ideal tracer there should be neither sources nor sinks in the system other than those adherent to the sought parameters. In practice we shall treat as a good tracer even a substance which has other sources or sinks if they can be properly accounted for, or if their influence is negligible within the measurement accuracy.

Before we proceed to our main subject it is necessary to define some other indispensable terms. Concentration is one of the most "self-evident" terms which has to be revised for dynamic systems. The resident concentration (C_R) expresses the mass of solute (Δm) per unit volume of fluid (ΔV) contained in a given element of the system at a given instant, t:

$$C_R(t) = \Delta m(t)/\Delta V \tag{1}$$

The flux concentration (C_F) expresses the ratio of the solute flux ($\Delta m/\Delta t$) to the volumetric fluid flux ($Q = \Delta V/\Delta t$) passing through a given cross-section:

$$C_F(t) = \frac{\Delta m(t)/\Delta t}{\Delta V/\Delta t} = \frac{\Delta m(t)}{Q \Delta t} \tag{2}$$

These two definitions are taken from Kreft and Zuber (1978), but we shall also use elegant and convenient to our purposes definitions introduced by Sa-

NOTATION

C_R	resident concentration of tracer defined by equation (1), expressed as mass or activity per ml, or pmc in TDC, or dpm/mmol Kr
C_F	flux concentration of tracer defined by equation (2), expressed as above
$C_{R,m}$	resident concentration in the mobile water, expressed as above
C_s	concentration of tracer in the stagnant water, expressed as above
C_{FR}	resident concentration resulting from flux injection
C_{FF}	flux concentration resulting from flux injection
C_I	concentration resulting from an unspecified impulse injection
C_{IFR}	C_{FR} concentration resulting from an instantaneous injection
C_{IFF}	C_{FF} concentration resulting from an instantaneous injection
C_i	mean tritium concentration in ith month, in TU
dpm	disintegrations per minute
D	dispersion coefficient, m^2 s^{-1}
D_m	coefficient of molecular diffusion, m^2 s^{-1}
D/v	dispersion constant characteristic for a given aquifer, m
D/vx	dispersion parameter (reciprocal of the Peclet number) characteristic for a given system, dimensionless
DM	dispersion model in general
DM–C_{FF}	dispersion model, case C_{FF}, i.e. the measurement is weighed by flow rates at the output
DM–C_{FR}	dispersion model, case C_{FR}, i.e. the measurement is averaged over the cross-section at the measuring site
$E(t)$	exit time distribution of water which entered the system at $t = 0$, here in year^{-1}
EM	exponential model
EPM	combined exponential-piston flow model
erfc(z)	$= 1 - \mathrm{erf}(z)$ where erf(z) is the tabulated error function
FSM	finite state mixing cell models
$g(t)$	system response function describing the exit time distribution of a tracer which entered the system at $t = 0$, here in year^{-1}
H	average height of water in the system, m
H_m	average height of mobile water in the system, m
k_d	distribution constant, ml g^{-1}
LM	linear model
LPM	combined linear-piston flow model
M	mass or activity of the tracer
n	fraction of space occupied by water in the system
n_m	fraction of space occupied by mobile water in the system
n_s	fraction of space occupied by stagnant water in the system
pmc	percent of modern carbon
PFM	piston flow model
P_i	amount of precipitation in ith month, m month^{-1}
Q	volumetric flow rate through the system, here in m^3 year^{-1}
q	concentration of tracer in the solid phase, mass or activity per gram
R	recharge rate, m year^{-1}
S	cross-section area of the system normal to flow lines
T	turnover time defined either as V_m/Q (equation (3a)), or V/Q (equation (22)), here in years
TDC	total dissolved carbon
t	time variable

NOTATION (continued)

t'	transit time variable
t_a	apparent time defined by equation (19)
t_e	time moment for which the tracer profile is measured in the infiltration studies
$\overline{t_t}$	mean transit time of tracer defined by equation (6)
$\overline{t_w}$	mean transit time of water equal to $T = V_m/Q$
x	length of the system measured along the flow lines from the recharge area to a given measuring point, or the space variable in equations (12), (17) and (18)
x_0	length of the recharge area measured along the flow lines
V	total volume of water in the system
V_m	volume of mobile water in the system
V_s	volume of stagnant water in the system
v	mean transit velocity of water in the system, $Q/n_m S$
v_a	apparent velocity of water defined by equation (20)
v_t	mean velocity of tracer, $v_t = x/\overline{t_t}$
v_f	Darcy velocity, $v_f = v n_m = v_a(n_m + n_s)$
α	ratio of the summer to winter infiltration coefficients
α_i	infiltration coefficient for ith month
β	fraction of old water component (i.e., without the radioisotope of interest)
$\delta(t)$	Dirac delta function defined for the time variable
η	ratio of the total volume of the system to the exponential flow volume or to the linear flow volume for the EPM or LPM, respectively
λ	radioactive decay constant, here in years^{-1}

fonov et al. (1979). Namely, C_R can be treated as a mean concentration obtained by weighting over a given cross-section of the system, whereas C_F is a mean concentration obtained by weighting by the volumetric flow rates through a given cross-section of the system. Examples of differences between C_R and C_F are given later in this chapter. The reader who would like to get acquainted with different aspects related to these two concepts of concentration is referred to works of Gardner et al. (1973) who consider flow in capillaries with negligible molecular diffusion, to Levenspiel and Turner (1970), and Levenspiel et al. (1970) who consider parallel systems, to a classic work of Brigham (1974) who considers the dead-end pore model, and to Kreft and Zuber (1978, 1979) who discuss in detail the dispersion model. Similar concepts are also well known in two-phase flow in pipes where usually differences between the so-called in situ and flowing concentrations result from the differences between the phase velocities.

It appears that the differences in observed concentrations result not only from the detection mode, i.e., not only from the measurement of C_R or C_F, but it also depends on how the tracer is injected into the system. In this chapter we shall limit our considerations to cases in which the tracer is injected proportionally to the volumetric flow rates, thus we shall deal with two cases which can be denoted as C_{FF} and C_{FR}, where the first subscript refers to the injection mode and the second one to the detection mode. A wider discussion of the injection-detection modes and the related transit time distributions of the tracer can be found in Gardner et al. (1973) and Kreft and Zuber (1978).

Now recalling the earlier remark about the injection of a tracer one realizes that definitions of an ideal injection and detection are also necessary. It is therefore proposed to define an ideal injection and an ideal detection as those which exactly correspond to the initial and boundary conditions of the applied model. In tracer experiments the model remains often unspecified and then an ideal injection and an ideal detection will be represented by the injection and detection corresponding to the case C_{FF} related to the whole system or to the investigated part of the system. It will be shown in the next section that these definitions lead to experiments in which the tracer represents the flux of the traced material.

It becomes clear that samples of the type C_F should be taken at the outflow(s) or from abstraction wells penetrating either the whole depth or the investigated part of the system. In the case of the abstraction well the samples are averaged (weighted) by flow rates, but we have to remember that even for a fully penetrating well only a part of the aquifer is observed. Samples of the type C_R are difficult to collect in environmental systems. In the unsaturated zone local samples of the type C_R can be extracted from the soil samples. In boreholes penetrating the saturated zone if samples are taken from consequent depths they will be locally of the type C_F. When properly weighted over the penetrated depths they will yield a value which may be treated as the mean C_R for the whole depth.

Individual local samples, either of the type C_F or C_R, which do not represent the whole system, or its given part, cannot be properly interpreted with the aid of lumped-parameter models. This statement may be generalized as follows: in any tracer experiment if the tracer is arbitrarily injected or measured the obtained results are doubtful.

SOME BASIC CONCEPTS AND THE MATHEMATICAL TOOL

In this work we shall be dealing with systems being hydrologically in a steady state. The question to which systems this approach is applicable will be discussed in the final section.

The turnover time (T) of a system is defined here as:

$$T = V_m/Q \tag{3a}$$

where Q is the volumetric flow rate through the system, and V_m is the volume of mobile water in the system. In some cases the turnover time may also be expressed as:

$$T = H_m/R \tag{3b}$$

where H_m is the mean height of the mobile water and R is the recharge rate expressed in units of water height per time unit. In a homogeneous system another relation may also be used, namely:

$$T = \frac{V_m}{Q} = \frac{Sn_m x}{Sn_m v} = \frac{x}{v} \tag{3c}$$

where x is the length of the system measured along the streamlines, v is the mean transit velocity of water (defined as $v = Q/n_m S$), n_m is the space fraction occupied by the mobile water, and S is the cross-section area normal to flow.

The exit age-distribution function, or the transit time distribution, $E(t)$, describes the exit time distribution of incompressible fluid elements of the system (water) which entered the system at a given $t = 0$. This function is normalized in such a way that:

$$\int_0^\infty E(t)\,dt = 1 \tag{4}$$

The mean age of water $(\overline{t_w})$ leaving the system, or the mean transit time, is according to the definition of the $E(t)$ given as:

$$\overline{t_w} = \int_0^\infty t\,E(t)\,dt = T, \tag{5}$$

which means that the mean age of water leaving the system is always equal to the turnover time.

Now, having defined terms related to the system we shall consider definitions and terms related to the behaviour of the tracer, which is supposed to supply information about the system. The mean transit time $(\overline{t_t})$ of a tracer is defined as:

$$\overline{t_t} = \frac{\int_0^\infty t C_I(t)\,dt}{\int_0^\infty C_I(t)\,dt} \tag{6}$$

where $C_I(t)$ is the tracer concentration observed at the measuring point as the result of an instantaneous injection at the entrance to the system at $t = 0$. Definition of equation (6) is applicable to any injection-detection mode. As mentioned earlier, Kreft and Zuber (1978) pointed out that $\overline{t_t}$ is equal to $\overline{t_w}$ only if a conservative, non-delayed tracer is both injected and measured in the flux, i.e., when in equation (6) $C_{IFF}(t)$ for an ideal tracer appears instead of the unspecified $C_I(t)$. When a radioisotope is used as a tracer the concentrations under the integrals in equation (6) have to be corrected for the decay by using $e^{\lambda t}$ factor, where λ is the radioactive decay constant. This factor converts the observed concentrations into those which would be observed for a non-decaying tracer.

The literature on the tracer method contains evident, and sometimes hidden, misconcepts and mistakes resulting from the lack of distinction between the injection-detection modes, which in turn result from the lack of distinction between the definitions of C_R and C_F. Some investigators, after finding that there are cases in which $\overline{t_w} \neq \overline{t_t}$, questioned the definition of the $E(t)$ function. Others came to a conclusion that the harmonic mean $(\overline{1/t_t})$ should be used instead of the arithmetic mean given by equation (6), and that the tracer in principle moves with a velocity different than that of the traced material. The problem is serious since many contributions to the theory and practice of the tracer method are not free of these and other related misconcepts.

An example confirming our statement that $\overline{t_t}$ differs from $\overline{t_w}$ if the tracer is not injected and measured in flux is discussed in the following section. A possible difference between these two quantities means that the tracer distribution does not necessarily follow the distribution of the traced mass. So, similarly to the $E(t)$ function describing the system, it is necessary to define a function describing the tracer distribution. The exit age-distribution function, or the transit time distribution of a conservative tracer describes the exit time distribution of particles of that tracer which entered the system at a given $t = 0$. This function, called sometimes the system response function, or more often the weighting function, $g(t)$, may be expressed as:

$$g(t) = \frac{C_I(t)}{\int_0^\infty C_I(t)\,dt} \qquad (7a)$$

or:

$$g(t) = \frac{C_I(t)\,Q}{M} \qquad (7b)$$

because the whole injected mass or activity (M) of the tracer has to appear at the outlet, i.e.:

$$M = Q \int_0^\infty C_I(t)\,dt \qquad (8)$$

Equation (7) serves for finding $g(t)$ both experimentally and theoretically. Theoretical solutions are obtained from the mass balance or transport equation for properly chosen initial and boundary conditions. However, in environmental groundwater systems the weighting function cannot be found experimentally because of technical difficulties in performing a proper experiment. Thus we have to rely on theoretical solutions and on experimental evidence gained in other fields. According to what was said earlier if $C_I(t)$ is represented by $C_{IFF}(t)$ for an ideal tracer, the weighting function is identical with the $E(t)$ function. In laboratory experiments, where T is known from the volumetric measurements, this property allows to test if a given tracer behaves like an ideal one for mass tracing. Thus, one may modify the Nir and Lewis' definition of an ideal tracer given in the preceding section. Namely: an ideal tracer for mass tracing is a substance that, when injected and measured in the flux, has the same response function as the traced material. The finding concerning the equality of the $g(t)$ and $E(t)$ functions for the cases C_{FF} and for an ideal tracer, is of a high importance if the tracer method is applied to dynamic systems. Whenever the experimenter is in doubt about the model, or is unable to formulate it, the measurement of the type C_{FF} (if possible) is free of the errors resulting from the differences between the $g(t)$ and $E(t)$ functions.

Equations (6) to (8) define several terms directly corresponding to experiments in which an impulse injection is applied. In such cases the counting of time may start at the moment of the injection. In environmental tracer investigations the injection is usually continuous and in consequence for each chosen time moment, t, it is necessary to take into account the entire past history of the input concentration. This can be done with the aid of the convolution integral, which in its simple form applicable to steady state systems reads:

$$C_{out}(t) = \int_0^\infty C_{in}(t-t') \exp(-\lambda t') g(t') \, dt' \qquad (9a)$$

or:

$$C_{out}(t) = \int_{-\infty}^t C_{in}(t') \exp[-\lambda(t-t')] g(t-t') \, dt' \qquad (9b)$$

where $C_{out}(t)$ and $C_{in}(t)$ are the output and input concentrations respectively, and $g(t')$ is defined by equation (7a) but for a condition of instantaneous injection in flux, i.e., for the cases C_{IFR} and C_{IFF} (Kreft and Zuber, 1978). In other words, equation (9) is applicable only to the cases corresponding to the injection in flux. Fortunately, the environmental radiotracers which enter groundwater bodies with precipitation are, in principle, injected proportionally to the volumetric flow rates by nature itself. This means that for a given precipitation, the higher infiltration rate at a given part of the recharge area, the higher amount of the tracer enters the system there. Doubts may arise in the case of ^{14}C which enters the groundwater systems via biogenic material with accompanying dissolution of inorganic carbonates. In heterogeneous recharge areas it may happen that the amount of total dissolved carbon (TDC) differs from site to site and thus the injection of ^{14}C is not necessarily proportional to the volumetric flow rates.

In equation (9a), t represents the calendar time whereas the integration is performed over the transit times. An example showing the meaning of both time scales and the computation of $C_{out}(t)$ for a very simple $g(t')$ function is given in Fig. 1-1. This example shows that $C_{out}(t)$ is the sum of concentrations which appear at the outlet at the time t after travelling with different transit times t'. The weighting of these particular concentrations is performed with

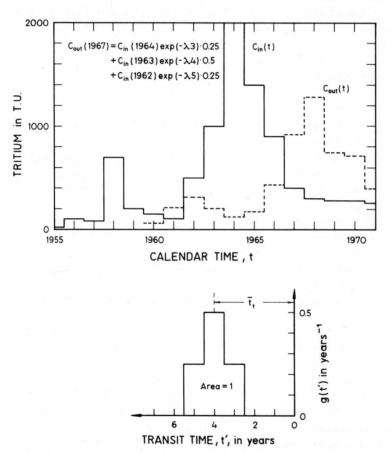

Fig. 1-1. An example showing the meaning of the time variable (calendar time here) and the transit time variable. The computation of $C_{out}(t)$ from a given $C_{in}(t)$ is performed using a simple $g(t')$ function of the binomial type. The values of the weighting function given in the summation in the upper part of the figure result from the assumed shape of the function given in the lower part of the figure. The position of the t' scale corresponds to the calculation of $C_{out}(t)$ for 1967. To calculate the output concentration for other values of t the t' scale is shifted and the computation repeated.

the aid of the $g(t')$ function which will be discussed in detail in the next section. The exponential term allows for the radioactive decay. Theoretically, the integration has to be performed for all possible transit times, i.e., from zero to infinity. In numerical calculations the integration limits have to be chosen in such a way that any their further change does not change the resulting $C_{out}(t)$.

In principle any form of the $g(t)$ function can be used in equation (9a) for finding the input-output relation. For instance, in chemical engineering even polynomials are often used. However, in tracer hydrology one is not interested in the input-output relations but in finding from them the physical parameters of the investigated system. Thus, the $g(t)$ function should have a clear physical meaning.

MODELS AND THEIR HYDROLOGICAL SIGNIFICANCE

Models in hydrology are usually used for computations in situations other than those expressed by experimental data. It means that a constructed model is tested or calibrated for a given set of data and next used to predict the response of a system in extrapolated situations for which no experimental data exist. Here the models will be used in a more restricted sense. Namely, from the input-output relations we shall be trying to determine some physical parameters of the system.

A mathematical model is an equation or formula in which pertinent physical principles are expressed (Snyder and Stall, 1965). Particular models will be defined by formulae describing the weighting functions of different shape. In general meaning by the model we shall understand here the type of the $g(t)$ function. In considering a given system, the type of the $g(t)$ function together with values of its parameters will be understood as the model of that system.

A number of lumped-parameter models exist which can be used for our purposes following a recent work of Małoszewski and Zuber (1982) who reviewed existing models and introduced several new ones. The weighting functions and the basic parameters of these models are given in Table 1-1.

Consider now the physical meaning and the characteristics of particular models. In the piston flow model (PFM) it is assumed that there are no flow lines of different velocities and that the hydrodynamical dispersion as well as molecular diffusion of the tracer are negligible. Thus the tracer moves from the recharge area with the mean velocity of water as if it were in a parcel. The weighting function is mathematically described by the Dirac delta function, which inserted into equation (9) gives:

$$C_{out}(t) = C_{in}(t-T) \exp(-\lambda T) \qquad (10)$$

for any $C_{in}(t)$. Equation (10) means that the tracer which entered at a given time $t-T$ leaves the system at the moment t with concentration decreased by the radioactive decay during the time span T. This equation, which describes a dynamic system, is mathematically equivalent to equation (11) describing the concentration of a radioisotope in a static water parcel separated since the recharge time or more exactly since the entry of the radioisotope, whereby:

$$C(t) = C(0) \exp(-\lambda t) \tag{11}$$

where t here is the radiometric age of water which may be thought of as defined by that equation. In the past this radiometric age was often used as the real age, T, of dynamic systems. This is true only for systems which can be approximated by PFM. The applicability of PFM can be judged by comparison with other, more realistic, models (see pp. 32 and 43).

The exponential model (EM) was introduced by Eriksson (1958) under the assumption that the exponential distribution of transit times (Fig. 1-2) corre-

TABLE 1-1

Models, their weighting functions, parameters, and the mean transit time of a conservative tracer

Model	Weighting functions, $g(t)$	Parameters[a]	Mean transit time of tracer, t_t
(1) Piston flow (PFM)	$\delta(t-T)$	T	T
(2) Exponential (EM) or good mixing	$T^{-1} \exp(-t/T)$	T	T
(3) Combined exponential and piston flow (EPM) or real system	$(T/\eta)^{-1} \exp(-\eta t/T + \eta - 1)$ for $t \geq T(1-\eta^{-1})$ 0 for $t < T(1-\eta^{-1})$	T, η	T
(4) Linear (LM)	$1/2T$ for $t \leq 2T$ 0 for $t > 2T$	T	T
(5) Combined linear and piston flow (LPM)	$\eta/2T$ for $T - T/\eta \leq t \leq T + T/\eta$, 0 for other t	T, η	T
(6) Dispersive flow (DM) (a) case C_{FF} (b) case C_{FR}	At^{-1} [b] $(2A - B)T^{-1}$ [c]	$T, D/vx$ $T, D/vx$	T $T(1 + D/vx)$

[a] T is the turnover time, D/vx is the dispersion parameter, η is the ratio of the total aquifer volume to the volume either with the exponential distribution of transit times (for the EPM) or with the linear distribution of transit times (for the LPM).
[b] $A = (4\pi tD/vxT)^{-1/2} \exp[-(1-t/T)^2 vxT/4Dt]$
[c] $B = (vx/2D) \exp(vx/D) \operatorname{erfc}[(1+t/T)(4Dt/vxT)^{-1/2}]$

sponds to a probable situation of decreasing permeability with the aquifer depth. Later Bredenkamp and Vogel (1970) in their consideration of an aquifer having the permeability and porosity exponentially decreasing with depth found a formula (their equation 4) equivalent to the exponential model for a constant C_{in} (equation 28 here). Similarly, Vogel (1970) found the same formula (his equation 4) for an aquifer with uniform permeability and porosity. In both cases the aquifer depth was assumed to be constant thus it is obvious that the hydraulic gradient has to be proportional to the distance from the watershed line to allow for the increasing water flow. The output concentration, $C_{out}(t)$, is measured in the outflow (spring, river) or in water from an abstraction well.

The exponential model is mathematically equivalent to the well known model of good mixing which is applicable to some lakes and industrial vessels. Many investigators were led astray by this property and rejected the applicability of EM to groundwater systems forgetting that the original concept of the model was based on the assumption of mixing at the sampling site without any mixing in the aquifer.

A more realistic model combines the properties of the models discussed

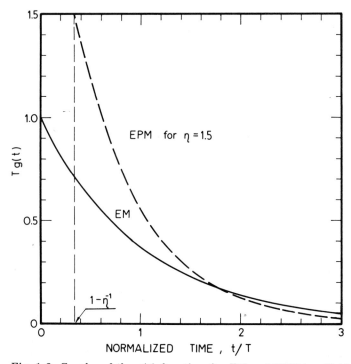

Fig. 1-2. Graphs of the $g(t)$ function for EM and EPM (see Table 1-1 for equations). The parameter η is the ratio of total volume of the system to the volume in which the exponential distribution of transit times exists.

above and is called shortly the exponential-piston model (EPM). In this model it is assumed that the aquifer consists of two parts in line, one with the exponential distribution of transit times, and another with the distribution, which may be approximated by the piston flow model. The weighting function of this model is given graphically in Fig. 1-2. The second parameter of this model, η, is equal to the total volume divided by the volume with the exponential distribution of transit times.

The linear model (LM) was derived similarly as EM but under an assumption that all the flow lines have the same velocity. The linear distribution of the transit times results from the different lengths of the flow lines (Eriksson, 1958).

The combination of LM with PFM gives similarly to EPM the linear-piston model (LPM). The weighting functions of LM and LPM are given in Fig. 1-3. The shape of these functions seems to be unnatural, and thus it is not astonishing that no practical application of LM or LPM is known.

The models presented above describe some idealized situations, whereas it is well recognized that the dispersion equation is the best mathematical formulation available for the description of macroscopic effects in the solute (tracer) transport in porous media, rivers, channels, and other natural or artificial systems. For our purposes the unidimensional form of this equation is of interest. It reads (for derivation see, e.g., Himmelblau and Bischoff, 1968):

$$D \frac{\partial^2 C}{\partial x^2} - v \frac{\partial C}{\partial x} = \frac{\partial C}{\partial t} \tag{12}$$

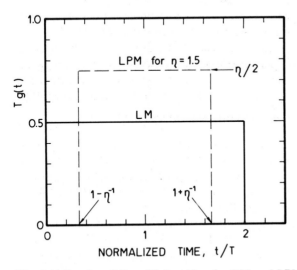

Fig. 1-3. Graphs of the $g(t)$ function for LM and LPM (see Table 1-1 for equations). The parameter η is the ratio of total volume of the system to the volume in which the linear distribution of transit time exists.

where D is the dispersion coefficient, which in general is a tensor, but in the unidimensional approximation is treated as a scalar, and x is here the spatial variable.

A misunderstanding is possible as a result of different possible applications of the dispersion equation. For instance, in pollutant movement studies the dispersion equation serves as a distributed parameter model, especially when numerical solutions are used. References to applications of the solutions to the dispersion equation as continuous models were given in the Introduction. Here, the unidimensional solutions are used as lumped-parameter models.

When solving equation (12) it is possible to allow for the injection-detection mode by a proper formulation of the initial and boundary conditions (Kreft and Zuber, 1978). As we stated earlier, only the solutions of the C_{IFR} and C_{IFF} types can be used as the $g(t)$ functions in equation (9a). Here we shall be using the solutions for a semi-infinite medium, neglecting a possible influence of the second boundary. In the author's opinion, in systems with dominant hydrodynamic dispersion, the solutions for semi-infinite media give a better approximation than the solutions obtained for finite systems. Some discussion of these problems can be found in Shamir and Harleman (1967), and Kreft and Zuber (1978). A number of laboratory studies (e.g., Rifai et al., 1956; Shamir and Harleman, 1967; Brigham, 1974; De Smedt and Wierenga, 1979) indicate that solutions obtained for semi-infinite media are also applicable to finite columns. However, laboratory experiments concern systems with low values of the dispersion parameter (D/vx). In such systems it is of little importance which boundary conditions are used. Unfortunately, there is a lack of experimental evidence for highly dispersive systems in this respect. The approach accepted here is based on intuitive concepts, discussed further in the section "Cases of a constant tracer input".

The analytical forms of the $g(t)$ functions of both DM models are given in Table 1-1. In the last column of the table it is shown that in the case C_{FR} the mean transit time of tracer is not equal to the mean transit time of water. All the other models given in Table 1-1 have their $g(t)$ functions equal to the $E(t)$ functions because they correspond to C_{FF} cases.

The dispersion model is the most flexible of all the models discussed here not only because of its applicability for both the C_{FF} and C_{FR} cases, but also because it permits the use of variety of possible shapes of the $g(t)$ function (Fig. 1-4). Undoubtly, the $g(t)$ functions of DM are more natural and cover a wider spectrum of shapes than other two-parameter models. More detailed discussion of this model is given on pp. 22 and 32.

In the past the so-called binominal model (Dincer et al., 1970; Davis et al., 1970) was used as a simplified form of DM. The binominal model, when used in its one- or two-parameter form, yields a symmetrical distribution of transit times, whereas the dispersion model in its correct form yields a nearly symmetrical distribution in the cases of low dispersion, and a highly asymmetrical distribution in the cases of high dispersion (Fig. 1-4). An example of the binomial model has been already shown in Fig. 1-1.

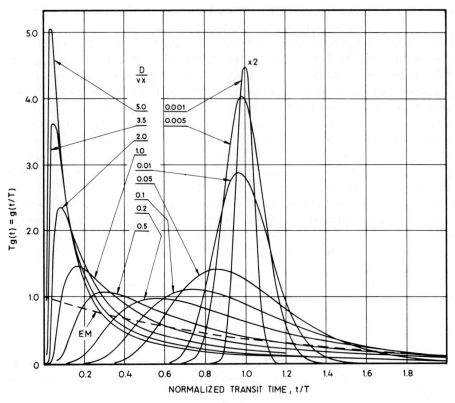

Fig. 1-4. Graphs of the $g(t)$ function for DM-C_{FF} (see Table 1-1 for equation), and different values of the dispersion parameter, D/vx (Lenda and Zuber, 1970). The $g(t)$ curve for EM is shown for comparison.

Fig. 1-5 presents several typical situations to which particular models are applicable. The piston flow model may be used to situations shown in the first part of Fig. 1-5 under the following conditions: (1) the width of the recharge zone has to be negligible in comparison with the distance to the sampling site, (2) the aquifer should be homogeneous, (3) the variations of the input concentration have to be slow. All these conditions are intuitive at this stage, but two of them will be confirmed further in the following section where the macroscopic and apparent dispersions are discussed, and in the section "Variable tracer input", where the output concentrations calculated with the aid of PFM are compared with the results obtained from other models for different input functions. The same situations can be treated with the aid of the dispersion model, without any restrictions and with additional possibility allowing for the sampling mode (compare cases indicated as c and e in Fig. 1-5 with the definitions of the C_{FF} and C_{FR} concentrations and the corresponding weighting functions). Now, it becomes clear that the piston flow model is an approximation of the dispersion model. In spite of all its drawbacks PFM may be convenient for fast and easy estimations.

The exponential model can be used for situations shown as a, b, and d in cross-section 2 of Fig. 1-5 providing the transit time through the unsaturated zone is negligible in comparison with the total transit time. This condition results from the shape of the weighting function (Fig. 1-2) in which the infinitesimally short transit time appears.

If the transit time through the unsaturated zone is not negligible, or if the abstraction well is screened at a certain depth below the water table (see case c in cross-section 2, Fig. 1-5), or finally if the aquifer is partly confined (cross-section 3, Fig. 1-5) the exponential-piston model and alternatively the dispersion model are applicable.

Similarly to the exponential model discussed above, cross-sections 4 and 5, Fig. 1-5, visualize situations to which the linear and linear-piston models are applicable.

According to Przewłocki and Yurtsever (1974) E. S. Simpson proposed the finite state mixing-cell models (briefly called the finite state models or FSM). These models are represented by a series (or even by two- or three-dimensional arrays) of boxes. Each box behaves like a good mixing model. The shape of the system and the flow distribution can be modelled by arranging the dimensions of boxes and their interconnections. Calculations of the theoretical $C_{out}(t)$ are performed by applying recursive equations. In spite of their simplicity these models are characterized by an unlimited number of fitting parameters, thus the values of physical parameters obtained from their use are doubtful.

The multi-box models (e.g. FSM) can represent situations in which mixing of several water components with different mean turnover times takes place. Unfortunately for groundwater systems there are usually no means for an independent determination of some of the parameters. Thus the number of fitting parameters is too large to obtain an unambiguous solution, and consequently the inverse problem remains unsolved. In other words, by applying multi-box models it is possible to obtain equally good fitting for a number of different sets of the parameters.

The interpretation of several case studies performed recently by Grabczak et al. (1984) suggests that in practice whenever the number of fitting parameters is larger than two, the inverse problem cannot be solved unambiguously.

There is one more problem of great importance related to the models. Models such as EM, LM, and some others not discussed here, are based on an assumption that the transit time distribution is related to the distribution of flow velocities. In other words, there is no exchange of tracer between flow lines of different velocities. Of course, the molecular diffusion always exists. However, in many systems its influence is negligible. In such cases the $g(t)$ and, consequently, $E(t)$ functions are related to the velocity distribution of flow lines. However, the dispersion model, owing to its flexibility, may serve

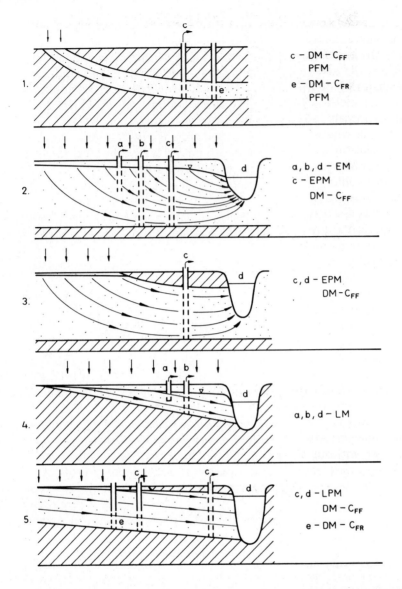

Fig. 1-5. Schematic situations showing examples of possible applicability of particular models (Małoszewski and Zuber, 1982). Cases *a*, *b*, *c*, and *d* correspond to sampling in outflowing or abstracted water, thus the sampling is averaged by the flow rates (case C_{FF}). Case *e* corresponds to samples taken separately at different depth (e.g., during drilling) and next averaged by the depth intervals (case C_{FR}). Cross-sections: 1. A confined aquifer with a narrow recharge area far from the sampling points. Sampling either in the pumped out water (case *c*) or averaged across the aquifer depth (case *e*). 2. An unconfined aquifer. Sampling including flow lines with infinitesimal short transit times is shown as cases *a*, *b*, and *d* to which the exponential model applies. In the case of deeper sam-

for describing either systems with no exchange of tracer between the flow lines or systems where the tracer distribution does not correspond to the velocity distribution. In the latter case the $E(t)$ function retains its properties given by equations (4) and (5) but cannot serve for finding the velocity distribution. The majority of investigators seem to treat these important properties of the tracer method as self-evident and do not even bother to mention them. Unfortunately, some other investigators treat the first case mentioned here as a rule whereas it is rather an exception. The reader who may have doubts in relation to the above statements is referred to the classic work of Taylor (1953) who considered a laminar flow through a capillary. In the case of a short capillary and a fast flow the tracer distribution only reflects the velocity distribution (and additionally the injection-detection modes, which were not discussed by Taylor). For a long capillary and a slow flow the tracer distribution is governed by the diffusion process and is related only to the mean flow velocity and the diffusion coefficient instead of the relation to the velocity profile.

Practice shows that sometimes it is impossible to obtain a good fit of the model to the experimental data without assuming that there are at least two water components: one with the radioisotope of interest, and another, older, without this radioisotope (in this older water the radioisotope has decayed completely since the recharge time). This effect can be allowed for by introducing an additonal fitting parameter, β, which is defined as the ratio of the old flow component to the total flow. Then the theoretical output concentration is obtained by applying an $(1-\beta)$ factor to the input concentration. In other words, the output concentration calculated without taking into account this effect has to be corrected by fitting the $(1-\beta)$ factor. An abstraction well penetrating two aquifers, one unconfined containing tritium, and another deeper without tritium, may serve as an example requiring the introduction of the β parameter for the interpretation of the tritium data.

→ pling (c) EM is not applicable and either the dispersion model or the exponential-piston flow model have to be used. Similarly if the infiltration time through the unsaturated zone is not negligibly short in comparison with the total turnover time, EM is not applicable to cases shown as a, b, and d. 3. An aquifer confined in the downstream part. The unconfined volume has an exponential distribution of transit times whereas the confined part is approximated by the piston flow. The parameter η in EPM represents the ratio of the total volume to the volume of the exponential part. 4. An unconfined aquifer with linearly increasing depth. Negligible infiltration time (see description of cross-section 2). 5. An aquifer with linearly increasing depth in the unconfined part. The downstream part is confined and can be approximated by piston flow (see description of cross-section 3). In the case of sampling averaged across the aquifer depth (e), the C_{FR} version of the dispersion model should be applied (cf. case e in cross-section 1).

DISPERSIVITY AND THE DISPERSION PARAMETER

Both the dispersion constant, D/v, more often called the dispersivity, and the dispersion parameter, D/vx, appearing in the dispersion model need more attention. The dispersivity is a common term for all those interested in mass transport in groundwaters. However, when solutions to the dispersion equation are treated as lumped-parameter models it is more convenient to use the dispersion parameter which is the reciprocal of the so-called Peclet number, well known in chemical engineering. In this section an attempt is made to explain the physical significance of this parameter in the lumped-parameter approach.

Laboratory experiments performed in granular homogeneous media yield the following relationship for the dispersion constant, D/v (e.g., see Perkins and Johnstone, 1963, or Fried and Combarnous, 1971):

$$D/v = D_m/\tau v + ad \quad \text{for } v \neq 0 \tag{13}$$

where D_m is the coefficient of molecular diffusion, τ is the tortuosity factor (Richardson, 1961), d is the mean grain diameter, and a is a constant. For most typical granular media this equation gives D/v values less than 1 cm. Artificial tracer experiments (Lenda and Zuber, 1970) as well as pollution movement studies (e.g., Małoszewski et al., 1980; Pinder, 1973; Robertson, 1974) show for granular aquifers values from a fraction of a metre to about 30 m, whereas for fractured and cavernous systems values from a few metres to more than 100 m are observed. The theoretical works of Saffman (1959) and de Josselin de Jong (1958) show that the dispersion constant is of the order of the length of an elementary capillary, obviously related to the mean grain diameter in unconsolidated media. Similarly, in heterogeneous media (as most aquifers are) we may expect that the dispersion constant will be of the order of the average length of heterogeneous zones. Zuber (1974) postulated that the dispersion constant may be thought of as a measure of the aquifer heterogeneity. In other words, instead of equation (13) we shall have:

$$D/v = D_m/\tau v + ad + b \tag{14}$$

where b is characteristic for a given aquifer. Its numerical value may depend on the scale of the experiment in relation to the scale of heterogeneities.

In environmental tracer investigations the situation is much more complex. Namely the weighting functions for DM are derived under an assumption that the injection to the system takes place at $x = 0$, whereas in fact it is extended over the recharge area. In such a situation, if the DM is applicable at all, it will yield an apparent value of the dispersion constant which may be expected to be of the order of the length of the recharge zone measured along the flow lines. In extreme cases (e.g., d in cross-section 2, Fig. 1-5) the mean distance from the recharge area to the drainage point is $x = 0.5x_0$, where x_0 is the length of the recharge zone. Thus the apparent dispersion parameter should not exceed 2 because $D/vx \leq x_0/x = 2$.

More exact estimation can be performed by comparing the $g(t)$ functions and their variances. The variance is defined as the second central moment, i.e., it describes the spread of the weighting function around the mean transit time. The variance of EM is T^2 whereas the variance of DM-C_{FF} is $(2D/vx)^2 T^2$. Obviously, these variances are equal for $D/vx = 0.5$. Inspection of the weighting function of DM-C_{FF} shows that for $D/vx = 0.5$ this function is relatively close to the $g(t)$ of EM (see Fig. 1-4). However, $D/vx = 0.5$ should not be treated as an extreme value. Later, when $C_{out}(t)$ for various radioisotopes is discussed, we shall see that depending on the input function and λ of a given radioisotope, the best agreement of DM with EM may also be obtained for the values of D/vx larger than 0.5.

All the considerations performed so far dealt with systems which were uniform in the horizontal direction, perpendicular to the flow lines. Possible behaviour of other systems can also be predicted. For instance, larger values of D/vx may probably appear in systems similar to those indicated as b and d in cross-section 2 of Fig. 1-5 if the recharge area is shaped in such a way that the flow lines with short transit times prevail over the flow lines with longer transit times (i.e., the amount of recharge close to the drainage prevails over that at a larger distance).

When an artificial tracer or pollutant is observed equation (14) describes the dispersion constant. Usually the value of b dominates over the other terms on the right-hand side of equation (14). The values of $D/v \cong b$ can be estimated from published case studies of systems similar to that of interest. With a decreasing scale of the experiment equation (14) may simplify into equation (13). Finally, for v approaching zero, $D \rightarrow D_m/\tau$. These last two cases are also applicable to the experimental situations in investigations of the infiltration rate in the unsaturated zone (see next section). The initial and boundary conditions used in the derivation of the $g(t)$ function for both the DM-C_{FF} and DM-C_{FR} cases correspond more exactly to the experimental situations occurring in the infiltration studies because the tracer appears in flux at $x = 0$ where x is here the depth measured from the land surface. Thus the apparent dispersion parameter does not appear in such studies at all. On the other hand the heterogeneity of the unsaturated zone usually does not contribute to the D/v value, i.e., $b = 0$, because at a given site the tracer movement takes place perpendicularly to the horizontal layering, and thus all the flow lines have macroscopically the same length and the same velocity.

The above discussion on the apparent dispersion parameter resulting from inadequate boundary conditions may create doubts whether the dispersion model is applicable to the interpretation of groundwater systems by the lumped-parameter approach. Some authors are of the opinion that in general the use of the dispersion models is not theoretically justified. Even in chemical engineering, where the boundary conditions are easier to describe mathematically, it is considered that "the only justification for the use of dispersion models is their success in representing the real process" (Himmelblau and Bischoff, 1968). However, for packed beds (e.g., soil columns or natural soils in filtration studies) or for pipelines with a turbulent or laminar flow (including rivers and karstic conduits) the use of the dispersion models is theoretically justified (e.g., see Rifai et al., 1956; Taylor, 1953, 1954; and Fischer, 1967). For other systems their theoretical foundation is not sound and the final judgement has to be left to the experimental evidence. Examples discussed on pp. 35 and 39 seem to show that the dispersion model yields parameters which are both internally consistent and consistent with other models.

INFILTRATION STUDIES

The unsaturated zone has attracted a lot of the attention among the researches interested either in the movement of fertilizers and pollutants or in studies of the infiltration process. Here we shall concentrate on the problems related to the possible use of the models discussed in the section "Models and their hydrogeological significance" for measuring the infiltration rate with the aid of environmental radioactive tracers. The tritium was the most promising among these tracers because of its high concentration peak observed around 1963 in the atmosphere. The hopes related to this tracer belong rather to the past as its atmospheric concentration is now slowly decreasing (e.g., see Figs. 1-1, 1-8 or 1-13). However, local increases of the tritium concentration observed from nuclear energy installations (e.g., see Weiss et al., 1979) open some further hopes for the applicability of the environmental tritium to the infiltration measurements.

The experimental conditions differ from those described in the preceding sections as usually at a given time, denoted here as t_e, the depth profile of the resident concentration $C_R(t_e, x)$, is measured by taking soil samples from different depths, x. Neglecting the disturbances which may be caused by ground heterogeneities in horizontal directions and by possible lateral flows resulting from the occurrence of perched water, the downward movement of a tracer in percolating water is similar to that in laboratory columns. A number of experiments showed that to such columns either the dispersion model, or its extended form called the dead-end pore model, are applicable. So, we have:

$$C_R(t_e, x) = \int_0^\infty C_{in}(t_e - t')\, [Q C_{IFR}(t', x)/M]\, \exp(-\lambda t')\, dt' \qquad (15)$$

where C_{IFR} is the unnormalized solution to the dispersion equation, or to the equations of the dead-end pore model discussed below, for the injection in flux (i.e., with infiltrating precipitation) and the measurement of the resident concentration. Q and M appear here as a result of the working definition of the $g(t)$ function (equation (8a)) and they cancel when a proper formula for the C_{IFR} is put into equation (15). For the dispersion model this formula reads:

$$C_{\text{IFR}}(t, x) = (M/Q)\{[(2v)/(4\pi Dt)^{1/2}]\exp[-(x-vt)^2/4Dt] - \\ - (v^2/2D)\exp(vx/D)\,\text{erfc}[(x+vt)/(4Dt)^{1/2}]\} \tag{16}$$

In this model, the dispersion coefficient (D) and the vertical transit velocity (v) are treated as fitting parameters. However, the model given by equation (16) was developed for a fully saturated medium with no pores containing stagnant water. Thus it is necessary to consider if, and when, the model is applicable to the unsaturated zone in which stagnant water exists. The dead-end pore model developed by Coats and Smith (1964) and followed by other investigators (e.g. Brigham, 1974; Gaudet et al., 1976) applies to media with stagnant water pores. Using notation consistent with other equations here it reads:

$$n_m D \frac{\partial^2 C_{R,m}}{\partial x^2} - n_m v \frac{\partial C_{R,m}}{\partial x} = n_m \frac{\partial C_{R,m}}{\partial t} + n_s \frac{\partial C_s}{\partial t} \tag{17a}$$

$$\frac{\partial C_s}{\partial t} = k(C_{R,m} - C_s) \tag{17b}$$

where n_s is the space fraction occupied by stagnant water, $C_{R,m}$ is the resident concentration in the mobile water, C_s is the concentration in the stagnant water, and k is the rate transfer constant. Equation (17b) describes the mass transfer of the tracer (solute) between the mobile and stagnant water, whereas possible interactions through the gaseous phase (e.g., for tritiated water) are neglected. In the models given by equations (17a) and (17b), or by their solutions, there are four non-disposable parameters: D, v, k, and n_s/m_s. Of these four parameters, v and n_s/n_m are of direct interest for determining the infiltration rate whereas k and D can be treated as auxiliary parameters, related to the behaviour of the tracer. The rate constant k is non-disposable because it depends not only on the molecular diffusion but also on the geometry of stagnant pores. For a more detailed discussion on the physical meaning of the rate transfer constant the reader is referred to Coats and Smith (1964). The dispersion coefficient is, in general, non-disposable. However, under favourable conditions this parameter can be relatively well predicted as will be discussed later in this section. The existence of four non-disposable parameters is little promising for solving the inverse problem. Thus we shall concentrate on two extreme cases in which the number of parameters is reduced.

There are several known solutions to equations (17a) and (17b). The solution of the C_{IFR} type, which is of interest to our considerations, was found by Villermeaux and Swaay (1969). However, two extreme cases can be obtained directly from equations (17a) and (17b). If the velocity of water is very low one may assume that the tracer concentration is equilibrated in the mobile and stagnant zones by the molecular diffusion process. In that case $C_{R,m} = C_s$ and equations (17a) and (17b) reduce to the following equation:

$$D \frac{\partial^2 C}{\partial x^2} - v \frac{\partial C}{\partial x} = \left(1 + \frac{n_s}{n_m}\right) \frac{\partial C}{\partial t} \tag{18}$$

If the real time, t, in equation (18) is replaced by the apparent time, t_a, given as:

$$t_a = \frac{t}{1 + n_s/n_m} \tag{19}$$

then equation (18) becomes mathematically equivalent to equation (12). In unidimensional flow, instead of the time transformation, one may transfer the mean velocity of water, v, into the apparent velocity, v_a, given as:

$$v_a = \frac{v}{1 + n_s/n_m} \tag{20}$$

By fitting the model in which v is meant as v_a (i.e., the C_{IFR} solution given by equation (16) but with a different meaning of v) one can easily find the recharge rate as:

$$R = Q/S = vn_m = v_a(n_s + n_m) \tag{21}$$

because $n_m + n_s$ is equal to the total moisture content measurable in situ (e.g., by neutron moisture method) or on samples.

Another simple situation is described by $k = 0$, which corresponds to the assumption that no tracer diffuses to the stagnant water. Even if this assumption is acceptable, and we hope that the velocity, v, found by fitting the model given by equation (16) represents the mean velocity of water, the recharge rate cannot be determined because the mobile water porosity, n_m, remains unknown.

Measurements of the concentration profile in the unsaturated zone can be treated as experiments on a small scale. Thus, in general, a low dispersion is to be expected. In the cases of low velocity the dispersion coefficient is then given as $D = D_m/\tau$ (see previous section), which means that the observed values will be even lower than the coefficient of molecular diffusion because the tortuosity factor, τ, is always larger than 1. In fact, values of D lower than D_m were observed in a number of infiltration experiments characterized by low velocities (e.g., Schmalz and Polzer, 1969; Smith et al., 1970; Allison and Hughes, 1974). The tortuosity factor for the saturated zone lies in the range of $1.3 - 1.5$ for unconsolidated media and in the range of $1.5 - 2.5$ for consolidated media (Richardson, 1961). Assuming a similar range of values for the unsaturated zone, the dispersion coefficient can be roughly estimated.

It has to be noted here that numerical differences between the flux solutions (type C_F) and concentration solutions (type C_R) disappear for low values of the dispersion parameter (e.g., see Kreft and Zuber, 1978). Thus, in the infiltration studies, when the assumption of $C_{R,m} = C_s$ is accepted, instead of equation (16), a simpler solution to the dispersion equation can be used. This solution can be found in any textbook on diffusion for the injection described by the Dirac delta function defined in space, i.e., for $\delta(x)$. According to the notation introduced by Kreft and Zuber (1978) and followed here, this solution is of the type C_{IRR}.

The main difficulties in the use of the environmental tritium method for the infiltration rate measurements result from: (a) seasonal variations of the tritium concentration, which make the exact estimation of $C_{in}(t)$ rather doubtful (see p. 33), (b) variable infiltration velocity, and (c) infiltration through the soil via short cuts in the form of cracks, root channels and animal burrows (e.g., see discussion in Allison et al., 1974).

Many of these basic difficulties as well as the computational problems resulting from the use of a complex formula under the convolution integral (equation (15)) may be overcome by the use of an artificial tracer injected below the root zone (e.g., see Blume et al., 1967, for the injection technique). This method is, however, beyond the scope of this work.

STAGNANT WATER IN THE SATURATED ŻONE

The dead-end pore model discussed in the previous section was originally developed for saturated media. The model is particularly useful for porous hard rocks where stagnant water pores occur. Usually the natural velocities are sufficiently low to assume that due to the molecular diffusion the concentrations in pores are equilibrated, i.e., $C_{R,m} = C_s$. In such cases the turnover time appearing in the weighting functions of Table 1-1, and in equation (4) has to be redefined:

$$T = \frac{V}{Q} = \frac{H}{R} = \frac{x}{v_a} \tag{22}$$

where V is the total volume of water, i.e., $V = V_m + V_s$. The apparent velocity found from equation (22) will be comparable with the transit velocity determined from the Darcy velocity, v_f, because the total porosity (in the meaning of open pores) is measurable, whereas the mobile water porosity remains unknown. Namely:

$$v_a = \frac{v_f}{n_m + n_s} \tag{23}$$

The problem of a proper definition and understanding what really is measured in a given experimental situation is not purely academic. The total volume in equation (22) and the total porosity in equation (23) should be understood in a microscopic meaning as those related to the open pores. Regional stagnant zones, as for instance a sedimentary pocket shown in Fig. 1-6 are not included into these definitions. The turnover time of the system shown in Fig. 1-6 will be defined for the microscopic total volume without the stagnant volume of the pocket. However, the movement of a tracer may be delayed due to its diffusion into the stagnant zone. In most cases this delay will be negligible because the limited area of the interface between the mobile and stagnant zones limits the transfer of the tracer to the stagnant sink. Notice, however, that water in pockets like that shown in Fig. 1-6, in spite of being radioisotopically dead, is hydraulically connected with the recharge area.

The mobile water porosity must not be identified with the specific yield, defined as the fraction of space occupied by water which can be drained

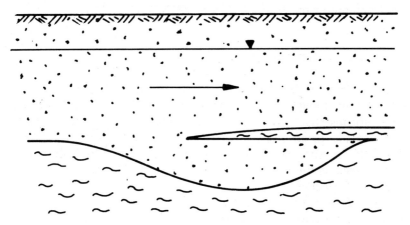

Fig. 1-6. Schematic representation of a radioisotopically dead zone which is, however, connected with the dynamic system.

under the force of gravity. It is well known that in fine-grained materials the specific yield is low, whereas the mobile water porosity is close or equal to the total porosity. Another misunderstanding may result from the conflicting terminology. Namely, the specific yield is often called the effective porosity, whereas in the tracer experiments this term is often used for the mobile water porosity.

As mentioned in the Introduction, fractured systems with a porous matrix cannot be correctly interpreted with the aid of models discussed in this paper. Neither is the above discussion of equations (22) and (23) applicable to such systems. In a fractured system with a porous matrix the mechanism of the delay of a tracer is, of course, similar to that described for systems with immobile water in the dead-end pores. The scale of the effect is, however, different because of different scales of the pertinent parameters. Namely, stagnant water in the dead-end pores occupies a volume which is a fraction of the mobile water volume. In the unsaturated zone (see previous section) the stagnant water volume increases but still remains a fraction of the mobile water volume (e.g., see De Smedt et al., 1981). Due to small dimensions of the dead-end pores the assumption of an instantaneous equalization of concentrations seems to be reasonable. However, in a fractured system with a porous matrix, the situation is quite different because the volume of water in the matrix may be much larger than that in the fractures, and the diffusion length in the matrix may be many times greater than that in the case of dead-end pores. Grisak and Pickens (1981) showed that in such a system the delay of a tracer may be orders of magnitude with respect to the velocity of water in the fractures. Usually the matrix is much less permeable than the fractured rock, and in consequence all the hydraulic observations are practically related to the fractured system only. Thus, the velocity of water determined from hydraulic observations and the apparent age of water determined with the aid of a lumped-parameter model are not comparable in such a case.

THE MOVEMENT OF DELAYED TRACERS

As mentioned earlier we would like to deal with ideal tracers, which do not undergo any other changes in the system than those adherent to the sought parameters. This means that for the water flow tracing the substances which undergo adsorption on, or exchange with the solid matrix are rather undesirable. Unfortunately, in the case of environmental radioisotopes the choice is not sufficiently free as the number of radioisotopes with adequate properties is very limited. Especially ^{14}C, which is an important tracer because of its suitable half-life time, is to be suspected of possible delay. Thus, some discussion of the models describing the macroscopic effects of the exchange with the solid matrix should be useful.

Substances which undergo reversible exchange with the solid matrix are removed for a certain time from the dynamic reservoir to the static one. If the process is governed by the linear adsorption isotherm (see definition of k_d below) and if the desorption process can be described by the same constants as the adsorption, the following model is applicable:

$$nD \frac{\partial^2 C_R}{\partial x^2} - nv \frac{\partial C_R}{\partial x} = n \frac{\partial C_R}{\partial t} + (1-n) \rho \frac{\partial q}{\partial t} \tag{24a}$$

and:

$$\frac{\partial q}{\partial t} = m(C_R - k_d^{-1} q) \tag{24b}$$

where ρ is the density of the rock material, $(1-n)\rho$ is the dry bulk density, q is the concentration of the tracer per mass unit of the rock material, m is the rate constant of the exchange reaction, k_d is the distribution constant defined as q^e/c_R^e where the superscript "e" means concentrations at the equilibrium stage for a given temperature. The other terms were defined earlier. The porosity and density appear in equation (24a) in order to obtain all the terms expressed in consistent units. This model was developed in gas chromatography and is commonly applied in the pollutant movement studies (e.g., Inoue and Kaufman, 1963; Robertson, 1974; Małoszewski et al., 1980). The reader interested in other possible models describing the behaviour of adsorbed substances is referred to Molinari and Rochon (1976).

Mathematically equations (24a) and (24b) are similar to the dead-end pore model discussed in the previous section. So, assuming a fast equilibration of C_R and q a time transformation is needed, or for the unidimensional flow the transformation of velocity, namely:

$$\text{Velocity of a delayed tracer,} \quad v_t = \frac{v}{1 + (1-n) n^{-1} \rho k_d} \tag{25}$$

As mentioned above, the adsorption model given by equation (24b) is generally accepted due to its easy applicability with the dispersion equation. Its simplified form resulting in equation (25), is applicable to any flow model. Equation (25) serves for estimating the velocity of a delayed substance in granular media if the distribution constant is obtained from a batch or column experiment (e.g., Inoue and Kaufman, 1963). For fissured and cavernous rocks the distribution constant is rather unmeasurable on samples.

Of the environmental radioisotopes used in groundwater investigations the radiocarbon is the most susceptible to the delay. Münnich et al. (1967) investigated the delay of ^{14}C in the form of CO_3^{2-} ions passing a column filled with 300-μm-diameter spheres of calcite. The delay factor obtained from column experiments was in a good agreement with a two-box model given as:

$$\frac{v}{v_t} = \frac{N_1 + N_2}{N_1} \qquad (26)$$

where N_1 is the number of carbon atoms in water contained in a unit volume of the aquifer, and N_2 the number of carbon atoms in the surface mono-molecular layer of solid carbonate contained in a unit volume of the aquifer. The observed delay factors were equal to 3 and 2 for 0.25 millimolar and 0.5 millimolar CO_3^{2-} solutions, respectively. For a porosity of 0.44 and density of 2,72 g cm^{-3} these delays correspond to the distribution constants (equation (24a)) of 0.56 and 0.28 ml g^{-1} respectively. Of course, such delays are enormous, but for well permeable media the surface of the mono-molecular layer available for the exchange process is at least an order of magnitude smaller than in the described experiments. Thus, the delay of ^{14}C caused by the exchange is negligible in most aquifers. More detailed discussion can be found in Münnich et al. (1967) and Münnich (1968). Here we shall note that the exchange of radiocarbon with the surface layer of the solid material influences its output concentration, because of the radioactive decay in this stagnant box, without any influence of the $\delta^{13}C$, providing a steady state condition has been reached. In such a case, the hydrochemical models based on the measurements of $\delta^{13}C$ may supply information on the initial dissolution of the organic carbon by the inorganic one, but cannot supply information if the exchange takes place.

The movement of tritiated water and ^{85}Kr can be treated as undelayed (see p. 4 for some exceptions in the case of tritiated water). Nothing is known about ^{32}Si.

CASES OF A CONSTANT TRACER INPUT

The cases of a constant tracer input can be solved analytically. In Fig. 1-7 relative concentrations are given as a function of the dimensionless turnover time, λT. Transfer to the real time is obtained for any radioisotope tracer by applying the adequate λ value. The $1/\lambda$ values expressed in years are: 17.9 for tritium ($t_{1/2}$ = 12.43 years), 8270 for ^{14}C, 290 for ^{39}Ar, and about 100 for ^{32}Si (Elmore et al., 1980; Kutschera et al., 1980). The graphs of Fig. 1-7 were calculated for PFM, EM, and LM from the following well-known formulae, which are easily obtainable from equation (9a):

$$C_{out}/C_{in} = \exp(-\lambda T) \qquad \text{for PFM} \qquad (27)$$

$$C_{out}/C_{in} = 1/(1 + \lambda T) \qquad \text{for EM} \qquad (28)$$

$$C_{out}/C_{in} = \frac{1-\exp(-2\lambda T)}{2\lambda T} \quad \text{for LM} \tag{29}$$

The concentration curves for the DM were calculated numerically directly from equation (9a).

It is clear from Fig. 1-7 that in the case of a constant C_{in} the turnover time can be determined only if the type of the one-parameter model is known. For two parameter models it is necessary to have a prior knowledge of the second parameter. The problem can be solved in principle by applying two radioisotope tracers with different λ values (Nir and Levis, 1975). Unfortunately, in hydrology the number of available tracers is not very promising.

Fig. 1-7 shows also that the PFM is applicable as an approximation for constant tracer input and low values of the D/vx parameter, say, $D/vx \leqslant 0.05$.

At first glance it may be astonishing that concentrations for large values of the D/vx parameter can be higher than those given by the exponential model (see Fig. 1-7). If the curves for the dispersion model were calculated for the $g(t)$ function derived for a finite system, then mathematically the case of D/vx approaching infinity would correspond to the mathematical model of good mixing, i.e., to the EM here. However, in the $g(t)$ functions derived for semi-

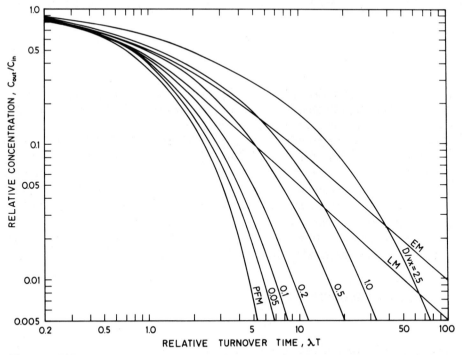

Fig. 1-7. Relative concentration versus relative turnover time, λT, for the case of a constant input of a radioactive tracer (Małoszewski and Zuber, 1982). EM = exponential model; LM = linear model; D/vx = dispersion parameter for the dispersion model; case C_{FF} (i.e. the measurement in the outflow).

finite media, a large value of D/vx parameter is not equivalent to a good mixing but simply means that there is a wide spectrum of transit times in the system. Such a wide spectrum, different than that for the EM, is easily imaginable. Thus, as mentioned in the section "Models and their hydrogeological significance", the solutions for semi-infinite media are chosen both for their simplicity and because in the author's opinion they give a better approximation of the real situations than the solutions derived under an assumption that the presence of the second boundary (e.g., an abstraction well or a spring) influences the dispersion in the system. If one assumes that at the boundary there is no dispersion (i.e., the second system is non-dispersive) then an unavoidable condition appears that there is no dispersion flux through the boundary, i.e., either all the flow lines reaching the drainage site have the same velocity or the same concentration. Such a condition is evidently unacceptable for natural systems.

Fig. 1-7 shows that the earlier discussed (p. 23) similarity of EM to DM with large values of the D/vx parameter also depends on the value of the turnover time.

It is generally accepted that radiocarbon measurements at two points along a given flow line allow to determine the age difference without knowing C_{in}, and consequently the flow velocity in the system. The formula obtained from equation (27) is:

$$T_2 - T_1 = \frac{1}{\lambda} \ln(C_1/C_2) \qquad (30)$$

It is clear that equation (30) is an approximation valid only for cases in which the PFM is applicable. Other models do not yield such simple relations and require more measuring points on a given streamline.

It may be interesting to note here that radioisotopes with a constant input are applicable as tracers for the age determination, and in turn by equation (3) for the mass transport, due to the existence of a sink (radioactive decay), which is adherent to the sought parameter (see the definition of an ideal tracer on p. 5). Other substances cannot serve as tracers for this purpose though they come under previous definitions of an ideal tracer. However, those other substances (e.g., Cl^-) serve as good tracers for other purposes, e.g., for determining the mixing ratio of different waters.

VARIABLE TRACER INPUT

^{3}H method

Seasonal variations of the tritium (^{3}H) concentration in precipitation cause serious difficulties in calculating the input function, $C_{in}(t)$. The yearly mean concentration weighted by the infiltration rates is

$$C_{in} = \sum_{i=1}^{12} C_i \alpha_i P_i / \sum_{i=1}^{12} \alpha_i P_i \tag{31}$$

where C_i, α_i, and P_i are the ^{3}H concentrations in precipitation, infiltration coefficients, and monthly precipitation for ith month, respectively. A rough simplification of equation (31) is usually applied. Namely, it is assumed that the infiltration coefficient in the summer months (α_s) is only a given fraction, α, of the winter coefficient (α_w). Putting $\alpha = \alpha_s/\alpha_w$ one obtains:

$$C_{in} = \frac{(\alpha \Sigma\, C_i P_i)_s + (\Sigma C_i P_i)_w}{(\alpha \Sigma P_i)_s + (\Sigma P_i)_w} \tag{32}$$

where subscripts "s" and "w" mean the summing over the summer and winter months, respectively. A further simplification results from assuming the same α value for each year. A new method for estimating α has been proposed recently by Grabczak et al. (1984). In that method the seasonal variations of the stable isotope content in precipitation are used to determine the contribution of the summer and winter infiltration to groundwater.

The input function is constructed by applying equation (32) to the known C_i and P_i data of each year, and for an assumed α value. In some cases equation (31) is applied if there is no surface run-off and α_i coefficients can be found from the actual evapotranspiration and precipitation data. The actual evapotranspiration is either estimated from pan-evaporimeter experiments (Andersen and Sevel, 1974) or by the use of an empirical formula for the potential evapotranspiration (Przewłocki, 1975). Monthly precipitation has to be measured in the recharge area, or can be taken from a nearby station. Monthly ^{3}H concentrations are known from publications of the IAEA (1965, 1970, 1971, 1973, 1975, and 1979) by taking data for the nearest station or eventually by applying correlated data of other stations (e.g., see Davis et al., 1967).

Present ^{3}H concentrations in precipitation at different observation stations are rather well correlated. However, in some areas, local short-time increases of the ^{3}H concentration are observed due to releases from nuclear installations. The expected future increase of the ^{3}H concentration caused by the development of nuclear power plants will be a result of releases at the earth surface; thus the distribution of ^{3}H in the atmosphere will probably be both less homogeneous geographically and more constant in time than at present.

Examples of typical ^{3}H input functions for two α values are shown in Fig. 1-8. Knowing the input function one may solve the direct problem, i.e., calculate the output concentrations for assumed models. Such results are helpful for estimating the potential applicability of the method and for planning the field work (e.g., in choosing sampling intervals). Examples of the solution of the direct problem for both versions of the dispersion model and for rather

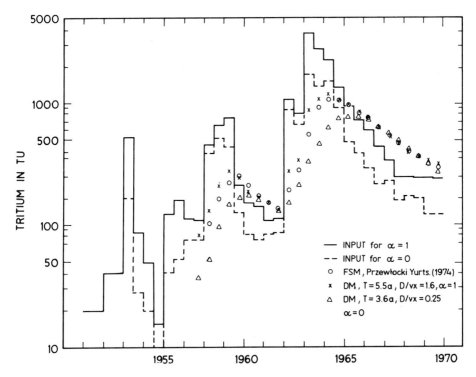

Fig. 1-8. An example showing the tritium input functions for the Modry Dul basin calculated for two extreme values of α (α is the ratio of summer to winter infiltration coefficients). The output concentrations are given for three models, which were fitted to the experimental data (not shown here) in the period 1966—1970. The FSM consisted of four boxes with individual recharges. The obtained turnover time was equal to 5.25 years for an assumed value of $\alpha = 1$. The dispersion model gave for $\alpha = 1$ a similar turnover time (5.5 years) whereas for $\alpha = 0$ the obtained turnover time was 3.6 years (after Małoszewski and Zuber, 1982, who reinterpreted original data of Dincer et al., 1970, and Przewłocki and Yurtsever, 1974; FSM, open circles).

extreme values of the dispersion parameter are given in Figs. 1-9 and 1-10. These figures show that in highly dispersive groundwater systems with turnover times not longer than about 50 years there is a tendency to the smoothing out of the ^3H concentrations. Systems with low values of the dispersion parameter are still interpretable.

The inverse problem, as mentioned earlier, consists in searching for the model of a given system for which the input and output concentrations are known. For this purpose the graphs representing the solutions to the direct problem can often be used. In such a case the graph which can be identified with the experimental data will represent the solution to the inverse problem. A more proper way is realized by searching for the best fit model. However, even the model of best fit is not necessarily the right one. In other words, the fitting procedure has to be used together with the geological knowledge, logic and intuition of the interpreter (Snyder and Stall, 1965). This means that all the available information should be used in selecting a proper type of the model prior to the fitting. If the selection is not possible prior to the fitting, and if more than one model give equally good fitting but with different values of the parameters, the selection has to be performed after the fitting.

Consider now again Fig. 1-8, where in addition to two typical input functions, the output concentrations calculated for three models are shown. The models were obtained by fitting to the experimental data (not shown here) collected in the time span of 1965—1969. All these models gave practically the same fitting. Were the observations performed for a longer time, there would be no difficulty in selecting the best fit model. The FSM is not recommended here because of a larger number of fitting parameters. Two versions of the dispersion model were obtained because fitting was performed for two extreme values of α. A more exact interpretation of this case would require either a longer record of the experimental data or a better estimate of the infiltration coefficient, as neither $\alpha = 0$ nor $\alpha = 1$ are realistic assumptions for the investigated system. However, the type of the model and the obtained mean values of T (4.5 years) and D/vx (approximately 1) seem to be consis-

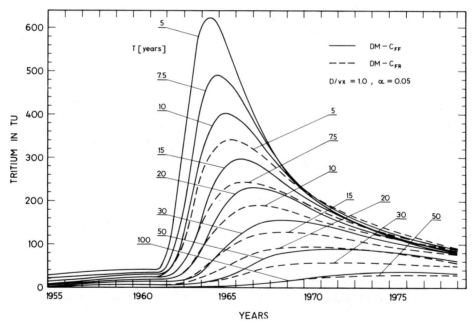

Fig. 1-9. An example of the tritium output concentration, in a region near Cracow (α taken as equal to 0.05), calculated for the dispersion model with a high value of the dispersion parameter (D/vx = 1.0). Note great differences between the case C_{FF} (sampling in outflowing water) and the case C_{FR} (sampling averaged over the aquifer depth).

tent with the geological knowledge of the system (a small drainage basin in the mountains) which may be approximated by the case given as in cross-section 2 of Fig. 1-5. Dincer et al. (1970) fitted the exponential and binomial models to the data of the Modry Dul basin for assumed $\alpha = 0$. The exponential model gave no satisfactory fitting whereas the binomial model gave the turn-over time relatively close to the value obtained from the DM for $\alpha = 0$, which is not astonishing if we recall our discussion on the applicability of the binomial model.

A very interesting study of a volcanic island, Cheju, Republic of Korea, was presented by Davis et al. (1970). The isotopic studies were aimed at finding if there is a chance for substantial groundwater resources. Many sites were sampled for ^3H and $\delta^{18}O$ measurements. The stable isotope measurements were helpful in determining the altitude of recharge, whereas ^3H served for estimating the turnover time. In the original work the binomial model was used for the determination of the turnover time. Małoszewski and Zuber (1982) using EPM reinterpreted two cases, which are shown in Fig. 1-11. Site 45 shows a short turnover time $T = 2.5$ years and the ratio of total volume to the exponential flow volume (η) equal to 1.7. The situation may be approximated by d in cross-section 3 of Fig. 1-5 though the actual flow takes place

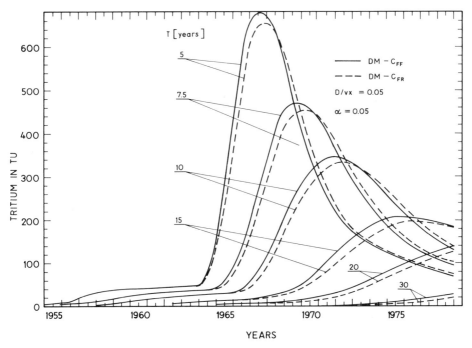

Fig. 1-10. An example of the tritium output concentration calculated for the same input function as in the case of Fig. 1-9, but for a low value of the dispersion parameter ($D/vx = 0.05$). Note low differences between the cases C_{FF} and C_{FR}.

in fissures and lava tunnels. This site is a medium altitude spring with the stable isotope composition showing the recharge area at somewhat higher altitude (Davis et al., 1970), thus the obtained model is consistent with other data and the obtained turnover time reliable.

Site 2 (Fig. 1-11) shows a long turnover time (21 years) and the η value of 1.1, which means that the main water body is in the recharge area, and that the water is led from there to the discharge site by a low-volume tunnel. This model is consistent with the stable isotope data (Davis et al., 1970) which show that this large coastal spring is recharged at a high altitude at the centre of the island. The exponential model ($\eta = 1.0$), which is also shown in Fig. 1-11 is inconsistent with the stable isotope data and has to be rejected in spite of reasonably good fitting.

Fig. 1-11 shows that in these two cases the EPM gives much better fitting than the binomial model, and thus one may expect that the obtained results are more reliable and at the same time more indicative as the parameter η has a clear physical meaning.

Another example was given by Przewłocki and Yurtsever (1974) and Przewłocki (1975) who interpreted the data of a small mountain spring in Grafendorf, Austria, with the aid of the FSM. The same data were reinter-

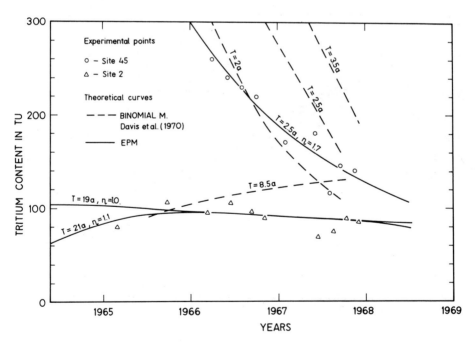

Fig. 1-11. Examples showing the solution of the inverse problem (parameters determined by fitting) for Cheju Island (reinterpretation after Małoszewski and Zuber, 1982, original data after Davis et al., 1970).

preted by Małoszewski and Zuber (1982) with the aid of EPM and DM (Fig. 1-12). At the first glance it may look strange that different types of models gave practically the same fitting. The use of a quantitative criterion of the best fit would not solve the problem because the scatter of the experimental points results most probably from the unsteady state, which is not taken into account in the models. However, one may notice that all the models gave similar values of the turnover time. On the other hand, other parameters of the models indicate that all the models describe a similar system. Namely, in FSM it was assumed that more than a half of the aquifer volume was in the recharge area, and next the flow took place through a number of cells without a direct recharge. EPM describes exactly the same situation ($\eta = 2.1$). DM is perhaps the least indicative in this respect but the low value of the dispersion parameter ($D/vx = 0.09$) is consistent with the findings of other models because recalling the discussion given on p. 23 this value shows a relatively short extend of the recharge area in comparison with the length of the aquifer.

FSM for Grafendorf was found using six cells. Recalling the discussion given in the section "Models and their hydrogeological significance" and applying general criteria on models one may state that models with the lowest

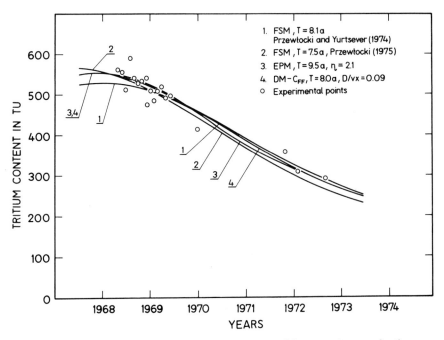

Fig. 1-12. An example showing how different models may give nearly the same fitting (Małoszewski and Zuber, 1982, original data from Przewlocki and Yurtsever, 1974, and Przewlocki, 1975, for a small mountain spring near Grafendorf, Austria).

number of fitting parameters give more reliable values of physical parameters than other models. In other words, in that example both EPM and DM are preferable to FSM.

It has also to be noted that the construction of FSM in the way described above requires a prior knowlegde on the structure of the system. In the case of EPM and DM the results are obtained without any additional assumptions and either are consistent with the known aquifer structure or supply new information.

^3H-^3He method

Torgersen et al. (1979) suggested that the ^{3}H-^{3}He method can be used for determining the water residence time in closed systems. The method involves the collection of about 10 grams of water. The dissolved gases are extracted and purified. ^{3}He, ^{4}He and Ne are then mass-spectrometrically measured to determine the concentrations in ml STP of gas per gram of water. The remaining water having been totally degassed is sealed in vacuo for ^{3}H analysis by the ^{3}He grow-in method. There may be five components of ^{3}He in groundwater: (1) ^{3}He of saturation — originating at the air-water interface in the infiltration area; (2) ^{3}He of oversaturation — resulting from air injections which are probably common in karstic systems (e.g, see Herzberd and Mazor, 1979) or even may happen in granular formations (Heaton and Vogel, 1981); (3) ^{3}He of mantle origin; (4) ^{3}He of crust origin; and (5) tritiugenic ^{3}He. Torgersen et al. (1979) describe how to find the last component in lake waters from the measured total ^{3}He, by making use of Ne and ^{4}He determinations, and by assuming that either mantle or crust components may be present, but not both. It still remains an open question if the determination of tritiugenic ^{3}He is possible with satisfactory accuracy in groundwaters. However, the ^{3}H-^{3}He method is worth considering owing to possibilities it offers. Here, in further considerations, the ^{3}He concentration will be used in the meaning of the tritiugenic ^{3}He concentration. Equation (9) can be rewritten for the parent ^{3}H in the form:

$$C_{^3H}(t) = \int_0^\infty C_{^3H,in}(t-t') \exp(-\lambda t') g(t') \, dt' \tag{33}$$

and for the daughter ^{3}He:

$$4.01 \times 10^{14} \, C_{^3He}(t) = \int_0^\infty C_{^3H,in}(t-t') [1 - \exp(-\lambda t')] g(t') \, dt' \tag{34}$$

where $C_{^3H,in}(t)$ is the variable tritium concentration at the entrance to the system expressed in tritium units (TU), and $C_{^3He}(t)$ is the ^{3}He concentration at the outlet expressed in ml STP per gram of water.

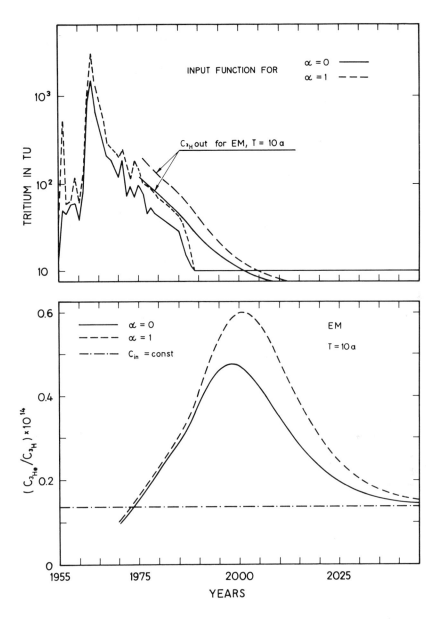

Fig. 1-13. A hypothetical example calculated for the exponential model (EM) with the turnover time of 10 years showing that the ^{3}H-^{3}He method will be more sensitive to the past tritium peak in the atmosphere in the near future than the ^{3}H method itself (Małoszewski and Zuber, 1983). Calculations performed for two extreme cases of the input concentrations $\alpha = 0$ means no summer infiltration, whereas $\alpha = 1$ means that infiltration coefficients in the summer and in the winter are equal. In the lower part the $C_{^3He}/C_{^3H}$ ratio is also shown for a hypothetical constant input concentration of 10 TU.

The ^{3}H-^{3}He method can be used in the form originally proposed; then the ratio of equation (34) to equation (33) is to be used for the theoretical curves, or in a more informative form (Małoszewski and Zuber, 1983) where the theoretical values of $C_{^3H}(t)$ and $C_{^3He}(t)$ are calculated separately and fitted individually to the experimental data. This latter method has the full advantages of the tritium method with the addition of the helium method, whereas the first version of the method may yield ambigous results because different models will often give the same ^{3}He/^{3}H ratio. This statement is rather self-evident because the same concentration ratio may result from an infinite number of different ^{3}He and ^{3}H concentrations. Małoszewski and Zuber (1983) showed this directly by calculating the output concentrations of ^{3}He and ^{3}H as well as their output ratio for a typical input concentration of ^{3}H and for different models.

The ^{3}H-^{3}He method has not been applied so far to groundwater systems. However, theoretical considerations of Małoszewski and Zuber (1983) show that this method is a very promising alternative to the tritium method for the next several tens of years, providing that the tritiugenic ^{3}He can be separated from the total ^{3}He. As an example may serve Fig. 1-13 where the ratio curves are shown for EM with a relatively short turnover time of 10 years. The ^{3}H-^{3}He method will give a strong peak in the concentration time curve, whereas the tritium method will give at the same time a concentration close to the concentration in the atmosphere.

^{85}Kr method

The ^{85}Kr concentration in the atmosphere results from nuclear power stations and plutonium productions for military purposes (see Chapter 2). Large scatter of the observed concentrations shows that there are local temporal variation of the ^{85}Kr activity (Rózański, 1979). However, yearly averages give relatively smooth input functions for both hemispheres (Grabczak et al., 1982). The input function for the northern hemisphere is shown in Fig. 1-14 together with examples of output concentrations for several models. The ^{85}Kr input function is based on measurements taken up to 1978 and extrapolated up to 1985. It is expected that in future a distinct increase of the ^{85}Kr concentration in the atmosphere will be observed again as the result of a wider development of nuclear power plants.

^{85}Kr as a gas is not a good tracer for surface waters or for systems having an easy contact with the atmospheric air. However, as a noble gas it should be an ideal tracer for confined aquifers providing no deep production of ^{85}Kr occurs. Its concentration is usually expressed as the number of desintegrations per millimole of Kr gas extracted from a water sample. Thus, its concentration practically does not depend on the solubility (i.e., temperature and partial pressure) in the recharge area. The present low concentrations of ^{85}Kr in most groundwaters require large samples (100—300 litres) and a complicated sampling procedure (Rózański and Florkowski, 1979). Further development of the measuring technique, and the expected increase in ^{85}Kr production and emission make this tracer a potential substitute for the tritium method.

The output concentrations for low values of the turnover time little depend on the model (e.g., curves for T = 5 years in Fig. 1-14). However, for higher values of the turnover time the differences between the models become more pronounced. Output concentrations calculated for different assumed models (Fig. 1-15) show that ^{85}Kr concentrations alone are not sufficient for a proper selection of the model. Thus this method should be used only as a complementary one to the tritium and ^{14}C methods.

^{14}C method

In hydrology the radiocarbon concentrations are usually reported in percent of modern carbon (pmc). 100 pmc means the radiocarbon concentration equal to that of 0.95 of the NBS standard in 1950 (for minor corrections of the standard see Stuiver and Polach, 1977). This concentration is supposed

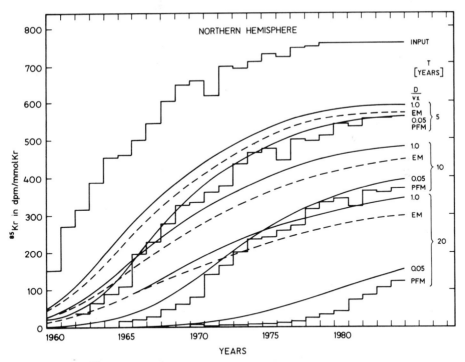

Fig. 1-14. The ^{85}Kr input function in the northern hemisphere and output concentrations calculated for three values of the turnover time, T, and for the following models: exponential (EM), piston flow (PFM), and dispersive with the dispersion parameter D/vx equal to 1.0 and 0.05 (Grabczak et al., 1982).

to represent that of plants in 1850, and is also used in hydrology as the so-called uncorrected initial concentration. However, the radiocarbon concentration oscillated in the past. The period of the oscillations was about 10,000 years with the last maximum (about 108.5 pmc) at about 6400 years B.P. and the last minimum (99 pmc) at about 1400 years B.P. (Damon et al., 1972; In Che Yang and Fairhall, 1972; Michael and Ralph, 1972.) Assuming that no other changes occurred in the last 50,000 years the mean concentration is 103.5 pmc. In the period of 1850—1950 the dilution of the radiocarbon by the antropogenic carbon (the Suess effect) prevailed over the natural tenden-

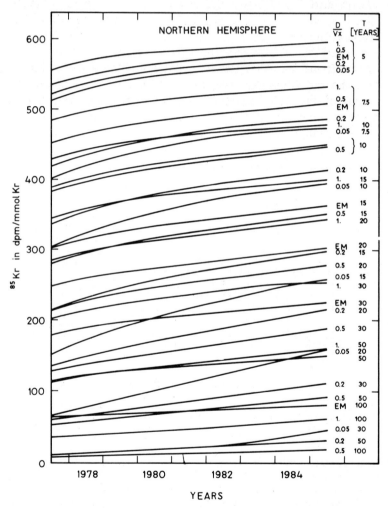

Fig. 1-15. The ^{85}Kr output concentrations calculated for the same input and models as in Fig. 1-14 but for a wider selection of turnover times (T) and dispersion parameters (D/vx).

cy and the ^{14}C concentration fell to about 98 pmc in 1940—1950. After 1950 an increase of the ^{14}C concentration in the atmosphere was observed as a result of nuclear bomb tests. Usually the radiocarbon concentrations are expressed in Δ^{14}C notation:

$$\Delta^{14}C, \%_{00} = \frac{A}{A_s}(1000 - 2\delta^{13}C + 50) - 1000 \qquad (35)$$

where A is the activity of the sample, A_s is the activity of the standard (0.95 of the NBS standard), and δ^{13}C is the ^{13}C concentration expressed as permil deviation from the PDB standard. Equation (35) physically means that the radiocarbon concentrations are expressed in permil deviations from the modern concentrations. The δ^{13}C term introduces corrections for ^{14}C fractionation effects, if δ^{13}C differs from the most common value for plants, i.e., $-25\%_{00}$.

Fig. 1-16 shows the radiocarbon concentrations in the atmosphere and some other materials of interest expressed in pmc. They were recalculated from the Δ^{14}C notation given in the original works into the pmc notation assuming δ^{13}C = $-25\%_{00}$, i.e., ^{14}C(pmc) = Δ^{14}C($\%_{00}$)/10 + 100. In other words

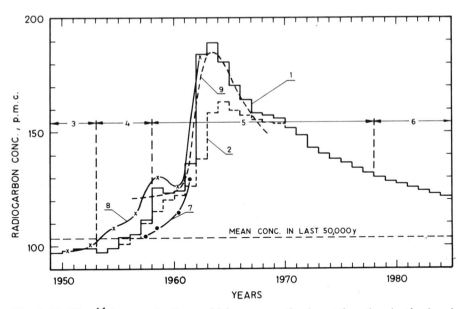

Fig. 1-16. The ^{14}C concentrations which serve as the input function in the bomb era expressed in pmc for δ^{13}C = $-25\%_{00}$: 1 = atmosphere in the northern hemisphere, 2 = atmosphere in the southern hemisphere, 3 = German wines (Institut für Radiohydrometrie, 1978), 4 = atmosphere (Walton et al., 1967), 5 = atmosphere (Levin et al., 1980), 6 = atmosphere exponentially extrapolated, 7 = soil CO_2 (Münnich and Roether, 1963), 8 = plants near Bonn (Tamers and Scharpenseel, 1970), 9 = oak near Hamburg (Berger, 1972).

the data of Fig. 1-16 are supposed to represent the input function for the recharge areas covered by plants with $\delta^{13}C = -25‰$.

The radiocarbon input and output concentrations are calculated here as if there were no changes due to the carbon hydrochemistry. In other words the concentrations are calculated as if the changes in radiocarbon content were caused only by the radioactive decay and given flow pattern in the system. However, the interpretation of ^{14}C in any groundwater system cannot be performed without taking into account the carbon hydrochemistry. In this respect the reader is referred to the most recent works of Wigley et al. (1978), Reardon and Fritz (1978), Fontes and Garnier (1979), and Mook (1980). In these works as well as in all the other related ones, the piston flow model was used for determining the age of water. The corrections resulting from a given model of the carbon hydrochemistry were applied to the initial carbon content. In such a case the initial activity is corrected to obtain a value which would be expected as a result of all processes except radioactive decay (Wigley et al., 1978). In order to apply more realistic flow models than PFM it is necessary to extend this definition of the so-called adjustment problem. Thus, the adjustment problem is to correct the initial activity of ^{14}C to the value which would be expected as a result of all processes except radio-

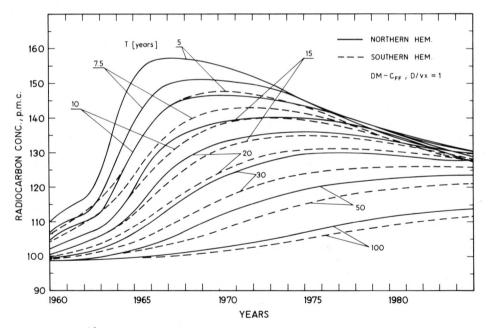

Fig. 1-17. The ^{14}C output concentrations in both hemispheres calculated for the dispersion model with a high value dispersion parameter (D/vx) and for sampling in the outflow (case C_{FF}). The concentrations are given for the total dissolved carbon as if there were no changes due to carbon hydrochemistry.

active decay and hydrodynamic effects. In other words, the input function or the calculated output concentrations have to be multiplied by the correction factor resulting from a given model of the ^{14}C hydrochemistry. However, it seems that a more convenient way, proposed by Grabczak et al. (1982) is to divide the experimental output concentrations by that factor and next to fit the uncorrected theoretical flow model to the corrected experimental data. In such a case the experimental output concentration is corrected to obtain the value which would be observed if there were no changes caused by the hydrochemical processes.

Figs. 1-17 and 1-18 show examples of the ^{14}C output concentrations for extreme values of the dispersion parameter. It is evident that in the case of a high dispersion the concentration exceeding the prebomb level does not necessarily mean that the mean turnover time is shorter than the duration of the bomb era. These figures also show that observations of changes in ^{14}C concentration should be very helpful in investigations of some systems having short turnover times.

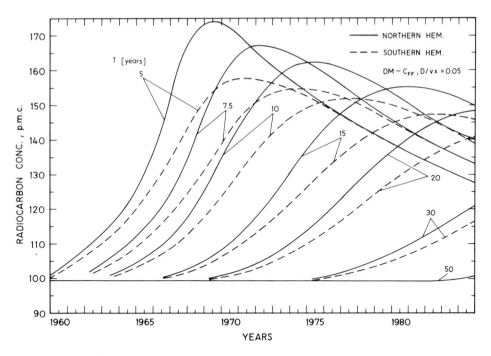

Fig. 1-18. The ^{14}C output concentrations in both hemispheres calculated for the dispersion model with a low value of the dispersion parameter (D/vx) and for sampling in the outflow (case C_{FF}). The concentrations are given for the total dissolved carbon as if there were no changes due to carbon hydrochemistry.

Other possibilities

In the previous subsections we discussed methods either routinely used (^{3}H, ^{14}C) or highly promising (^{85}Kr, ^{3}H-^{3}He). In some laboratories other possibilities are investigated. Among the radioisotopes under the consideration are: ^{32}Si, ^{39}Ar, ^{36}Cl, and ^{81}Kr.

Other non-radioactive environmental tracers can be used as complementary tools to the radioisotope tracers considered above. The most promising are freon-11 (CCl$_3$F), and freon-12 (CCl$_2$F$_2$), propelants whose concentrations in the atmosphere strongly increase due to the activity of man. Their applicability in hydrology has been considered by Thompson and Hayes (1979) and Grabczak et al. (1982).

COMBINED INTERPRETATION

The examples discussed on pp. 35—39 indicate that in practice the solution to the inverse problem is often ambiguous. There are two main sources of difficulties: the first usually results from a short record of the experimental data, whereas the second is related to the steady state approximation. To obtain meaningful results the experimenter cannot restrict himself to the radioisotope data. As mentioned, all the other available information should be included into considerations related to the selection of the right model from those of the best fit. There are no two identical natural groundwater systems in the world. Thus, each system has to be considered individually, though the knowledge of other known case studies (i.e., experience) is helpful in its interpretation. An example of the use of the stable isotope data for selecting a proper model was given earlier (Cheju Island, Site 2; see p. 38). Here we shall consider possibilities resulting from the existence of several radioisotopes.

From the considerations of the previous section it becomes clear that for systems having short turnover times several methods exist. Their combined use proposed by Grabczak et al. (1982) leads to a better selection of the model and consequently gives more reliability to the obtained results. These authors gave several examples of the combined interpretation involving ^{3}H, ^{85}Kr and ^{14}C. In spite of the preliminary character of their studies we shall consider here two of their examples in order to illustrate both the idea of the combined interpretation and the limitations encountered in observations of gaseous tracers. Table 1-2 contains data of a small karstic spring in a carbonate formation near Cracow. As no long record of the data was available, ^{85}Kr was used for the preliminary selection of several possible models (third column of Table 1-2) by making use of Figs. 1-14 and 1-15. Next, by making use of curves of the type shown in Figs. 1-9 and 1-10 and in Figs. 1-17 and 1-18 the expected concentrations of ^{3}H and ^{14}C were estimated, respectively. Their

TABLE 1-2

Environmental radioisotope data of the Eliseus Spring, Poland, and their interpretation (Grabczak et al., 1982)

Sampling date	^{85}Kr measured (dpm/mmol Kr)	Models[a] selected on the basis of ^{85}Kr	Tritium (TU)		^{14}C (pmc)	
			measured	expected from model[b]	corrected[c]	expected from model
7.02.79	381 ± 74	EM, T = 2 years β = 0.33	44 ± 2 45 ± 2 46 ± 2	47	108 ± 10	⩽ 122
17.02.79						
3.10.79						
		EM, T = 11.5 years β = 0		95		138
		DM, T = 9 years D/vx = 0.05 β = 0		90		154
		DM, T = 14 years D/vx = 1.0 β = 0		85		138

[a] T is the turnover time, D/vx is the dispersion parameter, and β is the ratio of non-radioactive water flow to the total flow for assumed mixing of two water components.
[b] In addition to the parameters given in the third column, the α coefficient was arbitrarily taken as equal to 0.05 (α is the ratio of summer to winter infiltration coefficients).
[c] Experimental output concentrations multiplied by $-25^0/_{00}/\delta^{13}$C, where δ^{13}C observed for TDC was equal to $-12.0^0/_{00}$.

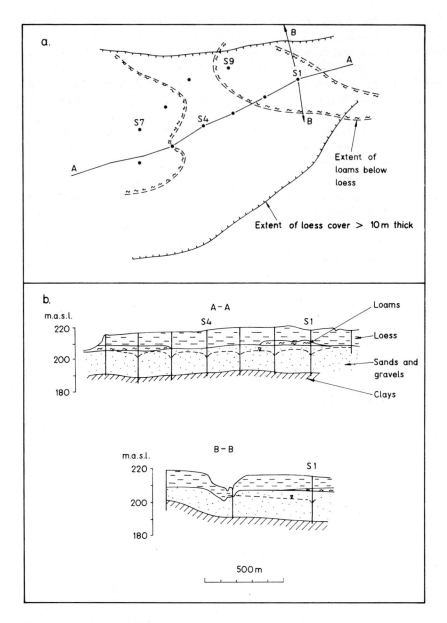

Fig. 1-19. Geological situation of a semiconfined aquifer for which the interpretation of the environmental radioisotope data is given in Fig. 1-20 and Table 1-3. (a) Location of the pumping wells, numbers are given for sampled wells; (b) geological cross-section.

comparison with the measured values shows that the first model listed in Table 1-2 should be selected. The ^{14}C concentrations were corrected with the aid of one of the simplest hydrochemical models, i.e., by dividing the experimental values by $-\delta^{13}C/25‰$ (see references in the section "^{14}C method"). This example seems to show that in spite of the lack of a long record of the output concentrations it is possible, under favourable conditions, to perform the interpretation by the combined use of several radioisotopes.

However, the preliminary interpretation of this case study is given here only as an example of the principles of the combined interpretation. The obtained results are rather doubtful because the concentration of ^{85}Kr in the karstic system under study may be influenced by possible contacts with the atmospheric air. Another source of errors in the estimation of the turnover time may result form the delay caused by the diffusion of tracers into the porous

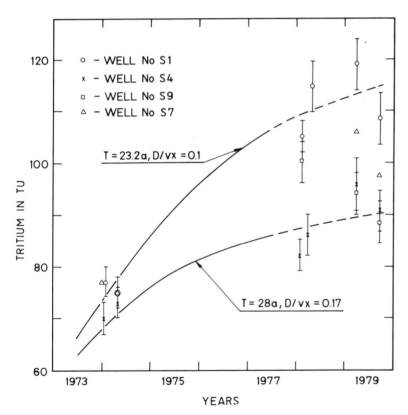

Fig. 1-20. The interpretation of tritium data of the aquifer shown in Fig. 1-19. Fitting procedure yielded the dispersion model whose parameters (turnover time, T, and the dispersion parameter, D/vx) are given in the figure. The ratio of summer to winter infiltration coefficients (α) was taken as equal to 0.05.

TABLE 1-3

Comparison of measured ^{14}C and ^{85}Kr concentrations with the expected values found by applying models fitted to the tritium data of Fig. 1-20; Leaky aquifer near Cracow, Poland (Grabczak et al., 1982)

Sampling site	Date	δ^{13}C (‰ vs. PDB)	^{14}C (pmc)			^{85}Kr (dpm/mmol Kr)	
			measured	corrected[a]	expected	measured	expected
Well S1	10.2.78	−15.8	77.4 ± 2.0	122 ± 10	125	208 ± 67	~90
	30.3.79	−13.7	54.6 ± 1.3 (?)	100 ± 10	128	—	—
	26.9.79	−15.5	81.0 ± 2.0	131 ± 10	131	—	—
Well S4	31.1.78	−15.3	77.8 ± 2.0	127 ± 10	119	172 ± 84	~70
	26.9.79	−15.6	76.0 ± 2.0	122 ± 10	122	—	—
Well S9	25.2.78	−17.5	78.8 ± 2.0	113 ± 10	122	247 ± 80	~80
	26.9.79	−17.4	80.6 ± 2.0	116 ± 10	126	—	—

[a] Experimental output concentrations multiplied by $-25‰/\delta^{13}C$.

matrix. As mentioned in the Introduction this effect has been pointed out recently by several investigators.

The second example is related to the investigations of abstraction wells in a sandy aquifer covered by a semipremeable loess layer (Fig. 1-19). Tritium data and the parameters of the fitted models are given in Fig. 1-20, whereas Table 1-3 contains ^{14}C and ^{85}Kr data compared with values estimated from these models. In this case of a sandy aquifer, the simple hydrochemical model works better than in the previous example and the ^{14}C data agree satisfactorily with the selected model. The measured ^{85}Kr values are about 2.5—3 times higher than those expected. Similarly first measurements of freon-11 indicated concentrations about three times higher than expected, whereas observed concentrations of the tritiugenic ^{3}He were two to three times lower than expected (Grabczak et al., 1982). These facts can be explained by the gaseous diffusion of ^{85}Kr and freon-11 from the atmosphere to the unsaturated zone whereas in the case of ^{3}He the concentration gradient was opposite because in that case the highest production rate of ^{3}He was in the unsaturated zone (the ^{3}H peak was there as it can be deduced from Fig. 1-20), and consequently ^{3}He was escaping to the atmosphere. This example is of great importance because it shows that whenever the infiltration time through the unsaturated zone is not negligible in comparison with the total turnover time a quantitative interpretation of the gaseous tracers will be hindered. In the case of tritium the influence of gaseous diffusion in the unsaturated zone is negligible because the main reservoir of the tracer is in the liquid form.

Recalling the discussion of the sections "Models and their hydrogeological significance" and "Dispersivity and the dispersion parameter", it may be interesting to note that the relatively low dispersion parameter (0.1—0.17) results in that case from a high contribution of the local infiltration process to the total value of the turnover time. This example also confirms our earlier statement that in the cases given in cross-section 2 of Fig. 1-5 the dispersion model should be applicable if the infiltration time is not negligibly short.

When interpreting concentrations of different tracers the same mathematical model of the flow system has to be applied to each of them. If the obtained results are inconsistent then another model is to be sought or the assumption revised. The obtained parameters and the type of the model have to be consistent with our knowledge of the geology of the investigated system. If they are not then another model has to be sought, or the geology studied more deeply.

The earlier remark related to a consistent interpretation of a given system, i.e., to applying the same model to different tracers, is particularly actual. Many investigators interpret ^{14}C with the aid of the PFM in the presence of tritium. Such a practice is unacceptable theoretically because tritium indicates water recharged in the bomb era, which means that the ^{14}C input was also variable. The PFM can be applied in systems with a low dispersion for a constant input (see Fig. 1-7) or for a monotonically changing input (see Fig.

1-14 for ^{85}Kr, or fig. 3 in Grabczak et al., 1982, for freon-11). For variable inputs, as it is the case for tritium and modern ^{14}C, the PFM does not even approximate properly systems with a low dispersion, as it can be seen by comparison of the theoretical input concentration of ^{14}C in Fig. 1-16 with the theoretical output concentrations curves in Fig. 1-18. For the PFM the output curves should follow the input curve shifted in time by a value of T years (the radioactive decay can be neglected because $T \ll t_{1/2}$). As two other arguments one may note that the presence of tritium suggests that the geological situation given in cross-section 1 of Fig. 1-5 is little probable and that there are no examples of a successful use of the PFM to the bomb-era tritium.

CONCLUDING REMARKS

The theoretical models and examples of their applications give an idea on how the environmental radioisotopes can be used for determining the parameters of groundwater systems. The considerations were limited to the steady state systems, i.e., to systems with a steady flow and volume of water. It is left to the experimenter's judgement, based on logic and intuition, which systems can be treated as being in a steady state, and how the values used are to be averaged if necessary. The time scale both for calculations and for observations has to be chosen in accordance with the steady state assumption. Examples given in the text, as well as many other examples of original works, indicate that even under this assumption the solution to the inverse problem remains often ambiguous. Thus, it is difficult to expect that the unsteady problems can be solved properly, because the number of fitting parameters increases, i.e., instead of a constant T, two variables appear, $Q(t)$ and $V(t)$. Obviously a long record of data is needed to obtain any meaningful result of $Q(t)$ and $V(t)$, or eventually in approximation of $T(t)$. Thus, the unsteady state approach cannot be applied meaningfully to the systems with long turnover times. The reader interested in the mathematical treatment of unsteady systems is referred to Lewis and Nir (1978).

It has to be kept in mind that systems which can be treated as being in a steady state at present, were not necessarily steady state in the past. Particular care is needed when interpreting the environmental radioisotopes having long half-life time, i.e., radiocarbon ($t_{1/2}$ = 5730 years) and in the future ^{36}Cl ($t_{1/2}$ = 305,000 years) or ^{81}Kr ($t_{1/2}$ = 210,000 years). The present measuring techniques available in most laboratories allow to measure the ^{14}C concentration down to 1 pmc. Thus the method covers theoretically, depending on the model applied (see Fig. 1-7), several tens of thousands of years at least. Even if a given system can be treated as a steady state in the Holocene (last 10,000 years), it is unacceptable to assume that the same recharge conditions existed through the last glaciation. In other words, the turnover times greater than 10,000 years have to be treated qualitatively only. However, syngenetic or paleoinfiltration waters can be interpreted with the aid of equation (11) (providing an existence of a proper radioisotope) in terms of the geological age if one can assume that the time period in which the water remained stagnant is much longer than the period in which water was in a dynamic state.

In spite of all these difficulties and limitations the environmental radioisotopes have proved their usefulness, especially in investigations of little known, or karst and fissured systems. General discussion on that subject was given in the Introduction. Here we may repeat our conviction that the mathematical models are both helpful in planning of field work and in interpretation of the field data for obtaining some basic parameters. They are also helpful in the understanding of the transport of soluble matter in the scale of the whole system. Thus they will be useful in the prediction of pollutant behaviour in the case of pollutants appearing in the whole recharge area.

ACKNOWLEDGEMENTS

The author is indebted to J. Ch. Fontes for the encouragement to this work and many helpful remarks and to A. Kreft for valuable discussions.

REFERENCES

Allison, G.B. and Hughes, M.W., 1974. Environmental tritium in the unsaturated zone: estimation of recharge to an unconfined aquifer. In: *Isotope Techniques in Groundwater Hydrology 1974, Vol. I*, IAEA, Vienna, pp. 57—72.
Andersen, L.J. and Sevel, T., 1974. Six years' environmental tritium profiles in the unsaturated and saturated zones. In: *Isotope Techniques in Groundwater Hydrology 1974, Vol. I*. IAEA, Vienna, pp. 3—20.
Berger, R., 1972. Tree ring calibration of radiocarbon dates. In: *Proceedings 8th International Conference on Radiocarbon Dating*. Royal Society of New Zealand, Wellington, pp. A97—A104.
Blume, H.P., Zimmermann, U. and Münnich, K.O., 1967. Tritium tagging of soil moisture: the water balance of forest soils. In: *Isotope and Radiation Techniques in Soil Physics and Irrigation Studies*. IAEA, Vienna, pp. 315—332.
Bredenkamp, D.B. and Vogel, J.C., 1970. Study of a dolomitic aquifer with carbon-14 and tritium. In: *Isotope Hydrology 1970*. IAEA, Vienna, pp. 349—372.
Brigham, W.E., 1974. Mixing equations in short laboratory cores. Soc. Pet. Eng. J., 14: 91—99.
Coats, K.H. and Smith, B.D., 1964. Dead-end pore volume and dispersion in porous media. Soc. Pet. Eng. J., 4: 73—84.
Damon, P., Long, A. and Wallick, E., 1972. Dendrochronology calibration of the C-14 time scale. In: *Proceedings 8th International Conference on Radiocarbon Dating*. Royal Society of New Zealand, Wellington, pp. A28—A44.
Davis, G.H., Dincer, T., Florkowski, T., Payne, B.R. and Gattinger, T., 1967. Seasonal variations in the tritium content of groundwaters of the Vienna basin. In: *Isotopes in Hydrology*. IAEA, Vienna, pp. 451—473.
Davis, G.H., Lee, Ch.K., Bradley, E. and Payne, B.R., 1970. Geohydrologic interpretations of a volcanic island from environmental isotopes. Water Resour. Res., 6: 99—109.

Davis, S.N. and De Wiest, R.J.M., 1967. *Hydrology*. Wiley, New York, N.Y., 463 pp.
De Josselin de Jong, G., 1958. Longitudinal and transverse diffusion in granular deposits. *Trans. Am. Geophys. Union*, 39: 67—74.
De Smedt, F. and Wieringa, P., 1979. Mass transfer in porous media with immobile water. *J. Hydrol.*, 41: 59—67.
De Smedt, F., Wierenga, P.J. and Van der Beken, A., 1981. *Theoretical and Experimental Study of Solute Movement through Porous Media with Mobile and Immobile Water*. Laboratory of Hydrology, Vrije Universiteit Brussel, Brussels, 219 pp.
Dinçer, T., Payne, B.R., Florkowski, T., Martinec, T. and Tongiorgi, E., 1970. Snowmelt runoff from measurements of tritium and oxygen-18, *Water Resour. Res.*, 6: 110—124.
Elmore, D., Anantaraman, N., Fulbright, H.W., Gove, H.E., Hans, H.S., Nishiizumi, K., Murell, M.T. and Honda, M., 1980. Half life of ^{32}Si using tandem accelerator mass spectrometry. *Phys. Rev. Lett.*, 45: 589—592.
Eriksson, E., 1958. The possible use of tritium for estimating groundwater storage. *Tellus*, 10: 472—478.
Fischer, H.B., 1967. The mechanics of dispersion in natural streams. *Proc. ASCE, J. Hydraul. Div.*, 93: 187—216.
Fontes, J.Ch. and Garnier, J.M., 1979. Determination of the initial ^{14}C activity of the total dissolved carbon: a review of the existing models and a new approach. *Water Resour. Res.*, 15: 369—413.
Fried, J.J. and Combarnous, M.A., 1971. Dispersion in porous media. *Adv. Hydrosci.*, 7: 169—282.
Gardner, R.P. and Ely, R.L., 1967. *Radioisotope Measurement Applications in Engineering*. Reinhold, New York, N.Y., 483 pp.
Gardner, R.P., Felder, R.M. and Dunn, T.S., 1973. Tracer concentration responses and moments for measurements of laminar flow in circular tubes. *Int. J. Appl. Radiat. Isot.*, 24: 253—270.
Gaudet, J.P., Jegat, H. and Vachaud, G., 1976. Etude du mécanisme des transferts d'eau et de soluté en zone non saturée avec prise en compte d'une fraction liquide immobile. *Houille Blanche*, 228: 243—249.
Geyh, M.A. and Backhaus, G., 1979. Hydrodynamics aspects of carbon-14 groundwater dating. In: *Isotope Hydrology 1978, Vol. II*. IAEA, Vienna, pp. 631—643.
Glueckauf, E., 1980. The movement of solutes through aqueous fissures in porous rock. *Harwell, Rep.*, AERE-R-9823.
Glueckauf, E., 1981. The movement of solutes through aqueous fissures in micro-porous rock during borehole experiments. *Harwell, Rep.*, AERE-R 10043.
Grabczak, J., Zuber, A., Małoszewski, P., Rózański, K., Weiss, W. and Šliwka, I., 1982. New mathematical models for the interpretation of environmental tracers in groundwaters and the combined use of tritium, C-14, Kr-85, He-3, and freon-11 methods. *Beitr. Geol. Schweiz, Hydrol. Ser.*, 28(2): 395—405; and Vodn. Resur. (in press) (in Russian).
Grabczak, J., Małoszewski, P., Rózański, K. and Zuber, A., 1984. Estimation of the tritium input function with the aid of stable isotopes. *Catena* (in press).
Grisak, G.E. and Pickens, J.F., 1980. Solute transport through fractured media, 1. The effect of matrix diffusion. *Water Resour. Res.*, 16: 719—730.
Grisak, G.E. and Pickens, J.F., 1981. An analytical solution for solute transport through fractured media with matrix diffusion. *J. Hydrol.*, 52: 47—57.
Grisak, G.E., Pickens, J.F. and Cherry, J.A., 1980. Solute transport through fractured media, 2. Column study of fractured till. *Water Resour. Res.*, 16: 731—739.
Heaton, T.H.E. and Vogel, J.C., 1981. "Excess air" in groundwater. *J. Hydrol.*, 50: 201—216.
Herzberd, O. and Mazor, E., 1979. Hydrological application of noble gases and tempera-

ture measurements in underground water systems: examples from Israel. *J. Hydrol.*, 41: 217—231.
Himmelblau, D.M. and Bischoff, K.B., 1968. *Process Analysis and Simulation: Deterministic Systems.* Wiley, New York, N.Y.
IAEA, 1965, 1970, 1971, 1973, 1975, and 1979. *IAEA Tech. Rep. Ser.*, 96, 117, 129, 147, 165, and 192.
In Che Yang, A. and Fairhall, A.W., 1972. Variations in natural radiocarbon during the last 11 millenia and geophysical mechanism for producing them. In: *Proceedings, 8th International Conference on Radiocarbon Dating.* Royal Society of New Zealand, Wellington, pp. A44—A58.
Inoue, Y. and Kaufman, W.J., 1963. Prediction of movement of radionuclides in solution through porous media. *Health Phys.*, 9: 705—715.
Institut für Radiohydrometrie, 1978. Jahresbericht, 1978. *Ges. Strahlen- Umweltforsch.*, Ber. R 169.
Kaufman, W.J. and Orlob, G.T., 1956a. Measuring ground water movement with radioactive and chemical tracers. *J. Am. Assoc. Waterworks*, 48: 559—572.
Kaufman, W.J. and Orlob, G.T., 1956b. An evaluation of groundwater tracers. *Trans. Am. Geophys. Union*, 37: 297—306.
Kreft, A. and Zuber, A., 1978. On the physical meaning of the dispersion equation and its solutions for different initial and boundary conditions. *Chem. Eng. Sci.*, 33: 1471—1480.
Kreft, A. and Zuber, A., 1979. On the use of the dispersion model of fluid flow. *Int. J. Appl. Radiat. Isot.*, 30: 705—708.
Kutschera, W., Henning, W., Paul, M., Smither, R.K., Stephenson, E.J., Yntema, J.L., Alburger, O.E., Cumming, J.B. and Harbottle, G., 1980. Measurements of ^{32}Si half life via accelerator mass spectrometry. *Phys. Rev. Lett.*, 45: 592—596.
Lenda, A. and Zuber, A., 1970. Tracer dispersion in groundwater experiments. In: *Isotope Hydrology 1970.* IAEA, Vienna, pp. 619—641.
Levenspiel, O. and Turner, J.C.R., 1970. The interpretation of residence time experiments. *Chem. Eng. Sci.*, 25: 1605—1609.
Levenspiel, O., Lai, B.W. and Chatlynne, C.Y., 1970. Tracer curves and the residence time distribution. *Chem. Eng. Sci.*, 25: 1611—1613.
Levin, I., Münnich, K.O. and Weiss, W., 1980. The effect of anthropogenic CO_2 and ^{14}C sources on the distribution of ^{14}C in the atmosphere. *Radiocarbon*, 22: 379—391.
Lewis, S. and Nir, A., 1978. On tracer theory in geophysical systems in steady and non-steady state, II. Non-steady state — theoretical introduction. *Tellus*, 30: 260—271.
Maloszewski, P., Witczak, S. and Zuber, A., 1980. Prediction of pollutant movement in groundwaters. In: *Nuclear Techniques in Groundwater Pollution Research.* IAEA, Vienna, pp. 61—81.
Małoszewski, P. and Zuber, A., 1982. Determining the turnover time of groundwater systems with the aid of environmental tracers, I. Models and their applicability. *J. Hydrol.*, 57: 207—231.
Małoszewski, P. and Zuber, A., 1983. Theoretical possibilities of the ^{3}H-^{3}He method in investigations of groundwater systems. *Catena*, 10: 189—198.
Micheal, H. and Ralph, K., 1972. Discussion of radiocarbon dates obtained from precisely dated sequoia and bristlecone pine samples. In: *Proceedings 8th International Conference on Radiocarbon Dating.* Royal Society of New Zealand, Wellington, pp. A11—A28.
Molinari, J. and Rochon, J., 1976. Mesure des paramètres de transport de l'eau et des substances en solution en zone saturée. *Houille Blanche*, 228: 223—242.
Mook, W.G., 1980. Carbon-14 in hydrological studies. In: P. Fritz and J.Ch. Fontes (Editors), *Handbook of Environmental Isotope Geochemistry, Vol. 1. The Terrestrial Environment,* A. Elsevier, Amsterdam, pp. 49—74.

Münnich, K.O., 1968. Isotopen-Datierung von Grundwasser. Naturwissenschaften, 4: 158—163.

Münnich, K.O. and Roether, W., 1963. A comparison of carbon-14 and tritium ages of groundwater. In: *Radioisotopes in Hydrology*. IAEA, Vienna, pp. 397—406.

Münnich, K.O., Roether, W. and Thilo, L., 1967. Dating of groundwater with tritium and ^{14}C. In: *Isotopes in Hydrology*. IAEA, Vienna, pp. 305—320.

Neretnieks, I., 1980. Diffusion in rock matrix — an important factor in radionuclide retardation? *J. Geophys. Res.*, 85: 4379—4397.

Nir, A. and Lewis, S., 1975. On tracer theory in geophysical systems in the steady and nonsteady state, I. *Tellus*, 27: 372—383.

Perkins, R.K. and Johnstone, O.C., 1963. A review of diffusion in porous media. *Soc. Pet. Eng. J.*, 3: 70—84.

Pinder, G.F., 1973. A Galerkin-finite element simulation of groundwater contamination on Long Island, New York. *Water Resour. Res.*, 9: 1657—1669.

Przewłocki, K., 1975. Hydrologic interpretation of the environmental isotope data in the Eastern Styrian Basin. *Steir. Beitr. Hydrol.*, 27: 85—133.

Przewłocki, K. and Yurtsever, Y., 1974. Some conceptual mathematical models and digital simulation approach in the use of tracers in hydrological systems. In: *Isotope Techniques in Groundwater Hydrology 1974, Vol. II.* IAEA, Vienna, pp. 425—450.

Reardon, E.J., and Fritz, P., 1978. Computer modelling of groundwater ^{13}C and ^{14}C isotope compositions. *J. Hydrol.*, 36: 201—224.

Richardson, J.G., 1961. Flow through porous media. In: V.L. Streeter (Editor), *Handbook of Fluid Dynamics*. McGraw-Hill, New York, N.Y. pp. 16-11 and 16-92.

Rifai, E.M.N., Kaufman, W.J. and Todd, D.K., 1956. Dispersion Phenomena in Laminar Flow through Porous Media. *Sanit. Eng. Res. Lab., Univ. Calif., Rep.*, 3.

Robertson, J.B., 1974. Application of digital modelling to the prediction of radioisotope migration in groundwater. In: *Isotope Techniques in Groundwater Hydrology, Vol. II.* IAEA, Vienna, pp. 451-478.

Różański, K., 1979. Krypton-85 in the atmosphere 1950—1977: a data review. *Environ. Int.*, 2: 139-143.

Różański, K. and Florkowski, T., 1979. Krypton-85 dating of groundwater. In: *Isotope Hydrology 1978, Vol. II.* IAEA, Vienna, pp. 949—961.

Saffman, P.G., 1959. Dispersion in flow through a network of capillaries. *Chem. Eng. Sci.*, 11: 125—129.

Safonov, M.S., Rozen, A.M. and Voskresienskij, N.M., 1979. Analysis of unidimensional flow model of mass transport by moving medium from the point of view of two possible ways of concentration averaging: over cross-section or by volumetric flow rate. *Teor. Osn. Khim. Tekhnol.*, 13: 17—23 (in Russian).

Schmalz, B.L. and Polzer, W.F., 1969. Tritiated water distribution in unsaturated soil. *Soil Sci.*, 108: 43—47.

Shamir, U.Y. and Harleman, D.R.F., 1967. Dispersion in layered porous media. *J. Hydraul. Div., ASCE*, 89: 67—85.

Smith, D.B., Wearn, P.L., Richards, H.J. and Rowe, P.C., 1970. Water movement in the unsaturated zone of high and low permeability strata by measuring natural tritium. In: *Isotope Hydrology 1970*. IAEA, Vienna, pp. 73—88.

Snyder, W.M., and Stall, J.B., 1965. Men, models, methods and machines in hydrologic analysis. *J. Hydraul. Div., ASCE*, 91: 85—99.

Stuiver, M. and Polach, H.A., 1977. Reporting of ^{14}C data. *Radiocarbon*, 19: 355—363.

Sudicky, A.E. and Frind, E.O., 1981. Carbon-14 dating of groundwater in confined aquifers: implications of aquitard diffusion. *Water Resour. Res.*, 17: 1060—1064.

Tamers, M.A. and Scharpenseel, H.W., 1970. Sequential sampling of radiocarbon in groundwater. In: *Isotope Hydrology 1970*. IAEA, Vienna, pp. 241—257.

Tang, D.H., Frind, E.O. and Sudicky, E.A., 1981. Contaminant transport in fractured porous media: analytical solution for a single fracture. *Water Resour. Res.*, 17: 555—564.
Taylor, G., 1953. Dispersion of soluble matter in solvent flowing slowly through a tube. *Proc. R. Soc. London, Ser. A*, 219: 186—203.
Taylor, G., 1954. The dispersion of matter in turbulent flow through a pipe. *Proc. R. Soc. London, Ser. A*, 223: 446—468.
Thompson, G.M. and Hayes, J.M., 1979. Trichlorofluoromethane in groundwater- a possible tracer and indicator of groundwater age. *Water Resour. Res.*, 15: 546—554.
Torgersen, T., Clarke, W.B. and Jenkins, W.J., 1979. The tritium-helium-3 method in hydrology. In: *Isotope Hydrology 1978, Vol. II*. IAEA, Vienna, pp. 917—930.
Villermeaux, J. and van Swaay, W.P.M., 1969. Modèle représentatif de la distribution des temps de séjour dans un réacteur semi-infini à dispersion axiale avec zones stagnantes. *Chem. Eng. Sci.*, 24: 1097—1111.
Vogel, J.C., 1970. Carbon-14 dating of groundwater. In: *Isotope Hydrology 1970*. IAEA, Vienna, pp. 225—240.
Walton, A., Baxter, M.S., Callow, W.J. and Baker, M.J., 1967. Carbon-14 concentrations in environmental materials and their temporal fluctuations. In: *Radioactive Dating and Methods of Low-Level Counting*. IAEA, Vienna, pp. 41—49.
Weiss, W., Bullacher, J., and Roether, W., 1979. Evidence of pulsed discharges of tritium from nuclear energy installations in Central European precipitation. In: *Behaviour of Tritium in the Environment*. IAEA, Vienna, pp. 17—30.
Wigley, T.M.L., Plummer, L.N. and Pearson, F.J., Jr., 1978. Mass transfer and carbon isotope evolution in natural water systems. *Geochim. Cosmochim. Acta*, 42: 1117—1139.
Zuber, A., 1974. Theoretical possibilities of the two-well pulse method. In: *Isotope Techniques in Groundwater Hydrology, Vol. II*. IAEA, Vienna, pp. 227—294.

Chapter 2

ISOTOPES IN CLOUD PHYSICS: MULTIPHASE AND MULTISTAGE CONDENSATION PROCESSES

JEAN JOUZEL

INTRODUCTION

Cloud physics is essentially devoted to the study of dynamical, thermodynamical and microphysical processes leading to the formation of different types of clouds and precipitations. This field is largely concerned with the transport of atmospheric water and with the physics of its phase changes. For that reason, the application of environmental isotopes in cloud physics is practically limited to the constitutive isotopes of the water molecules mainly $^1H_2^{18}O$, $^1H^2H^{16}O$ and $^1H^3H^{16}O$. $^1H_2^{17}O$ exists as a natural molecule (^{17}O is a stable isotope). Its behaviour, during the water cycle, is very similar to that of $^1H_2^{18}O$ (more abundant and easier to precisely determine), and its study is not of interest in cloud physics.

Radioelements produced by nuclear tests since 1952 are used in very specific studies related to cloud physics (Jaffé et al., 1954; Blifford et al., 1957; Bleeker et al., 1966). For example, these radioactive substances can provide informations about the scavenging of particulate matter from the atmosphere during precipitation and understanding of large-scale processes such as interhemispheric transport and stratospheric tropospheric exchanges. The application of environmental isotopes in these atmospheric studies, not directly connected with cloud physics, will not be considered in this review article.

The behaviour of ^2H and ^{18}O in the hydrosphere is intimately linked to the mechanism prevailing during cloud and precipitation formations. Isotope fractionation effects occur at each phase change except sublimation and melting of compact ice (Moser and Stichler, 1980). In the case of tritium, these isotopic effects are generally masked by mixing of water originating from different sources. However, this last isotope is very attractive as it represents the best conceivable tracer of the water molecule during its atmospheric cycle and thus it is a tool in the large-scale processes study. In cloud physics itself, its use is very limited (Ehhalt, 1967; Ostlund, 1968; Jouzel et al., 1975) and the main part of this review will be concerned with

^{2}H and ^{18}O studies. Note that despite the very limited use of tritium, the three isotopes appear very complementary in cloud physics, as in precipitation studies (Gat, 1980), groundwater hydrology (Fontes, 1980) and ice and snow investigations (Moser and Stichler, 1980).

After precipitation, some isotope redistribution may take place in the transition of snow to glacier ice especially in temperate glaciers (Moser and Stichler, 1980). However, primary isotopic distributions in solid precipitations (hail and snow) are generally well preserved since the diffusion coefficients of $^1H^2H^{16}O$, $^1H_2^{18}O$ and $^1H^3H^{16}O$ are very low in ice ($\simeq 10^{-14}$ m^2 s^{-1} at $T = -2°C$; Kuhn and Thürkauf, 1958). On the other hand, individual droplets and raindrops are considered isotopically homogeneous as diffusion coefficients are relatively high in liquid water ($\simeq 2.5 \times 10^{-9}$ m^2 s^{-1} at $T = 25°C$; Wang, 1965). Only hailstones keep isotopic records corresponding at each step of their growth which allow to deduce original information about it. Most of the studies dealing with application of isotopic methods to cloud physics are thus related to hailstone formation studies. This field has been continuously developed over the last twenty years and represents a large part of this review.

An increasing part of paleoclimatological research is based on the study of the ^{2}H and ^{18}O contents of paleoprecipitation. The isotopic signal can be directly recovered from paleowaters, as fluid inclusions in speleotherms (Schwarcz et al., 1976; Harmon and Schwarcz, 1981), lakes and groundwaters (Fontes, 1980) and ice and snow (Dansgaard et al., 1969, 1974; Lorius et al., 1979). Furthermore, it can be reconstructed from lake sediments (Létolle et al., 1980), tree-ring cellulose (Epstein and Yapp, 1976; Yapp and Epstein, 1977) and other organic materials. A quantitative interpretation of these data as climatological indicators needs a good knowledge and modelling of the fractionation effects occurring during the atmospheric water cycle and particularly in clouds. This paleoclimatological research represents a second field of interest for the development of isotopic studies in cloud physics. This aspect is discussed in the different review articles of this serie devoted to paleoclimatological applications of ^{2}H or ^{18}O determinations (Buchardt and Fritz, 1980; Moser and Stichler, 1980; Schwarcz, Chapter 7).

In the present review, the discussion on hailstone formation is preceded by a presentation of the basic principles governing the isotopic variation in clouds, the growth of individual elements and the models used. In the next section, a rapid survey is made of studies connected with the distribution of isotopes in atmospheric waters and with the use of isotopes in studies of precipitation other than hailstones.

BASIC PRINCIPLES

Units

The ^{2}H and ^{18}O concentrations are usually expressed with reference to the SMOW (Standard Mean Ocean Water). This theoretical standard has been defined rigorously with respect to the available secondary standard NBS-1 by Craig (1961a) as:

$$R\,^{18}O_{SMOW} = 1.008\, R\,^{18}O_{NBS\,1}$$

$$R\,^{2}H_{SMOW} = 1.050\, R\,^{2}H_{NBS\,1}$$

The results are expressed in δ versus SMOW defined as:

$$\delta = \frac{R_{sample} - R_{SMOW}}{R_{SMOW}}$$

R being the isotopic ratio (^{2}H/^{1}H or ^{18}O/^{16}O). The δ values are usually expressed in permil units.

At present, a new standard, called V-SMOW (Vienna Standard Mean Ocean Water), whose composition closely matches that of the defined SMOW is currently available from the International Atomic Energy Agency (IAEA) (Gonfiantini, 1978). This V-SMOW has the same ^{18}O/^{16}O ratio as the SMOW but a slightly lower ^{2}H/^{1}H ratio (0.2‰). The ^{2}H/^{1}H absolute ratio of V-SMOW is $(155.76 \pm 0.05) \times 10^{-6}$ (Hageman et al., 1970) and its absolute ^{18}O/^{16}O ratio is $(2005.2 \pm 0.45) \times 10^{-6}$ (Baertschi, 1976).

A second standard, called SLAP (Standard Light Antarctic Precipitation), has been calibrated. The recommended δ values for SLAP relative to V-SMOW are (Gonfiantini, 1978):

$\delta^{18}O$ (SLAP) = -55.5‰

$\delta^{2}H$ (SLAP) = -428‰

The tritium concentrations are usually expressed in tritium units (TU); 1 TU is equivalent to a ^{3}H/^{1}H ratio equal to 10^{-18}. A new tritium half-life equal to 12.430 has been recently evaluated (Unterweger et al., 1981) and recommended for the calculations of tritium concentrations (IAEA, 1981). In this new scale, 1 TU is equivalent to 3.193×10^{-12} Ci kg^{-1} (in water) and to 0.118 Bq kg^{-1} (in water).

Fractionation coefficients

The vapour pressure of the isotopic species of water, ^{1}H^{2}H^{16}O, ^{1}H$_2$^{18}O and $^1H^3H^{16}O$, are slightly different of those of $^1H_2^{16}O$. In the case of stable isotopes, these differences play an important role in the variations observed in the course of the atmospheric water cycle as they cause fractionation effects at vapour/liquid and vapour/solid phase changes.

Due to their lower vapour pressures, the condensed phases in equilibrium with vapour are enriched in heavy isotopes. The fractionation coefficient α is defined as the ratio of $^2H/^1H$ or $^{18}O/^{16}O$ in the condensed phase to the value of the same parameter in the vapour phase. The fractionation factor α is practically equal to the ratio of the corresponding vapour pressures of the pure isotopic molecules involved. A precise knowledge of the α values is necessary in cloud physics for the interpretation of experimental results as well as for the development of isotopic models, especially those used for the comparison of the ^2H and ^{18}O contents.

Many determinations have been made of the liquid/vapour fractionation coefficient at temperatures greater than 0°C either by determining the vapour pressures of both vapour and liquid (Jones, 1968; Jakli and Staschewski, 1977; Szapiro and Steckel, 1977) or from measurements of isotopic ratios in the two coexisting phases (Merlivat et al., 1963; Majoube, 1971a,b; Kakiuchu and Matsuo, 1979). A comparison of the different results has been recently given by Kakiuchu and Matsuo (1979).

Following Gat (1980) the best values describing the fractionation factors between liquid and vapour are those given by Majoube (1971a), where (T being the absolute temperature)

for ^{18}O:
$$\ln \alpha = \frac{1.137}{T^2} \times 10^3 - \frac{0.4156}{T} - 2.0667 \times 10^{-3} \tag{1}$$

and for ^2H:
$$\ln \alpha = \frac{24.844}{T^2} \times 10^3 - \frac{76.248}{T} + 52.612 \times 10^{-3} \tag{2}$$

Supercooled drops and droplets, commonly exist in natural clouds down to at least $-15°C$ (Rogers, 1979) and their presence is of a great importance in the physics of precipitation. Unfortunately, only one study deals with isotopic effects between liquid and vapour at temperatures lower than 0°C (Merlivat and Nief, 1967). A precise determination under similar conditions has never been done for ^{18}O. However, for comparison between the two isotopes it is preferable to use values determined by the same method for both isotopes. Thus, it is recommended to apply the previous formulae (1) and (2) extrapolated for temperatures below 0°C keeping in mind that experimental results have been obtained between 0°C and 100°C.

In the case of vapour/solid equilibrium, direct determinations have been performed by Merlivat and Nief (1967) for ^2H (between 0 and $-40°C$) and by Majoube (1971b) for ^{18}O (between 0 and $-34°C$). The use of formulae (3) and (4), which fit these experimental results is recommended.

For ^{18}O:
$$\ln \alpha = \frac{11.839}{T} - 28.224 \times 10^{-3} \tag{3}$$

For ^2H:
$$\ln \alpha = \frac{16288}{T^2} - 9.34 \times 10^{-2} \tag{4}$$

The ^2H and ^{18}O α values corresponding to equations (1) to (4), are given in Table 2-1 for temperatures of $20°$, $0°$ and $-20°C$. The $\alpha-1$ ratios reported in the last column give an idea of the isotopic effect expected for ^2H versus that expected for ^{18}O. One notes that, the equilibrium isotopic effect is 8—10 times higher for ^2H than for ^{18}O.

The fractionation effects occurring between liquid and solid phases are linked to variations of other physical properties such as heat capacities and latent heat of fusion.

From the knowledge of these parameters for the pure isotopic species, Weston (1955) deduced a ^2H fractionation factor of 1.0192 between liquid and ice. Several experimental determinations (Posey and Smith, 1957; Kuhn and Thürkauf, 1958; O'Neil, 1968; Arnasson, 1969), have been carried out. The complete compilation of the data given by Moser and Stichler (1980), shows that fractionation factors are close to 1.02 for ^2H and 1.003 for ^{18}O. The influence of solid liquid fractionation is limited in cloud physics since no fractionation is expected when droplets and raindrops freeze entirely (Friedman et al., 1964).

Molecular diffusivities

In the following, it will be shown that a kinetic isotopic effect due to the differences between the molecular diffusivities of $^1H_2{}^{16}O$, $^1H^2H^{16}O$, $^1H_2{}^{18}O$,

TABLE 2-1

Values of α^2H and $\alpha^{18}O$ at different temperatures

Temperature (°C)	Vapour/liquid equilibrium			Vapour/solid equilibrium		
	α^2H	$\alpha^{18}O$	$\frac{\alpha^2H-1}{\alpha^{18}O-1}$	α^2H	$\alpha^{18}O$	$\frac{\alpha^2H-1}{\alpha^{18}O-1}$
+20	1.0850	1.0098	8.7			
0	1.1123	1.0117	9.6	1.1330	1.0152	8.8
−20	1.1492	1.0141	10.6	1.1744	1.0187	9.2

in air, occurs during non-equilibrium processes. First theoretical estimation and experimental determination have been obtained respectively by Craig and Gordon (1965) and Ehhalt and Knott (1965). Merlivat (1978) has recently redetermined these coefficients and the values derived from the experimental data are:

$$D^1_{H^2H^{16}O}/D^1_{H_2^{16}O} = 0.9755 \text{ and } D^1_{H_2^{18}O}/D^1_{H_2^{16}O} = 0.9723$$

the ratios are practically independent of temperature.

The isotopic kinetic effect has thus almost the same value for ^2H and ^{18}O (the $D_i/D-1$ ratio is equal to 0.88). The difference between the relative importance for ^2H and ^{18}O of equilibrium and kinetic effects (8—10 compared to 0.88) is the basic reason of the different behaviour between the two isotopes during the atmospheric water cycle when non-equilibrium processes take place.

ISOTOPIC EFFECTS DURING GROWTH OF INDIVIDUAL ELEMENTS

Liquid phase formation

In a first stage, the liquid phase of clouds is composed of water droplets formed by condensation of water vapour on a condensation nucleus by heterogeneous nucleation. After this initial stage, growth of droplets and drops is mainly due to collision and coalescence. Larger elements fall faster than smaller ones and capture a fraction of those lying in their paths (Rogers, 1979).

From an isotopic view point, the condensation and evaporation processes involved during the formation and falling of droplets and raindrops have been analysed by Bolin (1958). He investigated the exchange of tritium between a freely falling waterdrop and its environment. Bolin's theoretical analysis is an extension of the theoretical and laboratory work of Kinzer and Gunn (1951) in which the authors studied the rate of evaporation of falling droplets. The same approach has been followed by Friedman et al. (1962), Booker (1965) and later by Jouzel (1975) and Stewart (1975) who derived the equation describing the variation in isotope content of a droplet or a drop when only vapour transfer processes (condensation and evaporation) are involved. In the δ notation, the variation of the liquid isotopic content versus time, $d\delta_L/dt$ is given by (Jouzel, 1975):

$$\frac{d\delta_L}{dt} = \frac{3\rho_s}{a^2 \rho_\varrho} \left\{ D_i f_i \left[S\left(1 + \delta_v - \frac{1+\delta_L}{\alpha}\right) + \frac{A(1+\delta_L)(S-1)}{\alpha} \right] - ADf(1+\delta_L)(S-1) \right\} \quad (5)$$

where a is the drop radius; ρ_ℓ is the volumic mass (kg m^{-3}) of the liquid water and ρ_s the specific humidity (kg m^{-3}) of the saturated water vapour above it at cloud temperature; f and D are the coefficients of ventilation (dimensionless) and diffusivity of water vapour in air (m^2 s^{-1}); f_i and D_i are the same coefficients relative to the isotopic species and δ_v is the isotope content of the vapour. S, the relative humidity, is defined as the ratio of the specific humidity of water vapour in the environment to ρ_s, $S-1$ being the degree of supersaturation. A is a dimensionless coefficient which takes into account the difference between the temperature of the liquid element and that of its environment. In the absence of supersaturation ($S = 1$), when the droplet or drop is neither losing nor gaining weight, and at constant isotopic content, integration of this equation leads to:

$$\delta_L = \alpha (1 + \delta_v) - 1 + [(1 + \delta_0) - \alpha (1 + \delta_v)] \exp(-t/t_a) \qquad (6)$$

where δ_0 is the liquid isotopic content at $t = 0$. The period t_a called adjustment time or relaxation time is:

$$t_a = \frac{a^2 \alpha \rho_\ell}{3 D_i f_i \rho_s} \qquad (7)$$

After t_a, the distance to isotopic equilibrium $(1 + \delta_L) - \alpha(1 + \delta_v)$, is $1/e$ times its initial value.

Isotopic equilibrium corresponds to the first term of equation (6):

$$\delta_L = \alpha (1 + \delta_v) - 1 \qquad (8)$$

Equation (6) shows that, in the previously given conditions, the liquid phase approaches isotopic equilibrium all the more quickly that t_a is smaller. Fig. 2-1 shows the values of the isotopic relaxation time as a function of the radius at different temperatures taking the ventilation coefficients experimentally determined by Kinzer and Gunn (1951). Values of t_a vary from a fraction of second for tiny droplets to more than half an hour for large drops at $-15°C$ (Fig. 2-1).

Applying equation (5) to a droplet evolving in a cloud where environmental conditions (air pressure, temperature and isotopic content of the water vapour) are continuously modified, Jouzel et al. (1975) have demonstrated that droplets with a radius of less than 30 µm can be considered at any time in isotopic equilibrium with the vapour phase whatever are the cloud conditions.

To give an idea of the isotopic modification affecting a drop if only the vapour transfer processes are considered, it is convenient to calculate the distance travelled during the adjustment time (Bolin, 1958; Friedman et al.,

1962; Woodcock and Friedman, 1963). As an example, this distance is equal to 2900 m for a radius of 1.5 mm (Friedman et al., 1962). However, this approach is only meaningful when the drops evolve in a saturated atmospheric profile without condensation or evaporation.

A noticeable part of the applications of isotopes in cloud physics is concerned with the relation existing between the two stable isotope contents, $\delta^2 H$ and $\delta^{18}O$ during the atmospheric water cycle. As this relationship is theoretically deduced from isotopic cloud models, this important point will be treated below. Included is a discussion of the problems connected with the isotopic modifications due to the evaporation of a drop during its fall to the ground in a subsaturated environment.

In the first part of this section only vapour transfers have been considered. The complete equation applicable when growth is due to vapour transfers and collection of droplets has been derived by Jouzel (1975). A collection term is added to the condensation term, $(d\delta_L/dt)_c$ given by equation (5) and the resulting equation is:

$$\frac{d\delta_L}{dt} = \left(\frac{d\delta_L}{dt}\right)_c + \frac{3EV_L W_g}{4a\rho_\varrho}(\delta_a - \delta_L) \tag{9}$$

where δ_a is the isotopic content of the collected elements, E the collection efficiency (the mass ratio of the amount of liquid water collected to that present in the volume swept by the droplet or drop under consideration), W_g is the mass of droplets per unit volume and V_L the fall speed of the element studied.

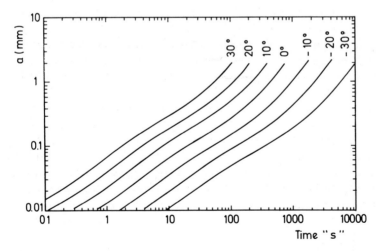

Fig. 2-1. Isotopic relaxation time, t_a, as a function of the drop radius for different temperatures and a pressure of 1013.25 mbar. t_a is given here for deuterium and calculated from equation (7).

In steady-state conditions of temperature, air pressure, and isotopic content of water vapour, for $S = 1$ and when the droplets collected are in equilibrium with the vapour, an isotopic relaxation time, t'_a, similar to t_a but relative to both condensation and coalescence processes has been defined (Jouzel et al., 1975). It is derived from equation (9) and given by:

$$t'_a = t_a \left(1 + \frac{EV_L W_g a \alpha}{4 D_i f_i \rho_s}\right)^{-1} \tag{10}$$

Depending upon the radius of the element and the environmental conditions of air pressure, temperature and liquid water content, t'_a can be up to fifteen times lower than t_a (for $a = 1$ mm). This indicates that the collection of droplets plays an important role in maintaining a drop relatively close to isotopic equilibrium during its evolution in a cloud. Under these conditions, the ^{2}H content of a single drop deviates by less than $10^0/_{00}$ from isotopic equilibrium during its ascent in a typical hailcloud. As shown in Fig. 2-5a this figure can be considerably higher when a population of drops is considered (Federer et al., 1982b), owing to the fact that in that case a drop grows so from smaller drops not in isotopic equilibrium.

Solid phase formation

Once a cloud extends to altitude above the 0°C isotherm, there is a chance that ice crystals will form. Two phase transitions can lead to ice formation: the complete freezing of a liquid droplet or the direct deposition (sublimation) of vapour onto a solid phase. Ice crystals usually begin to appear in a cloud when the temperature drops below about -15°C (Rogers, 1979). As previously stated, no isotope fractionation is expected when the first process takes place. Yet it is accepted that the ice formed by direct condensation of vapour, grows in isotopic equilibrium with this vapour (Dansgaard, 1961; Friedman et al., 1964; Aldaz and Deutsch, 1967; Miyake et al., 1968).

However, this growth includes a diffusion process of the vapour through a supersaturated environment. For this reason, Jouzel and Merlivat (1984) have suggested that a kinetic fractionation effect should control vapour deposition and have demonstrated the validity of this concept by laboratory experiment. A first estimate of the total fractionation factor between vapour and ice, α_{ci}, including both equilibrium and kinetic effects has been given by Jouzel et al. (1980). It is expressed:

$$\alpha_{ci} = \left[\frac{D}{D_i}\left(1 - \frac{1}{S_i}\right) + \frac{1}{\alpha_i S_i}\right]^{-1} \tag{11}$$

where D/D_i is the ratio of molecular diffusivities of H_2O and of the isotopic molecule ($^1H^2H^{16}O$ or $^1H_2^{18}O$), α_i, the equilibrium fractionation factor between solid and vapour (formulae (3) and (4)); S_i is the ratio of the specific humidities of the water vapour in the environment to that above the ice crystal in formation (corresponding to the saturation vapour pressure over ice at ice crystal temperature). This kinetic effect leads to isotopic enrichment in the ice below the equilibrium fractionation values α_i given in Table 2-1. Once embryos are formed, diffusional growth takes place in an environment essentially saturated with respect to vapour pressure over liquid, it is particularly interesting to determine α_{ci} under these conditions. At $-20°C$ ($S_i = 1.215$), α_{ci} is equal to 1.133 for ^2H and 1.010 for ^{18}O (instead of 1.174 and 1.019 for α_i).

A further aspect of the influence of fractionation during ice formation is connected with the phenomena occurring during hailstone growth and will be treated in the section "δ^2H-$\delta^{18}O$ relationship in hailstones".

ISOTOPIC CLOUD MODELS

Some general aspects of cloud formation

The mechanism of cloud formation is the cooling of moist air below its dewpoint, with the additional requirement for atmospheric clouds that the air also contains aerosol particles which can serve as centres of condensation. The cause of cooling can be radiative heat loss, advection over cold surfaces, mixing of air masses of different temperatures or expansion due to the vertical lifting. The associated rates of cooling can range from infinitesimal to an extreme of about $10°C$ per minute (Rogers and Vali, 1978).

There are three types of clouds: stratiform, cumuliform and cirriform which are used in combination to define ten main characteristic forms (a detailed description is beyond the scope of this review). The first two represent clouds formed respectively in stable and in convectively unstable atmospheres. The cirriform clouds are the ice clouds which are in general higher and more tenuous and less clearly reveal the kind of air motion which leads to their formation.

As a result of the different rates of cooling and in aerosol content cloud liquid water contents range from less than 0.01 g m^{-3} to 10 g m^{-3}. The most frequent values of water content found in all types of stratiform cloud fall within the narrow range of $0.05 - 0.25$ g m^{-3}. In small cumulus, the water content rarely exceeds 1 g m^{-3} and is generally much smaller. In the so-called cumulus congestus, much higher values are found and, over small localized regions, may approach the maximum theoretical value calculated on the assumption of adiabatic ascent of the air from the cloud base (Mason, 1971).

The release of latent heat of condensation is an important feedback mechanism between cloud formation and the dynamics of its movement. Growing clouds are sustained by upward air currents, which may vary in strength from a few centimetres per second in stratiform clouds up to a few tens of metres per second in large cumulonimbus.

Cumulus cloud development is conveniently described in terms of a parcel of air undergoing expansion while being lifted vertically. Its ultimate stage of development is the cumulonimbus in which hail and hailstones are generated. Very schematically, the condensed phase of a cumulonimbus cloud can be divided in four types of elements, cloud droplets, large drops, ice crystals and large ice particles (including hail and hailstones). The microphysical processes leading to the formation of these elements are drawn in Fig. 2-2. There is mixing of the ascending parcel with the outside air but in large clouds, a central zone unaffected by entrainment and referred to as updraft core (Chisholm, 1973) probably exists as sketched in Fig. 2-3.

It is more difficult to discuss stratiform clouds, as the point of origin and the subsequent path of an air parcel are less distinct than for cumulus. However conceptually at least, stratus can be treated in the same way as cumulus (Rogers and Vali, 1978). The same can be said for isotopic modelling: the microphysical processes which lead to fractionation effects are completely described in the more complete cumulonimbus.

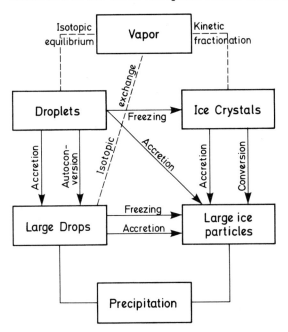

Fig. 2-2. Schematic diagram of the microphysical processes leading to the formation of the different types of elements in a cumulonimbus cloud (after Hirsch, 1971, with slight modifications). The processes giving rise to isotopic effects are drawn in dotted lines, whereas those taking place without isotopic modification are drawn in continuous lines. In this scheme, reverse processes such as melting and evaporation are neglected.

Isotopic modelling based on open system clouds

Early models are based on the assumption that clouds can be considered as open systems. The Rayleigh type, or simple distillation, model assumes that the condensed phase is immediately removed after its formation and that it leaves the air parcel in isotopic equilibrium with the vapour phase (Dansgaard, 1961, 1964; Friedman et al., 1964; Taylor, 1972). As stated above, the condensed phase is enriched in heavy isotopes versus the water vapour which generates it at a given point. This results in a continuous decrease of the heavy isotope content of the remaining water vapour and of the successive precipitation. For the vapour phase, this simple model writes:

$$\frac{d\delta_v}{1+\delta_v} = (\alpha - 1) \frac{dm_v}{m_v} \tag{12}$$

and the isotopic content (^2H or ^{18}O) of the liquid phase, δ_L, can be then formulated

$$\delta_L = \alpha (1 + \delta_0) \left(\frac{m_v}{m_0}\right)^{\overline{\alpha}-1} - 1 \tag{13}$$

where α is the fractionation factor at precipitation and $\overline{\alpha}$, its mean value between initial condensation and precipitation, m_0 and m_v are the mixing ratios of water vapour, defined as the amount of water vapour per kilogram

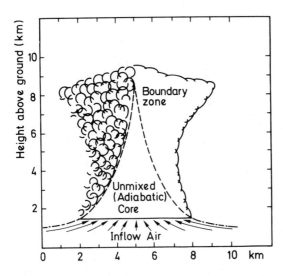

Fig. 2-3. Schematic representation of a cumulus-type cloud as described by Chilshom (1973).

of dry air, at initial condensation and precipitation respectively, δ_0 is the initial isotopic content of the vapour. As an example, $\delta_L{}^2H$ as a function of temperature is drawn in Fig. 2-4, for initial δ_0 and m_0, respectively, equal to $-133.5‰$ and 5.6 g of water per kg of air. In that case the mean deuterium temperature gradient between 3.5 and $-22°C$ is equal to $6‰\ °C^{-1}$.

Rigorously, this kind of open-system model cannot be applied to natural clouds as a condensed phase remains in the cloud. However, in view of the previous description of cloud formation, it is probably a good approach for stratiform clouds in which liquid water contents are indeed very low.

Isotopic modelling based on closed-system clouds

An opposite approach to determine the isotopic ratios of the different water phases coexisting in a cloud is to consider it as a closed system. This hypothesis is only valid if the condensed phase is composed of very tiny droplets or ice crystals which have no sufficient terminal fallspeed to leave the cloud and precipitate. This model is quoted in precipitation studies (Dansgaard, 1961, 1964; Eriksson, 1965) but has been especially applied to calculate the isotopic content of cloud droplets in a hailcloud (Facy et al., 1963; Macklin et al., 1970). There is then conservation of each isotopic specy ($^1H_2{}^{16}O$, $^1H^2H^{16}O$, $^1H_2{}^{18}O$), which leads to equation (14) (for $^1H_2{}^{16}O$) and (15) (for $^1H^2H^{16}O$ or $^1H_2{}^{18}O$):

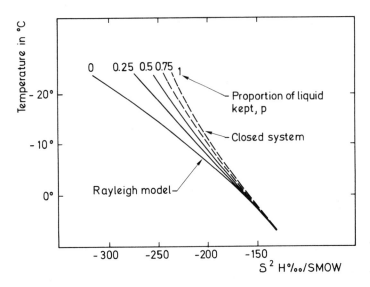

Fig. 2-4. Isotopic model including liquid phase only with possible removal of liquid at each step of the condensation (Jouzel, 1979). $p = 0$, corresponds to the classical Rayleigh model (no liquid kept), $p = 1$, corresponds to the closed system (no precipitation).

$$m_0 = m_L + m_v \tag{14}$$

$$m_0 \delta_0 = m_L \delta_L + m_v \delta_v \tag{15}$$

where the same notations as for equation (13), have been kept for α, δ_0, m_0, δ_v and m_v. m_L and δ_L, are the mixing ratio and the isotopic content of the cloud droplets.

It is also assumed that the cloud droplets are in isotopic equilibrium with the vapour at all times as they ascend, an assumption, which is, as shown above, reasonable and expressed by equation (8). Combining equations (8), (14) and (15), one obtains:

$$\delta_v = \frac{m_0 \delta_0 - (\alpha - 1) m_L}{m_v + \alpha m_L} \tag{16}$$

The relationships between the temperature in the updraft, the pressure and the altitude are obtained by following the adiabatic humidity curve from the cloud base. They allow to determine m_v and α, at each cloud level, and then to calculate δ_L from any initial values of m_0 and δ_0. This has been done, taking the same initial point as for the open-system model (Fig. 2-4). A similar but lower decrease of $\delta_L {}^2H$ with decreasing temperatures is predicted by this closed-system model. In the example presented in Fig. 2-4 the mean deuterium temperature gradient is equal to $3.5°/_{00}$ $°C^{-1}$ (between the cloud base and $-22°C$) instead of $6°/_{00}$ $°C^{-1}$ for the open system.

The validity of this model was studied by Jouzel et al. (1975), who defined six conditions necessary to use it to determine the isotopic content of condensed water for a storm:

(1) The condensed water belongs to the updraft core (Fig. 2-3).

(2) No liquid water accumulation zone, composed of large drops, does exist.

(3) The condensed water is made up of droplets and ice crystals of which at least 60% by weight have a radius below 30 μm and no more than 40% a radius between 30 and 100 μm.

(4) The contribution of the liquid phase to the growth of hailstones during the rise of a given volume of air is less than 4% of the water content value at the cloud base.

(5) The water content at the cloud base varies by less than 6% from its average value during the storm.

(6) The isotope content of the water at the cloud base varies by less than $0.07°/_{00}$ (for the deuterium) from its average value during the storm.

Recently, Jouzel et al. (1980), have pointed out that it is highly questionable whether these very stringent conditions are ever met in a hailstorm and they claimed the need for a more realistic isotopic model in that case.

Multistage isotopic models including liquid phase only

The importance of the presence of liquid water in a cloud on isotopic values, has been firstly taken into account by S. Epstein (unpublished lecture notes, quoted by Gat, 1980), and later by Craig and Gordon (1965), Miyake et al. (1968), Tzur (1971) and Merlivat and Jouzel (1979). In the case of a two-phase model, droplets and vapour in isotopic equilibrium, the following general equation has been derived taking into account the partial removal of precipitation (Merlivat and Jouzel, 1979):

$$\frac{d\delta_v}{1+\delta_v} = \frac{(\alpha-1)\,dm_v - m_L\,d\alpha}{m_v + \alpha m_L} \tag{17}$$

where the same notations as for equation (16) have been kept.

This equation (17) includes equations (13) and (16) (for open and closed systems). If no liquid water is kept within the cloud ($m_L = 0$) equation (17) becomes a Rayleigh distillation equation (equation (13)). If all the liquid is kept in the cloud, the integrated form of (17) becomes identical to equation (16).

From equation (17), δ_L, can be calculated all along the condensation process, if the amount of water leaving the cloud as precipitation at each step of the procedure is known. As an example, calculations have been carried out, assuming that, at each step, a constant proportion of the newly condensed water vapour, p, leaves the cloud (Jouzel, 1979). The importance of this parameter on the deuterium profile in the liquid water of a cloud is clearly shown in Fig. 2-4.

Isotopic models taking into account the entrainment of outside air

In all the previous studies, the air parcel was considered as a closed system for the air and mixing with surrounding air was not taken into account. However, this mixing is the main important process as far as tritium is concerned.

A tritium cloud model has been proposed by Ehhalt (1967) but only on a qualitative basis. The tritium presently observed has been produced by nuclear tests and mostly injected into the stratosphere. It mixes into the troposphere primarily during the spring. This downward mixing is reflected by the vertical gradient of the ^3H concentration in atmospheric water vapour. In a convective cloud rising in such an environment, a similar though weaker increase of the ^3H concentration in water with altitude is expected. The concentration in the cloud water will also vary laterally as the central region of fastest updraft is least influenced by mixing with outside air. The ^3H concentration is closer to that of ground level moisture in the central region than in the outer part of the cloud (called updraft core and boundary zone in Fig. 2-3). No quantitative model has been produced.

The influence of mixing on stable isotope distribution in convective clouds has been recently studied by Zaïr (1980) Federer et al. (1982a) and Grenier et al. (1983), from a simple cumulonimbus model similar to that sketched on Fig. 2-3. An increase of the deuterium temperature gradient is predicted but the effect is very slight (only a few permil through the entire depth of a typical hailcloud). The basic reason is that large hailclouds are not greatly influenced by mixing processes with surrounding air. It would be probably different for stratiform clouds in which, as stated above, these processes are more important but no study has been developed in this direction.

Multistage isotopic models including liquid and solid phases

The main problem with the above models is that they are too simplistic and, in particular, because only cloud droplets in isotopic equilibrium with water vapour are considered. Therefore, Jouzel et al. (1980) and Federer et al. (1982a) have derived an isotopic model based on a more realistic approach. As a basis, they used the steady-state model described by Hirsch (1971) which takes into account mixing with outside air, precipitation fallout and interactions between water vapour, cloud droplets, large drops, cloud ice and large ice particles. The scheme of the microphysical processes involved between these different phases corresponds to that reported in Fig. 2-2; the processes giving rise to isotopic effects are drawn in dotted lines whereas that taking place without isotopic modification are drawn in continuous lines. The reverse processes, evaporation and melting, have a negligible influence on the water mass balance and are not considered in this model.

To follow the isotopic content variations of the drops, both coalescence and condensation processes are taken into account, through an equation similar to (8) extended to a population of drops assuming a Marshall-Palmer drop distribution (Kessler, 1969).

The basic relation of this new isotopic model is that which allows to follow the evolution of the isotopic content of the water vapour. Keeping the same notations as previously used, it can be expressed as:

$$d\delta_v (m_v + \alpha m_L) = M' - V' -$$

$$(1 + \delta_v) \left[dm_v \left(1 - \frac{\alpha_{ci} m_i + \alpha m_L}{m_i + m_L}\right) + m_L d\alpha \right] \quad (18)$$

where M' and V' are the terms corresponding to the mixing with outside air and to the isotopic transfer between vapour and drops, m_i is the mixing ratio of cloud ice and α_{ci} the combined fractionation factor given by equation (11). This equation includes all the previous ones, equation (17) being

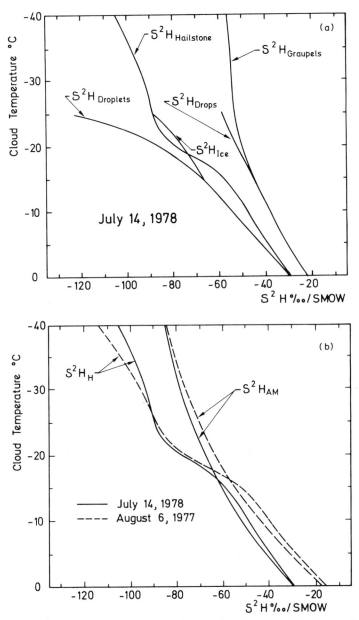

Fig. 2-5. (a) Isotopic model including liquid and solid phases (Jouzel et al., 1980; Federer et al., 1982a) with deuterium content variations of the droplets, drops ice and graupels as a function of cloud temperature. $\delta^2 H_{hailstone}$ is the deuterium content of the water collected by hailstones. (b) Isotopic model including liquid and solid phases. Comparison of the results with that obtained applying the classical adiabatic model (equation (16)) for water collected by hailstones. Calculations have been made for two different Switzerland hailstorms, the July 14 storm being very unstable and the August 6 one only marginally unstable.

obtained in the absence of cloud ice ($m_i = 0$), of mixing with outside air ($M' = 0$) and of isotopic transfer between vapour and drops ($V' = 0$).

It can be pointed out that this relation is very general and applicable to any cloud model provided that steady state conditions exist. Once δ_v determined, the isotopic contents of each condensed phase can be deduced as well as that of the water collected by an hailstone which is a mixture of these different phases.

As an illustration, the variation of the deuterium content versus temperature of a hailstorm in Switzerland is shown in Fig. 2-5a with the curves relative to droplets, drops, cloud ice and graupels (large ice particles) and the resulting one $\delta^2 H_{hailstone}$ deduced for the water collected by hailstones. As expected large drops, which are not in isotopic equilibrium with water vapour, are enriched in heavy isotopes with respect to cloud droplets at the same temperature.

In Fig. 2-5b, the $\delta^2 H$ of the water collected by hailstone is compared with the value of this parameter based on the classical adiabatic model (equation (16)). Results are presented for two different hailstorms which differ both in cloud base conditions and stability, the 14 July one being very unstable and the 6 August one only marginally unstable.

It is seen that the differences between the $\delta^2 H_H$ and $\delta^2 H_{AM}$ profiles are appreciable, especially at lower temperatures. At high temperatures the isotope content of hail is 1—3°/$_{00}$ higher, but below the point of intersection of $\delta^2 H_H$ and $\delta^2 H_{AM}$ at around $-17°C$, it becomes considerably lower (up to 20°/$_{00}$ at $-30°C$). The relatively higher isotope content of hail with respect to cloud water is due to the interaction with rainwater, which is seen to be substantially enriched with regard to the other species. Above the intersection, the difference then negative, increases very rapidly down to $-25°C$, owing to the rapid freezing of isotopically rich drops and their transformation to graupels, which are practically not collected by hailstones, and to the appearance of isotopically poor cloud ice both from freezing of cloud droplets and vapour deposition. This applies to the two, very different, hailstorms.

Although it would be necessary to pursue isotopic modelling of clouds from two- or three-dimensional and time-dependent cloud models, this new approach incorporates substantial improvements as compared to the isotopic cloud models previously used. In particular, the very general equation obtained (equation (18)), represents an important step forward for future work in this field.

Some considerations about isotopes distribution in clear air

The vertical gradients in atmospheric water vapour, remaining after a precipitation event, which globally leads to a depletion in heavy isotopes are firstly controlled by the above described microphysical processes taking place in the cloud during such precipitation events. Subsequently, these gradients can be modified by vertical and horizontal mixing possibly accompanied by vapour condensation, isotopic exchange between falling liquid precipitation and surrounding vapour, and by redistribution owing to evaporation of falling ice crystals (see below).

Eriksson (1965) has considered the effect of turbulence on the vertical isotopic gradients in water vapour of the troposphere. The equations can be solved for certain idealised cases. For the case of constant exchange coefficient and exponential decrease of specific humidity with height, the solution is given by (keeping the above notations):

$$\frac{d\delta_v}{1+\delta_v} = (\sqrt{\alpha}-1)\frac{dm_v}{m_v} \qquad (19)$$

where α is the prevailing fractionation factor in the condensation process. It can be pointed out that this relation is similar to that describing the Rayleigh process, α being replaced by $\sqrt{\alpha}$.

This solution is by no means general and Taylor (1968) has proposed a two-component mixture model to the troposphere. The turbulent exchange is assumed to mix vapour from the foot of the tropospheric column, effectively the region where the depletion of the isotopic concentration by condensation is observed not noticeable (see next section "Sampling of water vapour and condensed phases in the atmosphere") to greater heights where it mixes with vapour isotopically depleted to produce a detectable drop in isotopic concentration. The result of the mixture is to produce a resultant enrichment lying between 1 and $\overline{\alpha}$, the averaging of α over the condensation process. Taylor (1972) points out that in a region of intense horizontal mixing, little significance can be attached to the isotopic structure in the vertical. Besides, he predicts that the formation of ice crystals in high level cirrus clouds (around 500 mbar) and their subsequent evaporation after falling will tend to redistribute the heavier isotopes at the lower levels.

Using the available experimental data on the vertical deuterium distribution in atmospheric water vapor (next section), Rozanski and Sonntag (1982) drew attention on the fact that observed gradients are very generally steeper than that predicted by existing models. They proposed a multibox model including the isotope exchange between water vapour and cloud water and falling raindrops. This model reproduces the observed vertical profiles fairly well, the authors stating that the isotopic exchange is presumably responsible for the observed steep gradients. They also indicate that the fast increase of the deuterium content observed, in two cases near the tropopause suggests a strong source of isotopically enriched water vapor in the stratosphere which might be due to the chemical decomposition of methane gas.

The δ^2H-$\delta^{18}O$ relationship

The ^2H-^{18}O relationship is of a primary importance in all studies dealing with stable isotope in the water cycle (see Volume 1 of this serie).

At the cloud scale, this relationship can be derived from the previous models (Dansgaard, 1964; Taylor, 1972; Jouzel and Merlivat, 1979). Assuming that isotopic equilibrium fractionation prevails in atmospheric condensation processes; Taylor (1972) shows that the isotopic composition of the remaining vapour lies, in a δ^2H-$\delta^{18}O$ diagram on a straight line of a slope of about 8.8 $\delta_0(^2H)/\delta_0(^{18}O)$ where δ_0 is the isotopic content of the vapour at the initial stage of condensation. It is the same for isobaric and adiabatic coolings, in the case of instantaneous removal of condensate, and also in the case where the effective separation factor is altered by the presence of the liquid phase. The predicted slopes for mid-latitudes and tropical water vapour lies between 8.1 and 8.4. This prediction is in agreement with the experimental data obtained on tropospheric water vapours sampled in different European countries (Taylor, 1972; Jouzel, 1979).

At a global scale the interest is focused on precipitations which lie on a straight line of slope close to 8. This has been first observed by the Chicago group (Friedman, 1953). In 1961, Craig defined this relationship as the meteoric water line (MWL) whereby $\delta^2H = 8\ \delta^{18}O + 10$. It closely defines the isotopic composition of non-evaporated continental precipitations. The global validity of the value of 8 for the slope has been documented by the data of the IAEA network and even in the polar snow (Epstein et al., 1965; Jouzel et al., 1983). Theoretical modelling of this relationship for precipitation has been reviewed by Gat (1980) who concludes that, in spite of an apparent agreement there are a number of disturbing features and that one still lacks a satisfactorily global model to explain it.

An interesting step forward to a better understanding of the δ^2H-$\delta^{18}O$ relationship at a global scale comes from a study by Merlivat and Jouzel (1979) and Jouzel and Merlivat (1981, 1984). The first step of the model deals with the isotopic effects taking place during the evaporation from the ocean surfaces which were modelled through experiments conducted in a large air sea simulating facility (Merlivat and Coantic, 1975), Merlivat (1978a,b). The isotopic content of the water vapour leaving the ocean surface is expressed as a function of the sea surface temperature, of the relative humidity of the air masses formed and of the wind regime in the boundary layer close to the surface of the ocean.

In a first part, limited to the study of liquid precipitation (Merlivat and Jouzel, 1979), the water cycle is described by a cloud model allowing partial or total removal of precipitations (equation (16)) which permits the description of all the types of cloud systems encountered in nature. The atmospheric system, as a whole, is assumed to be in a steady state. The δ^2H-$\delta^{18}O$ relationship in the liquid phase leaving the clouds (down to a minimum value of $-20°C$) is demonstrated to correspond to the general linear form $\delta^2H = s_1\delta^{18}O + d_1$, for a large range of conditions closely resembling those encountered in nature. The slope s_1 and the intercept d_1, are shown to be very insensitive to cloud conditions such as the amount of liquid water kept in the cloud, although, as stated above (Fig. 2-4) this parameter strongly influences δ^2H or $\delta^{18}O$ contents. This linearity is a chance linked with the increase of the $\alpha-1$ ratios with decreasing temperature (Table 2-1), and the basic reason of the general law observed in the natural non-evaporated liquid precipitation. On the other hand, s_1 is demonstrated to depend upon the relative humidity of the air masses at the departure of the ocean surface, h, and of the ocean surface temperature, T_e, (Fig. 2-6), and d_1 only of h (Fig. 2-7). Through these results, the slope and the intercept of the MWL, are shown to correspond to $h = 81\%$ and $T_e = 26°C$. The similar equation given by Dansgaard (1964) yields $h = 79.5\%$ and $T_e = 25.4°C$. These figures appear very reasonable as the regions which are the main sources of water vapour over the oceans are the subtropics and the trade wind belts.

In a second step, Jouzel and Merlivat (1981, 1984) extended this global model taking into account the appearance of the solid phase using a simplified version of the isotopic model described by equation (18) in which the mixing

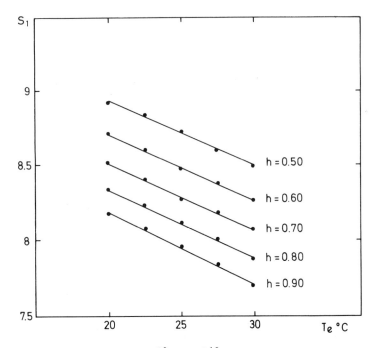

Fig. 2-6. Variation of s_1 ($\delta^2H = s_1\delta^{18}O + d_1$) in precipitation, as a function of the ocean surface temperature, T_e, for different values of the relative humidity, h, at the ocean surface (Merlivat and Jouzel, 1979).

and exchange terms M' and V' are neglected. In a first approach, the classical idea that snow is formed in isotopic equilibrium with the surrounding vapour has been applied, but the model failed to explain the δ^2H-$\delta^{18}O$ slope observed in snow. The calculated values lie around 6 instead of 8 as in the nature. Then, considering that snow is formed in supersaturated environment, the concept of a kinetic fractionation effect at vapour deposition on ice has been considered (equation (11)). It results an excellent agreement between experimental data and theoretical curves for very reasonable values of supersaturation. Another interesting point is that the ^{18}O temperature gradient is lowered of 40% when this kinetic effect is taken into account. This is very satisfactory to explain the low values (less than $1^0/_{00}$ $°C^{-1}$) measured in precipitation over the polar ice caps (Dansgaard, 1964), Picciotto et al. (1960), Lorius and Merlivat (1977).

This new model proves that the constancy of the δ^2H-$\delta^{18}O$ slope for solid precipitation around a value of 8 is not at all the demonstration, that atmospheric processes leading to their formation take place at isotopic equilibrium as very generally thought. On the contrary, this slope value results for solid phase from the existence of a kinetic effect at vapour deposition.

In a recent study, based on the above theoretical considerations, Jouzel et al. (1982), have shown that the deuterium excess ($d = \delta^2H - 8\,\delta^{18}O$) of paleowaters can be useful to obtain information about the variation of the relative humidity at the oceanic surface. A continuous d profile was ob-

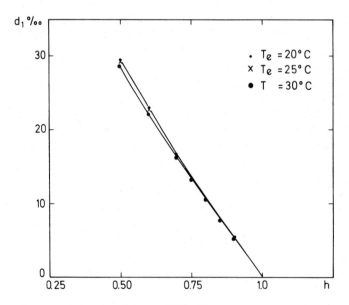

Fig. 2-7. Variation of $d_1 = s_1\delta^{18}O$-δ^2H in precipitation as a function of the relative humidity, at the ocean surface, h, for different ocean surface temperatures, T_e (after Merlivat and Jouzel, 1979).

tained from the 906 m deep Dome C Ice core (East Antarctica) spanning about 32,000 years. It exhibits significant variations over this period. In particular d was $4.5^0/_{00}$ lower around 18,000 B.P. than over the last 7500 years. This d decrease is attributed to a similar d change at the origin of precipitation (although the possibility that this decrease could partly result from modifications occurring during the atmospheric processes cannot be completely eliminated). It is interpreted (see Fig. 2-7 noting that d is very close to d_1) as reflecting higher relative humidity during the last glacial maximum (estimated $\sim$ 90%) over the oceanic areas providing moisture for Antarctic precipitation. From considerations about the evaporation flux, this interpretation is shown very consistent with the wind speed increase at this time suggested by large marine and continental inputs (Petit et al., 1981).

All global models assume that there is no vapour mixing between travelling and local air masses. This is no completely satisfactory as the exchange between the atmosphere and the sea is of a dominating importance over oceanic regions (Eriksson and Bolin, 1964) and that the reevaporation of water from over continents or lakes can also affect the composition of atmospheric moisture (Fontes et al., 1970; Gat, 1980). Recently, Rozanski et al. (1982) and Sonntag et al. (1981) have studied the factors controlling stable isotope distribution in modern European precipitation. They use a box model for moisture, taking into account evapotranspiration which is assumed to be isotopically identical with the precipitation, which in some cases may be a reasonable assumption (Zimmerman et al., 1967; Münnich, 1978). Applying a simple Rayleigh model (equation (12)) within each successive box, they claimed that the proposed model, provides, despite its rough character, an acceptable prediction of the observed variations in isotopic composition of precipitation over the main part of the European continent. Both continental and seasonal temperature effects observed in the stable isotope composition are fairly well reproduced, the calculations showing that the isotopic composition of local precipitation is primarily controlled by regional scale processes, i.e. by the water vapor transport patterns into the continent and by the average precipitation-evapotranspiration history of the air masses precipitating at a given place.

Future modelling of stable isotope repartition at a global scale needs both the understanding of isotopic transfers taking place in clouds, and that of mixing processes between air masses, and consequently, moisture of different origins. With this objective, a new approach of the isotopic modelling of the water cycle based on the use of General Circulation Models (GCM) of the atmosphere was proposed by Joussaume et al. (1983; an approach also suggested by Gedzelman and Lawrence, 1982). Joussaume et al. (1983) have incorporated a simulation of the $^1H^2H^{18}O$ and $^1H_2^{18}O$ cycles to the GCM of the Laboratoire de Météorologie Dynamique (France). These cycles are treated within the basic water cycle, taking into account the isotopic fractionations occurring at: (1) evaporation from the oceans, (2) formation of liquid and solid precipitation and (3) evaporation of rain under a cloud base. As a result, the $^1H^2H^{18}O$ and $H_2^{18}O$ content of the precipitation and of the atmospheric water vapour are predicted at every grid point. This simulation can be then compared to the well-documented distribution of the isotopic species in modern precipitation. Note also that with this formulation of the isotopic cycles, this GCM will be run for paleoconditions (for example, 18,000 B.P.) and that it is hoped in this way, to improve the theoretical basis of the isotopic method of paleoclimatic reconstitution.

The results of the first simulations of the water isotope cycles in the atmosphere using a GCM are now available for modern conditions (Joussaume et al., 1983, 1984). Despite some systematic biases (mainly in the polar areas), there is a good agreement between observed and predicted values; the $\delta^{18}O$-temperature, $\delta^{18}O$-precipitation and δ^2H-$\delta^{18}O$ relations are realistically simulated. As a whole, this approach appears very promising towards a better interpretation of isotopic data both in the climatologic and paleoclimatologic fields and will certainly be further developed in the next years.

Isotopic exchange during the fall of liquid precipitation

In the above models (except for the GCM simulation), isotopic modifications taking place during the fall of liquid precipitation and surrounding vapour are not considered. Isotopic exchange and evaporation affects both the δ^2H and $\delta^{18}O$ values and their relationship especially for precipitation falling through a dry atmosphere (Ehhalt et al., 1963). In the case of evaporation a kinetic fractionation effect occurs which is due to the differences between the molecular diffusivities of the different isotopic species of water vapour through air. Whereas for ^2H, the kinetic effect play only a minor role, in the case of ^{18}O the equilibrium and kinetic fractionation factors have about the same magnitude. The influence of evaporation on the isotopic content of drops has been the subject of both experimental and theoretical studies (Ehhalt et al., 1963; Dansgaard, 1964; Ehhalt and Knott, 1965; Stewart, 1974, 1975).

From a theoretical viewpoint, the ^2H and ^{18}O contents of a drop, can be followed by solving equation (5). This has been done by Stewart (1974, 1975), from a basically similar equation which accounts both for kinetic and equilibrium fractionation processes. It results an isotopic enrichment of the drop during its fall.

Applying equation (5) to ^2H and ^{18}O, for a drop initially in isotopic equilibrium with surrounding vapour (equation (8)), it becomes:

$$\frac{d\delta_L{}^2H}{d\delta_L{}^{18}O} = \frac{(1 + \delta_L{}^2H)\,[(D_i f_i/\alpha)_{^2H} - Df]}{(1 + \delta_L{}^{18}O)\,[(D_i f_i/\alpha)_{^{18}O} - Df]} \tag{20}$$

According to Stewart (1975), the products Df and $D_i f_i$ are well approximated by $k\ D^{0.58}$ and $k\ D_i^{0.58}$ with k independent of the molecule concerned leading to:

$$\frac{d\delta_L{}^2H}{d\delta_L{}^{18}O} = \frac{(1+\delta_L{}^2H)\left[\frac{1}{\alpha}\left(\frac{D_i}{D}\right)^{0.58}_{{}^2H}-1\right]}{(1+\delta_L{}^{18}O)\left[\frac{1}{\alpha}\left(\frac{D_i}{D}\right)^{0.58}_{{}^{18}O}-1\right]} \tag{21}$$

which can be calculated with the values of α and D_i/D given earlier in the sections "Fractionation coefficients" and "Molecular diffusivities". For a drop evaporating at 20°C, we obtain:

$$\frac{d\delta_L{}^2H}{d\delta_L{}^{18}O} = 3.6\ \frac{1+\delta_L{}^2H}{1+\delta_L{}^{18}O} \tag{22}$$

As $(1+\delta_L{}^2H)/(1+\delta_L{}^{18}O)$ generally lies between 0.9 and 1 for liquid precipitation (IAEA, 1981), it results that a drop evaporating without collecting droplets, evolves along a δ^2H-$\delta^{18}O$ line of slope close to 4 which is two times lower than that prevailing during equilibrium processes.

SAMPLING OF WATER VAPOUR AND CONDENSED PHASES IN THE ATMOSPHERE

The preceding sections are mainly concerned with theoretical aspects of isotopic fractionation effects occurring during the atmospheric water cycle. However, any model must be tested against field data and require the collection and analysis of atmospheric waters. In Volume 1 of this series, Gat (1980) presents the principal results obtained on tritium distribution in atmospheric waters. The purpose of this section is a review of the same kind of data available for ^2H and ^{18}O. Most measurements were made on water vapour sampled in clear air using aircrafts (Taylor, 1972; Ehhalt, 1974; Jouzel, 1979).

From water vapour samples collected on 20 aircraft flights, Taylor (1972) presents a very complete study of meteorological and theoretical aspects of the vertical ^2H, ^3H and ^{18}O distribution, up to 5 km over southwestern Germany during 1967 and 1968.

The variations found in the deuterium concentration are closely related to tropospheric stratification. Within the lowest layer of the troposphere, up to the fair-weather inversion (1—2 km), turbulence and convection smooth out the vertical gradients of isotopic concentration. Above the inversion, the

δ^2H and δ^{18}O vertical gradients are very much steeper and strong isotopic depletion is found (up to $-300^0/_{00}$ for δ^2H). The deuterium content of individual air masses above the inversion are related to water vapour mixing ratio through the equation:

$$\log(1 + \delta_v) = (\bar{\alpha} - 1) \log m_v/m_0 + \text{constant} \qquad (23)$$

This equation, given with our previously defined notations, corresponds to the Rayleigh condensation process (equation (11)). It is shown how this relationship is related to air mass origin and previous condensation history. The δ^2H and δ^{18}O values for 32 water samples lie along a straight line δ^2H = $(8.43 \pm 0.31) \delta^{18}$O + (16 ± 7). This slope value for water vapour, is in agreement with the above mentioned prediction of Taylor (1972) leading to a theoretical gradient of 8.1—8.4; note also measurements by Jouzel (1979) and Federer et al. (1982b) on a mountainside between 500 and 2100 m.

Ehhalt (1974) determined deuterium and tritium distribution over various locations of the U.S.A. and over the Pacific Ocean up to 13 km. Except in a few cases, a δ^2H decrease is observed in the troposphere. Unfortunately, these very well documented data, have not, untill now, give rise to detailed interpretation. As the data did not provide a single pattern, it is difficult to summarize them in a simple manner without taking into account the different meteorological conditions. Note, however, that the mean profiles served with that obtained by Ehhalt and Ostlünd (1970) and Taylor (1972) as a basis for the model developed by Rozanski and Sonntag (1982) (see p. 79).

For most of the profiles obtained during these different field experiments, the following general conclusions given by Taylor (1972) apply:
— The vertical gradients of isotopic concentration in water vapour in the lower troposphere, up to ca. 1—2 km, separated from the higher lying air by a temperature inversion are smoothed considerably by turbulence and convection.
— Above this ground layer, the isotopic gradients appears to be established by condensation and reevaporation processes.

To check the validity of isotopic cloud models, it is fundamental to collect simultaneously water vapour and condensed phases in clouds. In turn, the interpretation of the isotope distribution in the different cloud phases should also contribute to the understanding of the genesis of meteorological events. For the liquid phase, such efforts were made in hurricanes (Ostlünd, 1968; Ehhalt and Ostlünd, 1970) and rainstorms (Stewart, 1974).

Ehhalt and Ostlünd (1970) determined both horizontal variations and vertical profiles of the deuterium content for water vapour inside a hurricane cloud. The results exhibit large variations with radial distance and altitude. These authors compared the deuterium content of water vapour and of rainwater collected simultaneously and found a significant deviation from isotopic equilibrium. This clearly indicates that the basic assumption of isotopic equilibrium between liquid water and vapour is incorrect for a cumulus type rain and justifies the effort made to account for this disequilibrium in isotopic cloud models (Jouzel et al., 1980). However, a qualitative explanation was already given (Ehhalt and Ostlünd, 1970) leading to deduce that the updraft zone and the zone of falling rain were partly separated. According to these authors, this conclusion has important meteorological implications in the field of hurricane studies.

Stewart (1974) has experimentally, and as stated above theoretically, studied the raindrop evaporation and the origin of vapour during a rainstorm. He made a detailed investigation of the δ^2H-$\delta^{18}O$ relationship. Rain samples lie along a straight line of slope 4.4 ($\delta^2H = 4.4\ \delta^{18}O - 4.4$). The lowering of the slope versus that of the MWL, is attributed to evaporation and can be interpreted from equation (21). Moreover, the strength of the isotopic enrichment during the fall is related to the amount of evaporation of the drops. Using the values of relative humidity, temperature and deuterium content of the water vapour, measured during the sampling flight in the subcloud layer, Stewart deduced that the initial radius of the raindrops was 1.0—1.1 mm and that evaporation loss was about 35% of the initial condensate. Similar calculations were performed by Miyake et al. (1968) and Fontes (1976) showing that isotopic measurements thus appears suitable to estimate evaporation during rainfall which is normally very difficult to determine.

One can note, however, as shown by the IAEA network data, that evaporation does not significantly affect the global isotopic pattern in precipitation but locally the resulting change in the δ^2H-$\delta^{18}O$ slope is often used to characterize kinetic fractionation effects (IAEA, 1981).

Another way to get information about the isotope distribution in a cloud would be to use hailstones which keep a record of the isotopic composition of the water at the successive levels where they have grown. Indeed, isotope measurements in hailstones are essentially performed to get information about hailstone growth. Before reviewing this research field, we briefly present data about use of water isotopic species in the study of precipitation other than hailstones, i.e., essentially rain as isotopic studies in snow are practically only developed for paleoclimatological purposes.

A BRIEF SURVEY OF ISOTOPIC STUDIES IN RAIN AND SNOW

Studies based on isotopic measurements on rain and snow samples are very abundant. One can cite, works dealing with paleoclimatological research with the aim to document the δ^2H- or $\delta^{18}O$-temperature relationship (Picciotto et al., 1960; Aldaz and Deutsch, 1967; Kato, 1978; Smith et al., 1979; Siegenthaler and Oeschger, 1980; Weaver, 1981; Bromwich and Weaver, 1983), hydrological studies as reviewed by Gat (1980), Fontes (1980) and Moser and Stichler (1980) and numerous regional investigations as those developed by Gat and Dansgaard (1972), Schriber et al. (1977) and Salati et al. (1979).

On the other hand, isotopic measurements of rain and snow samples, have practically never been applied to study the processes leading to their formation. This kind of work is limited to the interpretation of data relative to fractionally collected rainwater (Dansgaard, 1953, 1961; Epstein, 1956; Ehhalt et al., 1963; Gambell and Friedman, 1965; Matsuo and Friedman, 1967; Miyake et al., 1968). The basic reason is that drops are isotopically homogeneous and thus do not contain information about the different steps of their growth and, moreover, that their mean isotopic content can be modified after their formation by evaporation processes taking place during their fall to the ground.

Dansgaard (1953) found that the ^{18}O content of rainshowers is constant throughout the precipitation this being confirmed from ^2H data during one shower studied by Matsuo and Friedman (1967). According to these authors, the constancy of the isotopic composition of a shower may be accounted for by, among other explanations, the thorough mixing in the air mass with high holdup of liquid water. Dansgaard (1961) and Ehhalt et al. (1963) found, however, that stable isotopes contents varied with time especially at the beginning of a shower where the precipitation intensity was low, and the constancy of the isotopic composition was observed only when the precipitation intensity was high. They attributed these changes to the evaporation during the rainfall and the difference in the altitude of raindrop formation. In the rainwater of cold-front origin, the heavy isotopes are generally enriched at the beginning and depleted at the end of precipitation (Epstein, 1956; Matsuo and Friedman, 1967). This is explained as resulting of the continuous depletion of water in the cloud leading to a parallel depletion in the heavy isotope content, theoretically demonstrated whatever is the amount of liquid water kept in the cloud (Fig. 2-4). On the other hand, Gedzelman and Lawrence (1982), found that sequential samples show a general increase of δ^2H with time. These changes are explained by means of a detailed meteorological analysis assuming that the precipitation derives from a Rayleigh condensation process in which all condensation in the air column directly above the sampling site is assumed to fall immediately to the ground.

One can also note the very complete study of Miyake et al. (1968) based on measurements performed on samples collected during ten precipitation events including a shower of snow pellets which grow by capturing cloud droplets and ice crystals. In that case, isotopic content increases with time. This is attributed to the fact that the largest snow pellets which are formed in the upper part of the cloud with a low isotopic content, arrive at the ground, according to their higher fall velocity, before the smaller ones grown in the lower part of the cloud with higher isotopic content.

At a larger time scale, Dansgaard (1964) drew attention on the so-called "amount effect", i.e.; the existence of a negative correlation between the isotopic values and the amount of monthly precipitation with low δ values in rainy months and high δ values in months with sparse rain. He noted that this "amount effect" is found all the year round at most tropical stations and in the summer time at mid latitudes. This author evoked three possible reasons to explain this effect, the first one, linked with the condensation processes itself (the δ values decrease when cooling proceeds whereas the amount of precipitation increases), the two others related to post-condensation processes (the enrichment by isotopic exchange with the surrounding vapor and by evaporation during the fall of rain being most pronounced for light rain). Recently, Yapp (1982) developed a simple quantitative model which predicts that a parametric correlation between the isotopic composition of precipitation and precipitation intensity can exist as a consequence of the condensation process when certain conditions are met such as the constancy of (1) the initial isotopic composition of the vapor, (2) the vertical rate of ascent of the precipitating air mass, and (3) the ratio of the condensed water to water vapor. The influence on the model of the variation of the vertical rate of ascent is discussed in view of the data relative to two tropical islands which may satisfy these conditions.

At a larger scale, it must be stressed that the ^{2}H concentration (or ^{18}O) can provide some additional information concerning the condensation history, e.g. whether a tropical mass has undergone condensation processes since it left its area of origin; to what extent a polar air mass has lost moisture during time spent over polar regions (Taylor, 1972). Lawrence and White (1981), White and Broecker (1981), Lawrence et al. (1982), Gedzelman and Lawrence (1982) and Covey and Schneider (1982), have, in this way, studied meteorological factors influencing deuterium content of precipitation, over North America. In particular, the existence of a relation between the deuterium content of precipitation and the paths followed by the storm in the Eastern U.S.A. was established (Lawrence et al., 1982). The same type of approach can be pursued using tritium which then acts as a radioactive tracer; as stated in the Introduction, a discussion of this kind of works is beyond the scope of this review.

THE STUDY OF HAILSTONE GROWTH MECHANISMS BY MEANS OF ISOTOPIC ANALYSES

Some general aspects of hailstone growth

In air that is sufficiently moist and unstable, convective clouds can grow to great heights, developed vigorous updrafts, and produce heavy rain, lightning and hail. These large severe storms may occur individually or more typically, in groups associated with synoptic scale fronts or mesoscale convergence areas (Rogers, 1979). Two important features of these cumulonimbus clouds are the existence of strong vertical updrafts in certain regions and the presence of supercooled droplets for air temperatures below 0°C. Under those environmental conditions hailstones are formed when either graupel particles or large frozen raindrops grow by accreting supercooled cloud droplets.

An important aspect of hail growth is the latent heat of fusion released when the accreted water freezes, leading to different hailstone growth regimes (Ludlam, 1958; Mason, 1971). Within the complex structure of hail, it is possible to distinguish three kind of deposits (Mason, 1971).

(1) Porous ice is formed when the deposited water freezes rapidly, mainly as individual droplets.

(2) Compact ice is formed when the droplets have time to spread over the surface and form a continuous film before freezing. This happens when the transfer of heat between the deposit and the environment is just sufficiently rapid to allow all of the accreted water to freeze and to maintain the surface at 0°C.

(3) Spongy ice is produced when the heat exchange between the hailstone and the environment is not sufficiently rapid to allow all of the deposited water to freeze. Only a fraction of the collected water freezes immediately and produces the skeletal framework of ice which retains the unfrozen water whereby the whole mixture is maintained at 0°C. Under these conditions, the hailstone may also collect ice crystals and snowflakes and grow rapidly in size.

Porous ice, corresponding to the dry growth regime, contain large numbers of tiny air bubbles giving the ice an opaque appearance. Compact and spongy ice, growing in the wet growth regime, with smaller concentrations of air bubbles, appear generally clear or transparent. During its lifetime, an hailstone may undergo alternate wet and dry growth as it passes through a cloud of varying temperature and liquid water content, thus developing the layered structure that is often observed. Besides hailstones reveal crystalline structure which is shown by examining an hailstone layer in polarized light. This crystalline structure is determined by several factors such as the concentrations, size and impact velocity of the supercooled droplets, the temperature of the air and hailstone surfaces and the detailed mechanisms of freezing (Mason, 1971). Both the layered and crystalline structures contain information about the mechanisms of hailstone growth and have been extensively used to increase our knowledge of hailstone formation.

Early isotopic studies on hailstones

As described in the previous sections, there are reasons for the variation of isotopic composition of the cloud condensed phases which participate in the hailstone growth. The idea of the existence of an "isotopic structure" in hailstones and that this structure could serve to reconstruct the hailstone history was born in the early sixties. It has first been proposed and applied by Facy et al. (1962, 1963).

The first measurements were made on hailstones gathered near Paris during a storm on April 9, 1961. They were immediately preserved in petroleum ether, below $-10°C$, to await experimental use. It is essential to avoid superficial fusions or transfers by evaporation to prevent modifications of the original isotopic composition. Refrigerated petroleum ether is effective in preventing this by immediately enclosing the hailstone and thus protecting it from possible sublimations or condensations.

In these first experiments, the hailstones were placed on loosely woven muslin and continuously sublimated. The resulting vapour was fed into a mass spectrometer which measured the deuterium concentration according to the techniques previously described by Nief and Botter (1959). The deuterium content as a function of the radius is reported in Fig. 2-8.

In their articles, Facy et al. (1962, 1963) give the principle of the reconstruction of the hailstone trajectory from the deuterium results. As a basis, they use the simple isotopic model described by equation (16) (closed system) which foresees a regular decrease of deuterium content of the cloud condensed phases with decreasing temperatures and thus with increasing altitudes (Fig. 2-4). From the shape of the curve reported in Fig. 2-8, the hailstone radius is seen to increase progressively with lower deuterium contents which means that the hailstone grows larger at lower temperatures (higher altitudes), at least from a 5-mm radius. The main growth of the hailstone took place on its way upward.

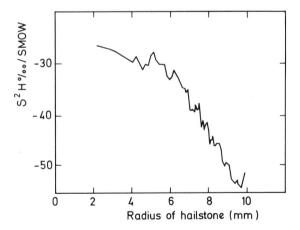

Fig. 2-8. Deuterium content of an hailstone collected near Paris on April 9, 1961, as a function of its radius (Facy et al., 1963). This specimen has been continuously sublimated.

Clearly, this first study was very encouraging and demonstrated the importance of the isotopic tool in hailstone studies. However, the authors pointed out the qualitative aspect of their interpretation and concluded that isotopic analyses should be supplemented by physical and meteorological information.

Merlivat et al. (1964a,b), have pursued this work by analysing three large hailstones (up to 70 mm in their larger dimension), collected in Minnesota and Iowa during May 1962. A new and more accurate sampling method was adopted. The hailstones were cut in thick slices (2—4 mm), from which well localized 25-mm³ samples were obtained. The deuterium results (Fig. 2-9) show that their growth started in the lower part of the updraft and that they underwent successive upwards and downwards movements before falling to the ground. They also proposed to estimate the unknown deuterium content, δ^2H_0, at the cloud base from the extreme values registered in the hailstones themselves by taking advantage from the fact that the hailstone growth is efficient only between the 0° and −40°C isotherms. Indeed, there are no droplets at temperatures lower than −40°C and the coalescence process is not efficient with the solid phase only. On the other hand, the influence of condensation processes is negligible during the fall to the ground. Thus, two estimations of δ^2H_0 were obtained from equation (16) and assuming that the isotopically richest and poorest layers are respectively formed at 0° and −40°C. The authors show that this procedure leads to two almost identical values (the difference is 4.5‰) and introduces an error on the deduced height of formation of 0.25 km only.

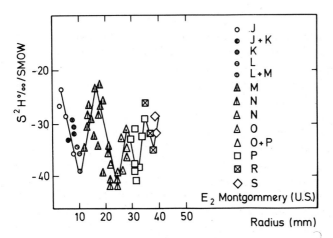

Fig. 2-9. Deuterium content of an hailstone collected in Minnesota during May 1962, as a function of its radius (Merlivat et al., 1964a). This hailstone has been cut in slices from which 25-mm³ samples were obtained.

An interesting observation is that the clear layers are isotopically lighter than the opaque ones. This is logical inasmuch as the first ones generally correspond to the wet growth regime and thus are expected to grow at relatively higher temperature than the second ones which are formed in the dry growth regime. This observation can be considered as a first qualitative check of the validity of the proposed interpretation.

Finally, these authors have proposed to determine the updraft speed U using the fact that $U = dZ/dt + V_L$, where V_L is the terminal fall speed of the hailstone, Z the altitude and t the time. dZ/dt is equal to $dZ/dR \cdot dR/dt$ and can be calculated from the trajectory, $Z = f(R)$ and the hailstone growth equation respectively. This method has led to updraft speeds lying between 100 and 300 km h^{-1} for the three studied hailstones which appear higher than those generally accepted in hailstorms.

Applying the same method, Bailey and Macklin (1965) have calculated the updraft profile corresponding to the first analysed hailstone by Facy et al. (1963). They pointed out that this profile disagrees with one deduced by Browning (1963) from radar observations of a severe hailstorm. As a possible explanation for this difference, they suggested that the isotopic content in a hailstone layer is not the same as that in the cloud droplets at the level at which the layer was formed. This particular point will be discussed in detail in the following section devoted to the $\delta^2 H$-$\delta^{18} O$ relationship in hailstones. Facy et al. (1965), have refuted this explanation but have indicated other possible causes of misinterpretation such as collection of large drops which are not at isotopic equilibrium and the shedding of liquid water from the hailstone surface.

Information deduced from tritium measurements

Following, deuterium studies, Ehhalt (1967) has shown that tritium measurements may provide additional information on the growth of hailstones. Deuterium and tritium measurements carried out on a large hailstone collected in Nebraska on July 5, 1965, are presented in Fig. 2-10. The two striking features of the tritium distribution are the very high content of the inner core (5 mm diameter) and a fairly abrupt increase at the 2-cm radius. Based on the previously described characteristics of the deuterium and tritium distribution in clouds, Ehhalt (1967) and Knight et al. (1975) have proposed a very detailed interpretation of these isotopic measurements. The high tritium value of the core strongly suggests that it was in contact with water from high altitudes. The abrupt increase at the 2-cm radium and its relatively constant value from there on, is explained as being caused by a mixture of cloud droplets and precipitation elements. The ^3H profile indicates that the water accumulated by the hailstone in the layer outside the 2-cm radius would consist of 60% precipitation and 40% cloud droplets.

The rapid decrease of tritium levels with time implies that this valuable tool of the late 60's becomes more difficult to handle in the seventies. This is primarily due to the small sample size which does not allow to perform electrolytic enrichment* and is the main reason of the very limited number of studies based on this isotope. Only one other study has been done on tritium content in five large hailstones (Jouzel et al., 1975). In any case, high tritium values are encountered in the hailstone embryos as shown in Fig. 2-11 which reports the tritium content of one of the hailstones collected in Oklahoma (U.S.A.) on April 26, 1971. In that case, the tritium values are fairly constant within the limits of experimental errors around a mean value of 80 TU (Fig. 2-11). This means that the hailstone was formed in the updraft core. Deuterium data support the hypothesis of a single growth episode with a rise followed by the fall to ground level.

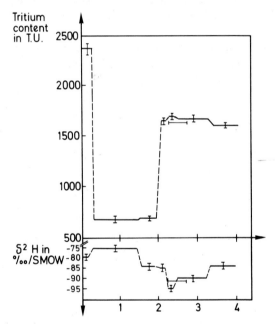

Fig. 2-10. Deuterium and tritium content of an hailstone collected in Nebraska on July 5, 1965, as a function of its radius (Ehhalt, 1967). Note the high tritium content of the core and the abrupt increase at a radius of 2 cm.

---

* Note that the new technique developed by Clarke et al. (1976), based on accumulation of ^{3}He and measurement by mass spectrometer, would theoretically allow to reach a detection limit of 20 TU for a 200-mg sample with accumulation times of 6 months to 1 year. This would be sufficiently low to deduce, even in the eighties; valuable information from tritium investigations in large hailstones.

The δ^2H-$\delta^{18}O$ relationship in hailstones

The first comparison of δ^2H and $\delta^{18}O$ values in hailstones was made by Majzoub et al. (1968) and is based on the analyses of two large hailstones. In the δ^2H vs $\delta^{18}O$ diagram, values are close to the MWL line for the inner layers, but the slope is smaller for the outer layers of the two specimens (3.3 and 6.2, respectively). This is interpreted as indication that the outer layers do not form in isotopic equilibrium with the surrounding water vapour.

From theoretical considerations, Bailey et al. (1969) have demonstrated that the δ^2H-$\delta^{18}O$ relationship could be modified during the freezing of supercooled droplets accreted by an hailstone. This is due to evaporation from the liquid during the freezing process, the hailstone surface temperature being higher than the air temperature. A subsequent δ^2H and $\delta^{18}O$ enrichment of the accreted ice is expected along a δ^2H-$\delta^{18}O$ line with a slope close to 4 as described above for evaporation processes and with maximum theoretical values of 6‰ in δ^2H and 1.5‰ in $\delta^{18}O$, reached in the wet growth regime.

Experimental evidence has been obtained from analyses of samples of accreted ice formed in an icing tunnel (Bailey et al., 1969). According to these authors, this effect appears sufficiently large to affect the interpre-

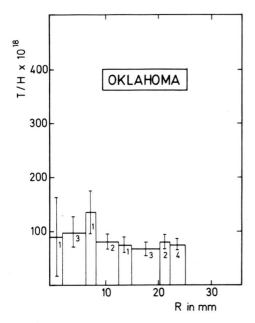

Fig. 2-11. Tritium content of an hailstone collected in Oklahoma on April 16, 1971, as a function of its radius. This hailstone is characterized by a fairly constant tritium value (Jouzel et al., 1975).

tation of isotopic analyses and explain a part of the differences between the clear and the opaque layers. However, δ^2H and $\delta^{18}O$ measurements made by Macklin et al. (1970) do not confirm that this process plays a role during the formation of natural hailstones.

Taking advantage of a new spectrometer (Hageman et al., 1978) allowing simultaneous determination of both isotopes on small samples (down to 20 mg of water), Jouzel and Merlivat (1978, 1980) and Jouzel et al. (1984) have shown that this effect can be observed on natural hailstones. A set of 78 samples from a large hailstone (9 cm in its largest dimension) collected in the Massif Central (France) on August 18, 1971, was analysed for δ^2H and $\delta^{18}O$. The experimental results are reported in Fig. 2-12, in the classical δ^2H-$\delta^{18}O$ diagram, each point representing the mean isotopic value of a layer. Cloud droplets curve (curve *1*) is calculated from an adiabatic cloud model assuming that it passes through the representative point of the isotopically poorer layer, which is shown to be formed around $-30°C$ and thus presumably not affected by evaporation processes. This figure shows a general shift of the other hailstone layers towards relatively higher $\delta^{18}O$ values as expected when evaporation occurs. The opaque embryo appears unaffected and the shift is higher for clear layers than for opaque ones. Using a modified ap-

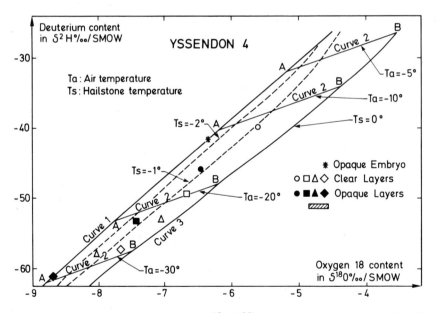

Fig. 2-12. Representative points in a δ^2H-$\delta^{18}O$ diagram of each layer of the Yssendon 4 hailstone collected in the Massif Central (France) on August 18, 1971, and theoretical curves (Jouzel and Merlivat, 1980). Curve *1* corresponds to cloud droplets (closed model), curves *2* to the evolution of a liquid film during its solidification and curve *3* to the upper limit of enrichment in the wet growth regime.

proach of the Bailey et al. (1969) treatment, taking into account continuous collection and shedding of liquid water, Jouzel and Merlivat (1980) have studied the isotopic behaviour of a growing film (wet growth). Its $\delta^2 H$-$\delta^{18}O$ content evolves along the curves 2 which are practically straight lines with a slope of 3.5 from A (isotopic equilibrium) to a value B, representing the maximum possible enrichment, which is very close to that previously calculated by Bailey et al. (1969).

In the dry growth regime, the calculated enrichment decreases with decreasing hailstone surface temperatures, T_s (dotted curves for $T_s = -1°C$ and $-2°C$), and becomes negligible where T_s is lower than $-2°C$. Thus, all the samples are expected between curves 1 and 3. This has been experimentally verified; clear layers formed at $T_s > 1°C$ and opaque ones at $T_s < 1°C$ in the hailstone studied (Fig. 2-12).

The isotopic content of the water at the moment of collection corresponds to the intersection of a straight line with a slope of 3.5 passing through the representative point of the sample with curve 1. This procedure allows us to deduce, from $\delta^2 H$ and $\delta^{18}O$ values, a corrected trajectory which takes into account the isotopic enrichment during the solidification process. This has been done by Jouzel and Merlivat (1980) for five hailstones, and the curves for two of them are presented in Figs. 2-13 and 2-14. Yssendon 4 is the hailstone studied above (Fig. 2-12) and Alberta E has been produced during the Alberta storm (Canada) previously studied by Jouzel et al. (1975). Even in the case where the maximum possible enrichment is almost reached for clear

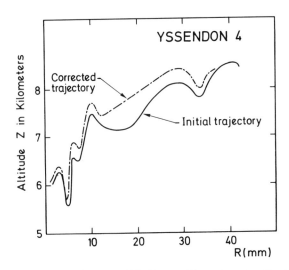

Fig. 2-13. Altitude of hailstone Yssendon 4 collected in the Massif Central (France) on August 18, 1971, as a function of its radius. The corrected trajectory, taking into account the isotopic modification during the solidification process, is obtained from the simultaneous $\delta^2 H$ and $\delta^{18}O$ determinations (Jouzel et al., 1980).

layers (Yssendon 4) the correction does not appear important (less than 0.5 km), and it becomes negligible for the Alberta stone (the enrichment being weaker). This study definitely shows that deuterium results are suitable for the reconstruction of hailstone trajectories even without this correction.

Recently (Federer et al., 1982b), combined δ^2H and $\delta^{18}O$ measurement were performed on a total of 86 hailstones collected in Switzerland. Concerning the embryos, the deviation from the equilibrium line is significantly larger for the frozen drops than for those classified as graupel which means than frozen drops evaporated more than graupels. This study shows also a significant difference between clear and opaque layers indicating that clear layers largely grow under non-equilibrium conditions with considerable evaporation from their surfaces.

The recycling process during hailstone formation

Large hailstones have fall speeds of 30 m s^{-1} or more, so any theory of their production requires that the cloud shall contain updraft of closely comparable speed in order to keep the stones suspended over periods of about 10 minutes or more. In its early stage of growth the fall speed increases rather slowly so that a strong steady updraft will carry it up through the supercooled region before it attains a large size. After descendence it therefore appears necessary to allow it to re-enter this zone to continue its growth

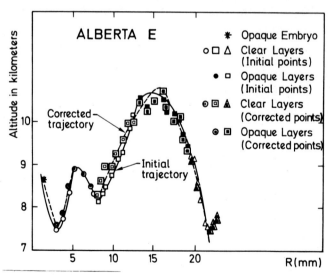

Fig. 2-14. Altitude of hailstone Alberta E, collected on August 7, 1971, as a function of its radius. The corrected trajectory is that deduced taking into account the isotopic modification during the solidification process (see Fig. 2-13).

(Mason, 1971). This "recycling" process has been suggested by Browning and Ludlam (1962) for severe hailstorms which may contain an almost steady strong updraft. An alternate means to prolong the lifetime of a hailstone in the storm is to suppose that the updraft is intermittent and that the hailstones repeatedly fall from a high level and then be carried up again without leaving the updraft core (Gokhale and Rao, 1969a,b; Gokhale, 1975).

The most interesting part of the isotopic study by Macklin et al. (1970) deals with this problem from the interpretation of the deuterium results. The deduced trajectory (Fig. 2-15) shows two successive ascents. According to these authors, the sharpness in the height transition confirms a recycling process of the nature envisaged by Browning and Ludlam (1962) with a descent outside the main body of the updraft.

From the same isotopic results Gokhale and Rao (1971), Gokhale (1975), concluded that this hailstone did not undergo Browning and Ludlam recycling process but that the entire growth is more likely to have occurred in the core of a pulsating and accelerating updraft. This was contradicted by Macklin et al. (1971) who stated, in their reply, that the sharpness of the boundary between the layers of different crystal sizes suggest an abrupt discontinuity.

Clearly in that case, deuterium results alone are not sufficient to choose between the two hailstone growth models. However the Gokhale and Rao's growth model has been well confirmed later on for three hailstones produced by a storm which arose on August 7, 1971, in the province of Alberta in Canada (Jouzel et al., 1975). The trajectories deduced from deuterium re-

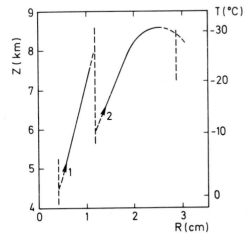

Fig. 2-15. Altitude of an hailstone collected as a function of its radius, showing two successive ascents (Macklin et al., 1970). After these authors, the sharpness in the height transition confirms a recycling process.

sults show successive upward and downward movements (Fig. 2-16). Recently (Jouzel and Merlivat, 1980), two other hailstones of the same storm have been analysed for δ^2H and $\delta^{18}O$, showing similar trajectories (Fig. 2-14). One notes that hailstone radius increases during the first descent in B and the second of C which excludes the interpretation that these falls have taken place outside the cloud. From tritium measurements the possibility of growth in the tritium-rich boundary zone (sketched in Fig. 2-3) is excluded. Thus, both isotope measurements suggest that the first descent of B and the second of C took place in the updraft core. The updraft velocity profile deduced from these curves indicates that the updraft was pulsating considerably. This finding appears in agreement with radar observations carried out during this storm (Renick et al., 1972) which revealed a multicellular character. This agreement gives an independent check on the validity of the interpretation of isotopic results at least on a qualitative basis.

One can note, at last, that from the combined δ^2H-$\delta^{18}O$ measurements quoted above, Federer et al. (1982b) suggest that some frozen drop embryos probably originate from melted and recirculated aggregates and graupels.

Estimate of the cloud liquid water content

At the limit between wet and dry growth regime, generally corresponding to opacity change, the amount of water collected by a hailstone per unit of

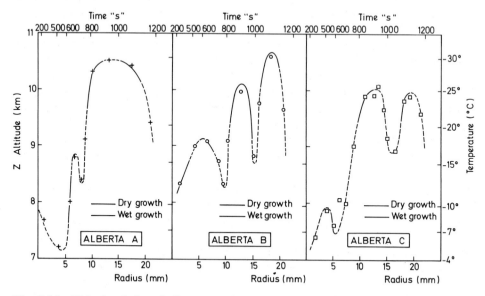

Fig. 2-16. Altitude of three hailstones collected in the province of Alberta (Canada) on August 7, 1971, as a function of their radius. Two, at least of the descents, the first of *B* and the second of *C*, have probably taken place in the updraft core.

time is just that which can be frozen. The corresponding amount of liquid water present in cloud (per unit of volume) at this time, is defined as the critical liquid water content, W_c. It can be calculated knowing the hailstone radius and pressure and temperature of the air (Macklin, 1963). It is interesting to compare W_c with the adiabatic values assumed in the isotopic closed model. This has been done for certain values of the hailstone radius corresponding approximately to transitions between layers (Macklin et al., 1970), giving then estimates of the actual liquid water content. There is a discrepancy by a factor of about 0.5—0.8 between the calculated W_c and the adiabatic values (Macklin et al., 1970). This discrepancy may be attributed to several processes as shedding of a part of the collected water, mixing of the main body of the updraft with dry air or growth of solid elements (ice crystals and hailstones). Macklin et al. (1970) point out that even accepting W_c values below the adiabatic ones does not affect significantly the inferred trajectory nor its implications. The same method was applied later on by Jouzel et al. (1975) giving similar results.

Recent investigations

Isotopic studies of hailstones undertaken since 1975 are usually combined with other methods of investigations of hailstone growth such as bubble structures, crystallographic observations, and radar measurements (Knight et al., 1975; Roos et al., 1976, 1977; Macklin, 1977; Macklin et al., 1977; Federer et al., 1978; Federer and Waldvogel, 1980; Zaïr, 1980; Grenier et al., 1983). On the other hand, the assumptions used to interpret the isotopic results in terms of trajectories are oversimplified according to many authors. For example, Knight et al. (1975) indicate that, although the description of the $\delta^2 H$ decrease in cloud droplets is qualitatively correct, such calculations cannot be quantitatively realistic. Therefore, they restrict themselves to more qualitative conclusions which require less strict and better justified assumptions. They analysed nine hailstones from Kansas (September 3, 1970) and Colorado (June 11, 1973) and found that even large hailstones can have simple trajectories with relatively little fluctuation in height. Furthermore, they did not go above the height of the $-25°C$ isotherm. In addition, it was found that in the case of wet growth of oblate hailstones significant lateral variation occurred in the deuterium content which implies movement of liquid water along the hailstone surface. This variation does not influence the previous conclusions but according to the authors would indicate that the long axis tends to be horizontal during the fall.

Macklin et al. (1977) combined the comparison of isotope distribution, crystallography and mode of occurrence of air bubbles in order to get a complete quantitative analysis and interpretation from the three independent methods. They include a determination of air and deposit temperatures, liquid water content and updraft velocity (see also Macklin, 1977). This removes some of the ambiguities in the interpretation of either isotopic or crystal data. Thus, they show from a complete study of eleven large hailstones (Colorado, August 15, 1974; Ohio, April 3, 1974 and Kansas, September 3, 1979) that the deuterium concentrations and the ambient temperatures deduced from the crystal structure generally show similar variations. Other studies showed similar agreement (Macklin et al., 1970; Knight et al., 1975; Roos et al., 1976, 1977; Federer et al., 1978, 1980, Zaïr, 1980). As there are completely different assumptions underlying the two types of analyses this general agreement is a new argument for the validity of the isotopic approach and of its interpretation at least on a qualitative basis. Knight et al. (1975) have stated that the change in deuterium content with radius is generally simpler than the change in either bubble content or crystal size. Thus isotope ratios are probably better indicators of relative growth altitude than the structural features. They conclude that this isotopic approach gives a better measure of real variability of trajectories and histories than do structural features and when combined with in cloud sampling and analysis, it would be indeed a very valuable quantitative tool.

An attempt to integrate the different approaches has been made over the last five years (1976—1981) in the Grossversuch IV program in Switzerland. The scope of this program is to provide a statistical and physical test of the Soviet method of hail suppression (Federer, 1977), a method based on rapid glaciation of the large supercooled drops in the so-called "accumulation zone" (Sulakvelidze, 1969). In the isotopic field, a particular effort has been pursued with a twofold purpose: firstly, to obtain more realistic profiles of the isotope content of the cloud water participating to the hailstone growth than the previously used adiabatic models; secondly, to determine more accurately than previously done the isotope content of the vapour feeding the cloud at its base. One of the final purposes was to specify, from δ^2H and $\delta^{18}O$ measurements, the temperature of formation of hailstone embryos which is one of the fundamental questions to solve, regarding the hailstone formation mechanisms, in this program mainly devoted to hail suppression.

The first objective has been achieved with the help of the multistage isotopic model including liquid and solid phases (see section "Isotopic cloud models" and Fig. 2-5a,b). Its application results in a much narrower range for the trajectories. It is then possible to reconcile the theoretical calculations with the large range of deuterium concentrations observed for three storms in one day (Roos et al., 1976, 1977) and to eliminate some contradictions which arose applying the previous closed model such as excessive updraft speeds and discrepancies between critical and actual cloud liquid water content.

Applying this model, Jouzel et al. (in preparation) have reinterpreted the isotopic data concerning five hailstones of the August 7, 1971, Alberta hailstorm. The multicellular nature of this storm is well confirmed but the growth temperature intervals range from —13 to —23°C (Fig. 2-17) showing a considerable reduction if compared to that obtained applying the "closed system" model (Fig. 2-16). The estimated wind speed ranges from 15 to 42 m s^{-1}, very consistent with information derived from radar observations giving confidence in the interpretation of isotopic data. In their article, Jouzel et al. also discussed the results obtained on six hailstones collected in the Massif Central region (France; August 18 and 19, 1971) and fully developed the theoretical aspects dealing with the isotopic modifications of the collected water both in the wet and dry growth regimes. For the eleven discussed hailstone trajectories, the most common growth temperatures are for the layers between −15° and −25°C and, remarkably, the embryo source region is situated around −15°C for all the three storms. At last, examination of individual trajectories put forward some new interesting points as preferential growth of clear layers during ascents (and conversely of opaque layers during descents) and the evidence that large clear layers can possibly grow over a large range of continuously changing environmental conditions.

The isotope content of vapour at cloud base is usually determined from the extreme isotopic values measured in hailstones (Merlivat et al., 1964a), or more recently from crystallographic analysis knowing that there are no large crystals for air temperatures lower than −24°C and no small crystals at temperatures higher than −15°C (Federer et al., 1980). To check the validity of these assumptions water vapour has been sampled in and near the Swiss experimental area between 430 and 2100 m above mean sea level in typical

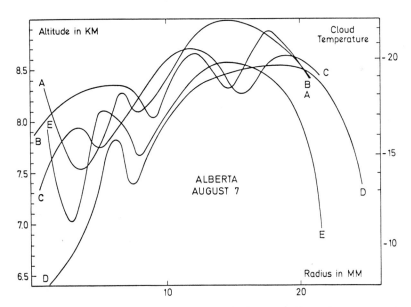

Fig. 2-17. Trajectories of the five Alberta hailstones collected on August 7, 1971, deduced from application of the Isotopic Cloud Model developed by Jouzel et al. (1980) and Federer et al. (1982a).

pre-storm conditions (Federer et al., 1980). It was shown (Federer et al., 1982b) that the vertical, horizontal and temporal variations of δ^2H_0 in the experimental area can be as large as 10—15‰ even on days with no precipitation. In the samples taken in the inflow of thunderstorms, however, the vertical gradient was substantially reduced. On August 7, 1979, water vapour was sampled at two altitudes (1500 and 2100 m) just before the storm from which hailstones were studied, the two measurements yielded almost the same value of δ^2H_0 which was adopted leading to trajectories consistent with the radar pattern and crystallography.

Direct sampling from aircraft of water vapor in the storm inflow was undertaken by Knight et al. (1981) for three storms. Unfortunately, convincing experimental results were only obtained on the June 22, 1976, storm in northeast Colorado, where five samples were collected in a period of 110 minutes. The deuterium results showed variations of 13‰. In this large multicell storm there was a very substantial diversity of air being sampled which can explain that the inflow deuterium values also varied widely. The study by Knight et al. (1981) is a serious qualification to the direct use of the deuterium method to deduce quantitatively hailstone growth trajectories in such big multicell hailstorms.

Very generally, the value for δ^2H_0 cannot be obtained from direct sampling and the crystallographic method described above is therefore recommended.

Considering now the trajectories deduced in these two recent studies which dealt with respectively 10 (Knight et al., 1981) and 86 (Federer et al., 1982b) hailstones, they support the generally narrow range of growth temperatures of hailstones. The Knight et al. results lead to place most hailstone growth in the $-10°$ to $-25°C$ temperature range with (as stated in almost all the previous studies) hailstone embryo formation usually at a lower level than the growth of the bulk of the hailstones. The overall growth range for the 86 hailstones studied by Federer et al. (1982b) lies between $-2.5°$ and $-30.5°C$ but hailstones grow mainly between $-15°$ and $-25°C$ in a wet mode. The trajectories of the large hailstones are surprisingly flat, indicating that an approximate balance is maintained between updraft and the increasing terminal velocities of the growing hailstones.

CONCLUSION

The previous section has demonstrated the importance of isotopic determinations in the hailstone growth research and in the study of the microphysics and of the dynamics of hailstorms, through estimates of the cloud liquid water content and of the updraft velocity profiles. Most of the studies undertaken in this field of research are based on ^2H determinations and subsequent reconstruction of hailstones trajectories. However, the investigations including ^3H and ^{18}O measurements have shown that the three isotopes bring very complementary information: ^{3}H allowing to specify the zone of formation of hailstones in the storm and ^{18}O leading to a better knowledge of growth regimes.

Most of the possibilities to check the validity of the transformation of the deuterium results in trajectories have been explored, and with success if we keep in mind the limitations of the different methods. There is a very general agreement for accepting this interpretation, at least on a qualitative basis, and that isotope ratios are probably better indicators of relative growth altitude than structural features. From a quantitative point of view, the simple closed-system model appears unrealistic but an important step forward has been the development of a new model (Jouzel et al., 1980; Federer et al., 1982a) which opens the possibility for a realistic quantitative interpretation of the isotopic results, notwithstanding the fact that efforts have to be pursued in the field of isotopic cloud modelling.

It may be noted that other methods can be used in the research on the growth of hailstones such as studies of water-insoluble particles (Rozinski, 1966, 1967; Rozinski et al., 1976, 1979), chemical analysis of hailstone slice residues (Vittori et al., 1969) and distribution of micro-organisms (Vittori et al., 1973).

The isotopic method is a very specific tool as it concerns the water itself. On the other hand, it has been seen in this section, that its application is greatly facilitated when it is combined with other studies such as crystallographic and bubble structure observations and radar measurements.

REFERENCES

Aldaz, L. and Deutsch, W., 1967. On a relationship between air temperature and oxygen isotope ratio of snow and firn in the South Pole region. *Earth Planet. Sci. Lett.*, 3: 267—274.
Arnasson, B., 1969. Equilibrium constant for the fractionation of deuterium between ice and water. *J. Phys. Chem.*, 73: 3491—3494.
Baerstchi, P., 1976. Absolute ^{18}O content of Standard Mean Ocean Water. *Earth Planet. Sci. Lett.*, 31: 341—344.
Bailey, I.H. and Macklin, W.C., 1965. On the study of the formation of hailstones by means of isotopic analysis. *J. Geophys. Res.*, 70: 493—497.
Bailey, I.H., Hulston, J.R., Macklin, W.C. and Stewart, J.R., 1969. On the isotopic composition of hailstones. *J. Atmos. Sci.*, 26: 689—694.
Bleeker, W., Dansgaard, W. and Lablans, W.N., 1966. Some remarks on simultaneous measurements of particulate contaminants including radioactivity and isotopic composition of precipitation. *Tellus*, 18: 773—785.
Blifford, I.H., Patterson, R.L., Jr., Lockhart, L.B., Jr. and Baus, R.A., 1957. On radioactive hailstones. *Bull. Am. Meteorol. Soc.*, 38: 139—141.
Bolin, B., 1958. On the use of tritium as a tracer for water in nature. *Proc. 2nd U.N. Conf., Peaceful Uses of Atomic Energy, Geneva*, 18: 336—344.
Booker, D.V., 1965. Exchange between water droplets and tritiated water vapour. *Q. J. R. Meteorol. Soc.*, 91: 73—79.

Bromwich, D.H. and Weaver, C.J., 1983. Distance from main moisture source is principal control on $\delta^{18}O$ of snow in coastal Antarctica. Nature, 301: 145—147.
Browning, K.A., 1963. The growth of hail within a steady updraft. Q. J. R. Meteorol. Soc., 89: 490—506.
Browning, K.A. and Ludlam, F.H., 1962. Airflow in convective storms. Q. J. R. Meteorol. Soc., 88: 117—135.
Buchardt, B. and Fritz, P., 1980. Environmental isotopes as environmental and climatological indicators. In: P. Fritz and J.Ch. Fontes (Editors), Handbook of Environmental Isotope Geochemistry, 1. The Terrestrial Environment, A. Elsevier, Amsterdam, pp. 473—500.
Chisholm, A.J., 1973. Alberta hailstorms, 1. Radar case studies models. Meteorol. Monogr., 14(36): 37—95.
Clarke, W.B., Jenkins, W.J. and Top, Z., 1976. Determination of tritium by mass spectrometric measurement of ^{3}He. Int. J. Radiat. Isot., 27: 515—522.
Craig, H., 1961a. Standard for reporting concentrations of deuterium and oxygen-18 in natural water. Science, 133: 1833—1834.
Craig, H., 1961b. Isotopic variations in meteoric waters. Science, 133: 1702—1703.
Craig, H. and Gordon, L., 1965. Deuterium and oxygen 18 variation in the ocean and marine atmosphère. In: E. Tongiorgi (Editor), Stable Isotopes in Oceanographic Studies and Paleotemperatures, Spoleto 1965. CNR, Pisa, pp. 9—130.
Covey, K. and Schneider, S., 1982. Isotopic composition of snow calculated from the trajectories of air parcels in storms. Symp., Milankovitch and Climate, Lamont Doherty Geological Observatory, Palisades, N.Y., 30 November—3 December 1982 (presented paper).
Dansgaard, W., 1953. The abundance of ^{18}O in atmospheric water and water vapour. Tellus, 5: 461—469.
Dansgaard, W., 1961. The isotopic composition of natural waters. Medd. Grønl., 165(2): 1—120.
Dansgaard, W., 1964. Stable isotopes in precipitation. Tellus, 16: 436—468.
Dansgaard, W., Johnsen, S.J., Moeller, J. and Langway, C.C., Jr., 1969. One thousand centuries of climatic record from Camp Century on the Greenland ice sheet. Science, 166: 377—381.
Dansgaard, W., Johnsen, S.J., Reeh, N., Gundestrup, N., Clausen, H.B. and Hammer, G.U., 1974. Climatic changes, Norsemen and modern man. Nature, 255: 24—28.
Ehhalt, D.H., 1967. Deuterium and tritium content of hailstones: additional information on their growth. Trans. Am. Geophys. Union, 48: 106—107.
Ehhalt, D.H., 1971. Vertical profiles and transport of HTO in the troposphere. J. Geophys. Res., 76: 75—84.
Ehhalt, D.H., 1974. Vertical profiles of HTO, HDO and H_2O in the troposphere. NCAR Tech. Note, NCAR-TN/STR100: 131 pp.
Ehhalt, D.H. and Knott, K., 1965. Kinetische Isotopentrennung bei der Verdampfung von Wasser, Tellus, 17: 389—397.
Ehhalt, D.H. and Ostlünd, G., 1970. Deuterium in Hurricane Faith 1966. J. Geophys. Res., 75: 2323—2327.
Ehhalt, D.H., Knott, K., Nagel, J.F. and Vogel, J.C., 1963. Deuterium and oxygen-18 in rain water. J. Geophys. Res., 68: 3775—3780.
Epstein, S., 1956. Variations of the $^{18}O/^{16}O$ ratios of fresh water and ice. Natl. Acad. Sci., Nucl. Sci. Ser., Rep., 19: 20—25.
Epstein, S. and Yapp, C.J., 1976. Climatic implications of the D/H ratio of hydrogen in C-H groups in tree cellulose. Earth Planet. Sci. Lett., 30: 252—261.
Epstein, S., Sharp, R.P. and Gow., A.J., 1965. Six years record of oxygen and hydrogen isotope variations in South Pole firn. J. Geophys. Res., 70: 1809—1814.

Eriksson, E., 1965. Deuterium and oxygen-18 in precipitation and other natural waters: some theoretical considerations. *Tellus*, 4: 498—512.

Eriksson, E. and Bolin, B., 1964. Oxygen-18, deuterium and tritium in natural waters and their relation to the global circulation of water. In: *Proceedings of the Second Conference on Radioactive Falloup from Nuclear Weapon Tests, Germantown, November 3—6, 1964*, pp. 675—686.

Facy, L., Merlivat, L., Nief, G. and Roth, E., 1962. Etude de la formation de la grêle par une méthode d'analyse isotopique. *J. Méc. Phys. Atmos.*, 2: 67—77.

Facy, L., Merlivat, L., Nief, G. and Roth, E., 1963. The study of the formation of a hailstone by means of isotopic analysis. *J. Geophys. Res.*, 68: 3841—3848.

Facy, L., Merlivat, L., Nief, G. and Roth, E., 1965. Reply to the comment of I.H. Bailey and W.C. Macklin. *J. Geophys. Res.*, 70: 498—499.

Federer, B., 1977. Methods and Results of Hail Suppression in the European Area. *Am. Meteorol. Soc., Meteorol. Monogr.*, 38.

Federer, B. and Waldvogel, A., 1978. Time-resolved hailstone analyses and radar structure of Swiss storms. *Q. J. R. Meteorol. Soc.*, 104: 69—90.

Federer, B., Jouzel, J. and Waldvogel, A., 1978. Hailstone trajectories determined from crystallography deuterium content and radar back scattering. *Pure Appl. Geophys.*, 116: 111—129.

Federer, B., Thalmann, B., Oesch, A., Brichet, N., Waldvogel, A., Jouzel, J. and Merlivat, L., 1980. The origin of hailstones embryo deduced from isotopic measurements. *Proc., 8th Int. Conf. on Cloud Physics, Clermont-Ferrand*. LAMP, Clermont-Ferrand.

Federer, B., Brichet, N. and Jouzel, J., 1982a, Stable isotope in hailstones, I. The isotopic cloud model. *J. Atmos. Sci.*, 39: 1323—1335.

Federer, B., Thalmann, B. and Jouzel, J., 1982b. Stable isotopes in hailstones, II. Embryo and hailstone growth in different storms. *J. Atmos. Sci.*, 39: 1336—1355.

Fontes, J.Ch., 1976. *Isotopes du milieu et cycles des eaux naturelles: quelques aspects*. Thèse de Doctorat ès Sciences, Université Pierre et Marie Curie, Paris, 218 pp.

Fontes, J.Ch., 1980. Environmental isotopes in groundwater hydrology. In: P. Fritz and J.Ch. Fontes (Editors), *Handbook of Environmental Isotope Geochemistry, 1. The Terrestrial Environment, A*. Elsevier, Amsterdam, pp. 75—140.

Fontes, J.Ch., Gonfiantini, R. and Roche, M.A., 1970. Deuterium et oxygen-18 dans les eaux du lac Tchad. In: *Isotope Hydrology*. IAEA, Vienna, pp. 387—404.

Friedman, I., 1953. Deuterium content of natural water and other substances. *Geochim. Cosmochim. Acta*, 4: 89—103.

Friedman, I., Machta, L. and Soller, R., 1962. Water vapour exchange between a water droplet and its environment. *J. Geophys. Res.*, 67: 2761—2766.

Friedman, I., Redfield, A.C., Schoen, B. and Harris, J., 1964. The variations of the deuterium content of natural waters in the hydrologic cycle. *Rev. Geophys.*, 2: 177—224.

Gambell, A.W. and Freidman, I., 1965. Note on the aerial variation of deuterium/hydrogen ratios in rainfall for a single storm event. *J. Appl. Meteorol.*, 4: 533—535.

Gat, J.R., 1980. The isotopes of hydrogen and oxygen in precipitation. In: P. Fritz and J.Ch. Fontes (Editors), *Handbook of Environmental Isotope Geochemistry, 1. The Terrestrial Environment, A*. Elsevier, Amsterdam, pp. 21—48.

Gat, J.R. and Dansgaard, W., 1972. Stable isotope survey of the fresh water occurrences in Israel and the northern Jordan rift valley. *J. Hydrol.*, 16: 177—212.

Gedzelman, S.D. and Lawrence, J.R., 1982. The isotopic composition of cyclonic precipitation. J. Appl. Meteorol., 21: 1335—1404.

Gokhale, N.R., 1975. *Hailstorms and Hailstones Growth*. State University of New York Press, New York, N.Y, 465 pp.

Gokhale, N.R. and Rao, K.M., 1969a. Theory of hailgrowth. *J. Rech. Atmos.*, 4: 153—178.

Gokhale, N.R. and Rao, K.M., 1969b. Accumulation of large hydrometeors in the upper part of an intensive updraft in a cumulonimbus. In: *Proceedings of the Sixth Conference on Severe Local Storms, Chicago, Ill.* pp. 59—62 (preprint).

Gokhale, N.R. and Rao, K.M., 1971. Comments on "The analysis of hailstone" by W.C. Macklin, L. Merlivat and C.M. Stevenson. *Q. J. R. Meteorol., Soc.*, 97: 575—576.

Gonfiantini, R., 1978. Standards for stable isotope measurements in natural compounds. *Nature*, 271: 534—536.

Grenier, J.C., Admirat, P. and Zaïr, S., 1983. Hailstone growth trajectories in the dynamic evolution of a moderate hailstorm. *J. Appl. Meteorol.*, 22: 1008—1021.

Hageman, R. and Lohez, P., 1978. Twin mass spectrometers for simultaneous isotopic analysis of hydrogen and oxygen in water. In: N.R. Daly (Editor), *Advances in Mass-Spectrometry*, 7. Heyden, pp. 504—507.

Hageman, R., Nief, G. and Roth, E., 1970. Absolute isotopic scale for deuterium analysis of natural waters. Absolute D/H ratio for SMOW. *Tellus*, 22: 712—715.

Harmon, R.S. and Schwarcz, H.P., 1981. Changes of ^{2}H and ^{18}O enrichment of meteoric water and pleistocene glaciation. *Nature*, 290: 125—128.

Hirsch, J.H., 1971. Computer modelling of cumulus clouds during project cloud catcher. *S. Dak. Sch. Mines Technol., Rep.*, 71—77.

IAEA (International Atomic Energy Agency), 1981. *Statistical Treatment of Environmental Isotope Data in Precipitation. IAEA Tech. Rep. Ser.*, 206.

Jaffé, G., Wittmann, W. and Bates, C.C., 1954. Radioactive hailstones in the district of Columbia area: a brief explanation. *Geol. Soc. Am. Bull.*, 35: 245—249.

Jakli, G. and Staschewski, D., 1977. Vapour pressures of H_2 ^{18}O ice ($-50°$ to $0°$C) and H_2 ^{18}O water ($0°$ to $170°$C). *J. Chem. Soc., Faraday Trans.*, 1(73): 1505—1509.

Jones, W.M., 1968. Vapour pressures of tritium oxide and deuterium oxide. Interpretation of the isotope effects. *J. Chem. Phys.*, 48: 297—214.

Joussaume, S., Jouzel, J. and Sadourny, R., 1983. Simulation of the HDO and H_2 ^{18}O cycles for present day and glacial age climates, using an atmospheric general circulation model. *Symp., Ice and Climate Modelling, Evanston, June 1983.*

Joussaume, S., Jouzel, J. and Sadourny, R., 1984. Water isotope cycles in the atmosphere: first simulation using a general circulation model (submitted to *Nature*).

Jouzel, J., 1975. Complementarité des mesures de deutérium et de tritium pour l'étude de la formation des grêlons. *Note C.E.A.*, N-1833: 1—60.

Jouzel, J., 1979. Teneurs isotopiques de la vapeur d'eau tropospherique. Mise au point d'un système de prélèvement embarquable. *J. Rech. Atmos.*, 13: 261—269.

Jouzel, J. and Merlivat, L., 1978. Deuterium and ^{18}O in hailstones. *Trans. Am. Geophys. Union*, 89: 285.

Jouzel, J. and Merlivat, L., 1980. Growth regime of hailstones as deduced from simultaneous deuterium and oxygen-18 measurements. In: *Proceedings of the Eighth International Conference on Cloud Physics, Clermont-Ferrand.* LAMP, Clermond-Ferrand.

Jouzel, J. and Merlivat, L., 1981. A global isotopic model to interpret deuterium and oxygen-18 variations in rain and snow. *Symp., Variations in the Global Water Budget, Oxford.* Preprints, p. 54.

Jouzel, J. and Merlivat, L., 1984. Deuterium and oxygen-18 in precipitation: modeling of the isotopic effects at snow formation. *J. Geophys. Res.* (in press).

Jouzel, J., Merlivat, L. and Roth, E., 1975. Isotopic study of hail. *J. Geophys. Res.*, 80: 5015—5030.

Jouzel, J., Brichet, N., Thalmann, B. and Federer, B., 1980. A numerical cloud model to interpret the isotope content of hailstones. In: *Proceedings of the Eighth Conference on Cloud Physics, Clermont-Ferrand.* LAMP, Clermond-Ferrand.

Jouzel, J., Merlivat, L. and Lorius, C., 1982. Deuterium excess in an East Antarctic ice core suggests higher relative humidity at the oceanic surface during the last glacial maximum. *Nature*, 299: 688—691.

Jouzel, J., Merlivat, L., Petit, J.R. and Lorius, C., 1983. Climatic information over the last century deduced from a detailed isotopic record in the South Pole snow. *J. Geophys. Res.*, 88: 2693—2703.

Jouzel, J., Merlivat, L. and Federer, B., 1984. Isotopic study of hail: the SD-δ^{18}O relationship and the growth history of large hailstones (submitted to *Q.J.R. Meteorol. Soc.*).

Kakiuchu, M. and Matsuo, S., 1979. Direct measurements of D/H and ^{18}O/^{16}O fractionation factors between vapour and liquid water in the temperature range from 10 to 40°C. *Geochem. J.* 13: 307—311.

Kato, K., 1978. Factors controlling oxygen isotopic composition of fallen snow in Antarctica. *Science*, 272: 46—48.

Kessler, E., 1969. On the distribution and continuity of water substance in atmospheric circulation. *Meteorol. Hong.*, 10: 84 pp.

Kinzer, G.D. and Gunn, 1951. The evaporation temperature and thermal relaxation time of freely falling water drops. *J. Meteorol.*, 8: 71—83.

Knight, C.A., Ehhalt, D.H., Roper, N. and Knight, N.C., 1975. Radial and tangential variation of deuterium in hailstones. *J. Atmos. Sci.*, 32: 1990—2000.

Knight, C.A., Knight, N.C. and Kime, K.A., 1981. Deuterium contents of storm inflow and hailstone growth layers. *J. Atmos. Sci.*, 38: 2485—2499.

Kuhn, W. and Thürkauf, M., 1958. Isotopentrennung beim Gefrieren von Wasser und Diffusionskonstanten von D und ^{18}O im Eis. *Helv. Chim. Acta*, 41: 938—971.

Lawrence, J.R. and White, J.W.C., 1981. Meteorological factors influencing the hydrogen isotope composition of precipitation in Northeastern United States. *Variations in the Global Water Budget, Oxford.* Preprints, p. 65.

Lawrence, J.R., Gedzelman, S.D., White, J.W.C., Smiley, D. and Lazov, P., 1982. Storm trajectories in eastern U.S.: D/H isotopic composition of precipitation. *Nature*, 296: 638—640.

Létolle, R., Olive, P. and Lemaille, E., 1980. ^{18}O variations in lacustrines carbonates from the pre-alps during the late or postglacial period. *Meet. Advisory Groups on the Variation of the Isotopic Composition of Water, Vienna.*

Lorius, C. and Merlivat, L., 1977. Distribution of mean surface stable isotope values in East Antartica; observed changes with depth in a coastal area. In: *Isotopes and Impurities in Snow and Ice, Proc. Grenoble Symp., August—September 1975.* IASH Publ., 118: 127—137.

Lorius, C., Merlivat, L., Jouzel, J. and Pourchet, M., 1979. A 30,000 year isotope record from Antarctic ice. *Nature*, 280: 644—648.

Ludlam, F.H., 1958. The hail problem. *Nubila*, 1: 7—27.

Macklin, W.C., 1963. Heat transfer from hailstones. *Q. J. R. Meteorol. Soc.*, 89: 360—369.

Macklin, W.C., 1977. The characteristics of natural hailstones and their interpretation. In: G.B. Foote and C.A. Knight (Editors), *Hail: A Review of Hail Science and Hail Suppression. Am. Meteorol. Soc., Meteorol. Monogr.*, 38: 65—86.

Macklin, W.C., Merlivat, L. and Stevenson, C.M., 1970. The analysis of a hailstone. *Q. J. R. Meteorol. Soc.*, 96: 472—486.

Macklin, W.C., Merlivat, L. and Stevenson, C.M., 1971. Reply to the comment of N.R. Gokhale and K.M. Rao. *Q. J. R. Meteorol. Soc.*, 97: 576—577.

Macklin, W.C., Knight, C.A., Moore, H.E., Knight, N.C., Pollock, W.H., Carras, J.N. and Thwaites, S., 1977. Isotopic, crystal and air bubble structure of hailstones. *J. Atmos. Sci.*, 34: 961—967.

Majoube, M., 1971a. Fractionnement en oxygene 18 et en deuterium entre l'eau et sa vapeur. *J. Chim. Phys.*, 10: 1423—1436.

Majoube, M., 1971b. Fractionnement en oxygène 18 entre la glace et la vapeur d'eau. *J. Chem. Phys.*, 68: 625—636.

Majzoub, M., Nief, G. and Roth, E., 1968. Variations and comparisons of deuterium and oxygen-18 concentrations in hailstones. In: *Proceedings of the International Conference on Cloud Physics, August 26—30, Toronto, Ont.* American Meteorological Society, pp. 450—454.

Mandrioli, P., Puppi, G.L., Bagni, N. and Prodi, F., 1973. Distribution of microorganisms in hailstones. *Nature*, 246: 416—417.

Mason, B.J., 1971. *The Physics of Clouds.* Clarendon Press, Oxford, 671 pp.

Matsuo, S. and Friedman, I., 1967. Deuterium content in fractionally collected oceanwater, *J. Geophys. Res.*, 72: 6374—6376.

Merlivat, L., 1977. Contribution à l'étude du mécanisme de transfert de masse à l'interface air-eau par une méthode isotopique. Comparaison des coefficients de frottement et d'évaporation. *J. Rech. Oceanogr.*, 4: 1—14.

Merlivat, L., 1978a. Molecular diffusivities of water H_2 ^{16}O, $HD^{16}O$ and H_1 ^{18}O in gases. *J. Chem. Phys.*, 69: 2864—2871.

Merlivat, L., 1978b. The dependence of bulk evaporation coefficients on air-water interfacial condition as determined by the isotopic method. *J. Geophys. Res.*, 83: 2977—2980.

Merlivat, L. and Nief, G., 1967. Fractionnement isotopique lors des changements d'état solide-vapeur et liquide-vapeur de l'eau à des températures inférieures à $0°C$. *Tellus*, 19: 122—127.

Merlivat, L. and Coantic, M., 1975. Study of mass transfer at the air-water interface by an isotopic method. *J. Geophys. Res.*, 80: 3453—3464.

Merlivat, L. and Jouzel, J., 1979. Global climatic interpretation of the deuterium-oxygen-18 relationship for precipitation. *J. Geophys. Res.*, 84: 5029—5033.

Merlivat, L., Botter, R. and Nief, G., 1963. Fractionnement isotopique au cours de la distillation de l'eau. *J. Chim. Phys.*, 60: 56—59.

Merlivat, L., Nief, G. and Roth, E., 1964a. Formation de la grêle et fractionnement isotopique du deutérium. *Abh. Dtsch. Akad. Wissensch., Berlin*, 7: 839—853.

Merlivat, L., Nief, G. and Roth, E., 1964b. Etude de la formation de la grêle par une méthode isotopique. *C.R. Acad. Sci. Paris*, 258: 6500—6502.

Miyake, Y., Matsubaya, O. and Niskihara, 1968. An isotopic study on meteoric precipitation. *Pap. Meteorol. Geophys.*, 19: 243—266.

Moser, H. and Stichler, W., 1980. Environmental isotopes in ice and snow. In: P. Fritz and J.Ch. Fontes, (Editors), *Handbook of Environmental Isotope Geochemistry, 1. The Terrestrial Environment, A.* Elsevier, Amsterdam, pp. 141—178.

Munnich, K.O., 1978. Soil-water plant relationship. *Symp., Study and Management of Water Resources in Arid and Semi-arid Regions, Ahmedabad, March 1—4.*

Nief, G. and Botter, R., 1959. In: J. Waldron (Editor), *Advances in Mass Spectrometry.* Pergamon, Oxford, 515 pp.

O'Neil, J., 1968. Hydrogen and oxygen isotope fractionation between ice and water. *J. Chem. Phys.*, 72: 3683—3684.

Ostlünd, H.G., 1968. Hurricane tritium, II. Air-sea exchange of water in Betsy 1965. *Tellus*, 20: 577—594.

Petit, J.R., Briat, M. and Royer, A., 1981. Ice age aerosol content from East Antarctic ice core samples and past wind strength. *Nature*, 293: 391—394.

Picciotto, E., De Maere, X. and Friedman, I., 1960. Isotope composition and temperature of formation of Antartic snows. *Nature*, 187: 857—859.

Posey, J.C. and Smith, H.A., 1951. The equilibrium distribution of light and heavy waters in a freezing mixture. *J. Am. Chem. Soc.*, 79: 555.

Renick, J.H., Chilshom, A.J. and Summers, P.W., 1972. The seedability of multicell and supercell hailstorms using droppable pyrotechnic flaves. *3rd Conf. on Weather Modification, Rapid City, S. Dak.* (preprint).

Rogers, R.R., 1979. A short course in cloud physics. *Int. Ser. Nat. Philos.*, 96: 235 pp.

Rogers, R.R. and Vali, 1978. Recent developments in meteorological physics. *Phys. Rep.*, 48: 177 pp.

Roos, D. v.s.d., Vogel, S.C. and Carte, A.E., 1976. Hailstone growth. *Int. Conf. on Cloud Physics, Boulder, Colo.* Preprints, pp. 240—245.

Roos, D. v.s.d., Schooling, H. and Vogel, J.C., 1977. Deuterium in hailstones collected on 29 November 1972. *Q. J. R. Meteorol. Soc.*, 103: 751—768.

Rozanski, K. and Sonntag, C., 1982. Vertical distribution of deuterium in atmospheric water vapour. *Tellus*, 34: 135—141.

Rozanski, K., Sonntag, C. and Münnich, K.O., 1982. Factors controlling stable isotope composition of modern precipitation. *Tellus*, 34: 142—150.

Rozinski, J., 1966. Solid water-insoluble particles in hailstones and their geophysical significance. *J. Appl. Meteorol.*, 52: 481—492.

Rozinski, J., 1967. Insoluble particles in hail and rain. *J. Appl. Meteorol.*, 6: 1066—1074.

Rozinski, J., Browning, K.A., Langer, G. and Nagamoto, C.T., 1976. On the distribution of water-insoluble aerosol particles in hailstones and its possible value as an indication of hail growth history. *J. Atmos. Sci.*, 33: 530—536.

Rozinski, J., Knight, C.A., Nagamoto, C.T., Morgan, M.G., Knight, N.C. and Prodi, F., 1979. Further studies of large water insoluble particles within hailstones. *J. Atmos. Sci.*, 36: 882—891.

Salati, E., Dall'Olio, A., Matsui, E. and Gat, J.R., 1979. Recycling of water in the Amazon Basin: an isotopic study. *Water Resour. Res.*, 15: 1250—1258.

Schriber, G., Stauffer, B. and Müller, F., 1977. $^{18}O/^{16}O$, $^{2}H/^{1}H$ and ^{3}H measurements on precipitation and air moisture samples from the North Water area. In: *Isotopes and Impurities in Ice. Proc. Grenoble Symp., August—September 1975.* IAHS Publ., 118: 182—187.

Schwarcz, H.P., Harmon, R.S., Thompson, P. and Ford, D.C., 1976. Stable isotope studies of fluid inclusions in speleothems and their paleoclimatic significance. *Geochim. Cosmochim. Acta*, 40: 657—665.

Siegenthaler, U. and Oeschger, H., 1980. Correlation of ^{18}O in precipitation with temperature and altitude, *Nature*, 285: 314—317.

Smith, G.I., Friedman, I., Klieforth, H. and Hardcastle, K., 1979. Areal distribution of deuterium in eastern California precipitation 1968—1969. *J. Appl. Meteorol.*, 18: 172—188.

Sonntag, C., Rozanski, K., Münnich, K.O. and Jacob, H., 1981. Spatial variation of deuterium and ^{18}O in European precipitation and groundwater, a simple water vapour model of this continental effect in D and ^{18}O. *Variations in the Global Water Budget, Oxford.* Preprints, pp. 145—146.

Stewart, M.K., 1974. *Stable isotopes in raindrops and lakes: laboratory and natural studies.* Ph.D. Thesis, University of Pensylvania, Philadelphia, Pa., 110 pp.

Stewart, M.K., 1975. Stable isotope fractionation due to evaporation and isotopic exchange of falling water drops: application to atmospheric processes and evaporation of lakes. *J. Geophys. Res.*, 80: 1138—1146.

Sulakvelidze, G.K., 1969. *Rainstorms and Hail.* Gidrometeoizdat, Leningrad, 1967 (translat by Israël Program for Scientific Translations, Jerusalem).

Szapiro, S. and Steckel, F., 1967. Physical properties of heavy oxygen water, 2. Vapour pressure. *Trans. Faraday Soc.*, 63: 883—894.

Taylor, C.B., 1968. *Die Isotopenzusammensetzung des atmosphärischen Wasserdampfs im Höhenbereich 500 bis 5000 m über Festland,* Doctoral Thesis, University of Heidelberg, Heidelberg, 54 pp.

Taylor, C.B., 1972. The vertical variations of the isotopic concentrations of tropospheric water vapour over continental Europe and their relationship to tropospheric structure. *N.Z. Dep. Sci. Ind. Res., Inst. Nucl. Sci., Rep.*, INS-R-107: 45 pp.

Tzur, Y., 1971. *Isotopic composition of water vapour.* Ph.D. Thesis, Feinberg Graduate School at the Weizmann Institute, Rehovot.

Unterweger, M.P., Coursey, B.M., Schima, F.J. and Mann, W.B., 1980. Preparation and calibration of the 1978 National Bureau of Standards tritiated water standards. *Int. J. Appl. Radiat. Isot.*, 31: 611—614.

Vittori, O., Prodi, F., Morgan, G. and Cesari, G., 1969. Natural tracer distribution in hailstones. *J. Atmos. Sci.*, 26: 148—152.

Wang, J.H., 1965. *J. Phys. Chem. Ithaca*, 69: 4412.

Weaver, C.J., 1981. *Control exorted by sea ice extent and large scale circulation over oxygen isotopic ratios in coastal Antarctic precipitation.* M.S. Thesis, Atmospheric Sciences Program, Ohio State University, Columbus, Ohio (unpublished).

Weston, R.E., Jr., 1955. Hydrogen isotope fractionation between ice and water. *Geochim. Cosmochim. Acta*, 8: 281—284.

White, J.W.C. and Broecker, W.S., 1981. Sources of air vapour in northern America as indicated by stable hydrogen and oxygen isotope ratios. *Variations in the Global Water Budget, Oxford.* Preprints, p. 165.

Woodcock, A.H. and Friedman, I., 1963. The deuterium content of raindrops. *J. Geophys. Res.*, 68: 4477—4483.

Yapp, C.J., 1982. A model for the relationships between precipitation, D/H ratios and precipitation intensity. *J. Geophys. Res.*, 87: 9614—9620.

Yapp, C.J. and Epstein, S., 1977. Climatic implications of D/H ratios of meteoric water over North America (9500—22,000 B.P.) as inferred from ancient wood cellulose C-H hydrogen. *Earth Planet. Sci. Lett.*, 34: 333—350.

Zaïr, S., 1980. *Analyse cristallographique et isotopique des grêlons; application à l'orage du 6 août 1978 dans la région du Napf.* Thèse 3ème cycle, Grenoble, 101 pp.

Zimmernan, U., Ehhalt, D. and Münnich, K.O., 1967. Soil water movement and evapotranspiration: change in the isotopic composition of the water. In: *Isotopes in Hydrology.* IAEA, Vienna, pp. 567—585.

Chapter 3

ENVIRONMENTAL ISOTOPES IN LAKE STUDIES

ROBERTO GONFIANTINI

INTRODUCTION

Lakes are systems which, although complex as all natural systems, are often well suited to study with the environmental isotopes. The only prerequisite is that isotope content variations occur, as a consequence of natural processes, between the various parts of the lake and/or between the lake water and other waters in the environment, i.e. runoff and groundwater: we shall see that this condition is often fulfilled. In addition, with respect to other hydrological systems, lakes have the advantage of being in general accessible at all points for sampling and for in situ measurement of physical and chemical parameters.

The investigations which more frequently are carried out in lakes with environmental isotopes are dealing with the water dynamics and with the water balance; recently, also studies on gas exchange between lake water and atmosphere have been undertaken. The isotopes which have been used mostly in these studies are ^2H, ^{18}O, ^3H and, to a much less extent, ^3He and ^{222}Rn. The status of art of the application of isotope techniques to limnological studies has been reviewed in 1977 in a meeting organised by IAEA, the proceedings of which give a fairly comprehensive account of the results achieved in this field (IAEA, 1979). Other recent reviews of the environmental isotope applications to lake studies are those of Pearson and Coplen (1978) and of Gat (1981b). Pearson and Coplen discuss also nitrogen and sulphur isotopes, which are not included in this chapter.

With respect to other tracers, isotopes are often more conservative (apart from radioactive decay, which is easy to correct for) and therefore enable us to separate physical processes from chemical and biological ones which provide sources or sinks for various substances. Inparticular, the isotopes incorporated into the water molecules, i.e. ^3H, ^2H and ^{18}O, are almost ideal tracers to investigate certain lake parameters.

The discussion and the few selected examples reported in this chapter summarize some of the most important results achieved in lake studies with environmental isotopes and they should be sufficient to give a fairly good idea of the potentiality and of the limitations of the method to hydrologists and limnologists.

The stable isotope fractionations occurring during evaporation are first discussed, as this is perhaps the most important and the most used isotopic tracing occurring in lakes due to natural processes. Its application to water balance studies of lakes is also shown.

The lake dynamics investigations constitute the second category of applications discussed, including the use of the new methods based on ^{3}He deriving from tritium decay.

Other researches are briefly described in the last section of this chapter, which, however, does not include studies on lake sediments; these are in fact discussed in another chapter of this volume.

Definition of terms

A brief definition of some terms used in this chapter is given below for the reader who is not fully familiar with the limnological terminology.

In a lake which is not vertically well mixed because of differences in density established by temperature and/or salt concentration changes, the upper, less dense region is called the *epilimnion:* this is generally fairly turbolent and warmer with respect to the deep, more dense and relatively undisturbed region, called the *hypolimnion*. The intermediate region of relatively rapid change of density is called the *metalimnion*. The plane of maximum density gradient with depth is the *pycnocline*, which is called the *thermocline* when it corresponds also to the maximum vertical temperature decrease with depth responsible for the parallel increase of water density. A thermocline is established in most lakes during summer, which constitutes an efficient barrier against vertical mixing.

The *overturn* is the annual mixing process which lakes normally undergo in autumn, when the epilimnion slowly cools down to a temperature equal to that of hypolimnion.

Some lakes are permanently stratified and are called *meromictic*, and *meromixis* is the related phenomenon. The deeper, stable region of these lakes is called *monimolimnion*, while *mixolimnion* is the upper region through which most of the water flows (Hutchinson, 1957).

ISOTOPIC FRACTIONATIONS DURING EVAPORATION OF WATER

The Craig-Gordon model of evaporation

During evaporation in a natural environment, the water vapour released from a liquid surface is depleted in heavy isotopes with respect to the liquid, because the isotopic species of water $^1H^2H^{16}O$ and $^1H_2^{18}O$ are, broadly speaking, less volatile than the lighter, most abundant one, $^1H_2^{16}O$. The physical processes which are responsible for the isotopic fractionations accompanying evaporation are rather complex, and they can be better described and understood by subdividing the whole evaporation process in steps, as firstly proposed by Craig and Gordon (1965) and subsequently adopted by everybody.

(1) Water vapour in isotopic equilibrium with the liquid phase surface is first released. At the water-air interface there is therefore a "virtually saturated sublayer" (Gat, 1981a) of water vapour, depleted in heavy isotopes with respect to the liquid water by the equilibrium fractionation factor.

(2) The vapour is then migrating away from the liquid-air interface, and it crosses a region of the atmosphere in which the transport is governed by the molecular diffusion (Craig and Gordon, 1965; Brutsaert, 1965, 1975; Merlivat and Coantic, 1975). This gives rise to a further depletion in heavy isotopes, due to the fact that the diffusivities in air of $^1H^2H^{16}O$ and $^1H_2^{18}O$ are lower than that of $^1H_2^{16}O$.

(3) The vapour then reaches a fully turbolent region, in which isotopic fractionations during transport do not occur. Here the water vapour liberated by the liquid surface mixes with the vapour deriving from other sources already present in the atmosphere.

(4) The water vapour of the atmospheric turbolent region also penetrates the molecular diffusion layer to reach and to condensate on the liquid surface. This process is usually called molecular exchange of the liquid with the atmospheric vapour.

The above mechanism of evaporation and of vapour migration in the atmosphere above the liquid water explains qualitatively the isotopic behaviour of water bodies evaporating in natural conditions, and in particular: (1) the deviation of the isotopic composition from the line characteristic of non-evaporated fresh water (the so-called meteoric water line) in a diagram ($\delta^{18}O$, δ^2H), because the molecular diffusion adds a non-equilibrium fractionation; and (2) the limited isotopic enrichment which can be attained, as a consequence of the molecular exchange with the atmospheric vapour. Observations of these facts were first reported by Craig et al. (1963) and by Gonfiantini (1965): the latter showed that waters initially isotopically different reach the same isotopic composition in the final stages of evaporation in the same environment, because of the isotopic exchange with the atmospheric water vapour.

By adopting the Craig-Gordon model, the net water vapour flux across the molecular diffusion layer, corresponding to the evaporation rate E, is proportional to the vapour concentration difference at the boundaries:

$$E = (C_s - C_a)/\rho = C_s(1-h)/\rho \tag{1}$$

where ρ is a resistance coefficient, C_s is the saturation concentration of vapour at the liquid-air interface, C_a is the vapour concentration in the turbulent atmospheric region, and $h = C_a/C_s$ is the relative humidity of the latter normalized to the liquid surface temperature.

For the rare isotopic species $^1H^2H^{18}O$ or $^1H_2^{18}O$ (denoted by the subscript "i") the evaporation rate is given by:

$$E_i = (C_s R_s - C_a R_a)/\rho_i = C_s \left(\frac{R}{\alpha} - hR_a\right)/\rho_i \qquad (2)$$

where R, R_s and R_a are respectively the isotopic ratios ($^2H/^1H$ or $^{18}O/^{16}O$) of the surface layer of the liquid water, of the saturated vapour at the interface, and of the atmospheric vapour in the turbulent region, and $\alpha = R/R_s$ is the equilibrium fractionation factor between liquid and vapour.

It should be noted that R corresponds to the isotopic ratio of the whole water column, only if this is vertically well mixed. If this does not occur, and the surface water is actually enriched in heavy isotopes as a consequence of the removal of isotopically lighter vapour, then a term $\rho_{i,L}$ should be introduced in equation (2), to take into account the resistance to vertical transport in the liquid phase (Craig and Gordon, 1965; Gat, 1970). However, this correction can be neglected in most of the natural cases, being vertical mixing relatively fast, and R can be assumed to represent the isotopic composition of the whole liquid water column. In the case of a stratified lake, R represents the isotopic composition of water in the epilimnion.

Isotopic composition of an evaporating water body

Let us now consider a drying up water body, without inflow and outflow, and from which the water is removed only by evaporation. In this case $E = -dN/dt$ and $E_i = -dN_i/dt$, where N and N_i are, respectively, the number of molecules present in the liquid phase of the most common isotopic species and of the rare one. Dividing the equation (2) by (1) one obtains:

$$\frac{dN_i}{dN} = \frac{\frac{R}{\alpha} - hR_a}{(1-h)\frac{\rho_i}{\rho}} \qquad (3)$$

Considering that $R = N_i/N$ and $f = N/N_0$ (fraction of remaining water), equation (3) may be rewritten as:

$$\frac{dR}{d\ln f} = \frac{\frac{R}{\alpha} - hR_a}{(1-h)\frac{\rho_i}{\rho}} \qquad (4)$$

The term $\Delta\epsilon$ is now introduced, defined as (Craig and Gordon, 1965):

$$\Delta\epsilon = (1-h)\left(\frac{\rho_i}{\rho}-1\right) \qquad (5)$$

This term, proportional to the evaporation rate (proportional to $1-h$) and to the relative difference of transport resistance in air between isotopic molecules, is usually called kinetic or excess enrichment factor: in fact, it represents the additional isotopic enrichment introduced by the transport of vapour across the molecular diffusion layer of the atmosphere, which sums up with the equilibrium enrichment factor ϵ at the interface to produce the total effective enrichment.

Making use of equation (5), and introducing the more common δ units ($\delta = R/R_{\text{reference}} - 1$), equation (4) becomes:

$$\frac{d\delta}{d\ln f} = \frac{h(\delta - \delta_a) - (\delta + 1)(\Delta\epsilon + \epsilon/\alpha)}{1 - h + \Delta\epsilon} \qquad (6)$$

being $\epsilon = \alpha - 1$. Equation (6) describes the variation of the isotopic composition of an evaporating water body as the residual liquid fraction decreases.

Assuming that the evaporation conditions remain unchanged, that is that h, δ_a, $\Delta\epsilon$, ϵ and α remain constant, equation (6) can be easily integrated:

$$\delta = \left(\delta_0 - \frac{A}{B}\right) f^B + \frac{A}{B} \qquad (7)$$

where δ_0 is the initial isotopic composition of water and:

$$A = \frac{h\delta_a + \Delta\epsilon + \epsilon/\alpha}{1 - h + \Delta\epsilon} \qquad (8)$$

$$B = \frac{h - \Delta\epsilon - \epsilon/\alpha}{1 - h + \Delta\epsilon} \qquad (9)$$

From equation (7) it can be seen that A/B is the isotopic composition which a water reaches in its final evaporation stages when f approaches zero. In natural cases A/B corresponds invariably to a maximum value of heavy isotope enrichment. In addition, A/B does not depend on δ_0, and this explains why waters with different initial isotopic composition tend to the same δ value during evaporation, as in experiments already mentioned (Gonfiantini, 1965).

An example of the variation of the isotopic composition of an evaporating water as a function of its remaining fraction f and for various values of relative humidity is given in Fig. 3-1. The values of δ_0 and δ_a, although arbitrary, are however realistic for low- to middle-latitude coastal sites.

From Fig. 3-1 it can be seen that the $\delta(f)$ curves have the concavity downwards for values of h greater than about 50% and upwards for smaller h val-

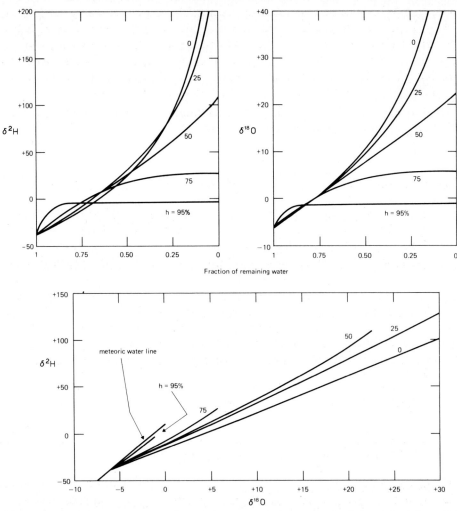

Fig. 3-1. Stable isotope composition of an evaporating water body as a function of the fraction of remaining water, computed for different relative humidities. At $h = 0$ the heavy isotope enrichment obeys a law similar to that of the Rayleigh distillation. The initial isotopic composition of water is assumed $\delta^2 H = -38$ and $\delta^{18}O = -6$. The isotopic composition of the atmospheric vapour is assumed $\delta^2 H = -86$ and $\delta^{18}O = -12$. The lower diagramme shows the change of the slope of the $\delta^2 H$-$\delta^{18}O$ relationship for various relative humidities.

ues. This corresponds to the change of the exponent B of f from values greater than 1 to less than 1, or also to the change in sign of $d^2\delta/df^2$, taking into account that $\Delta\epsilon$ and ϵ/α are usually small in comparison with h and $1-h$.

If $h \gg 50\%$ the isotopic composition rapidly approaches the limiting value A/B and then remains practically constant, simulating the occurrence of an isotopic steady state in the final evaporation stage. This has been actually observed in most evaporation experiments since the first reported by Craig et al. (1963), in which the relative humidity was always high. On the other hand, some studies carried out in low relative humidity conditions have shown the continuously faster increase of δ values shown in Fig. 3-1 (Fontes and Gonfiantini, 1967a; Gonfiantini et al., 1973).

Finally, when $h = 0$ (dry atmosphere), equation (7) reduces to:

$$\frac{\delta + 1}{\delta_0 + 1} = f^{-(\Delta\epsilon + \epsilon/\alpha)/(1 + \Delta\epsilon)} \tag{10}$$

This relationship is a form of the Rayleigh distillation equation, applied in this case to the distillation of mixtures of isotopic molecules of water. When $h = 0$, the heavy isotope enrichment of the remaining water is theoretically unlimited.

From equation (7) it can also be seen that the isotopic values of an evaporating water fall in a $\delta^{18}O$-δ^2H diagram into a curve very close to a straight line, the slight deflections from which become apparent only at low f values (Fig. 3-1); deflections which are, however, practically impossible to detect in evaporation experiments. It can be shown also from equation (7) that the slope $S = d\delta^2H/d\delta^{18}O$ of these lines decreases with decreasing relative humidity: for instance, at h close to 100 the slope is approximately 8, and drops to 3.7 at h close to 0 with the values of other parameters adopted to construct Fig. 3-1.

The isotopic fractionation factors

Let us now summarize the data available on the isotopic fractionation and enrichment factors between liquid water and vapour.

The equilibrium fractionation factor α

The water vapour in thermodynamic equilibrium with liquid water is with respect to the latter slightly depleted in heavy isotopes, due to the difference of the hydrogen bond energy in the liquid phase between isotopic molecules. This results in turn in a difference in the vapour pressure of the isotopic species of water. To give an idea of it, the heavy isotope depletion of vapour with respect to liquid water is about 80‰ for ^2H and 10‰ for ^{18}O at room temperature. The enrichment factor slightly decreases with increasing temperature, at a rate of about 1.2‰ °C^{-1} for ^{2}H and of 0.1‰ °C^{-1} for ^{18}O (at room temperature).

For ^{18}O, the fractionation factor α (defined, as said before, as R_{liquid}/R_{vapour}) decreases monotonously with increasing temperature until it becomes equal to one at the critical temperature of water (374.1°C). For ^{2}H, α decreases monotonously up to 220°C, where it becomes equal to one. Above this temperature, α ^{2}H is smaller than one (the vapour is therefore enriched in deuterium with respect to the liquid), it reaches a minimum value of about 0.996 at 275°C, and then increases to reach the value of one at the critical temperature (see data plots by Friedman and O'Neil, 1977).

The equilibrium fractionation factors have been determined experimentally by a number of authors. In general the results are sufficiently consistent each other, at least in the temperature range of interest for this discussion, and the differences are not really important, although sometimes definitely greater than the experimental error. Although it is practically impossible to critically assess the quality of each determination, it should be said, however, that the values reported by Majoube (1971) have been perhaps those most frequently used in recent years.

TABLE 3-1

Equilibrium fractionation factors in the system liquid water—water vapour. Values of coefficients of equation (T in Kelvin):

$$\ln \alpha = AT^{-2} + BT^{-1} + C$$

A	B	C	ϵ (‰) at 0°C	ϵ (‰) at 20°C	Temperature range of validity (°C)	Notes
$^{18}O/^{16}O$						
1534	−3.206	+0.00264	11.53	9.60	0—100	(1)
1137	−0.4156	−0.00207	11.71	9.79	0—100	(2)
4568.3	−23.755	+0.03713	11.46	9.30	0—50	(3)
5970.2	−32.801	+0.05223	(12.24)	9.86	10—40	(4)
$^2H/^1H$						
24844	−76.248	+0.05261	112.3	85.0	0—100	(2)
2408	+64.55	−0.1687	(105.0)	82.8	10—40	(4)

Notes:
(1) From Bottinga and Craig (1969), obtained by fitting experimental data of Baertschi and Thürkauf (1960) and of Craig et al. (1963).
(2) From Majoube (1971).
(3) Obtained by fitting the experimental data on vapour pressure of Jakli and Stachewski (1977) in the temperature range 0—50°C.
(4) From Kakiuchi and Matsuo (1979).

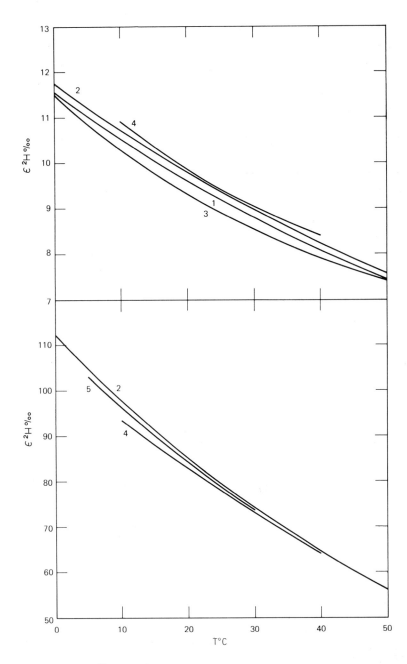

Fig. 3-2. The ^{18}O and ^{2}H equilibrium enrichment factor ϵ between liquid water and vapour in the temperature range 0—50°C. *1* = Bottinga and Craig, 1969; *2* = Majoube, 1971; *3* = Jakli and Stachewski, 1977; *4* = Kakiuchi and Matsuo, 1979. Equations fitting these curves are given in Table 3-1.

The variation of the equilibrium fractionation factors with the temperature can be represented by an expression of the type:

$$\ln \alpha = AT^{-2} + BT^{-1} + C$$

where T is the absolute temperature in Kelvin and A, B and C are coefficients obtained by fitting series of experimental data. The values of these coefficients from some recent determinations are reported in Table 3-1, together with the values of $\epsilon^0/_{00}$ calculated at 0 and at 20°C for comparison. The variations of $\epsilon^0/_{00}$ with the temperature are also shown in Fig. 3-2.

Finally, it should be noted that the equilibrium fractionation factor between liquid water and vapour can be defined in two ways, which are mathematically inverse: (1) $\alpha^+ = R_{\text{liquid}}/R_{\text{vapour}}$, where the superscript "+" refers to the fact that $\alpha^+ > 1$ at room temperature; or (2) $\alpha^* = R_{\text{vapour}}/R_{\text{liquid}} = 1/\alpha^+$, in which case obviously $\alpha^* < 1$. The choice between α^+ and α^* is generally made so as to simplify somewhat the mathematical expressions, and this depends on the type of problems dealt with: for instance for evaporation problems α^* is frequently preferred. In spite of this, here the use of a α^+ (called simply α) has been adopted, in order to avoid confusion between α^+ and α^*, and because the data on isotopic fractionation factors reported in the literature generally refer to the definition which produces $\alpha > 1$. The isotopic equilibrium enrichment factor ϵ is therefore equal to $\alpha - 1$ in this text.

The kinetic enrichment factor

As we have seen (equation (5)), the term $\Delta\epsilon$ is proportional to the moisture deficit $(1 - h)$ of the atmosphere and to the relative difference of transport resistance in air between isotopic molecules $(\rho_i/\rho - 1)$. On theoretical grounds and on the basis of preliminary experiments by Dansgaard (1961), Craig et al. (1963) and Gat and Craig (1966), it was found that the term ρ_i/ρ for oxygen isotopes was close to $(D/D_i)^n$, where $n = 0.5$ and D and D_i the molecular diffusion coefficients in air of $^1H_2^{16}O$ and $^1H_2^{18}O$ respectively (Craig and Gordon, 1965). The so-called turbolence parameter n can in principle range between 0 and 1, depending on the evaporation conditions; the value of 0.5., however, appears to represent reasonably well the conditions more frequently encountered in nature (see also Stewart, 1975).

The D/D_i ratio in air was calculated by Craig and Gordon (1965) by means of the following expression obtained from the kinetic theory of gas of low concentration diffusing in other gas:

$$\frac{D}{D_i} = \left(\frac{M_i (M + 29)}{M (M_i + 29)} \right)^{1/2} \tag{11}$$

where M and M_i are the molecular weights of $^1H_2^{16}O$ and of $^1H_2^{18}O$ (or

$^1H^2H^{16}O$), respectively, and 29 is the mean molecular weight of air. For oxygen isotopes ($M = 18$ and $M_i = 20$) equation (11) gives $D/D_i = 1.0324$ and $\rho_i/\rho = 1.0160$, while for hydrogen isotopes ($M_i = 19$) one obtains $D/D_i = 1.0166$ and $\rho_i/\rho = 1.0083$: this latter value is not in agreement with the experimental data.

Evaporation pan experiments were carried out by Gat (1970) to evaluate the kinetic enrichment factor to be used for the isotope balance study of Lake Tiberias (Isreal). The results showed that the coefficient $\Delta\epsilon/(1-h)$ was generally ranging for ^{18}O between 13 and 16‰, but at night values up to 21‰ were possible. This was attributed to the difference between the day windy and the night windstill conditions, the latter implying an increase of the parameter n. Less conclusive were the deuterium measurements.

Recently, Merlivat (1978) has determined experimentally the ratios of the diffusion coefficients of the isotopic species of water vapour in air at 21°C, and has found: 1.0285 ± 0.0008 for $D(^1H_2^{16}O)/D(^1H_2^{18}O)$ and 1.0251 ± 0.0010 for $D(^1H_2^{16}O)/D(^1H^2H^{16}O)$: while the first of these values if not too distant from that calculated with (11), the second is definitely quite different.

The square root of Merlivat's D/D_i values give the values of 1.0141 ± 0.0004 for $\rho(^1H_2^{18}O)/\rho(^1H_2^{16}O)$ and of 1.0125 ± 0.0005 for $\rho(^1H^2H^{16}O)/\rho(^1H_2^{16}O)$, which are in very good agreement with those found by Vogt (1976) in wind tunnel experiments: 1.0143 ± 0.0004 and 1.0124 ± 0.0017 respectively (for these determinations see also Münnich, 1979).

In conclusion, the kinetic enrichment factor $\Delta\epsilon$ for the evaporation conditions most frequently occurring in nature can be evaluated adequately with the following relationships:

$$\Delta\epsilon^{18}O‰ = 14.2\,(1-h)$$
$$\Delta\epsilon^2H‰ = 12.5\,(1-h)$$
(12)

The dissolved salt effects

The three effects of dissolved salts on the isotopic composition of an evaporating water are:

(1) The dissolved salt decreases the thermodynamic activity of the water and also its evaporation rate:

(2) The ions co-ordinate in their hydration sphere water molecules having an isotopic composition and an evaporation rate which may be different from that of "free" water. This effect is different for each salt:

(3) Certain salts reach the concentration of saturation in the course of evaporation, and crystallize with water of crystallization, thus removing from the system water with an isotopic composition in general different from that of the remaining liquid water.

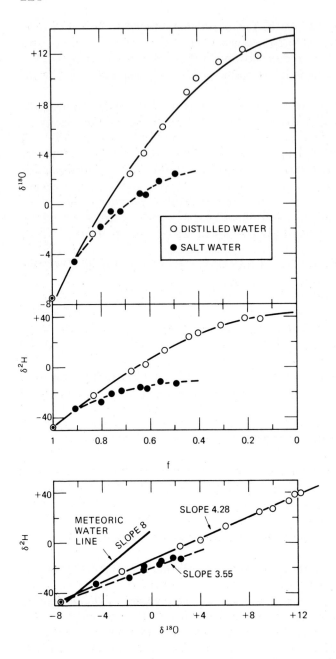

Fig. 3-3. Isotopic composition of distilled water and of NaCl saturated solution, both evaporating in the same environment, as a function of the remaining water fraction f (from Gonfiantini, 1965). The salt water enrichment is less than that of distilled water due to the lower water activity (a_w = 0.75).

For convenience, these effects will be discussed separately although they occur together (especially the first two). On the other hand, the most common salt occurring at relatively high concentrations in natural waters, NaCl, produces only negligible effects of the second type listed above, and no effect of the third type.

Effects due to changes of the thermodynamic activity of water

This is the most important effect produced by dissolved salts on the isotopic composition of evaporating waters. Practically, the dissolved salts decrease the thermodynamic activity of water, that is the saturated vapour concentration, which becomes $a_w C_s$, where $a_w < 1$ being the activity coefficient of water. The relative humidity of air then becomes h/a_w normalized to the temperature and activity of water. Therefore, the term h/a_w should be introduced in equations (1) to (6) and (12) in place of h to account for the occurrence of dissolved salts.

The activity coefficient of water is significantly lower than 1 only at relatively high salt concentrations: for instance, for a NaCl content of 1 g/l a_w is 0.9994, and becomes equal to 0.99 at a concentration of 17.5 g/l. For seawater, the activity coefficient is about 0.98. Therefore, the salt effect on evaporating water activity can be neglected in most natural cases. The effect, however, may become important for saline lakes and sebkhas, and for drying up lakes in the last evaporation stages.

The simplest case to consider is that in which a_w (and salt concentration) remains constant during evaporation, with a value of a_w considerably lower than 1. This occurs, for instance, for saturated solutions, in which the water removal by evaporation is counter-balanced by the salt precipitation. In this case all the equations written so far remain valid, except for the substitution of h with h/a_w.

Fig. 3-3 shows the isotopic composition of distilled water ($a_w = 1$) and of water saturated with NaCl ($a_w = 0.75$) evaporating in the same environment (Gonfiantini, 1965). This experiment shows that salt water becomes less enriched in heavy isotopes than distilled water, as it should be expected, because the first is virtually evaporating at higher relative humidity ($h/0.75$).

More complex, but also more common in nature, is the case of non-saturated salt water, in which during evaporation the salt concentration increases and in parallel the water activity decreases. Under these circumstances the heavy isotope content of water increases at the beginning, reaches a maximum value and then decreases: such a trend, which can be theoretically predicted, has been observed and explained in seawater evaporation experiments (Fig. 3-4, Gonfiantini, 1965; Lloyd, 1966; Fontes, 1966) and for a drying-up saline lake (Fontes and Gonfiantini, 1967a). Recently, the problem of evaporating seawater has been also rediscussed by Jauzein (1982).

The activity coefficient as a function of the salt molality M can be conveniently represented by a relationship of the type:

$$a_w = DM^2 + EM + G \tag{13a}$$

where D, E and G are constants which can be obtained by fitting the experimental values of the water activity coefficient. If evaporation is the only process removing water from the system, then an expression similar to (12) can be used to represent a_w as a function of the remaining water fraction f, being $f = M_0/M$:

$$a_w = D'f^{-2} + E'F^{-1} + G \tag{13b}$$

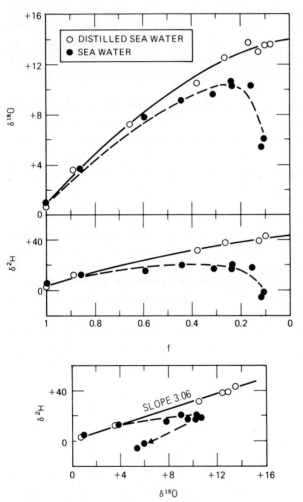

Fig. 3-4. Isotopic composition of distilled seawater and of seawater, both evaporating in the same environment, as a function of the remaining water fraction f (from Gonfiantini, 1965). The seawater enrichment reverses its trend in the final evaporation stages as a consequence of the continuous decrease of the water activity.

This equation is valid up to the value of f corresponding to salt saturation: afterwards a_w remains constant.

The water activity coefficients at different salt concentration are tabulated by Robinson and Stokes (1959) for some of the most common salts. For instance, for a sodium chloride solution having an initial activity of 0.98 (similar to that of seawater) we obtain:

$$a_w = -0.000543\, f^{-2} - 0.018521\, f^{-1} + 0.99931$$

which is valid approximately up to $f = 0.1$ (saturation).

Using an expression similar to (13b) to represent a_w, equation (6) can be integrated by steps by introducing h/a_w in place of h (also in evaluating $\Delta\epsilon$ which is a function of h).

Effects due to the hydration of ions

The water molecules entering the hydration sphere of ions have often an isotopic composition which is different from that of the "free" water, as firstly shown by Taube (1954) for oxygen isotopes. The experimental evidence of this is given, for oxygen isotopes, by the difference in isotopic composition between the CO_2 equilibrated with the pure water and the CO_2 equilibrated with a salt solution prepared with the same water.

Evaporation experiments show that also hydrogen isotope fractionations occur in ion hydration. This was anticipated by Gonfiantini (1965) to explain the different hydrogen isotope behaviour of pure water and of water saturated with sodium chloride during evaporation under Rayleigh conditions (while no difference occurs for oxygen isotopes).

The effect of isotopic fractionation in ion hydration on the isotopic composition of evaporating waters has been fully demonstrated by Sofer and Gat (1975). Equation (2) should be written as:

$$E_i = C_s \left(\frac{\Gamma R}{\alpha} - \frac{h R_a}{a_w} \right) / \rho_i$$

where a_w is the water activity coefficient as discussed above and:

$$\Gamma = \frac{55.56}{Mn(\alpha_H - 1) + 55.56} = \frac{R_F}{R} \tag{14}$$

where $\alpha_H = R_B/R_F$ is the hydration fractionation factor, i.e. the ratio between the isotopic ratios of water bound in the hydration sphere of ions (subscript B) and of free water (subscript F); M is the salt molality and n is the hydration number, i.e. the number of water molecules coordinated by ions in their hydration sphere.

Equation (6) then becomes:

$$\frac{d\delta}{d\ln f} = \frac{\dfrac{h}{a_w}(\delta - \delta_a) - (\delta + 1)\left(\Delta\epsilon + \dfrac{\alpha - \Gamma}{\alpha}\right)}{1 - \dfrac{h}{a_w} + \Delta\epsilon} \qquad (15a)$$

which describes the isotopic composition during evaporation of any water body. If the salt content is negligible, then $a_w = 1$ and $\Gamma = 1$ and the equation reduces to (6). In equation (15a), the kinetic enrichment factor $\Delta\epsilon$ is now:

$$\Delta\epsilon = K(1 - h/a_w)$$

where K is 14.2‰ for ^{18}O and 12.5‰ for ^2H.

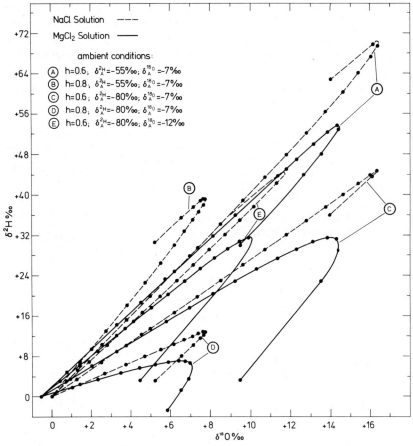

Fig. 3-5. Computed $\delta^{18}O$-δ^2H curves for NaCl and MgCl$_2$ solutions evaporating under different ambient conditions (from Sofer and Gat, 1975). Initial salt concentration: 0.5 molal.

For saturated solutions, equation (15a) can be easily integrated being a_w and Γ constant. The expression obtained is formally identical to (7) but where A and B have now the values:

$$A = \frac{\dfrac{h}{a_w}\delta_a + \Delta\epsilon + \dfrac{\alpha - \Gamma}{\alpha}}{1 - h + \Delta\epsilon}$$

$$B = \frac{\dfrac{h}{a_w} - \Delta\epsilon - \dfrac{\alpha - \Gamma}{\alpha}}{1 - h + \Delta\epsilon}$$

(15b)

For non-saturated solutions, expressions similar to (13) and to (14) should be introduced in (15a) respectively for a_w, Γ and for evaluating $\Delta\epsilon$. The molality M becomes equal to M_0/f, being M_0 the initial molality.

The hydration fractionation factor α_H has been determined for a number of electrolytes, but the results obtained by different authors sometimes do not agree too well. The fractionation values for some common electrolytes are reported in Table 3-2. Truesdell (1974) has also determined the variations

TABLE 3-2

Fractionation factors between water bound in the hydration sphere of ions and free water (α_H) and between free water and the whole water at molality equal to 1 (Γ ($M = 1$))

Isotope (and temperature)	Electrolyte	$n\,(\alpha_H - 1)$ (‰)	$\Gamma\,(M = 1)$	References
$^{18}O/^{16}O$ (25°C)	NaCl	0	1	Taube, 1954; Götz and Heinzinger, 1973
	KCl	−9.0	1.00016	Sofer and Gat, 1972
	KCl	−1.9	1.00003	Götz and Heinzinger, 1973
	CaCl$_2$	26.0	0.99953	Sofer and Gat, 1972
	MgCl$_2$	61.5	0.99889	Sofer and Gat, 1972
	MgCl$_2$	66.7	0.99880	Taube, 1954
$^2H/^1H$ (20°C)	Na$_2$SO$_4$, K$_2$SO$_4$	−28	1.0005	Stewart and Friedman, 1975
	NaCl, KCl	134	0.9976	Stewart and Friedman, 1975
	NaCl	22	0.9996	Sofer and Gat, 1975
	CaCl$_2$	262	0.9953	Stewart and Friedman, 1975
	CaCl$_2$	341	0.9939	Sofer and Gat, 1975
	MgCl$_2$	493	0.9912	Stewart and Friedman, 1975
	MgCl$_2$	285	0.9949	Sofer and Gat, 1975
	Na$_2$CO$_3$	279	0.9950	Stewart and Friedman, 1975
	MgSO$_4$	347	0.9938	Stewart and Friedman, 1975

TABLE 3-3

Isotopic fractionation factors α_C between water of crystallization and mother water[a]

Salt	Isotope	α_C	$T(°C)$	Reference
Mirabilite	$^{18}O/^{16}O$	1.0020	0	Stewart, 1974
$Na_2SO_4 \cdot 10H_2O$	$^2H/^1H$	1.019	0	Stewart, 1974
Gypsum	$^{18}O/^{16}O$	1.0040	17—37	Gonfiantini and Fontes, 1963
$CaSO_4 \cdot 2H_2O$	$^2H/^1H$	0.985	17—57	Fontes and Gonfiantini, 1967b
Gaylussite $Na_2CO_3 \cdot CaCO_3 \cdot 5H_2O$	$^2H/^1H$	0.987	18—25	Matsuo et al., 1972

[a] Values of $\alpha_C > 1$ indicate that the heavy isotope is concentrated in the water of crystallization with respect to the mother water; vice versa when $\alpha_C < 1$.

of α_H with temperature, especially at high temperatures, but the discussion of his results are beyond the scope of this chapter.

Curves reporting the isotopic composition of evaporation solutions of NaCl and of $MgCl_2$ for various values of relative humidity have been computed by Sofer and Gat (1975). In a $\delta^{18}O$-δ^2H plot, the curves show the characteristic hook shape, which is mainly due to the isotopic fractionation in the hydration sphere of ions (Fig. 3-5).

Effects due to water of crystallization of salts

This effect is in general a minor one, and becomes important only in the very last stages of evaporation, when dissolved salts reach the saturation concentration and start to crystallize. Some deposited salts contain water of crystallization which generally has an isotopic composition different from that of the mother water, thus affecting the value of f and of δ of the residual water in competition with the evaporation process.

The fractionation factors between water of crystallization and mother water for some salts commonly formed during evaporation of natural waters are reported in Table 3-3.

WATER BALANCE STUDIES OF LAKES

Stable isotope balance equations

The water balance and the stable isotope balance of a lake during the time dt are given by:

$$\frac{dV}{dt} = I - Q - E \qquad (16)$$

$$\frac{Vd\delta + \delta dV}{dt} = I\delta_I - Q\delta_Q - E\delta_E \qquad (17)$$

where V is the volume of the lake water having a mean isotopic composition δ, I, Q and E are the inflow, the outflow and the net evaporation rates, and δ_I, δ_Q and δ_E are the respective mean isotopic compositions.

The mean isotopic composition of the inflow is given by:

$$\delta_I = \Sigma \delta_i I_i / \Sigma I_i$$

where the sums include all types of inflow, i.e. precipitation, runoff and groundwater. For well mixed lakes, the isotopic composition of the liquid outflows (which include effluent(s) and bottom infiltration) is equal to that of the lake water, i.e. $\delta_Q = \delta$, because no isotopic fractionation occurs. The isotopic composition of the water removed by evaporation from a well mixed lake is:

$$\delta_E = \frac{1}{1-h+\Delta\epsilon}\left(\frac{\delta-\epsilon}{\alpha} - h\delta_a - \Delta\epsilon\right)$$

as it can be shown using equation (6).

Introducing these expressions in equation (17) we obtain:

$$\frac{Vd\delta + \delta dV}{dt} = I\delta_I - Q\delta - \frac{E}{1-h+\Delta\epsilon}\left(\frac{\delta-\epsilon}{\alpha} - h\delta_a - \Delta\epsilon\right) \qquad (18)$$

Equation (18) is valid for well mixed lakes.

Let us first consider the case of lakes the volume of which does not change significantly with time. In this case we have $dV/dt = 0$ and $I = Q + E$. By putting $E/I = x$ and $Q/I = 1 - x$, being x the fraction of the inflowing water lost by evaporation, and $V/I = T$ mean residence time of water in the lake, equation (18) becomes:

$$\frac{d\delta}{dt} = -\frac{1}{T}[(1 + Bx)\delta - \delta_I - Ax] \qquad (19)$$

where A and B are the same as defined on p. 117, equations (8) and (9). In the case of lakes with long residence time, having therefore a rather slow or negligible response to the short duration (daily to seasonal) variations of the inflow-outflow-evaporation rates and conditions, we can assume that δ_I, δ_a, T, x, h, α, ϵ and $\Delta\epsilon$ remain constant and have values corresponding to their long term average values. With this assumption equation (19) can be integrated:

$$\delta = \frac{\delta_I + Ax}{1 + Bx} + \left(\delta_0 - \frac{\delta_I + Ax}{1 + Bx}\right) \exp[-(1 + Bx)t/T] \qquad (20)$$

δ tends with time to a steady value δ_S:

$$\delta_S = \frac{\delta_I + Ax}{1 + Bx} = \frac{\delta_I (1 - h + \Delta\epsilon) + x(h\delta_a + \Delta\epsilon + \epsilon/\alpha)}{(1-x)(1-h+\Delta\epsilon) + x/\alpha} \qquad (21)$$

For the so-called terminal lakes, in which evaporation exactly compensates inflow ($x = 1$), the steady value of the isotopic composition is:

$$\delta_S^T = \alpha\delta_I(1 - h + \Delta\epsilon) + \alpha h\delta_a + \alpha\Delta\epsilon + \epsilon \qquad (22)$$

This is the maximum enrichment in heavy isotopes which can be reached by a constant volume lake*. On the contrary when the evaporation term is negligible for the lake water balance ($x = 0$), δ_S becomes equal to δ_I, as it should be expected. It is also interesting to note that at $h = 1$ (saturated vapour pressure) we obtain from (21) $\delta_S = \alpha\delta_a + \epsilon$, i.e. vapour and liquid are in isotopic equilibrium. ($\Delta\epsilon$ vanishes at $h = 1$, as from equations (12)).

Fig. 3-6 shows the steady-state isotopic composition δ_S of constant volume lakes as a function of x for different values of relative humidity.

Equation (21) can be used to evaluate x if all other variables can be measured or evaluated:

$$x = \frac{\delta_S - \delta_I}{A - \delta_s B} = \frac{(\delta_S - \delta_I)(1 - h + \Delta\epsilon)}{(\delta_S + 1)(\Delta\epsilon + \epsilon/\alpha) + h(\delta_a - \delta_S)} \qquad (23)$$

However, the error affecting this evaluation is generally rather high, because it is difficult to determine or evaluate some of the variables on the right-hand side term of (23) with sufficient accuracy, as will be seen also in the next section. In particular, the relative humidity is difficult to evaluate even if continuous measurements are available in stations on the lake shore, because h increases across the lake in the wind direction, in a manner which depends on meteorological conditions and it is difficult to predict.

The same applies to δ_a: the isotopic composition of the atmospheric moisture varies across the lake as the contribution of lake vapour increases. The influence of this factor on the stable isotope content of atmospheric vapour has been shown for Lake Geneva (Fontes and Gonfiantini, 1970) and on tri-

---

* Sometimes terminal lakes can be highly saline (in theory they should reach salt saturation), and therefore corrective terms should be introduced into (22) to take into account salt effects on the isotopic composition of water, as discussed in the section "The dissolved salt effects" (pp. 123—130).

tium content for Lake Tahoe (Östlund and Berry, 1970) and for Perch Lake (Brown and Barry, 1979). Also, the increase of relative humidity and the change of isotopic composition of the water vapour in the atmosphere above the lake can be responsible for the difference in slope observed in the $\delta^{18}O$-δ^2H correlation of Lake Chad (slope 5.2) with respect to that of river waters (slope 3.9), both being submitted to intense evaporation (Fontes et al., 1970).

Let us now consider the case of well mixed lakes the volume variations of which are not negligible. If these variations are relatively slow, regular and unidirectional, and if the system is sufficiently large to smooth short duration variations of δ_I, δ_a, α, ϵ, $\Delta\epsilon$, h, I, Q and E then equation (18) can be easily integrated using for the above parameters their mean values over the integration period. Equation (18) can be rewritten as:

$$\frac{d\delta}{d\ln f} = \frac{1}{I - Q - E} [I(\delta_I - \delta) + E(A - \delta B)] \qquad (24)$$

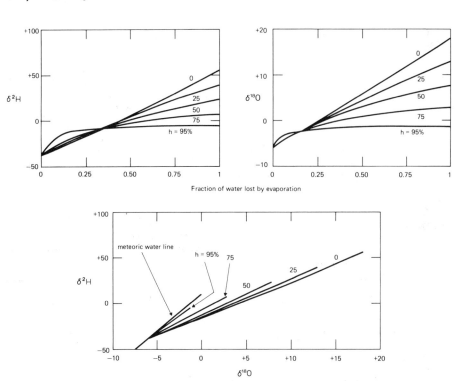

Fig. 3-6. Heavy isotope enrichment for constant volume, well mixed lakes as a function of the fraction of water lost by evaporation with respect to the total inflow. The computations are made for various relative humidities. The assumed isotopic values for the inflow are: $\delta^2H = -38$ and $\delta^{18}O = -6$; and for the atmospheric vapour: $\delta^2H = -86$ and $\delta^{18}O = -12$. The lower diagramme shows the change of slope of the δ^2H-$\delta^{18}O$ relationshps with different relative humidities.

where $f = V/V_0$ (residual fraction of water in the lake) and $df = dV/V_0 = (I - Q - E)dt/V_0$. Again A and B are those defined by equations (8) and (9).

It should be noted that in principle df/dt can also be positive, when $I > Q + E$: this corresponds to a volume increase and implies that $f > 1$.

The integration of equation (24) produces:

$$\delta = \left(\delta_0 - \frac{\delta_I + Ax}{1 + Bx}\right) f^{-(1+Bx)/(1-x-y)} + \frac{\delta_I + Ax}{1 + Bx} \qquad (25)$$

where $x = E/I$ (fraction of inflow water lost by evaporation) and $y = Q/I$ (fraction lost by isotopically non-fractionating outflows).

Note that equation (25) reduces to equation (7) if $I = Q = 0$, i.e. the lake is drying up due to evaporation only. If $I = 0$ and Q is not negligible (for instance, this would be the case of a pond formed by flood and then desiccating by both evaporation and water infiltration in the soil), equation (25) becomes:

$$\delta = \left(\delta_0 - \frac{A}{B}\right) f^{Bz} + \frac{A}{B} \qquad (26)$$

where $z = E/(E + Q)$.

For lakes showing fast volume variations with rate changing in time, equation (18) or (24) should be integrated by steps, each one representing a short time interval during which the inflow-outflow-evaporation conditions can be assumed reasonably steady. Similar lakes are rather common, being the volume variations connected with the seasonal cycle of precipitation and of drought, and also with the exploitation rate for artificial reservoirs. Lakes of this type, however, are often not well mixed.

The water balance of not well mixed lakes is in general difficult to assess with isotopes as well as with any other tracer. However, by analysing samples from a dense network it is possible to obtain an idea of the degree of mixing (which is a useful information per se) and decide whether it is possible to use mean values for water balance evaluations with a reasonable degree of accuracy. In the extreme case of a shallow lake in which regular isotopic variations are observed along the direction of the water flows, a possible approach is to disregard mixing and to consider each sample as representative of a different evaporation stage of the lake water (Fontes et al., 1970; Dinçer et al., 1979a).

Chemical analysis are always a useful complement to isotope studies of lakes. For instance simple conductivity measurements, which can be carried out "in situ", give immediately an idea of the degree of mixing and of evaporation.

Considering the salt balance, for well mixed lakes not changing in volume the fraction x of water lost by evaporation is given by:

$$x = 1 - M_I/M_S \tag{27}$$

where M_I is the mean salt molality of the inflow and M_S is that of the lake at the steady state (corresponding to the isotopic composition δ_s). For terminal lakes ($x = 1$), M_S^T should be much higher than M_I (theoretically M_S^T should correspond to the salt saturation).

For well mixed lakes with variable volume the fraction of residual water is:

$$f = \left(\frac{M_0 (1-x) - M_I}{M (1-x) - M_I} \right)^{1-y/(1-x)} \tag{28}$$

being M and M_0 the salt concentration at the time t and the initial one. This expression reduces to:

$$f = M_0/M \tag{29}$$

when only evaporation removes water from the system and there is no inflow ($I = Q = 0$). When $I = 0$ and $E/(E + Q) = z$ expression (28) becomes:

$$f = (M_0/M)^{1/z} \tag{30}$$

Equations (27) to (30) are valid until the salt saturation is reached, and assuming that the chemical composition is essentially conservative: only concentrations in water can vary, but not the relative proportions of dissolved compounds.

Equations (27) to (30) when used in conjunction with equations (19) to (26) may allow to evaluate x, y and z, the hydrological balance terms of a lake. In general, salt concentration or electrical conductivity can be used instead of molality in equations (27) to (30).

Applications

Lake Bracciano, Italy

The lake occupies an ancient volcanic crater at about 30 km northwest of Rome. It has the following major morphometric parameters: altitude 164 m a.s.l.; surface 57.5 km², volume 4.95×10^9 m³; maximum depth 160 m; catchment basin 149 km². The lake has a small effluent, the Arrone River, and in addition its water is used to supply the old aqueduct of "Acqua Paola" built in the 17th century. Effluent and aqueduct remove together about 40×10^6 m³ a⁻¹. Evaporation from the lake is evaluated to about 55×10^6 m³ a⁻¹. The turnover time of water is therefore about 50 years.

The ¹⁸O isotopic values are (Gonfiantini et al., 1962): $\delta_S = + 0.6‰$ (measurements performed in the years 1960—1961, which also showed that the lake is well mixed); $\delta_I = -6.0°/_{00}$ (observed in a spring and in a creek, as well as in many groundwater samples); $\delta_a = -13°/_{00}$ (estimated, considering the vicinity of the sea); $\alpha = 1.01025$ (and thus $\epsilon = 10.25°/_{00}$), calculated from Majoube's equation at $15°C$. For the relative humidity we assume the value of $h = 0.75$: this also fix the value of $\Delta\epsilon$ to $3.55°/_{00}$.

By introducing the above values in equation (23) we obtain: $x = 0.48$, to be compared with the value of 0.58 obtained from the hydrological balance. By assuming for h a value of 0.80, the computed value of x becomes 0.63.

If now we assume the following errors: $0.2°/_{00}$ for δ_S (which is easy to determine and therefore quite well known), $0.5°/_{00}$ for δ_I, $1°/_{00}$ for δ_a and 0.05 for the relative humidity (which also implies an error of $0.2°/_{00}$ on $\Delta\epsilon$), and neglecting the errors attached to α and ϵ (which in any case are minor), we obtain for x an error of ± 0.145. This error is rather high (about 30% of the computed evaporation), most probably higher than that affecting evaluations based on other methods. In particular the errors associated with δ_a and h are responsible for about 90% of the error on x. In fact, these two parameters are the most difficult to measure or to evaluate with reasonable accuracy, especially because of their changes across the lake.

Lake Neusiedl, Austria

The evaluation of the evaporation rate of this lake from stable isotope data has been attempted by Zimmermann and Ehhalt (1970).

Lake Neusiedl is large (about 250 km^2 of surface area) but very shallow (mean depth about 1 m), and consequently the mixing is rather poor: this, however, was not the major difficulty encountered in the isotope investigation, because an adequate network of sampling stations allowed a rather precise evaluation of the lake isotopic composition and of ist variations in time and space. Also volume variations were taken into account.

In establishing the isotope balance, hydrogen isotopes were preferred because the kinetic contribution to the evaporation fractionation factor is relatively much smaller than for oxygen isotopes: in fact, the term $\Delta\epsilon$ was neglected in the computations. Even so, however, it was concluded, perhaps too pessimistically, that "it can be difficult to determine the evaporation rate of a lake with an accuracy better than 50%". An evaluation of the different terms which contribute to this error, showed that the term related to the isotopic composition of the water removed by evaporation is by far the most important. This term depends mainly on the moisture content of atmosphere and on its isotopic composition, which are thus again indicated as the parameters most responsible of errors in assessing the isotopic balance of lakes. In particular, Zimmermann and Ehhalt (1970) also determined the isotopic composition of the water vapour in the atmosphere over the lake, and found that the δ^2H mean monthly values would differ up to $10°/_{00}$ in different locations: this gives an idea of the difficulty of keeping within acceptable limits the error attached to δ_a.

Three Turkish lakes

Sometimes, part of the difficulties discussed in previous examples can be removed. Dinçer (1968) reported the isotopic composition of three Turkish lakes (Lake Burdur, Lake Egridir and Lake Beysehir) in the southern part of the Anatolian plateau, which can be considered in climatic and environmental conditions practically identical. Among these, Lake Burdur can be considered a terminal lake, being its liquid water losses negligible as indicated by its relatively high salt content (23 g/l). This salinity is, however, still too low to influence significantly the isotopic fractionation processes: therefore, equation (22) can be used, which introduced in equations (21) and (23) gives:

$$\delta_S = \delta_I + \frac{x(\delta_S^T - \delta_I)}{\alpha(1-x)(1-h+\Delta\epsilon)+x} \quad (31)$$

$$x = 1 - \frac{\delta_S^T - \delta_S}{\delta_S^T - \delta_S + \alpha(\delta_S - \delta_I)(1-h+\Delta\epsilon)}$$

This equation does not contain the term δ_a relative to the isotopic composition of the atmospheric vapour, which is certainly the most difficult to evaluate with a good accuracy.

Table 3-4 shows the values of x obtained for lakes Beysehir and Egridir

TABLE 3-4

Evaluation of the water lost by evaporation (x) by three Turkish lakes (Dinçer, 1968)[a]

	^{18}O	^2H
$\alpha (= 1 + \epsilon)$	1.01053	1.0952
$\Delta\epsilon$ (%0)	5.68 ± 0.71	5.0 ± 0.7
δ_I (%0)	−8.75 ± 0.5	−55.0 ± 4.0
δ_S^T (Lake Burdur)	+1.60 ± 0.50	−2.0 ± 3.0
Lake Behysehir		
δ_S (%0)	−0.85 ± 0.50	−14.7 ± 3.0
x (%)	56.9 ± 8.9	58.5 ± 10.3
Lake Egridir		
δ_S (%0)	−2.36 ± 0.50	−21.4 ± 3.0
x (%)	39.8 ± 6.8	43.4 ± 8.3
δ_a (%0) (computed from L. Burdur value)	−18.28 ± 2.16	−119.1 ± 8.3

[a] Mean temperature: 12°C; relative humidity: 60 ± 5%.

using lake Burdur's isotopic values for δ_S^T. It can be seen that the error on x, although still rather high (but probably comparable with that given by non-isotopic methods), is, however, considerably lower than that computed for Lake Bracciano, in spite of the fact that for the latter the error on δ_S was lower. Table 3-4 includes also the δ_a values computed from Lake Burdur isotopic composition, which appear quite reasonable for such an environment.

Use of evaporation pans

Evaporation pans have been used to evaluate some difficult terms of the isotope balance equations. Gat (1970) used them to evaluate the term $\Delta\epsilon$, in connection with a study on Lake Tiberias, as already discussed on p. 123.

Recently, Stolf et al. (1979) used evaporation pans to evaluate the subsurface outflow from the artificial reservoir of Quebra Unhas in the arid northeastern Brazil. The study was simplified by the fact that the inflow was negligible, and therefore the reservoir was behaving essentially as a leaky evaporation pan. Equation (7) was used to fit the $\delta^{18}O$ values of the pan and to find the numerical values of A and B, which then were introduced in equation (26) to compute the value of z of the reservoir. The result was that 2/3 of water was lost by evaporation, and 1/3 by subsurface outflow. Chloride data indicated independently that the subsurface losses were only 1/4, which was considered in reasonable agreement with the result given by ^{18}O, due to the error attached to each of these evaluations.

Evaporation pans, however, should be used with caution, because their evaporation rate is usually higher (up to 30%) than in lakes, and because of the different isotopic "inertia" with respect to lakes: short duration changes of the evaporation conditions can affect the isotopic composition of the pan water, due to its small volume, and not significantly that of the lake.

The fast response of pans to changes of the parameters governing evaporation, already pointed out by Gat (difference between day and night values), has been further confirmed by Allison et al. (1979). These authors showed that it is impossible to evaluate the atmospheric relative humidity and its isotopic composition with an acceptable accuracy from pan data when the meteorological conditions are unstable. In addition, the value of the term A/B which is obtained by extrapolation of the δ values to $f = 0$ is particularly critical, even if the evaporation conditions are stable. In order to reduce the error associated with this term, Allison et al. propose the use of two spiked pans of equal dimensions, one having the initial heavy isotope content lower than A/B and the other higher. By combining the two equations (7) relative to the two pans we obtain:

$$\frac{A}{B} = \frac{\delta_1 \delta_{0,2} - \delta_{0,1} \delta_2}{\delta_1 - \delta_{0,1} - (\delta_2 - \delta_{0,2})}$$

Two ephemeral lakes in Sahara

The isotopic behaviour of two small ephemeral lakes, formed by the flood of December 1965 of the Saura River in western Sahara, has been investigated by Fontes and Gonfiantini (1967a). The first (Guelta Gara Dibah) is a pond in proximity of Beni Abbes; the second, much larger and salted (Sebkha el Melah, salt lake), is about 120 km to the southeast, between Kerzaz and Fum el Kheneg. They have been probably the first lakes to be studied with isotopes in a very arid environment.

The lake water attained a strong enrichment in heavy isotopes due to evaporation at low relative humidity. The maximum δ values observed were + 31.3 for ^{18}O and + 129 for ^2H in the Guelta; in the Sebkha the enrichment was slightly lower, due to the effect of dissolved salts.

The daily average evaporation conditions were roughly steady for several months, at least until April 1966, when the Guelta dried up. The isotopic values of the Guelta fit equations similar to (7) with the following values of the parameters* (Fig. 3-7):

	^{18}O	^2H
$A/B = \delta_S$	45.6 ± 0.9	205 ± 6
B	0.35 ± 0.02	0.32 ± 0.03

The value of B is that producing the maximum correlation coefficient between the δ and f^B pairs measured at different times. Errors represent 1σ. The fraction f was computed from the sodium content, and losses by infiltration were considered negligible. Using equation (9) we can then compute the average relative humidity and we obtain: 0.28 ± 0.02 from ^{18}O and 0.31 ± 0.02 from ^2H data. These values of h seem to be reasonable when compared with the available observations, which show daily variations of the relative humidity from 70 to 10% in parallel to temperature variations from 5 to 35°C.

Assuming that the isotopic composition of water before evaporation started is given by the interception of the Guelta line with the global meteoric water line ($\delta^2H = 8\ \delta^{18}O + 10$) in a $\delta^{18}O$-δ^2H diagram, one obtains for this $-7.0 \pm 0.5‰$ for ^{18}O and $-46 \pm 4‰$ for ^2H. Extrapolating back to these values, we obtain for f_0 a value of 1.74 ± 0.2, which means that already 43 ± 6% of the water was lost by evaporation during run-off and in the lake before the study started. Applying the same equations to the Sebkha, the water lost by evaporation before the beginning of the study is 25 ± 6%.

---

* The values reported here are slightly different from those of Fontes and Gonfiantini (1967a), because the data treatment is also slightly different. The conclusions, however, remain unchanged.

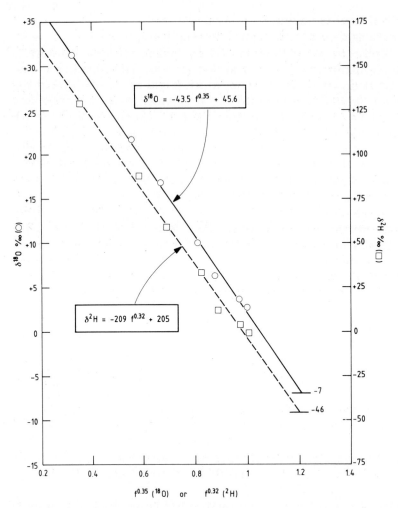

Fig. 3-7. Variation of the isotopic composition of an ephemeral pond in the Sahara (Guelta Gara Dibah, Beni Abbes, Algeria), submitted to evaporation under extremely arid conditions, as a function of the residual water fraction f raised to the exponent B, the value of which mainly depends on relative humidity (equation (9)) (adapted from Fontes and Gonfiantini, 1967a). The value of B has been evaluated from the isotopic data, and it is that which produces the maximum correlation coefficient between δ and f^B pairs. The linear correlation is extrapolated to the isotopic values of the meteoric water ($\delta^2 H = -46$ and $\delta^{18} O = -7$), to evaluate the fraction of water evaporated before the beginning of the study.

Two young artificial lakes in Germany

In 1970 Zimmermann (1978, 1979) started the study of the isotopic composition of two small artificial lakes near Heidelberg, the Wiesensee and the Waidsee, created, respectively, only two years and one year before as recreation resorts. The lakes are about 11 m deep and have no surface inflow and outflow: the water turnover is ensured by groundwater.

One of the most interesting features of the study was that the lakes in 1974 had not yet reached the steady state, and both the stable isotope composition and the volume of water were still changing. In addition, seasonal variations were overimposed on the main trend of the isotopic composition towards the steady value (Fig. 3-8), due to the summer stratification of the lakes, and to a less extent also to the seasonal change of the evaporation/inflow ratio.

In its approach to the isotope balance of the two lakes, Zimmerman disregarded the volume variations, considered small and uninfluential (although the mean yearly volume decrease of the two lakes is about 6%). Therefore, equation (20) is used to find out by trails the value of δ_S which produces the best correlation between $\ln(\delta_S - \delta)$ and t (Fig. 3-9), in a manner analogous to that adopted by Fontes and Gonfiantini (1967a). The slope of the correlation provides the turnover time of the two lakes. The values of A and B are computed independently, because h is known from field measurements and δ_a directly from isotopic analyses of atmospheric water vapour samples re-

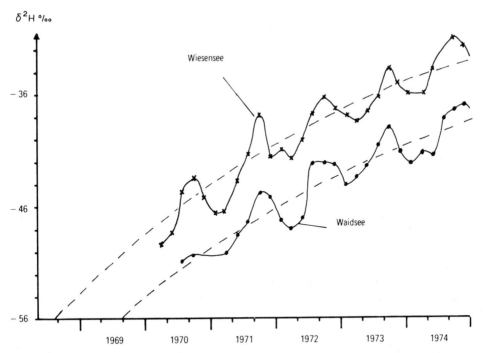

Fig. 3-8. Variations in time of the isotopic composition of two young artificial lakes near Heidelberg (from Zimmermann, 1979). Seasonal variations are superimposed on the main enrichment trend towards a steady-state value.

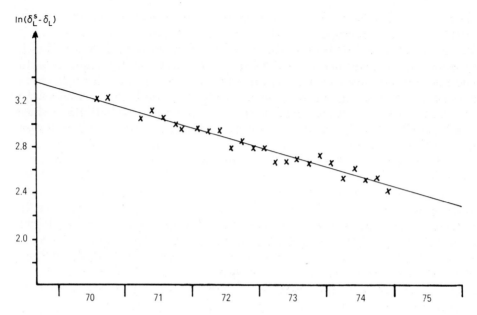

Fig. 3-9. Change with time of the hydrogen isotopic composition of a young artificial lake (Waidsee), expressed as $\ln(\delta_L^S - \delta_L)$, where δ is the lake water value and δ_L^S is the steady-state value towards which the lake tends (from Zimmermann, 1979).

gularly collected. It is therefore possible to compute the water lost by evaporation, and from this the terms related to subsurface inflow and outflow.

The error attached to these evaluations is rather high, often not less than 30%. In spite of this, however, "the stable isotope method is suitable for determining subsurface inflow and outflow, as there is no other method that allows a determination of these components with comparable accuracy" (Zimmermann, 1979).

The Okavango swamp, Botswana

The Okavango River originates in Angola and flows in the southeast direction. After entering Botswana it forms a large delta of about 20,000 km², the half of which is occupied by the swamp. Here intense evaporation and plant transpiration — the swamp is largely covered by reeds and other aquatic vegetation — remove most of the water. Only 3% of the inflow (10.5 × 10⁹ m³ a⁻¹) is left, which is collected by the Boteti River and transported to the Makgadikgadi salt pan.

In studying with isotopes the hydrological balance of the swamp, Dinçer et al. (1979a) disregarded horizontal mixing and assumed that each sample was representing, isotopically and chemically, a different stage of evaporation of the same initial water. The losses to groundwater were considered negligible.

An equation similar to (26) was used, in which z represents now the fraction of water lost by evaporation with respect to the total evapo-transpiration losses. The fraction of residual water was computed from the chemical conductivity using equation (29), assuming that the plant transpiration process, which is not isotopically fractionating, removes from the system only water and not salt. In this way, it was possible to show that transpiration losses are important especially in summer, when they account for 25—75% of the total losses. These estimates are, of course, affected by a large error, the evaluation of which was not attempted by Dinçer et al. In spite of this, isotopes appear to be the only tool available to evaluate the evaporation/evapo-transpiration ratio.

Other lakes

Stable isotopes have been used to study the water balance of a number of other lakes. Lake Kinneret in Israel (called also Sea of Galilee or Lake Tiberias) has been investigated first by Gat (1970) and later by Lewis (1979). The evaporation rate was evaluated by the latter making use of a two-boxes model (corresponding to epilimnion and hypolimnion) and of a periodic function to describe evaporation. The uncertainty of this estimate is about 30%, which perhaps could be reduced to 15% if some of the hydrological and isotopic parameters would be better known. An error analysis shows that the main sources of errors are the isotopic composition of the atmospheric vapour and the non-equilibrium fractionation factor between vapour and liquid water. The latter strongly depends on relative humidity, which is also not easily measureable over all the lake.

Lake Titicaca has been studied by Fontes et al. (1979a), who concluded on the basis of ^{18}O results that subsurface losses were small, if any. However, the error associated with this estimate is rather high: a subsurface loss of about one third of the inflow would still be compatible with the isotopic data. Lake Titicaca (8100 km^2 of surface area) also shows a remarkable homogeneity of the isotopic composition: for instance, in September 1976 the mean $\delta^{18}O$ was -4.42 with $\sigma_1 = 0.18$ for 74 samples collected at various depths from 27 stations all over the lake.

Matsuo et al. (1979) have investigated the water balance of the Hakone Caldera (6.9 km^2) in Japan using hydrogen and oxygen isotopes and dissolved chloride. Six equations with six unknowns are obtained, three for the lake and three for the catchment area, assuming a steady-state model: in fact, the lake does not show any significant isotopic composition variation in time, as well as groundwater and the Hayakawa River. The computation indicates that 58% of the inflow derives from groundwater and 42% from precipitation, while 12% of the outflow is due to evaporation and 88% to surface drainage (including exploitation for irrigation purposes). However, no attempt has been made to evaluate the errors attached to these figures.

The steady-state model has been used also for Lake Schwerin, in the German Democratic Republic, by Hübner et al. (1979), who found that 19% the outflow is due to evaporation. Also in this case no error evaluation is given.

Friedman and Redfield (1971) have discussed the deuterium and the salt balance of the lakes of the Lower Grand Coulee, Washington, U.S.A. The salinity of these lakes, previously high due to evaporation, has lowered considerably during the sixties due to inflow of irrigaton water from the Columbia Basin Irrigation Project. In some lakes, this has been accompanied by a decrease of the deuterium content, as it should be expected as a consequence of the decreased fraction of water removed by evaporation (Lake Lenore, Alkali Lake). In other lakes the isotopic and salt balance is less satisfactory, mainly because of the complexity of the system: for instance, in the Blue Lake a drop in chloride content from 33 to less than 10 ppm is accompanied by a slight rise of the deuterium content, which passes from -110 to $-105‰$. The peculiar feature of this study, however, is that a so-called "effective fractionation factor" α' of empirical origin is used in the isotopic balance model, and the exchange with the atmospheric vapour is disregarded: the model is therefore based on a Rayleigh distillation process. It is also assumed that α' has a value close to that of the equilibrium fractionation factor. This approach had already been discussed by Friedman et al. (1964) — who also reported the deuterium content of a number of lakes in the U.S.A. — but it should be considered less rigorous than that presented here in the section "Stable isotope balance equations". In fact, this approach is an approximation which is valid only in the initial evaporation stages.

In a recent paper, Zuber (1983) examines the water balance of Lake Chala, a small crater lake at the border between Kenya and Tanzania, by using stable isotope date obtained and in part published by Payne (1970). For the subsurface outflow of Lake Chala, Zuber finds a value of 3.9×10^6 m^3 a^{-1}, which is only about the half of the value obtained by Payne (1970) on the basis of an injected tritium experiment. It is interesting to note, however, that Payne has recently revised his tritium data interpretation on Lake Chala, especially because he had the opportunity to measure again the tritium content of a lake sample collected in 1977, that is, nine years after the end of the first study. With the new data, including a revised estimation of the precipitation inflow, Payne (1983) finds now a value of 4.14×10^6 m^3 a^{-1} for the subsurface outflow, which is in excellent agreement with that found by Zuber. However, no evaluation of the error attached to their estimates has been attempted by Payne and by Zuber.

Very saline lakes (sebkhas)

Very saline lakes often occur in arid regions, especially in coastal areas and in structural depressions, where they are formed by the discharge of groundwater systems. In Arab countries they are called sebkhas, and this name is becoming more and more of common use everywhere.

As discussed in the section "The dissolved salt effects", salts dissolved in high concentration in water give rise to a number of isotopic effects during evaporation, which are now rather well known, especially for the most common salts. Therefore, the isotopic balance equations should be modified to take into account these effects.

If the lake salinity does not change, then the modifications are simple: in practice, the set of equations (19) to (26) is still valid, but for the constants A and B one should use equations (15b) instead of (8) and (9). Conditions of steady salinity occur mainly in lakes in which volume variations are negligible and in drying up lakes when they reach the salt saturation or when they attain a value of a_w such that $h/a_w = 1$: in these conditions the net evaporation becomes equal to zero, but the isotopic exchange with the atmospheric water vapour continues.

For lakes in which as a consequence of evaporation the salt content of water is increasing and the volume is decreasing, equation (24) with A and B defined by (15b) should be integrated by steps, because a_w and Γ are not constant and change with the salt concentration (equations (13) and (14)). The fraction of residual water can be expressed as a function of salinity by using equations (27) to (30).

Balance studies of very saline lakes based on isotopic data are rather limited in number. Among other things, the salt effects on the isotopic composition of water, although well understood, increase, however, considerably the errors on the balance terms computed from isotopic data.

Gat (1979, 1980) has reported the case of three sebkhas in Israel. The isotopic composition of these sebkhas showed strong fluctuations in the period January to March as a consequence of winter rains, but then, with the dry season, the ^{18}O content was regularly increasing to reach a steady value. Using the $\delta^{18}O$ of the most enriched sebkha 51 as index, it was computed that in sebkha Fjndim evaporation approximately compensates inflow ($x \simeq 1$), while in sebkha G'vaoth inflow was significantly higher than evaporation ($2 > x > 1$).

Fontes et al. (1979b) studied Lake Asal in the Republic of Djibouti. This lake at 155 m below sea level and at about 15 km from the sea, is fed mainly by seawater. The salt concentration attains the saturation, and large deposits of sodium chloride are formed. As a consequence of the high salinity the ^2H and ^{18}O contents of the lake water are only modestly enriched: + 12 and + 3.7, respectively, to be compared with + 6 and + 0.7 of the inflowing water. Fontes et al. computed that approximately 15—20% of the inflow was lost by the lake through subsurface leakages. This figure was also supported by the lithium and sulphate balance. However, an evaluation of the errors of all the parameters used in the isotopic computation indicates that actually the water fraction removed by subsurface losses may range between 0 and 50%.

The desiccation of the brines of Owens Lake in southern California, U.S.A., after an exceptional flood occurred in early 1969 was investigated by Friedman et al. (1976). The δ^2H value of the flooding water was about — 120. The δ^2H

value of the lake water was -83 in February 1970 and raised steadily to -20 in November, it dropped to -45 in January 1971 due to rain and runoff and then it raised again to about 0 in summer 1971. The dissolved salt — ranging from 136 g/l in February 1970 to 470 in August 1971 — follows a similar history, somewhat more complicated by the precipitation-dissolution processes. The Rayleigh distillation model obviously cannot account for the isotopic values observed in the last evaporation stages, and the authors are therefore induced to admit the role of the isotopic exchanges with the atmospheric vapour and of the salinity on the evaporation process.

WATER MIXING STUDIES

General remarks

The isotopic composition of water — like the chemical composition and the temperature — is often not uniform throughout lakes, and therefore it is conceivable to use the isotopic differences naturally established and their variations with time to investigate the mixing processes and rates.

Shallow lakes and swamps are in general vertically well mixed, but poorly mixed horizontally. In these lakes, the isotopic composition of the inflowing water (streams, groundwater) is gradually modified by evaporation and by exchange with the atmospheric moisture as the distance from the inflow points increases. The extent of the isotopic variation depends on many parameters, including temperature and atmospheric relative humidity (which governs the evaporation rate), the depth and the mean residence time of water in the lake, the degree of horizontal mixing, etc. In extreme cases, horizontal mixing can be almost neglected, and the lake can be treated as a stream which evaporates during it flows (Gat, 1981b).

In deep lakes, the horizontal mixing is generally rather fast, and it is mainly due to turbulent mass transport under the wind stirring action. Therefore, deep lakes are often horizontally well mixed, with the exception of restricted areas around the inflow points. The mixing rate, however, decreases with increasing depth; in stratified lakes, the horizontal mixing rate is much greater in the epilimnion, which is the lake region exposed to the wind action, than in the hypolimnion. For instance, experiments by Quay et al. (1979, 1980) by using injected tritiated water as tracer, have shown that in a small Canadian lake the horizontal eddy diffusion coefficient was 1350 cm^2 s^{-1} in the epilimnion, and only 18 and 14 cm^2 s^{-1}, respectively, at the thermocline depth and in the hypolimnion.

Vertical mixing in deep lakes is comparatively rather slow, with the obvious exception of the overturn period. In addition, the thermocline established in summer or permanently occurring as in meromictic lakes, is an efficient barrier hampering mass transport between epilimnion and hypolimnion. For instance, in the experiment by Quay et al. mentioned above, the vertical diffusion coefficient across the thermocline was evaluated to $(5.0 \pm 0.3) \times 10^{-5}$ cm^2 s^{-1}, i.e. several orders of magnitude lower than the horizontal one. This value results from the sum of two almost equally important terms related to the eddy diffusion and to the molecular diffusion, the latter being for tritiated water equal to 1.8×10^{-5} cm^2 s^{-1} at 15°C (Wang et al., 1953). Clearly in all other cases, the contribution of molecular diffusion to water transport is negligible with respect to the turbulent component.

The main environmental isotopes which are used in studying water mixing in lakes are ^{18}O and ^{2}H, ^{3}H and its daughter product ^{3}He, and ^{222}Rn. The examples described in the following section should give to the reader a sufficient idea of the applicability of the different isotopic methods.

Applications

Lake Chad, Africa

Lake Chad is a large, shallow lake in the heart of the tropical Africa. Its surface is about 20,000 km^2 and its depth, seasonally variable, seldom exceeds 2 m. Evaporation from the lake is strong, due to the semi-arid climatic conditions. The Chari River, having its watershed in the southern part of the Chad Basin where precipitation is high, is responsible for most of the water inflow (83%) into the lake. The Chari River discharge is mainly concentrated in the September-December period, which accounts for about 3/4 of the total annual discharge.

The stable isotope study of the Lake Chad system has been carried out by Fontes et al. (1970). During the flood discharge, the δ^{18}O of the Chari River ranges from -3 to -5, i.e. it is considerably depleted in heavy isotopes with respect to the lake water enriched by evaporation. Therefore, the penetration of the river water into the lake can be followed by means of the isotopic composition.

Because of its shallow depth and of the occurrence of sand islets and of reeds, Lake Chad is not well mixed, both isotopically and chemically. By plotting the δ^{18}O value of the water samples collected all over the lake versus their electrical conductivity, two different correlations can be identified (Fig. 3-10) corresponding to the two main regions of the lake: the southern basin (where the Chari River is flowing in) and the northern basin, which are separated by the so-called Grande Barrière, a system of sand and reed islets crossing the lake in the west-east direction. This indicates that the water dynamics is different in the two basins.

In the southern basin the isotopic and the chemical composition of water follow the usual correlation which can be expected from an evaporation process, with reduced or even negligible horizontal mixing, even at the bottom of lake branches as for instance at Daguil and at Hadjer el Hamis.

In the northern basin, the δ^{18}O-conductivity relationship seems to be almost linear, indicating the occurrence of horizontal mixing overimposed on the evaporation process. In addition, the different ^{18}O-conductivity relationships indicate that the return of water from the northern to the southern basin is practically negligible.

The whole set of isotopic and chemical data of Lake Chad further indicates that the mechanism which regulates the salinity of water — always rather low in spite of the lack of any surface effluent from the lake — should consist of diffuse seepage to groundwater all around the lake, rather than leakages concentrated in limited section of the lake shore. Isotopic studies conducted on

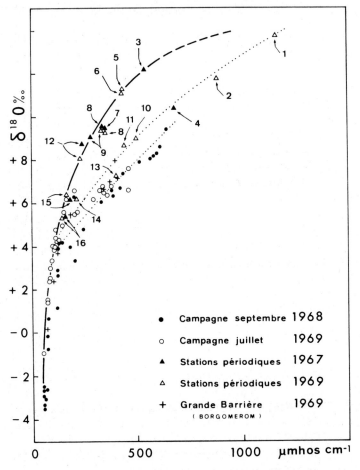

Fig. 3-10. Isotopic composition versus conductivity in Lake Chad water (from Fontes et al., 1970). The numbers identify stations on the lake. Stations on the full line are in the southern zone of the lake, those on the dotted lines are the northern zone. Values from open waters tend to plot along straight lines because the mixing is more efficient.

groundwater samples are also in agreement with this conclusion (Fontes et al., 1970, and unpublished data).

The Lake Chad data, however, should perhaps be partially re-elaborated, in order to investigate the influence of plant transpiration on the ^{18}O-conductivity relationship (see p. 142 on the Okawango swamp).

Lake Kainji, Nigeria

An almost unique case is that of Lake Kainji in Nigeria, in which large isotopic variations occurring seasonally in the inflowing water have been used to investigate the mixing patterns (Dinçer et al., 1979b; Zimmermann et al., 1976).

Lake Kainji is a large artificial lake created in 1968 after the construction of the Kainji dam on the Niger River. The lake is about 115 km long and has a maximum width of about 30 km, an average depth of 14 m, an average surface area of 1100 km^2 and volume ranging from 6×10^9 m^3 in August to 15×10^9 m^3 in January-March.

The Niger River has two flood periods in the year: the white flood (so-called because the water is turbid due to the considerable load of clayey suspended sediments) which occurs from August to November; and the black flood (so-called because the suspended sediments have been already deposited and therefore the water is dark-blue) which occurs from January to March. The white flood originates from precipitation in the region immediately to the north of the lake, while the black flood derives from precipitation in the upper basin of the Niger in Mali, Ivory Coast and Guinea (that is about 2500 km apart from the lake which also explains the delay with respect to the white flood) and flows through the large swamp area between Segou and Timbuktu in Mali, where it deposits the suspended sediments and is submitted to evaporation. As a consequence of their different history, white flood and black flood have also different stable isotope composition: the first has a relatively low heavy isotope content (δ^2H up to -38 in August 1970), while the second is significantly enriched by evaporation (δ^2H about -12 in February 1971). The maximum heavy isotope enrichment in the inflowing water is observed in June-July, when, however, the discharge of the Niger River is at the minimum. In the lake outflow, the minimum and the maximum δ-values are delayed of about two months with respect to the inflow (Fig. 3-11).

Zimmermann et al. (1976) introduced a so-called "pseudo-mixing index", K, to evaluate the degree of mixing:

$$K = \frac{\overline{I} \cdot \Delta t (\overline{\delta}_I - \overline{\delta}_O)}{\Delta \delta_O \cdot V}$$

where $\overline{\delta}_I$ and $\overline{\delta}_O$ are the measured mean isotopic composition of inflow and

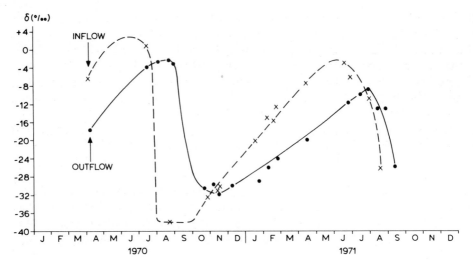

Fig. 3-11. Deuterium content of inflow and outflow of Lake Kainji, Nigeria, from April 1970 to October 1971 (from Dinçer et al., 1979b).

outflow, respectively, during the time interval Δt, $\Delta \delta_O$ is the corresponding variation of the outflow isotopic composition, $\overline{I}$ is the average inflow rate and V is the mean lake volume. The value of the index K ranges from about 10% in May-July, indicating poor mixing, to 80—100% in January-March, indicating that the lake is well mixed during high filling stages.

Isotopic analyses performed at various lake sites show significant horizontal variations especially at the beginning of the white flood, when the lake is at its minimum level: the water movement across the lake is approaching the piston flow. Later on, the horizontal variations become almost negligible, until the arrival of the black flood, which also initially displaces the lake water without complete mixing: this is reached only in the final section of the lake.

River water penetration in Lake Geneva, Switzerland-France

An application to some extent similar to those discussed in previous sections, is the study of the penetration in the Lake Geneva (surface area 582 km^2, volume 89 × 10^9 m^3, maximum depth 309 m) of the water of the two major tributaries, the Rhône and the Dranse rivers. For this purpose Meybeck et al. (1970) used the tritium content variations, which was the only tool available to differentiate waters with similar physico-chemical characteristics.

The study was carried out in 1968. At that time, the tritium concentration in the runoff was higher than that of the lake, and therefore it was possible to identify the river water and their mixing patterns within the lake and the zones of bottom waters less affected by the tributaries variations. In particular, the occurence of three water layers, associated with the three main water currents, was observed: the first layer, at 0—50 m depth, showing rapid tritium variations influenced by wind and by tributaries; the second, at 50—100 m depth, in which large, regular tritium variations occur originated from the annual floods of the major tributaries; and the third, at 150—300 m depth, where variations are small and slow, due to turbidity currents.

The application of the tritium method to this problem would be much more difficult nowadays, because the tritium content differences between tributaries and lake water are now much smaller than during the 60's. The possibility of using the stable isotope method should perhaps be considered especially for identifying the Rhône water penetration, which has a mean $\delta^{18}O$ content significantly lower than that of the lake: -14.9 versus -12.3. On the contrary, the Dranse water, with a mean $\delta^{18}O = -11.3$, is slightly enriched in heavy isotopes with respect to the lake. These values were observed in the period October 1963 to June 1964, during which the $\delta^{18}O$ of the Rhône River ranged from -15.4 to -14.2, and that of the Dranse from -11.6 to -10.3 (unpublished measurements by B. Blavoux, J.Ch. Fontes and R. Gonfiantini).

Lake Tahoe and Crater Lake, U.S.A.

Lake Tahoe is a large, deep lake (surface area 499 km^2, maximum depth 501 m) at high elevation (1897 m a.s.l.) in the Sierra Nevada between California and Nevada. Due to depth, vertical gradients are extremely small in the hypolimnion and practically unaffected by seasonal climatic changes. The study of the vertical mixing is therefore difficult.

Samples collected along a vertical profile in August 1973 showed tritium content and alcalinity considerably uniform (Imboden et al. 1977; Imboden, 1979a). This confirmed that a complete overturn had taken place in the lake at the end of the previous winter, as also indicated by Paerl et al. (1975) on the basis of nitrate concentration. Computations of the tritium content variation with time assuming different vertical mixing models led to the conclusion that at least another complete overturn should have occurred in Lake Tahoe in the period 1964—1968. It is also possible that, in spite of its depth, this lake mixes completely most of the years.

Crater Lake, Oregon, has elevation (1882 m a.s.l.) and depth (maximum 590 m) similar to Lake Tahoe, but it is considerably smaller in size (53.2 km^2). Most of the water input derives from precipitation: 78.5% of the catchment basin is in fact constituted by the lake surface.

In August 1967, the vertical tritium concentration was considerably uniform in the hypolimnion (mean value 24 TU), whereas its value was about 30 TU in the epilimnion (Simpson, 1970) as a consequence of the fact that at that time the tritium content of the inflowing water (precipitation, runoff, atmospheric moisture) was still considerably higher than that of the lake. Also in the case of Crater Lake, however, tritium indicated that complete vertical mixing is a frequent event — perhaps annual — in spite of the considerable depth.

A tritium balance study was attemped for both Lake Tahoe and Crater Lake, in order to evaluate the fraction of tritium deriving from exchange with the atmospheric moisture. Imboden et al. (1977) estimated that 68% of the tritium present in Lake Tahoe was deriving from the atmospheric vapour in August 1973. Simpson (1970) concluded that the contribution of atmospheric vapour to the tritium content of Crater Lake was about 30% in August 1967.

Lake Tanganyika and Lake Malawi, Eastern Africa

The values of some major morphometric, hydrological, chemical and isotopic parameters of these two large and deep meromictic lakes of the African Rift are reported in Table 3-5. Both of them show a rather peculiar feature: the epilimnion is depleted in heavy isotopes with respect to the hypolimnion, as also shown in Fig. 3-12 (Craig, 1974; Gonfiantini et al., 1979). According to Craig, this reflects a past, different water balance due to a drier, colder regime of higher evaporation. The water level record of Lake Malawi, which shows a minimum value in the period 1910—1916 at about 5 m below the

TABLE 3-5

Mean values of some major morphometric, hydrological, chemical and isotopic parameters of Lakes Tanganyika and Malawi (Craig, 1974; Gonfiantini et al., 1979)

	Lake Tanganyika	Lake Malawi
Altitude (m a.s.l.)	773	471
Surface, (m^2)	3.26×10^{10}	2.88×10^{10}
Volume (m^3)	1.89×10^{13}	8.4×10^{12}
Maximum depth (m)	1470	785
Mean precipitation (m a^{-1})	0.90	1.35
Mean runoff inflow (m a^{-1})	0.44	0.66
Mean runoff outflow (m a^{-1})	0.08	0.33
Epilimnion		
Temperature (°C)	26.9	25.4
Conductivity (μS cm^{-1})		248
Chloride (ppm)	26.6	
Tritium (TU)	6.4	4.6
δD (°/oo)	23.8	11.4
δ^{18}O (°/oo)	3.53	1.65
Hypolimnion		
Temperature (°C)	23.3	22.7
Conductivity (μS cm^{-1})		265
Chloride (ppm)	27.8	
Tritium (TU)	0.2	2.0
δD (°/oo)	27.8	13.5
δ^{18}O (°/oo)	4.19	2.08

mean level of recent years, appears in agreement with the above hypothesis. The successive increase of the lake level is a consequence of the increase of the mean yearly precipitation in the area. This increase has been about 200 mm since the beginning of the century.

The stable isotope content difference between epilimnion and hypolimnion is however much less in Lake Malawi than in Lake Tanganyika, indicating that in the first the vertical mixing, although slow, occurs most probably at a rate which is significantly faster than in the second lake. This conclusion is fully confirmed by the tritium content, which was practically zero in the hypolimnion of Lake Tanganyika in February 1973 (Craig, 1974), while it was 2 TU, i.e. well above zero, in Lake Malawi hypolimnion in June 1976.

By using a three-box model, the annual rate of mixing between epilimnion and hypolimnion of Lake Malawi has been evaluated to 5% (Gonfiantini et al., 1979): this mixing, however, is likely to occur mainly in the southern end of the lake, where, according to Eccles (1974), sinking of surface chilled water takes place. Clearly, isotopic profiles in a network of lake stations would indicate where vertical mixing mainly occurs.

An evaluation of the tritium balance indicates that 85% of tritium present

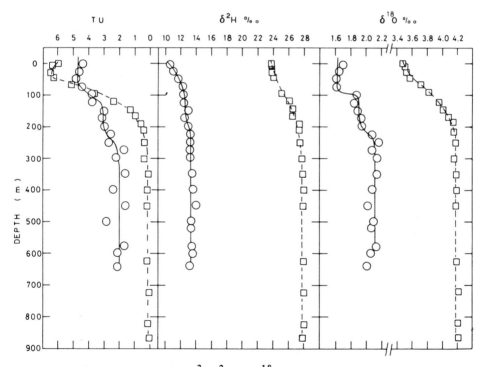

Fig. 3-12. Variation with depth of ^3H, ^2H and ^{18}O in Lakes Tanganyika (□) and Malawi (o). Values are from Craig (1974) and Gonfiantini et al. (1979).

in Lake Tanganyika in February 1973 and 60% of that of Lake Malawi derives from exchange with the atmospheric vapour (Gonfiantini et al., 1979).

The ^3H-^3He method applied to Lakes Erie, Huron and Ontario, U.S.A.-Canada

In recent years, the measurement of the ^3H content of water and of its daughter product ^3He has become possible by mass spectrometry. With this technique it is possible to determine the "age" of a parcel of water considered as a closed system by means of the expression:

$$t = \frac{1}{\lambda} \ln \left(\frac{[^3He^*]}{C_T} + 1 \right) \tag{32}$$

where $[^3He^*]$ is the excess of dissolved ^3He over the equilibrium value with atmospheric helium (with respect to which the $^3He/^4He$ of dissolved helium is lower by 1.4%), C_T is the tritium concentration of water, and λ is the decay constant of tritium (1.527×10^{-4} day^{-1}, corresponding to a half-life of 12.43 years). $[^3He^*]$ and C_T should be obviously expressed in the same units, usually in atoms g^{-1} of water. The method is strictly analogous to that of potassium-argon for rock dating.

The overall accuracy of the ^3H-^3He age determinations is ± 4 days. The correction for atmospheric helium is based on ^4He concentration measurements, and possibly also on dissolved neon (which has no radioactive precursors) to check whether part of the ^4He is radiogenic. The limnological application of the ^3H-^3He technique has been thoroughly discussed by Torgensen (1977).

The ^3H-^3He method has been first used by Torgensen et al. (1977) to evaluate the "age" of hypolimnetic water in Lakes Erie, Huron and Ontario. The ^3H-^3He age represents the time elapsed since the hypolimnetic water has been segregated from the contact with the atmosphere, i.e. since the moment in which the summer thermal stratification of the lake has stopped the loss to the atmosphere of the ^3He excess produced by tritium decay in deep waters. This loss is supposed to occur during the overturn period, when the ^3H-^3He watch is set to zero.

Fig. 3-13 illustrates the results obtained by Torgensen et al. (1977) at station 12B on Lake Huron at two different dates in summer 1974. Those of June, when the thermocline is in its initial stage of development, show that vertical mixing is still occurring, as the surface water has a ^3H-^3He age of about 30 days and deep water of 50—60 days. In other stations the age values are lower, indicating that most probably at station 12B the ^3He excess loss of deep water has not been completed during the overturn. In addition, the lack of age homogeneity among stations indicates that also horizontal transport could have been investigated with the ^3H-^3He method if more profiles were available.

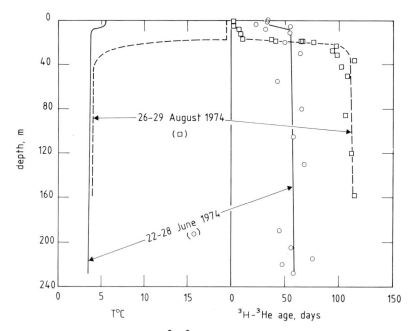

Fig. 3-13. Temperature and ^{3}H-^{3}He ages versus depth at station 12B on Lake Huron. When the thermocline is still in its initial stage of development, values are more scattered in the hypolimnion and some He is still migrating to the epilimnion. Values are from Torgensen et al. (1977).

Two months later, at the end of August, the thermocline is well developed in Lake Huron, and the separation between epilimnion and hypolimnion much more efficient. In the epilimnion the age of water approaches zero, while in the hypolimnion is about 110 days at station 12B. At other stations the results are now much more consistent with those of 12B. Intermediate age values are obtained from samples collected in the thermocline.

Analogous results are shown by Lakes Erie and Ontario. In the latter, the ^{3}H-^{3}He age of hypolimnetic water was about 160 days in mid October 1974.

In meromictic lakes, the ^{3}H-^{3}He age of the monimolimnion increases with depth and can reach several years at the bottom (9 years at 47 m depth in the Green Lake, New York, U.S.A.; Torgensen et al., 1978).

Vertical transport can also be studied with the ^{3}H-^{3}He method, as shown by Torgensen et al. (1977). With a box model, the following equation can be used to obtain the vertical eddy diffusion coefficient k_v as a function of depth z:

$$\frac{\partial [^3He^*]}{\partial t} = A_0 + k_v \frac{\partial^2 [^3He^*]}{\partial z^2} \tag{33}$$

where $A_0 = \lambda C_t$ is the in-situ ^{3}He production rate from tritium decay. If $k_v = 0$, each layer of water acts as a closed system retaining all the ^{3}He produced by tritium decay: this is the ideal case for the application of ^{3}H-^{3}He watch.

In steady state conditions, $\partial[^3\text{He}^*]/\partial t = 0$, and equation (33) can be integrated to produce:

$$[^3\text{He}^*] - [^3\text{He}^*]_0 = \frac{A_0 z}{k_v} (z_m - z/2) \tag{34}$$

where $[^3\text{He}^*]_0$ is the ^{3}He excess at the surface and z_m is the depth of the box.

Equation (34) can be used to describe the vertical eddy diffusion in lake epilimnion. Torgensen et al. (1977) found the following values of k_v: 0.6 cm^2 s^{-1} for Lake Erie (but the lower epilimnetic layer of this lake gives 0.22 cm^2 s^{-1}); 1.2 cm^2 s^{-1} for Lake Huron; and 3.6 cm^2 s^{-1} for Lake Ontario.

Also evaluations of k_v in the thermocline are possible, although more difficult.

Finally, it is also interesting to not that the ^{3}He/^{4}He ratio in lakes has been used as a tool for uranium prospecting (Clarke et al., 1977; Top and Clarke, 1981).

The ^{222}Rn method: Greifensee, Switzerland

^{222}Rn (half-life 3.83 days) and his parent ^{226}Ra (half-life 1620 years) belong to the ^{238}U radioactive family. ^{222}Rn is released into lake water by the bottom sediments and it is also produced by the dissolved ^{226}Ra. The measurement technique was firstly developed to investigate the near bottom eddy diffusion in the ocean (Broecker et al., 1968; Chung and Craig, 1972) but then it has also found application in limnology.

The first lake studied was the Greifensee, a small lake near Zurich, having a surface area of 8.6 km^2 and a maximum depth of 32 m. Measurements of ^{222}Rn and ^{226}Ra activities were performed monthly in the period April to November 1975 along two vertical profiles (Imboden and Emerson, 1978; Imboden, 1979b).

If the lake bottom is flat and the radon release from sediments spacially uniform (conditions which are more frequently fulfilled in the ocean and in large lakes in stations far from shore), the radon activity profile can be interpreted as due to one-dimensional (vertical) eddy diffusion flux:

$$\frac{\partial [\text{Rn}^*]}{\partial t} = \frac{\partial}{\partial z} \left(k_z \frac{\partial [\text{Rn}^*]}{\partial z} \right) - \lambda [\text{Rn}^*] \tag{35}$$

where [Rn*] is the excess ^{222}Rn activity (corresponding to the total ^{222}Rn activity minus the dissolved ^{226}Ra activity), z is the distance from the sediment-water interface (lake bottom), k_z is the vertical eddy diffusion coefficient and $\lambda = 0.181$ day^{-1} is the decay rate constant of ^{222}Rn. Assuming steady-state conditions, $\partial[\text{Rn}^*]/\partial t = 0$, and k_z independent from depth, the integration of equation (35) produces:

$$[\text{Rn}^*] = [\text{Rn}^*]_0 \exp[-z(\lambda/k_z)^{1/2}] \tag{36}$$

where $[\text{Rn}^*]_0 = F/(\lambda k_z)^{1/2}$ is the excess ^{222}Rn activity at the sediment-water interface, and F is the ^{222}Rn flux through the interface. F depends essentially on the ^{226}Ra content of the sediments and it can be experimentally determined.

Equation (36) fits the ^{222}Rn results at station A (approximately in a central position in Greifensee) with a value of k_z equal to 0.20 ± 0.025 cm^2 s^{-1} for the period May-August. At station B, closer to shore, equation (36) cannot be applied.

The one-dimensional flux conditions are in fact seldom fulfilled in lakes. In most cases, the influence of the horizontal transport cannot be neglected, and therefore a two-dimensional steady state equation should be used to describe the radon distribution:

$$\frac{\partial}{\partial x}\left(k_x \frac{\partial[\text{Rn}^*]}{\partial x}\right) + \frac{\partial}{\partial z}\left(k_z \frac{\partial[\text{Rn}^*]}{\partial z}\right) - \lambda[\text{Rn}^*] = 0 \tag{37}$$

The values of k_x and k_z reported in Table 3-6 have been obtained by using equation (37). It can be seen that with this more rigorous approach, the value of k_z at station A is not significantly different from that evaluated disregarding horizontal transport. Fig. 3-14 shows the ^{222}Rn data fitting for May 1975.

Lake Constance, Central Europe

Lake Constance is a large, deep lake (surface area 5.38×10^8 m^2, volume 4.85×10^{10} m^3, maximum depth 252 m) in Central Europe, shared between

TABLE 3-6

Vertical and horizontal eddy diffusion rates in the Greifensee from ^{222}Rn (Imboden and Emerson, 1978; Imboden, 1979b)

Period	k_z (cm^2 s^{-1})	k_x (cm^2 s^{-1})
April 1975	0.2 ± 0.05	200—2000
May to August 1975	0.15 ± 0.03	100—1000
September to November 1975:		
above 26 m depth		300—1000
below 26 m depth	0.05 ± 0.02	10— 200

Germany, Switzerland and Austria. The tritium study of Lake Constance initiated in 1963, which was almost an ideal starting date, being this the year in which the tritium content of precipitation reached its maximum value in the northern hemisphere, before to start to decline as a consequence of the thermonuclear moratorium. The results of the isotopic study of Lake Constance have been reported by Weiss et al. (1978, 1979) and by Weiss (1979).

From 1963 to the end of 1966 the tritium content of the shallow water was higher than that of the deep water; these two values were, however, much closer in the late winter and in the early spring, that is during the overturn. In the same period, the tritium content of the Alpenrhein, that is, of the Rhine River flowing into the lake, was also higher than that of the lake, while the tritium content of the Seerhein, that is, of the Rhine River flowing out of the lake, was practically identical to that of the lake shallow water (Fig. 3-15).

From 1967, the tritium contents of the shallow and of the deep lake water and of the Seerhein become practically identical, while that of the Alpenrhein becomes lower.

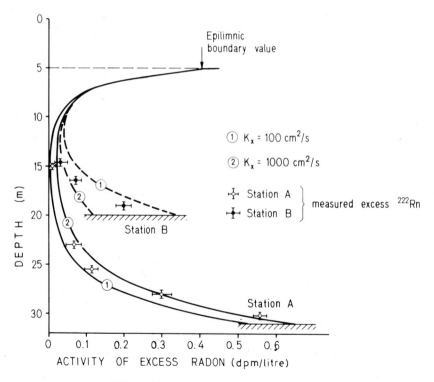

Fig. 3-14. Excess ^{222}Rn activity versus depth in Greifensee in May 1975 and computed profiles (from Imboden, 1979b).

From temperature data, the epilimnion of Lake Constance has a volume which approximately corresponds to 17% of that of the hypolimnion. A two-box model of the lake based on tritium data, indicates that every year another 15% of the deep water is exchanged with the shallow water during the period of stratification. Therefore, a renewal rate of 32% per year is estimated for the lake deep water, being the overturn and mixing during stagnation almost equally efficient. The mixing during stagnation (corresponding to a vertical diffusivity of 1 cm² s⁻¹) does not occur mainly by diffusivity across the thermocline, but it is largely due to the inflow of cold river water which goes down to depths corresponding more or less to that of the thermocline and thus renews the deep water directly, as observed in other non-isotopic studies.

The ^{3}H-^{3}He method has also been used to investigate the dynamics of Lake Constance. The ages ranged from few months in shallow water to about 15 months in the hypolimnion, in the period from 9 November 1976 (thermo-

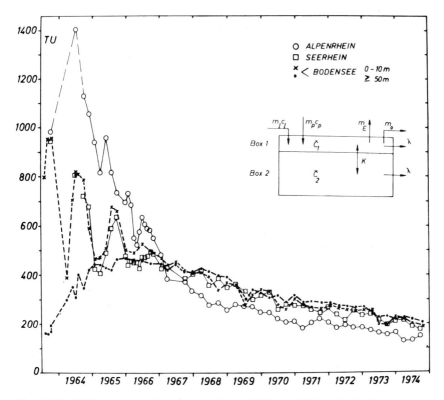

Fig. 3-15. Tritium content variations from 1963 to 1974 of Lake Constance, of the inflowing water (Alpenrhein) and of the outflowing water (Seerhein) (from Weiss, 1979). The sharp peak of Alpenrhein at the beginning of 1964 is due to the large thermonuclear tests made in the atmosphere in 1962. Since 1964, the tritium content is decreasing as a consequence of the thermonuclear moratorium.

cline well developed) to 13 January 1977 (thermocline disappeared). In November, the water age in the thermocline was about 8 months. These results demonstrate once more the complexity of the overturn process, which is also shown by the variability with depth of ^{3}He excess when the lake is not thermally stratified. The fact that in deep waters ages up to 15 months can be attained indicates lack of complete vertical mixing and that the ^{3}He excess is not fully lost during the yearly overturn.

The ^{222}Rn method has also been applied to Lake Constance to evaluate the vertical eddy diffusion in the hypolimnion (Weiss et al., 1978). Using the one-dimensional steady state model (equation (36)), a value ranging from 0.1 to 0.3 cm^2 s^{-1} was found for k_z in September 1977.

Other lakes

In Lake Kinneret, Israel, shallow water is usually depleted in heavy isotopes with respect to deep water in spring and in most of summer. Then it becomes enriched by evaporation at the end of summer up to the late fall, when the overturn occurs (Gat, 1970). The isotopic composition of hypolimnetic water is obviously much more constant; however the exceptionally large floods occurred in the Jordan River during the 1968—69 winter caused a decrease of about 0.7‰ of the hypolimnion ^{18}O content, which then required about five drier years to recover the old value (Lewis, 1979).

A small crater lake in southern Italy — Lago Piccolo near Monticchio, surface area 0.16 km^2, depth 38 m — has been studied by Mongelli et al. (1975). Shallow water (up to a depth of approximately 10 m) shows large seasonal variations in both temperature and stable isotopes, while water below 15 m depth is much more constant in time. A peculiarity of this lake is that the temperature increases from about 7.5°C at 15 m to about 10°C at the bottom, while the δ^{18}O values decrease about 0.5—1‰ in the same depth interval. This is due to the recharge which mainly takes place by springs at the lake bottom. The lake, therefore, is never well mixed, and the two water bodies, above 15 m and below, are always well distinct and identifiable on the basis of their isotopic composition.

In Lake Neusiedl, Austria, the tritium content variations help in locating inflows of groundwater, and also indicate that groundwater contribution is greater than previously estimated (Rank and Schroll, 1979).

A particular case is that of Lake Vanda, Antarctica, permanently covered by an ice layer about 4 m thick. Below this layer the temperature is about 5°C in the depth interval between 5 and 10 m, about 8°C between 18 and 37 m, and it rises up to 25°C at a depth of 67m. The stratification of water is maintained by the strong salt content increase with depth: it is also probable, for this reason, that the deeper water has been isolated from contact with the atmosphere since a long time. A variation in heavy isotope content accompanies the stratification in temperature and salinity and it suggests that the upper and lower layers of water were formed at different times from different sources of glacial melt water (Ragotzkie and Friedman, 1965).

MISCELLANEOUS STUDIES

Relationships between lakes and groundwater

The enrichment in heavy isotopes frequently occurring in lake water as a consequence of evaporation has been used several times to investigate the relationships between lakes and groundwater. The study mentioned earlier on Lake Bracciano, Italy, was actually initiated in order to verify the lake contribution to the shallow aquifer recharge, which was found negligible (Gonfiantini et al., 1962).

A similar investigation was undertaken in southern Turkey, in order to verify whether some large karst springs flowing from the foothill of the Taurus Mountains were fed in an appreciable proportion by lakes on the Anatolian Plateau (Dinçer and Payne, 1971). The contribution of Lake Egridir to the base flow of the Koprucay River was found to be 31% on the basis of the deuterium content, but the error attached to this value was not given: a reasonable estimate would be probably $(31 \pm 10)\%$. Other springs did not show any appreciable fraction of lake water.

Also some springs in the vicinities of Lake Chala, Kenya, were believed to be fed by the lake, the water losses of which by seepage are considerable. However, in a study which involved also the use of injected tritium to asses the volume and the water balance of the lake, Payne (1970) was able to show on the basis of the stable isotope composition that the lake contribution to the springs was negligible.

In most cases, lake water seems to contribute in a significant proportion to groundwater recharge only in areas immediately adjacent to lakes. This was observed, for instance, in the shallow groundwater of the Lake Chad polders (Fontes et al., 1970) and in wells close to Ram Lake, a small crater lake in Israel (Mazor, 1976).

In a small lake in the Federal Republic of Germany, Stichler and Moser (1979) were able to show the direction of the subsurface outflow by means of the stable isotope content variations in the surrounding groundwater. Wells located at a distance ranging from a few meters to about 200 m from the lake, showed that water was flowing mostly to the north, in which direction the more enriched deuterium values were observed. The subsurface outflow was smaller but still significant to the west and to the east, and practically negligible to the south. The seasonal variations of the groundwater isotopic composition, reflecting those occurring in the lake, decrease in amplitude as the distance increases, and show also a phase shift, indicating a groundwater velocity in the north direction of about 0.5—1 m per day.

Exchanges with the atmosphere

The effects of the exchanges with the atmospheric moisture on the isotopic composition of an evaporating water body have already been discussed in detail. Also the fraction of tritium derived from the isotopic exchange with the atmospheric vapour in the tritium budget of lakes can be evaluated, as shown in examples already reported (Lake Kinneret — Gat, 1970; Crater Lake — Simpson, 1970; Lake Tahoe — Imboden et al., 1977; Lakes Tanganyika and Malawi — Gonfiantini et al., 1979).

Laboratory experiments to investigate the exchange rate and mechanism have been made by several authors by using stable isotopes, which have been by far the most used isotopic tool in evaporation studies. The decrease of tritium concentration from a spiked water by exchange with atmospheric vapour has been investigated in laboratory and in field experiments by Horton et al. (1971) and by Prantl (1974).

Environmental ^{222}Rn is generally not suitable to investigate gas exchange of lakes with the atmosphere. In fact, ^{222}Ra is not only produced by dissolved ^{226}Ra, but it also introduced in lakes by groundwater and by release from bottom sediments. Therefore the activity ratio between ^{222}Rn and ^{226}Ra cannot be related in a simple manner with the exchange rate with the atmosphere. This difficulty can be overcome by injecting ^{226}Ra in lakes in order to artificially increase its concentration of about two orders of magnitude and to make negligible the ^{226}Rn produced by other sources. The method has been used in small Canadian lakes by Emerson et al. (1973) and Emerson (1975), who also discuss the theoretical background.

Weiss et al. (1978) proposed to use the environmental ^{3}He to investigate the gas exchange with the atmosphere of Lake Constance. As already shown in this chapter, ^{3}He, produced by tritium decay, is dissolved in lake water in excess with respect to equilibrium with the atmosphere, and accumulates mainly in the hypolimnion during the summer stratification of lakes. After the late autumn overturn, the ^{3}He concentration gradient with depth depends on the gas exchange rate with the atmosphere and provides an evaluation of it.

The CO_2 exchange with air can also be investigated with the help of the carbon isotopes. Broecker and Walton (1959) used ^{14}C to evaluate the rate of atmospheric CO_2 dissolution in American lakes (Pyramid Lake, Walker Lake, Mono Lake, Nevada-California), which was found to range from 10^{-2} to 10^{-1} mol m^{-2} day^{-1}. A similar range of values was also found by Killey and Fritz (1979) in Perch Lake, Ontario, Canada, by means of the ^{13}C/^{12}C ratio. The method is based essentially on the difference in isotopic composition between the CO_2 released into the lake by the bottom organic sediments and the atmospheric CO_2, and on the resulting vertical gradient of the dissolved inorganic carbon (DIC) concentration and isotopic composition.

Of particular interest are the ^{13}C and concentration profiles of DIC with depth in Lake Tanganyika (Craig, 1974). At station 1 the trend is the following:

(1) In the first 30 m the DIC is about 5.88 mmol l^{-1} and the δ^{13}C is + 1.45

(versus PDB): this value practically corresponds to that of isotopic equilibrium with the atmospheric CO_2.

(2) Between the depth of 30—200 m the DIC content increases rapidly to about 6.66 mmol l^{-1} and the $\delta^{13}C$ decreases to about + 0.1. There is here a production of CO_2 from oxidation of organic matter, for which the oxygen probably derives from dissolved nitrate: the content of the latter, in fact, shows a maximum at 100 m and practically drops to zero below 200 m.

(3) In the hypolimnion, from 200 m depth to bottom the DIC steadily increases to 6.84 mmol l^{-1} and also the $\delta^{13}C$ to the value of + 0.7 (these bottom values are actually extrapolated from the existing data, which are available only to a depth of 867 m).

The hypolimnion profile is believed to be a mixing profile, with a flux of CO_2 from the bottom having a $\delta^{13}C$ = + 28. Such a value is, however, much heavier than those currently observed in natural compounds. A process which could explain it is the formation of CH_4 from CO_2 by a fermentation process: The carbon isotope fractionation between CO_2 and methane is in fact very large (more than 70°/$_{oo}$ at room temperature and in equilibrium conditions, and even larger if kinetic effects occur) and therefore the residual CO_2 can become enriched in ^{13}C. An alternative hypothesis is that the hypolimnion profile is a transient one, resulting from mixing with a remnant of water having a $\delta^{13}C$ value of DIC of about + 0.7.

REFERENCES

Allison, G.B., Brown, R.M. and Fritz, P., 1979. Evaluation of water balance parameters from isotopic measurements in evaporation pans. In: *Isotopes in Lake Studies*. IAEA, Vienna, pp. 21—32.

Baertschi, P. and Thürkauf, M., 1960. Isotopie-Effect fur die Trennung der Sauerstoff Isotopen ^{16}O and ^{18}O bei der Rektifikation von leichtem und schwerem Wasser. *Helv. Chim. Acta*, 43: 80—89.

Bottinga, Y. and Craig, H., 1969. Oxygen isotope fractionation between CO_2 and water and the isotopic composition of marine atmosphere. *Earth Planet. Sci. Lett.*, 5: 285—295.

Broecker, W.S. and Walton, A., 1959. The geochemistry of ^{14}C in freshwater systems. *Geochim. Cosmochim. Acta*, 23: 15—38.

Broecker, W.S., Cromwell, J. and Li, Y.H., 1968. Rates of vertical eddy diffusion near the ocean floor based on measurements of the distribution of excess ^{222}Rn. *Earth Planet. Sci. Lett.*, 5: 101—105.

Brown, R.M. and Barry, P.J., 1979. A review of HTO evaporation studies at Chalk River Nuclear Laboratories. In: *Isotopes in Lake Studies*. IAEA, Vienna, pp. 73—86.

Brutsaert, W., 1965. A model for evaporation as a molecular diffusion process into a turbolent atmosphere. *J. Geophys. Res.*, 70: 5017—5024.

Brutsaert, W., 1975. A theory for local evaporation (or heat transfer) from rough and smooth surfaces at ground level. *Water Resour. Res.*, 11: 543—550.

Chung Y. and Craig, H., 1972. Excess-radon and temperature profiles from the eastern equatorial Pacific. *Earth Planet. Sci. Lett.*, 14: 55—64.

Clarke, W.B., Top, Z., Beavan, A.P and Gandhi, S.S. 1977. Dissolved helium in lakes: uranium prospecting in the Precambrian terrain of central Labrador. *Econ. Geol.*, 72: 233—242.

Craig, H., 1974. Lake Tanganyika geochemical and hydrographic study: 1973 Expedition. *Scripps Inst. Oceanogr., Univ. Calif., SIO Ref. Ser.*, 75—5: 83.

Craig, H. and Gordon, L.I., 1965. Deuterium and oxygen-18 variations in the ocean and the marine atmosphere. In: E. Tongiorgi (Editor), *Stable Isotopes in Oceanographic Studies and Paleotemperatures*. C.N.R., Laboratorio di Geologia Nucleare, Pisa, pp. 9—130.

Craig, H., Gordon, L.I. and Horibe, Y., 1963. Isotopic exchange effects in the evaporation of water, 1. Low temperature experimental results. *J. Geophys. Res.*, 68: 5079—5087.

Dansgaard, W., 1961. The isotopic composition of natural waters. *Medd. Groenl.*, 165 (2): 1—120.

Dinçer, T., 1968. The use of oxygen-18 and deuterium concentrations in the water balance of lakes. *Water Resour. Res.*, 4: 1289—1305.

Dinçer, T. and Payne, B.R., 1971. An environmental isotope study of the south-western Karst region of Turkey. *J. Hydrol.*, 14: 233—258.

Dinçer, T., Hutton, L.G. and Kupee, B.B.J., 1979a. Study, using stable isotopes, of flow distribution, surface-groundwater relations and evapotranspiration in the Okawango Swamp, Botswana. In: *Isotope Hydrology 1978, Vol. I*. IAEA, Vienna, pp. 3—26.

Dinçer, T., Zimmermann, U., Baumann, U., Imevbore, A.M.A., Henderson, F. and Adeniji, H.A., 1979b. Study of the mixing pattern of Lake Kainji using stable isotopes. In: *Isotopes in Lake Studies*. IAEA, Vienna, pp. 219—225.

Eccles, D.H., 1974. An outline of the physical limnology of Lake Malawi (Lake Nyasa). *Limnol. Oceanogr.*, 19: 730—742.

Emerson, S., 1975. Gas exhange rates is small Canadian shield lakes. *Limnol. Oceanogr.*, 20: 754—761.

Emerson, S., Broecker, W. and Schindler, D.W., 1973. Gas exchange rates in a small lake as determined by the radon method. *J. Fish. Res. Board Can.*, 30: 1475—1484.

Fontes, J.Ch., 1966. Intérêt en géologie d'une étude isotopique de l'évaporation. Cas de l'eau de mer. *C.R. Acad. Sci. Paris*, 263: 1950—1953.

Fontes, J.Ch. and Gonfiantini, R., 1967a. Comportment isotopique au cours de l'évaporation de deux bassins sahariens. *Earth Planet. Sci. Lett.*, 3: 258—266 and 386.

Fontes, J.Ch. and Gonfiantini, R., 1967b. Fractionnement isotopique de l'hydrogène dans l'eau de cristallisation du gypse. *C.R. Acad. Sci. Paris*, 265: 4—6.

Fontes, J.Ch. and Gonfiantini, R., 1970. Composition isotopique et origine de la vapeur d'eau atmosphérique dans la région du Lac Leman. *Earth Planet. Sci. Lett.*, 7: 325—329.

Fontes, J.Ch., Gonfiantini, R. and Roche, M.A., 1970. Deutérium et oxygène-18 dans les eaux du lac Tchad. In: *Isotope Hydrology 1970*. IAEA, Vienna, pp. 387—404.

Fontes, J.Ch., Boulange, B., Carmouze, J.P. and Florkowski, T., 1979a. Preliminary oxygen-18 and deuterium study of the dynamics of Lake Titicaca. In: *Isotopes in Lake Studies*. IAEA, Vienna, pp. 145—150.

Fontes, J.Ch., Florkowski, T., Pouchan, P. and Zuppi, G.M., 1979b. Preliminary isotopic study of Lake Asal system (Republic of Djibouti). In: *Isotopes in Lake Studies*. IAEA, Vienna, pp. 163—174.

Friedman, I. and Redfield, A.C., 1971. A model of the hydrology of the lakes of Lower Grand Coulee, Washington, *Water Resour. Res.*, 7: 874—898.

Friedman, I. and O'Neil, J.R., 1977. Compilation·of stable isotope fractionation factors of geochemical interest. In: M. Fleischer (Editor), *Data of Geochemistry. U.S. Geol. Surv. Prof. Pap.*. 440-KK.

Friedman, I., Redfield, A.C., Schoen, B. and Harris, J., 1964. The variation of the deuterium content of natural waters in the hydrologic cycle. *Rev. Geophys.*, 2: 177—224.

Friedman, I., Smith, G.I. and Hardcastle, K.G., 1976. Studies of Quaternary saline lakes, II. Isotopic and compositional changes during desiccation of the brines in Owens Lake, California, 1969—1971. *Geochim. Cosmochim. Acta*, 40: 501—511.
Gat, J.R., 1970. Environmental isotope balance of Lake Tiberias. In: *Isotopes in Hydrology 1970.* IAEA, Vienna, pp. 109—127.
Gat, J.R., 1979. Isotope hydrology of very saline surface waters. In: *Isotopes in Lake Studies.* IAEA, Vienna, pp. 151—162.
Gat, J.R., 1980. Isotope hydrology of very saline lakes. In: A. Nissenbaum (Editor), *Hypersaline Brines and Evaporitic Environments.* Elsevier, Amsterdam, pp. 1—7.
Gat, J.R., 1981a. Isotopic fractionation. In: J.R. Gat and R. Gonfiantini (Editors), *Stable Isotope Hydrology — Deuterium and Oxygen-18 in the Water Cycle.* IAEA, Vienna, pp. 21—33.
Gat, J.R., 1981b. Lakes. In: J.R. Gat and R. Gonfiantini (Editors), *Stable Isotope Hydrology — Deuterium and Oxygen-18 in the Water Cycle.* IAEA, Vienna, pp. 203—221.
Gat, J.R. and Craig, H., 1966. Characteristics of the air-sea interface determined from isotope transport studies. *Trans. Am. Geophys. Union*, 47: 115.
Gonfiantini, R., 1965. Effetti isotopici nell'evaporazione di acque salate. *Atti Soc. Toscana Sci. Nat. Pisa, Ser. A*, 72: 550—569.
Gonfiantini, R. and Fontes, J.Ch., 1963. Oxygen isotopic fractionation in the water of crystallisation of gypsum. *Nature*, 200: 644—646.
Gonfiantini, R., Togliatti, V. and Tongiorgi, E., 1962. Il rapporto $^{18}O/^{16}O$ nell'acqua del lago di Bracciano e delle falde a sud-est del lago. *Not. Com. Naz. Energ. Nucl. (Italy)*, 8, (6): 39—45.
Gonfiantini, R., Borsi, S., Ferrara, G. and Panichi, C., 1973. Isotopic composition of waters from the Danakil Depression. *Earth Planet. Sci. Lett.*, 18: 13—21.
Gonfiantini, R., Zuppi, G.M., Eccles, D.H. and Ferro, W., 1979. Isotope investigation of Lake Malawi. In: *Isotopes in Lake Studies.* IAEA, Vienna, pp. 195—207.
Götz, D. and Heinzinger, K., 1973. Sauerstoffisotopieeffekte und Hydratstruktur von Alkalihalogenid-Lösungen in H_2O und D_2O. *Z. Naturforsch.*, 28a: 137—141.
Horton, J.H., Corey, J.C. and Wallace, R.M., 1971. Tritium loss from water exposed to the atmosphere. *Environ. Sci. Technol.*, 5: 338—343.
Hübner, H., Richter, W. and Kowski, P., 1979. Studies on relationships between surface water and surrounding groundwater at Lake Schwerin (German Democratic Republic) using environmental isotopes. In: *Isotopes in Lake Studies.* IAEA, Vienna, pp. 95—102.
Hutchinson, G.E., 1957. *A Treatise on Limnology Vol. I.* John Wiley and Sons, New York, N.Y., 1015 pp.
Imboden, D.M., 1979a. Complete mixing in Lake Tahoe, California-Nevada, traced by tritium. In: *Isotopes in Lake Studies.* IAEA, Vienna, pp. 209—212.
Imboden, D.M., 1979b. Natural radon as a limnological tracer for the study of vertical and horizontal eddy diffusion. In: *Isotopes in Lake Studies.* IAEA, Vienna, pp. 213—218.
Imboden, D.M. and Emerson, S., 1978. Natural radon and phosphorus as limnologic tracers: horizontal and vertical eddy diffusion in Greifensee. *Limnol. Oceanogr.* 23: 77—90.
Imboden, D.M., Weiss, R.F., Craig, H., Michel. R.L. and Goldman, C.R., 1977. Lake Tahoe geochemical study, 1. Lake chemistry and tritium mixing study. *Limnol. Oceanogr.*, 22: 1039—1051.
International Atomic Energy Agency, 1979. *Isotopes in Lake Studies. Proceedings of an Advisory Group Meeting on the Application of Nuclear Techniques to the Study of Lake Dynamics, Vienna, 1977.* IAEA, Vienna, 290 pp.
Jakli, G. and Staschewski, D., 1977. Vapour pressure of $H_2^{18}O$ ice (-50 to $0°C$) and $H_2^{18}O$ water (0 to $170°C$). *J. Chem. Soc., Faraday Trans. 1*, 73: 1505—1509.

Jauzein, A., 1982. Deutérium et oxygène 18 dans les saumures: modelisation et implications sédimentologiques. *C.R. Acad. Sci. Paris*, 294: 663—668.

Kakiuchi, M. and Matsuo, S., 1979. Direct measurements of D/H and $^{18}O/^{16}O$ fractionation factors between vapor and liquid water in the temperature range from 10 to 40°C. *Geochem., J.*, 13: 307—311.

Killey, R.W.D. and Fritz, P., 1979. Carbon dioxide flux across a lake-atmosphere interface as determined by carbon isotope data. In: *Isotopes in Lake Studies*. IAEA, Vienna, pp. 245—249.

Lewis, S., 1979. Environmental isotope balance of Lake Kinneret as a tool in evaporation rate estimation. In: *Isotopes in Lake Studies*. IAEA, Vienna, pp. 33—35.

Lloyd, R.M., 1966. Oxygen isotope enrichment of seawater by evaporation. *Geochim. Cosmochim. Acta*, 30: 801—819.

Majoube, M., 1971. Fractionnement en oxygène-18 et en deutérium entre l'eau et sa vapeur. *J. Chim. Phys.*, 197: 1423—1436.

Matsuo, S., Friedman, I. and Smith, G.I., 1972. Studies of Quaternary saline lakes, I. Hydrogen isotope fractionation in saline minerals. *Geochim. Cosmochim. Acta*, 36: 427—435.

Matsuo, S., Kusakabe, M., Niwano, M., Hirano T. and Oki, Y., 1979. Water budget in the Hakone caldera using hydrogen and oxygen isotope ratios. In: *Isotopes in Lake Studies*. IAEA, Vienna, pp. 131—144.

Mazor, E., 1976. The Ram crater lake. A note on the revival of a 2000-year old groundwater tracing experiment. In: *Interpretation of Environmental Isotope and Hydrochemical Data in Groundwater Hydrology*. IAEA, Vienna, pp. 179—181.

Merlivat, L., 1978. Molecular diffusivities of $H_2^{16}O$, $HD^{16}O$ and $H_2^{18}O$ in gases. *J. Chem. Phys.*, 69: 2864—2871.

Merlivat, L. and Coantic, M., 1975. Study of mass transfer at the air-water interface by an isotopic method. *J. Geophys. Res.*, 80: 3455—3464.

Meybeck, M., Hubert, P., Martin, J.M. and Olive, Ph., 1970. Etude par le tritium du mélange des eaux en milieu lacustre et estuarien. In: *Isotope Hydrology 1970*. IAEA, Vienna, pp. 523—541.

Mongelli, F., Panichi, C. and Tongiorgi, E., 1975. Studio termico ed isotopico dei craterilaghi di Monticchio (Lucania). *Arch. Oceanogr. Limnol.*, 18: 167—188.

Münnich, K.O., 1979. Report on gas exchange and evaporation studies. In: *Isotopes in Lake Studies*. IAEA, Vienna, pp. 233—243.

Östlund, H.G. and Berry, E.X., 1970. Modification of atmospheric tritium and water vapour by Lake Tahoe. *Tellus*, 22: 463—465.

Paerl, H.W., Richards, R.C., Leonard, R.L. and Goldman, C.R., 1975. Seasonal nitrate cycling as evidence for complete vertical mixing in Lake Tahoe, California-Nevada. *Limnol. Oceanogr.*, 20: 1—8.

Payne, B.R., 1970. Water balance of Lake Chala and its relation to groundwater from tritium and stable isotope data. *J. Hydrol.*, 11: 47—58.

Payne, B.R., 1983. Radioisotopes for the estimation of the water balance of lakes and reservoirs. In: *Tracer Methods in Isotope Hydrology*. IAEA Tech. Doc., 291: 157—163.

Pearson, F.J., Jr., Coplen, T.B., 1978. Stable isotope studies of lakes. In: A. Lerman (Editor), *Lakes: Chemistry, Geology, Physics*. Springer-Verlag, New York, N.Y., pp. 325—336.

Prantl, F.A., 1974. Isotope balances for hydrological systems, I. Isotope evaporation studies with tritiated water in the laboratory. *Pure Appl. Geophys.*, 112: 209—218.

Quay, P.D., Broecker, W.S., Hesslein, R.H., Fee, E.J. and Schindler, D.W., 1979. Whole lake tritium spikes to measure horizontal and vertical mixing rates. In: *Isotopes in Lake Studies*. IAEA, Vienna, pp. 175—193.

Quay, P.D., Broecker, W.S., Hesslein, R.H. and Schindler, D.W., 1980. Vertical diffusion rates determined by tritium tracer experiments in the thermocline and hypolimnion of two lakes. *Limnol. Oceanogr.*, 25: 201—218.

Ragotzkie, R.A. and Friedman, I., 1965. Low deuterium content of Lake Vanda, Antarctica. *Science*, 148: 1226—1227.

Rank, D. and Schroll, E., 1979. Test for the applicability of combined nuclear and geochemical methods in relation to the water balance of Lake Neusiedl, Austria. In: *Isotopes in Lake Studies*. IAEA, Vienna, pp. 121—130.

Robinson, R.A. and Stokes, R.H., 1959. *Electrolyte Solutions*. Butterworths, London, 559 pp.

Simpson, H.J., 1970. Tritium in Crater Lake, Oregon. *J. Geophys. Res.*, 75: 5195—5207.

Sofer, Z. and Gat, J.R., 1972. Activities and concentrations of oxygen-18 in concentrated aqueous salt solutions: analytical and geophysical implications. *Earth Planet. Sci. Lett.*, 15: 232—238.

Sofer, Z. and Gat, J.R., 1975. The isotope composition of evaporating brines: effect on the isotopic activity ratio in saline solutions. *Earth Planet. Sci. Lett.*, 26: 179—186.

Stewart, M.K., 1974. Hydrogen and oxygen isotope fractionation during crystallization of mirabilite and ice. *Geochim. Cosmochim. Acta*, 38: 167—172.

Stewart, M.K., 1975. Stable isotope fractionation due to evaporation and isotopic exchange of falling waterdrops: application to atmospheric processes and evaporation of lakes. *J. Geophys. Res.*, 80: 1133—1146.

Stewart, M.K. and Friedman, I., 1975. Deuterium fractionation between aqueous salt solutions and water vapour. *J. Geophys. Res.*, 80: 3812—3818.

Stichler, W. and Moser, H., 1979. An example of exchange between lake and groundwater. In: *Isotopes in Lake Studies*. IAEA, Vienna, pp. 115—119.

Stolf, R., De Menezes Leal, J., Fritz, P. and Salati, E., 1979. Water budget of a dam in the semi-arid Northeast of Brazil based on oxygen-18 and chlorine contents. In: *Isotopes in Lake Studies*, IAEA, Vienna, pp. 57—66.

Taube, H., 1954. Use of oxygen isotope effects in the study of hydration of ions. *J. Phys. Chem.*, 58: 523—528.

Top, Z. and Clarke, W.B., 1981. Dissolved helium isotopes and tritium in lakes: further results for uranium prospecting in Central Labrador. *Econ. Geol.*, 76: 2018—2031.

Torgersen, T., 1977. *Limnologic studies using the tritium-helium 3-tracer pair: a survey evaluation of the method*. Ph.D. Thesis, Columbia University of New York, N.Y.

Torgersen, T., Top, Z., Clarke, W.B., Jenkins, W.J. and Broecker, W.S., 1977. A new method for physical limnology. Tritium-helium-3 ages: results for Lakes Erie, Huron, and Ontario. *Limnol. Oceanogr.*, 22: 181—193.

Torgersen, T., Clarke, W.B. and Jenkins, W.J., 1978. The tritium/helium-3 method in hydrology. In: *Isotope Hydrology 1978*, Vol. II. IAEA, Vienna, pp. 917—930.

Truesdell, A.H., 1974. Oxygen isotope activities and concentrations in aqueous salt solutions at elevated temperatures: consequences for isotope geochemistry. *Earth Planet. Sci. Lett.*, 23: 387—396.

Vogt, H.J., 1976. *Isotopentrennung bei der Verdampfung von Wasser*. Staatsexamensarbeit, Universität Heidelberg, 78 pp.

Wang, J.H., Robinson, C.V. and Edelman, I.S., 1953. Self-diffusion and structure of liquid water III. Measurement of the self-diffusion of liquid water with ^2H, ^3H, and ^{18}O as tracers. *J. Am. Chem. Soc.*, 75: 466—470.

Weiss, W., 1979. Tritium and helium-3 studies in Lake Constance. In: *Isotopes in Lake Studies*. IAEA, Vienna, pp. 227—231.

Weiss, W., Fisher, K.-H., Kromer, B., Roether, W., Lehn, H., Clarke, W.B. and Top, Z., 1978. Gas exchange with the atmosphere and internal mixing of Lake Constance (Obersee). *Verh. Ges. Ökol., Kiel 1977*, pp. 153—191.

Weiss, W., Lehn, H., Münnich, K.O. and Fischer, K.H., 1979. On the deep water turnover of Lake Constance. *Arch. Hydrobiol.*, 86: 405—422.

Zimmermann, U., 1978. Isotopenhydrologie von Baggerseen: Bestimmung des unterirdischen Zu-bzw. Abfluss und der Evaporation mit Hilfe der naturlichen Deuterium-bzw. Sauerstoff-18 Gehalts des Seewassers. *Steir. Beitr. Hydrogeol.*, 30: 139—167.

Zimmermann, U., 1979. Determination by stable isotopes of underground inflow and outflow and evaporation of young artificial groundwater lakes. In: *Isotopes in Lake Studies*. IAEA, Vienna, pp. 87—94.

Zimmermann, U. and Ehhalt, D., 1970. Stable isotopes in the study of the water balance of Lake Neusiedl, Austria. In: *Isotope Hydrology 1970*. IAEA, Vienna, pp. 129—138.

Zimmermann, U., Baumann, U., Imevbore, A.M.A., Henderson, F. and Adeniji, H.A., 1976. Study of the mixing pattern of Lake Kainji using stable isotopes. *Catena*, 3: 63—76.

Zuber, A., 1983. On the environmental isotope method for determining the water balance components of some lakes. *J. Hydrol.*, 61: 409—427.

Chapter 4

ENVIRONMENTAL ISOTOPE AND ANTHROPOGENIC TRACERS OF RECENT LAKE SEDIMENTATION

W.R. SCHELL and R.S. BARNES

INTRODUCTION

A tracer is any substance which appears in the water-sediment system and which can be measured and used to trace normal biogeochemical processes. An effective tracer of lake sedimentation is one that makes it possible to follow the dynamic behavior of allochthonous and autochthonous particles by measuring the concentration distribution in space and/or time. Tracers are said to be environmental if they are not intentionally added by man for the purpose of a dynamic study; they are said to be artificial if they are deliberately added by man in the form of artificial radioactive tracers, activation isotope tracers, fluorescence tracers, and/or heavy metals tracers.

We shall use the term environmental tracers in this chapter to mean both the natural radioactive isotope tracers such as ^{210}Pb and ^{14}C as well as the global fallout radioactive isotope tracers produced by nuclear detonations such as ^{137}Cs and ^{239}Pu, ^{240}Pu. In addition to these radioactive tracers, we have included the heavy metals tracers such as Pb, Cd, and Cu which are environmental pollutants and have been widely dispersed; these tracers also can be considered as environmental tracers. To fulfill the criteria of a quantitative tracer of lake sedimentation, a substance must not undergo transformation when subjected to dynamic physical or chemical action and it must behave in the same way as the sediment when subjected to the same dynamic action.

The purpose of this chapter then is to collect, identify, and evaluate the available information on tracers of lake sedimentation and to illustrate, by example, applications of the tracer tool to solving geochemical problems in fresh water systems. These problems include cycling of nutrients and metals, reconstruction of the history of processes such as deforestation in the watershed, climatic changes and pollution as well as the sedimentation rates in lakes.

BIOGEOCHEMICAL CYCLING IN LAKES

The major pathways and reservoirs of materials cycling through the ecosystem are illustrated in Fig. 4-1 (after Oldfield, 1977). Materials are exchanged via a number of biological, chemical, and physical processes between the four major reservoirs consisting of the atmosphere, biosphere, hydrosphere, and lithosphere. The compartment or reservoir within which a particular material resides the longest is generally regarded as a "sink" for that material even though such a sink may be, on a geological time basis, a source as well. The water columns of lakes, as a subset of the hydrosphere are conventionally viewed as transient reservoirs for the materials which enter them; lake sediments on the other hand represent a lithospheric sink of considerable magnitude. At the present time, lakes form about 1% of the earth's continental surface and thus are important sites of sediment accumulation since they intercept much of the sediment transported by rivers (Collinson, 1978).

In discussing biogeochemical cycling of elements within a lake it is important to distinguish between those materials that originate within the lake itself — the autochthonous materials — and those that originate externally to the lake and are brought into the lake by advective or diffusive processes which are termed allochthonous materials.

Materials entering the water column are subjected to continual physical, and biological interactions with their surroundings until they are ultimately removed from the water column to a more permanent sink by flushing, evaporation or sedimentation. It is this last pathway to sediment that we will attempt to examine and quantify by means of isotopic and elemental tracer techniques in this chapter. Of all the removal processes, sedimentation is the

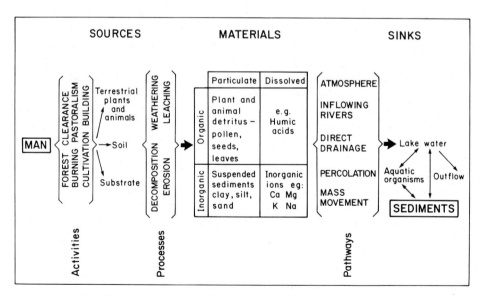

Fig. 4-1. A simplified partial model of lacustrine sedimentation in a lake-watershed ecosystem strongly influenced by man (Oldfield, 1977).

only one which can provide contemporary information as well as give some insight into the nature of past events which have left their imprints on the lake and its sediments.

The many complex and diverse chemical and biological interactions that materials undergo in the water column are beyond the scope of the present chapter and the reader is referred to Hutchinson (1957, 1975), Collinson (1978) and Lerman (1978). From the standpoint of sedimentation, these processes may be viewed as being a series of reservoir boxes from which sedimentation is the output quantity of interest. One's ability to relate such a simplified approach to the actual material dynamics in the water column itself is a difficult task; the success of the relationships derived depend markedly on both the specific tracer used and the nature of the lacustrine system which is studied. Using the reservoir box analogy it is, however, still possible to measure the rates of influx or efflux of a tracer, and by coupling this with direct measurements of the total quantity of the tracer within the box, to derive an average residence time for the tracer with the reservoir box.

ENVIRONMENTAL ISOTOPE AND ELEMENT TRACERS

Artificial changes in the environment due to man's activities have increased in recent years due to technological developments and to population pressures on the global ecosystem. Natural changes have occurred over geologic time due to climatic variations and to tectonic activity. Oceans and lakes provide repositories where samples can be collected to measure these changes through the use of certain tracers. An accurate time base is needed where a determination can be made of the rate at which the deposition of material has occurred. Environmental and artificial tracers can provide this time base.

Lake sediments preserve a record of processes which have occurred in the overlying water column as well as events which have occurred in the watershed. To reconstruct the history of processes such as deforestation, climatic changes and pollution, a knowledge of the sedimentation rate is necessary. This has been accomplished by counting varves, by palynological (pollen) methods or by independently dating some sediment anomaly (volcanic ash, landslides, glaciers). Radioactive dating of organic-rich sediment by the ^{14}C method has been used to give sedimentation rates over the last one to fifty millenia but does not extend to the present time (Stuiver, 1967).

Recent advances in measuring environmental radionuclides have given two promising methods for determining sedimentation rates over the past 150 years, the ^{137}Cs method (0—30 years), and the ^{210}Pb method (0—150 years). In addition, the input of trace metals and trace organic compounds as pollutants as a result of the industrial revolution have produced additional tracers which can also be used to trace the recent geochemical processes in lakes.

Application of the fallout radionuclides, particularly ^{137}Cs ($t_{1/2}$ = 30 years) to measuring recent sedimentation rates in lakes and reservoirs has been shown to be particularly useful (Pennington et al., 1973; Richey et al., 1973; Robbins and Edgington, 1975; Ashley and Moritz, 1979). In fresh water, cesium is preferentially adsorbed or fixed to the micaceous component of the sediment (Francis and Brinkley, 1976). Thus, the variation in ^{137}Cs content in the stratified layers can be compared with the local ^{137}Cs deposition record to determine the sedimentation rate. By comparing the fallout pattern for ^{137}Cs with the pattern measured in the sediment core profile, accurate dating of the core sections can be made (Lerman and Lietzke, 1975).

In contrast to bomb debris, the ^{210}Pb ($t_{1/2}$ = 22.35 years) tracer results from the radioactive decay of ^{226}Ra ($t_{1/2}$ = 1620 years) and ^{222}Rn ($t_{1/2}$ = 3.8 days) which emanates as a gas from the earth's crust and surface waters into the atmosphere. After formation in the atmosphere, ^{210}Pb becomes attached to the aerosol particles and subsequently is deposited on the earth's surface (Nevissi et al., 1974; Schell, 1977). This direct input of unsupported ^{210}Pb (i.e., the amount not in equilibrium with its parent ^{226}Ra) to the lakes and reservoirs is the result of wet and dry fallout onto these bodies of water. The ^{210}Pb dating technique has been applied primarily to the deposition of sediments in large lakes with relatively low sedimentation rates such as the Laurentian Great Lakes and to a few smaller lakes with high sedimentation rates. By using the ^{210}Pb sedimentation rates in conjunction with the profile of the trace metal and/or trace organic compound concentrations in the sediment, it is possible to calculate removal rates of trace constituents from the overlying water column. If trace metal concentrations in the overlying water column are also known, average trace metal residence times may also be calculated as discussed below.

Sample collection and processing

Because of the importance placed on the upper layers of sediment, the devices used for collecting these samples must be designed so that negligible mixing or losses occur in sampling. Hand-collected cores, piston cores, and box cores are probably the most reliable collection devices; gravity core sampling is a convenient collection method but may not necessarily give reliable samples. For soft lake sediments a piston corer designed by Donaldson (1967) has been found satisfactory; for marine sediments, the box corer has been used with success. A few gravity cores have been taken and used successfully when the water overlying the sediment in the core sample is clear, which indicates that little or no mixing has occurred. This latter method is not the recommended way to collect undisturbed cores.

After collection in the field the sediment cores are usually cut into 0.5-, 1.0- or 2.0-cm sections and placed in polyethylene bags. The best field procedure is to extrude cores slowly upwards removing the required thickness into a sample bag which is labeled and stored in a cold box to avoid bacterial decomposition until processed in the laboratory. For ^{137}Cs measurements, 25—50 g dry sediment is usually required for analysis by a calibrated low background Ge(Li) detection system. The gamma spectra obtained is also used for the analysis of ^{226}Ra, by counting its decay product, ^{214}Bi which is also required in the ^{210}Pb dating method. For ^{210}Pb measurements, a 2- to 3-g aliquot of each core section is spiked with 10.0- to 50.0-dpm ^{208}Po (a chemical yield tracer), leached with concentrated HNO_3 and HCl, and centrifuged to remove the undissolved sediment. The leachate is then evaporated to dryness and the residue converted to chlorides (Schell et al., 1973; Bruland, 1974). Alternatively, after the ^{208}Po tracer addition, the sediment sample is transferred into a quartz tube and heated to 550—600°C where the polonium is volatilized and condensed onto moist glass wool (Eakins and Morrison, 1976; El-Daoushy, 1978). The glass wool is then leached with HCl and made to 0.3 M HCl. In the former method, interferences in the counting of polonium caused by the presence of iron in the samples which also plates out are avoided by adding a few hundred milligrams of ascorbic acid to the solution to complex the iron. A silver disc, coated on one side with varnish, is placed in the solution causing the polonium to plate-out spontaneously. After eight hours, the plate is taken out of the solution, washed with distilled water and acetone and placed in a low background silicon surface barrier detector system for measuring the alpha particle spectrum.

The ^{137}Cs tracer method

Significant levels of ^{137}Cs were produced in nuclear weapons tests and have now been deposited on the earth as fallout. The maximum deposition occurs in both hemispheres at the latitudinal band of 30—60°. Since the tests were mainly in the northern hemisphere, the fallout of ^{137}Cs to the present time is approximately 4 times greater at comparable northern versus southern latitudes. The amount of ^{137}Cs deposited on the earth has been measured at precipitation stations at many locations and these data may be used to determine the cumulative input of cesium (see HASL reports). Cesium entering the terrestrial aqueous system has been found to sorb strongly onto clay particles. Adsorption to surface sites which are in exchange equilibrium with other cations also occurs with mixed sedimentary minerals. For illite and mica minerals in particular, cesium uptake occurs at lattice sites where particularly strong bonding occurs. In "hard" water systems competition between cesium and other cations takes place with cation substitution at the surface exchangeable sites; however, even in such high ionic strength water, the lattice sites still retain cesium. These bonding properties of cesium to minerals permits the use of this fallout produced radionuclide to trace sediment particles in a wide variety of systems.

An example of ^{137}Cs input is shown in Fig. 4-2 where estimates have been made for Lake Michigan, the English Lake district and for the Canadian West

Coast (Pennington et al., 1973; Ashley and Moritz, 1979). The profiles of ^{137}Cs deposition are shown for Lake Windermere, England and for Pitt Lake in the Fraser River delta, Canada, in Fig. 4-2. Both stations show the bomb fallout peak of 1963. The similarity in the integrated deposition of ^{137}Cs from the precipitation and from the sediment cores shows that the major input is at the lake surface rather than by deposition on the catchment basin with subsequent transport to the lakes. If redistribution and "focusing" in deep basins occurred, the integrated amount of ^{137}Cs would be greater in the sediment than in precipitation. The profiles also suggest that mixing by bottom fauna and/or by diffusion in the pore waters is limited. Two time periods can be derived from the measurements of ^{137}Cs in the core sections to determine sedimentation rates: (a) the depth where the 1963 peak is found, and (b) the depth where the ^{137}Cs is first detected, 1954.

The ^{210}Pb tracer method

The use of decay products of naturally occurring ^{226}Ra for sediment geochronologies was first outlined by Goldberg (1963). The decay chain of ^{226}Ra proceeds through the several radionuclides with short half-lives to long-lived ^{210}Pb as shown in Fig. 4-3. The basic principle of ^{210}Pb dating is that radon gas, ^{222}Rn, is emitted to the atmosphere from the lithosphere, surface waters, and airborne dust, and there decays to ^{210}Pb. This change in phase isolates radon from its radium precursor and destroys the initial secular equilibrium. A new secular equilibrium is then reached by the radon and its decay prod-

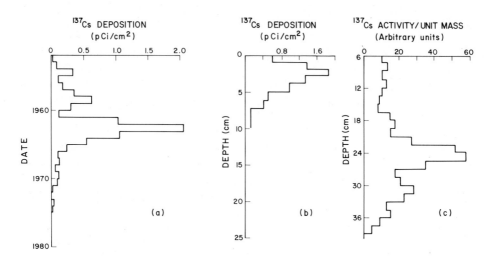

Fig. 4-2. Time history of ^{137}Cs: (a) input from bomb fallout; (b) deposition profile at Lake Windermere, England; and (c) deposition profile at Pitt Lake, British Columbia, Canada.

ucts with the longest-lived product ^{210}Pb becoming the predominant radionuclide. Radon remains in the atmosphere until it decays, whereas ^{210}Pb and other ^{222}Rn daughter products rapidly become attached to atmospheric aerosols and are removed by dry fallout and precipitation. It has been estimated that the residence time of ^{210}Pb in the lower atmosphere may be less than a week as compared to 30 days in the upper troposphere (Poet et al., 1972; Nevissi et al., 1974). This continuous source of ^{210}Pb provides a widespread flux to land and water surfaces.

Recent studies have shown that the global atmospheric distribution of ^{210}Pb at coastal and marine stations (Fig. 4-4) fits a distribution pattern in precipitation similar to nuclear weapons-produced ^{137}Cs and ^{90}Sr with mid-latitude peaks in the concentration (Schell, 1977). The terrestrial source (^{210}Pb) and the stratospheric source (^{90}Sr) have similar fallout patterns. Most atmospheric aerosols from either terrestrial or stratospheric sources are removed from the moist layer of the troposphere near the zonal belt of injection. However, terrestrial ^{210}Pb or its parent ^{222}Rn must somehow be transported through the lower moist layers into the upper dry layers of the troposphere for meridional transport and deposition at mid-latitudes. It is suggested that part of the ^{222}Rn which emanates from the land areas is mixed rapidly upward into the dry upper trosposphere/lower stratosphere and there decays to ^{210}Pb.

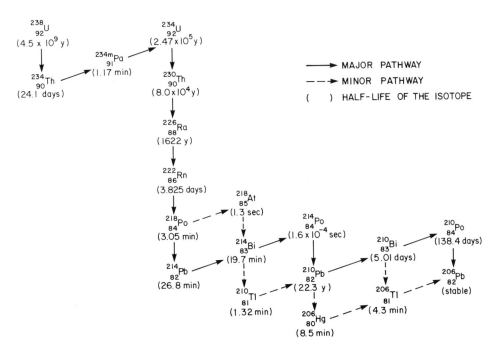

Fig. 4-3. Decay scheme for ^{210}Pb from its parent ^{238}U.

The deposition of ^{210}Pb on the surface ocean is reflected by the atmospheric input by precipitation and distribution as also shown in Fig. 4-4. Similarly, at lakes where prevailing winds are from over the ocean and where continental land masses are small, e.g., at islands and west coasts of continents, a similar latitudinal input from precipitation could be expected. However, at mid to eastern regions of continents, the atmospheric input of ^{210}Pb to lakes would be higher, more irregular and would depend strongly on local or regional terrestrial rock and soil compositions. This factor is shown in the atmospheric ^{210}Pb data summarized by Rangarajan et al. (1976) and by Krisnaswami and Lal (1978).

The predominant chemical forms of lead interacting with the terrestrial aqueous system, i.e., at pH 5.5—8.5, are PbO and PbCO$_3$. Organic-rich soil horizons are efficient in retaining ^{210}Pb from atmospheric precipitation which

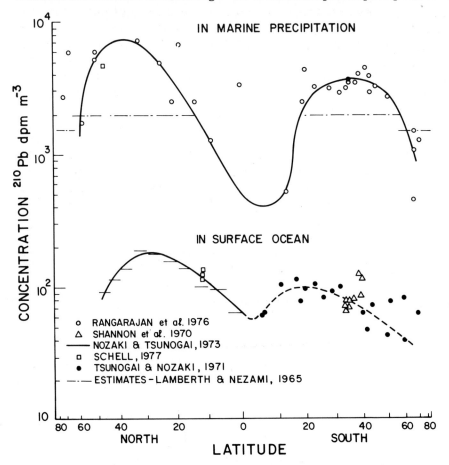

Fig. 4-4. Latitudinal distribution of ^{210}Pb in marine precipitation and in surface ocean waters (Schell, 1977).

subsequently may be transported to rivers mainly by soil erosion (Lewis, 1977). Upon entering the river systems, the ^{210}Pb is incorporated into suspended particulate matter and removed rapidly from the water (Goldberg, 1963). It is also removed rapidly in lakes and estuarine systems since it follows the natural detrital particles. However, most of the ^{210}Pb entering lakes comes directly from atmospheric input across the atmosphere-water interface and descends to the bottom with the suspended particulate matter.

Formation of sediments is a continuous process and the rates of both sediment and ^{210}Pb deposition may be constant over long periods of time. For dating purposes, it must be assumed that the ^{210}Pb is scavenged by suspended particulate matter and accumulated with the lake sediments. From a knowledge of the present atmospheric ^{210}Pb input to a region, the age of a sediment horizon which has been isolated from the contemporary input can be determined. By measuring the ^{210}Pb concentrations in several different sediment horizons, the rates of sediment accumulation can be calculated. However, because ^{210}Pb is also formed in situ from the small but finite amounts of terrigenous ^{226}Ra present in the sediment, this supported amount must be subtracted from the total concentration to obtain the unsupported ^{210}Pb. The unsupported ^{210}Pb represents the amount added to the water column from atmospheric sources.

The unsupported ^{210}Pb radioactivity in the sediment decreases by a factor of two every 22.35 years (the half-life of ^{210}Pb). When the logarithm of the unsupported ^{210}Pb activity (disintegrations per minute per gram, dpm g^{-1}) is plotted as a function of the total mass of dry sediment accumulated above a sediment core section, a log-linear relationship is found. Typical sedimentation profiles of ^{210}Pb concentrations with depths are shown in Fig. 4-5. This

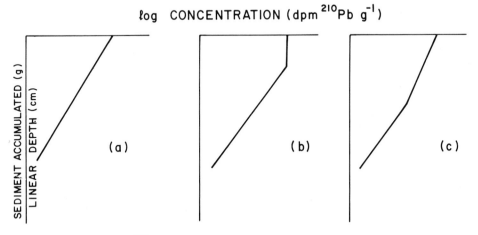

Fig. 4-5. Examples of ^{210}Pb concentration profile in lake cores showing the rhythmic accumulation of sediment: (a) simple lake system, (b) profile where biological mixing occurs in the upper layers, and (c) profile where the sedimentation rate has changed.

log-linear proportionality function represents the sedimentation rate only if the independent variable is the mass of overlying sediment deposited. This arises because of the compaction of the sediment due to the mass of overlying material which produces an approximately exponential decrease in porosity. Thus, the bulk sediment density increases with increasing depth in the sediment until a constant compaction is reached.

The linear sedimentation rate calculated from a plot of log ^{210}Pb concentration versus depth does not give the real particle-by-particle accumulation rate over time because of this compaction which must be corrected through measurements of in-situ density. The in-situ density is found by measuring the water content and the dry weight per unit volume, d, in each of the sediment core sections. The sediment porosity, P, (i.e., the fraction of the sediment volume occupied by voids which are filled with water) is determined by weighing the core sections before and after drying. Thus, the in-situ density ρ of each of the core sections is found by:

$$\rho = d(1 - P) \tag{1}$$

By using the density corrections, the real sedimentation rate can be calculated from a plot of log ^{210}Pb concentration versus sediment accumulation in g cm^{-2}.

Often the profiles of unsupported ^{210}Pb measurements obtained from the logarithmic plot of the unsupported activity as a function of the overlying mass of dry sediment accumulated show variations which are outside the analytical errors expected from the measurement of radioactive decay. Possible reasons for these discrepancies are:

(1) Disturbance or incomplete recovery of the upper layers of sediment by the coring device.

(2) In-situ mixing of the upper sediment layers due to burrowing organisms, gas bubbles, etc. (Fig. 4-5b).

(3) Post-depositional mobility.

(4) Presence of foreign material such as wood chips, pebbles, shells, etc.

(5) Actual changes in sedimentation rate (Fig. 4-5c).

(6) Non-equilibrium of ^{210}Po with ^{210}Pb in the upper segment of cores with high sedimentation rates.

The ^{210}Pb concentration in samples can be measured by alpha or beta and possibly by gamma, radioactivity counting. ^{210}Pb decays by emission of low-energy beta particles (0.061 MeV maximum) and gamma rays of 0.047 MeV, 4.1% into ^{210}Bi, as shown in Fig. 4-3, which decays by emission of high-energy beta particles (1.160 MeV maximum) into ^{210}Po which decays by alpha particles (5.305 MeV) to ^{206}Pb, a stable isotope. In principle, any of these three isotopes could be measured and the values related to ^{210}Pb since the half-lives of ^{210}Bi ($t_{1/2}$ = 5.0 days) and ^{210}Po ($t_{1/2}$ = 138 days) are short compared to ^{210}Pb. The beta counting methods are outlined in Bruland

(1974) and Koide et al. (1971), and the alpha counting procedures in Schell et al. (1973) and Schell (1977). The authors prefer the alpha counting methods since a chemical yield tracer is measured on each sample and the background counting rate for the silicon surface barrier detection system used in alpha detection is much less than that used for the Geiger or proportional counting of beta particles. This lower background permits greater sensitivity of the method.

The change in concentration of ^{210}Pb by decay with time, t, varies as:

$$-dN/dt = \lambda N \qquad (2)$$

where N = disintegration rate of ^{210}Pb, and λ = decay constant (0.0311 a^{-1})

By making the assumption that the ^{210}Pb once deposited with the sediment remains firmly fixed, the time of deposition can be determined from the decay. Thus, the activity of ^{210}Pb measured at any depth is a function of the ^{210}Pb deposited from the water column (unsupported ^{210}Pb) and the ^{210}Pb which is derived from ^{226}Ra (supported ^{210}Pb). Thus, the excess or unsupported ^{210}Pb is used to determine sedimentation rates of particulate matter from the water as $N_{excess} = N_{(total)} - N_{(from\ ^{226}Ra)}$, $N_t - N_{Ra}$ where "t" stands for total ^{210}Pb and "Ra" for ^{210}Pb coming from ^{226}Ra. The age at any depth, z, is found by integrating equation (2):

$$N_{(excess)} = N_t\, e^{-\lambda t} - N_{Ra}\, e^{-\lambda t} \qquad (3)$$

$$t_{(age)} = \frac{1}{\lambda} \ln(N_0/N) \qquad (4)$$

The sedimentation rate (s) is found from (4) by replacing t (age) by the quotient z/s and rearranging:

$$s = \frac{z\lambda}{\ln(N_0/N)} \qquad (5)$$

The average rate of input of ^{210}Pb, r, to a particular sampling location in a lake can be determined by integrating over its lifetime, the total unsupported ^{210}Pb, N (dpm g^{-1}) found in weighed sections of a core profile (g) of a given area A (cm^2) and by multiplying by the decay constant as:

$$r = (N/A) \cdot \lambda \text{ dpm g}^{-1} \text{ cm}^{-2} \text{ a}^{-1} \qquad (6)$$

Because of its short half-life, the unsupported ^{210}Pb is often found in lake cores at depths of less than 60 cm (a sedimentation rate of less than 0.3 cm a^{-1}); at this depth, samples can be collected by standard coring devices. Since most of the ^{210}Pb comes from atmospheric deposition onto the surface of the lake and the residence time of particulate matter which scavenges the ^{210}Pb is usually short, this average value is often the regional atmospheric ^{210}Pb input from terrestrial and atmospheric sources. This value can be compared with the amount collected from atmospheric rain/dust collectors and thus may be used as a first estimate of the regional ^{210}Pb input to the lakes, as will be discussed later.

Models

Profiles of unsupported ^{210}Pb activities in sediment cores may be used to establish sedimentary geochronologies by several different numerical models. Because the dates so derived are very model sensitive, it is necessary to consider carefully the applicability of each model to the experimental observations as well as the history of each watershed. There are two basic models to be considered: the constant initial concentration (CIC) model and the constant rate of supply (CRS) model. Both models assume that there is a constant input of unsupported ^{210}Pb to the lake and a constant residence time of ^{210}Pb in the lake water column, which, when taken together, amounts to saying that there is a constant flux of ^{210}Pb to the sediment. Additionally, both models assume that there is no migration of the ^{210}Pb within the sediment or loss from the sediment to the overlying water. At this point the two models diverge abruptly, as the CIC model makes the very crucial assumption that in each stage of sediment accumulation the *initial concentration* of unsupported ^{210}Pb was constant, regardless of any change in the rate of bulk sediment accumulation. The resulting corollary is that the unsupported ^{210}Pb profile must decrease monotonically with depth (time) in an undisturbed core. This implies that either the sedimentation rate remains constant, or that, if the rate of bulk sediment accumulation changes, the unsupported ^{210}Pb concentration changes in a same manner; this factor conflicts with the initial assumption of a constant flux of unsupported ^{210}Pb.

Despite such restrictions or inconsistencies in logic, the CIC model has been extensively employed (cf. Krishnaswami et al., 1971; Petit, 1974; Robbins and Edgington, 1975; Pennington et al., 1976). Awareness of the potential limitations of the CIC model has, however, led a few researchers, notably Appleby and Oldfield (1978) and Oldfield et al. (1978), to investigate the alternative CRS model. Although the CRS model, proposed originally by Goldberg (1963), is somewhat more complex, it is at least theoretically capable of handling change in bulk sedimentation rates so long as the rate of supply of unsupported ^{210}Pb to the sediment is constant. In comparing the two techniques, Oldfield et al. (1978) found that the CIC model was not compatible with ^{210}Pb profiles in sediments at sites which had experienced rapid acceleration in sediment accumulation in recent time. They also noted that in all cases where such an effect is believed to be present, the CIC model will underestimate the true age of the sediment below the onset of the acceleration in rate; the corollary of which is that the CIC model will overestimate the rate of sedimentation during periods of acceleration. Acceleration in sedimentation rates results in inflection points in the ^{210}Pb profiles when the log of the unsupported ^{210}Pb activity is plotted as a function of depth or dry mass of sediment accumulated. The unsupported ^{210}Pb profiles for the sediment cores from the lakes studied by Barnes et al. (1979) and Birch et al. (1980) clearly indicate that such inflections may be expected generally in lakes which have undergone any substantial degree of interaction with modern man.

The consequences of using the two different models to explain the physical process of net sedimentation, for example, are shown for sample core locations in Lake Washington (Figs. 4-6 to 4-9). Fig. 4-7 shows the concentration of unsupported ^{210}Pb as a function of depth and mass accumulated in the core profile; the scale for dry mass accumulated is a linear function

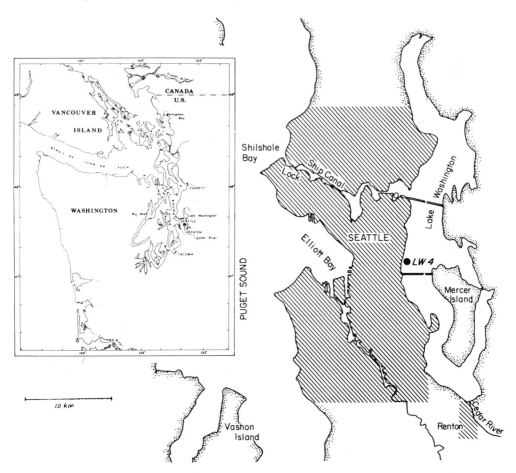

Fig. 4-6. Location of sampling station in Lake Washington.

whereas, because of porosity and possible changes in sedimentary material, the depth scale in the core is not linear. The average sedimentation rates using the CIC model have been determined for five time regions of core 104C.

Fig. 4-8 shows a histogram of the activity deposited in dpm (dpm ^{210}Pb g^{-1} of dry sediment in each 1-cm section) as a function of depth in (a) core 53C, collected in 1973 and (b) core 104C, collected in 1979 at the 63-m-deep

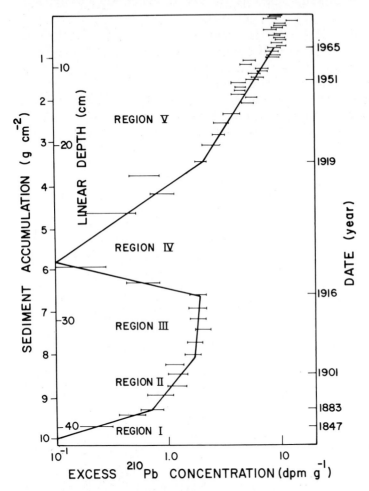

Fig. 4-7. Unsupported ^{210}Pb profile as a function of mass of sediment deposited at the 63-m deep station in Lake Washington. Measurements were made on each 0.5-cm section for 0 to 20 cm and each 1-cm section for the 20- to 40-cm depth intervals. Assuming the constant input of concentration of ^{210}Pb (CIC model) the age of the sediment layers were calculated from decay, the sediment deposited in each region calculated and the errors determined from "best fit" of the counting data by least-squares methods as: region I, 119 ± 18; region II, 525 ± 108; region III, 3940 ± 3139; region IV, 10,603 ± 1791; and region V 623 ± 44 g m^{-2} a^{-1}.

trench of the lake; the total ^{210}Pb deposited is the same for both cores at 245 dpm. For comparison with the CIC model, Fig. 4-9 shows the CRS model where the integrated deposition of ^{210}Pb in the cores from Fig. 4-8 is used. The total integrated deposition of ^{210}Pb is the numerical sum of the unsupported ^{210}Pb from 0 to 34 cm for 53C and 0 to 40 cm for 104C; i.e., to the depth where the cumulative activity is negligible. This total value is used as the initial input N_0; the value at any depth then, is the difference between the total input and the cumulative input at that depth, N. Using these values for N_0 and N, the age at that depth is calculated from equation (4). The instantaneous time sedimentation rates are then calculated from the grams deposited per interval of time (year) per unit area of the core for each of the sections and are shown graphically in Fig. 4-9.

The inflection points shown in Fig. 4-7 at 9.2, 8.1, 6.6, 5.9 and 3.5 g cm^{-2} *imply* changes in sedimentation rates between regions I, II, III, IV, V according to the standard, or CIC model interpretation. The average sedi-

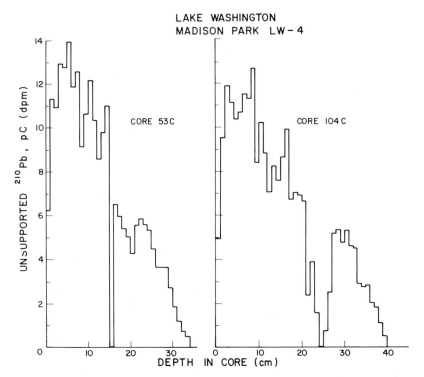

Fig. 4-8. Histograms of deposited activity of ^{210}Pb (dpm ^{210}Pb g^{-1} per gram dry sediment in each 1-cm section) with depth in two cores collected in 1973 (53C) and 1979 (104C) at the 63-m deep trench station in Lake Washington. The integration of the total unsupported ^{210}Pb from the two cores is the same even though the cores were collected and processed at different times.

mentation rates derived by the CIC model for regions I through V are compared with the point-by-point sedimentation rates obtained by the CRS model interpretations in Fig. 4-9. It can be seen that the CIC and CRS models give approximately the same sedimentation rates at the region I time interval. However, during the initial increase in sedimentation rates, region II, i.e. 1880—1905, the CIC model over-estimates the average sedimentation rates by a factor of 1.5 relative to the CRS model. From 1905 to 1916, region III, the CIC model over-estimates the average sedimentation rates by a factor of two relative to the CRS model. From 1916 to 1920, region IV, the CIC model gives a factor of 10 more than the CRS model. From 1920 to 1932, the CIC model gives a factor of 10 less than the CRS model. The two models compare favorably after 1935, region V. During the last time period the CRS model shows large variations (a factor of 2) in the sedimentation rates. The time of maximum sedimentation rate occurs between 1920 and 1926; small additional peaks of high sedimentation rates appear to be at 1947 and 1957.

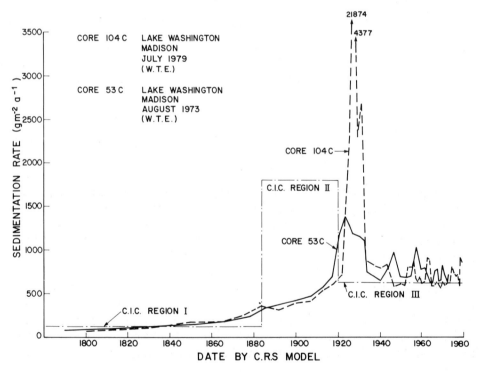

Fig. 4-9. Instantaneous sedimentation rate as a function of year of deposition determined using the CRS model for two cores collected in Lake Washington. The average sedimentation rates for the regions of core 104C determined by the CIC model is shown for comparison. It appears that the CRS model gives the time for the maximum sedimentation rate at 1926 or seven years later than that inferred by using the CIC model.

These instantaneous values may possibly represent input variations in ^{210}Pb or sediment but they cannot be observed with the CIC model. The reasons for the changes are discussed in the section "Comparisons of anthropogenic and ^{210}Pb dating of Lake Washington".

Trace element or trace organic compounds method

Certain trace elements or trace organic compounds also serve as tracers of lake sedimentation. Conversely, by measuring trace elements or trace organic compound concentrations and their time sequence in a sediment core profile, the pollution history of the lake can be reconstructed. The input of these trace constituents may be from both point and non-point sources. For example, the discharge of sewage effluent into a lake and the consequent input of a trace metal such as lead is a direct or point source, whereas atmospheric input of lead from the combustion of gasoline or from coal burning for home heating would be a diffuse or non-point source. Once in the water, the deposition of particulates containing the trace metals may be near the shore initially with subsequent resuspension by the lake currents to deeper zones where current transport is limited. It is in these permanent accumulation zones that the time history of the processes occurring in the water column can be inferred from measurement of sediment cores (Thomas, 1972; Aston et al., 1973; Förstner and Muller, 1974; Tschopp, 1977; Stumm and Baccini, 1978).

APPLICATIONS

Variation from the average deposition rates

Low surface values of ^{210}Pb found in profiles of marine sediments from Baja California, and the Santa Barbara Basin have been attributed to mobilization of ^{210}Pb; the low ^{210}Pb values found in Lake Tahoe and Lake Titicaca sediments were attributed to incomplete recovery of the upper sediment horizons (Koide et al., 1973). Later it was determined that the surface ^{210}Pb anomaly in the Santa Barbara Basin probably was due to a higher sedimentation rate occurring at the present time and the possibility of post-depositional migration of ^{210}Pb was unlikely (Krishnaswami et al., 1971). It has been suggested that ^{210}Pb mobility in Lake Michigan sediments was probably insignificant and that post-depositional redistribution by physical mixing or bioturbation could account for the observed anomalies (Robbins and Edgington, 1975).

In the ^{210}Pb lake sediment geochronologies reported by the aforementioned papers, it was possible to derive only a single average sedimentation rate over the last 80—100 years; no direct indication from the ^{210}Pb measurements that anthropogenic influences may have affected the sedimentation rates over this time interval were found. This failure to observe the fine structure of sediment accumulation, which is certainly possible in the sedimentation rate analysis, may be due to several factors including lake size and sampling frequency within the core. Large lake systems are inherently less sensitive to local watershed disturbances that might tend to alter sedimentation rates at the deep basin. This inherent sensitivity in observing changes in sedimentation rates is due to the distance between the sediment source and the deep basin as well as to the physical dispersion of a given sediment source over a relatively large sediment accumulating area. The sampling frequency in the core involves the vertical thickness of each sediment horizon being measured relative to the rate of sediment deposition at that location. In the case of Lake Mendota and Trout Lake, two of the smaller lakes examined, the thickness of the sediment sections was 5 cm and only every other 5-cm section was measured (Koide et al., 1973). If a minimum of three points is necessary to define an average sedimentation rate over a period of time, then it is evident that for these lakes the minimum time period for which a sedimentation rate could be obtained corresponds to the deposition of 25 cm of material, or 40 years at the reported sedimentation rate of 0.6 cm a^{-1} (Koide et al., 1973). In both cores from a pond in Belgium, and cores from Lake Biwa and Lake Shinji, Japan, measured sections were used whose thicknesses corresponded to as much as 40 years of deposition, and which at best, represented a 36-year average sedimentation rate (Petit, 1974, Matsumoto 1975a, b).

In larger lakes with generally lower sedimentation rates, an analogous problem is encountered even if the sediment core is dated at consecutive 2-cm intervals. In Lakes Titicaca and Tahoe, for example, the 2-cm incremental samples used represented about 60 years (Koide et al., 1973). Lake Michigan cores were measured at 1-cm intervals but with the observed linear sedimentation rate of 0.03—0.05 cm a^{-1}, three consecutive 1-cm sampling intervals still amounted to 60—100 years of deposition (Robbins and Edgington, 1975). Thus, no evidence of fine structure in the ^{210}Pb sedimentation rates of lakes based on ^{210}Pb was identified. Recent evidence for different sedimentation rates of lakes will be discussed below.

Comparisons of anthropogenic and ^{210}Pb dating of Lake Washington

The Lake Washington ^{210}Pb dating results are based on cores taken in the deep trench (62—65 m) off Madison Park by W.T. Edmondson (University of Washington). The location for samples is shown in Fig. 4-6. As is shown in Figs. 4-7 and 4-9, the sedimentation rate in Lake Washington began to increase rapidly about 1880. Between 1786 and 1882, the sedimentation rate averaged 119 g m^{-2} a^{-1} or about 0.6 mm a^{-1}. This linear rate is in agreement with ^{14}C results on sediment cores evaluated by Gould and Budinger (1958) which indicated a sediment accumulation rate of about 0.7 mm a^{-1} since the Mt. Mazama ash layer was deposited in the lake some 6700 years ago.

By 1900 the sedimentation rate was almost triple the natural background rate and still increasing. Between 1880 and 1900 the city of Seattle had grown from a small waterfront town on Elliot Bay, with a population of 3500 in 1880, to a large and very rapidly growing city of 81,000 in 1900. The suburbs had reached Lake Washington and were spreading along the lake's shore. This growth and development accelerated even more in the next decade when the population tripled to 237,000. By about 1915 the sedimentation rate had increased to 525 g m^{-2} a^{-1}. The end of the 1880—1915 interval marked a dramatic change in Lake Washington's watershed, hydrology and history. In 1916, the lake level was lowered 2.7 m with the opening the Lake Washington Ship Canal and the H.M. Chittenden U.S. Government Locks. Now the lake had a westward outlet; its former outlet was the Black River which passed through the city of Renton at the extreme southern end of the lake and into the Duwamish River. The Cedar River which had entered the Black River just below the lake was permanently diverted into the lake to provide water for operating the new locks. By this action, the hydraulic residence time in the lake was almost halved to 2.35 years (R.S. Barnes, unpublished data) and the fluvial sediments of a major river entered the lake. The estimated amount of sediment input from the river is 4-5 $\times$ 10^{7} kg a^{-1} (Crecelius, 1975). This input gives about 200 g m^{-2} a^{-1} at the deepest area of the lake, Madison Park — 63 m.

The lowering of the lake left its mark in the sediments as the former littoral sands were eroded back into the lake forming a silt band (Edmondson and Allison, 1970). This band is found at the 19- to 21-cm horizon in the Madison Park core (63 m deep), is present at depths of 12—18 cm in the central and northern profundal sediments and is found at greater depths in cores taken near the mouth of the Cedar River.

The silt band is primarily responsible for the very high sedimentation rates occurring between 1916 and 1920 which average about 10,603 g m^{-2} a^{-1}. Since that time the sedimentation rate at the Madison Park site has decreased to an average of about 623 g m^{-2} a^{-1}. Whether the severe anthropogenic eutrophication of the lake which culminated about 1963, or the subsequent large-scale sewage diversion (1963—1968), have effected the sedimentation rates cannot be determined at this time. Any direct organic contribution of effluent to the sediment is very slight since the sediment containing 5—7% carbon is primarily inorganic; however, indirect contributions to the sediment accumulation via an increase in productivity of silicous diatoms could be significant; a decrease in sediment accumulation from biogenic silica also might have resulted from the removal of nutrients by sewage diversion.

Based on the known cultural history of the area, the initial acceleration in sedimentation rate coincides with deforestation and land clearing. However, its duration and increase in intensity from 1900 to 1915, which was well after logging was completed, suggests that not only the initial deforestation resulted in higher sedimentation rates, but other activities following land clearing tended to prolong the erosion. The clearcutting practices in the watershed at that time were carried out primarily using oxen and cordoroy "skid rows", and later, railroads, neither of which are as land erosive as the extensive roadbuilding and ancillary activities associated with the more modern logging practices. The construction of roads for the automobile shortly after the turn of the century may have contributed to ultimately prolonging the

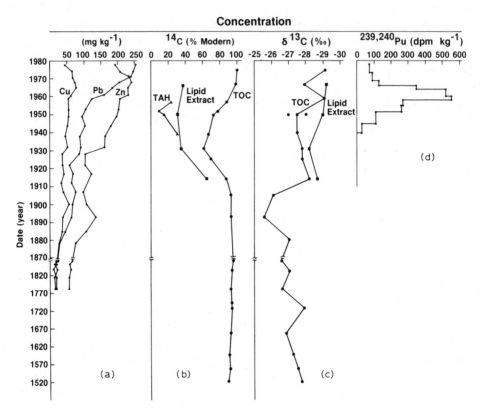

Fig. 4-10. Concentration profiles of trace metals, organic carbon and plutonium in Lake Washington as a function of year of sediment deposited at the 63-m deep station. (a) Concentrations of Cu, Pb, and Zn with time illustrating anthropogenic sources which contribute to the increases. (b) Changes in the organic ^{14}C concentration with time using 1950 wood as standard for 100% modern. The decrease from the natural biogenic level of ~93% modern (400 years old) before 1900 to the minimum in 1930 shows the fossil fuel dilution. TOC is total organic carbon and TAH is total aromatic hydrocarbon. (c) Profiles of the $^{13}C/^{12}C$ isotopic ratio as a function of depth based on the PDB carbonate standard of $\delta^{13}C$. The feature of increasing $\delta^{13}C$ values during the time period 1890—1915 may indicate the contribution of inorganic carbonates to the lake from land erosion at this time. (d) Concentrations of plutonium with time illustrating the fallout from nuclear weapons tests which deposited in the lake. The maximum in the early 1960s is at the correct time but the presence of plutonium below the 1952 layer shows mixing or diffusion.

high erosion rate of the surrounding soils which were deposited in the lake. The relatively uniform sedimentation rate since about 1932 suggests that stabilization between erosion and soil conservation in the watershed has been reached.

Trace metal profiles of lead, copper and zinc in Lake Washington sediments, shown in Fig. 4-10a, have been found by Spyridakis and Barnes (1978) to be similar throughout the central deep basin. These elements show increasing enrichment towards the surface of the sediment from about 1880 to 1968 when the copper and zinc input decreased in response to sewage diversion. The accelerating sedimentation rate during 1890—1922 did not appear to dilute greatly the actual zinc, copper, and lead concentrations in the sedimentary material. This suggests that the input sources of these metals either increased with the increasing sedimentation rate or that the higher land erosion brought proportionately increased levels of contaminated soil into the lake. The increase in lead, copper, and zinc concentrations on top of a relatively constant sedimentation rate since 1932 suggest that real fluctuations in anthropogenic sources occurred and that accelerated erosion prior to 1932 probably diluted these sources of metals.

Spyridakis and Barnes (1978) showed that three principal sources of lead to the lake over the past 100 years have existed. These are coal burning, the ASARCO smelter in Tacoma and the combustion of tetraethyl lead in gasoline since 1923. Coal and gasoline combustion are general area sources, whereas the smelter is a point source from which a plume dispersion by atmospheric processes occurs during the 35-km distance between the smelter stack and the lake. Because of the real point source identity, the smelter was assumed by Crecelius and Piper (1973) to be the dominant source term for lead between its construction in 1890 and its conversion to copper refining in 1913. Crecelius and Piper (1973) showed that the lake currently receives about 2×10^3 kg a^{-1} of arsenic (a by-product of copper smelting), all of which is due to the smelter. Of the current stack emission rate estimated at 3×10^5 kg a^{-1}, 30—40% is arsenic and 20—30% is lead. Thus, based on an As/Pb ratio of 7/5, the sediments of the lake must receive at least 1400 kg a^{-1} of lead from current smelter emissions. It was assumed that before 1913, even more lead was released to the atmosphere.

However, Spyridakis and Barnes (1978) presents historical information and identified that a stack height of 93.3 m rather than 173.7 m would reduce the long range transport before 1917 and suggested that coal combustion, rather than the smelter, was the major source of lead to the lake before 1917. Since coal has been mined extensively in the Puget Sound Basin and was used by industry and for home heating before the use of petroleum, the proposal has merit. Crecelius and Piper (1973) estimated that lead emissions from coal combustion were about 4×10^3 kg a^{-1} in the area in 1910. Using the difference in the values given by Kolde (1956) for coal production and for coal export in Washington as being locally consumed, and by assuming

50 mg g^{-1} lead in the coal, Barnes et al. (1979) found that the lead from coal combustion in fly ash could have amounted to 15 t in 1890, 31 t in 1900, 43 t in 1905, 65 t in 1910 and 75 t in 1918. Shapiro et al. (1971) found high concentrations of spherical and crystalline particles at depths of 16—24 cm at a station (SBP) located 4 km from the mouth of the Cedar River. These particles were rich in P (26%), Fe (29%), Si (33%), and Al (7%) which is consistent with fly ash from coal combustion. The change in energy use from coal to petroleum occurred rapidly after 1918. Since the chimneys were generally close to the ground for coal combustion, the many sources would introduce lead to the atmosphere, much like the automobile emissions of today. Thus, the lead enrichment of Lake Washington sediments before 1920 was thought to be due primarily to local coal combustion and not to the smelter.

After 1920 the lead enrichment was due primarily to the tetraethyl lead combustion by the automobile. Barnes and Schell (1973) show that the lead concentrations in Lake Washington sediments since 1920 increase in proportion to the gasoline consumption trends for King County. The concentrations of lead in lake sediments have been increasing at a rate of 6% per year from 1946 to 1972. Spyridakis and Barnes (1978) showed that during the gasoline shortage of 1973—1974, the lead concentrations in sediments decreased. Schell and Barnes (1974) also showed a lowering in the rate of lead concentration increase with time in the lake sediments which occurred between 1941 and 1945, i.e., during the time of World War II gas rationing.

Wahlen and Thompson (1980) using measurements of ^{137}Cs and ^{210}Pb in sediment cores from three lakes in New York State to fit the time history, estimated the magnitude of the change in the flux of trace metals and terrigenous materials due to man's impact on these watersheds. The increase in Pb from Lakes Champlain and Sylvan over the past 100 years was found to be 3.7 and 7.0 times, respectively; for Lake Canadarago a 20% increase in Pb input has occurred over the past 20 years. Littoral transport of sediment has been identified by the excess ^{210}Pb and ^{137}Cs inventories found in the cores over that expected from fallout onto the lake surface.

Other recent studies of the use of the tracers ^{210}Pb, ^{137}Cs, ^{239,240}Pu and ^{14}C and the CRS model in sediment chronology have been completed by Appleby et al. (1979), Battarbee et al. (1980), and Oldfield et al. (1980). In these studies man induced changes in sedimentation rates due to watershed erosion have been documented and the use of the CRS model appears to be better than the CIC model for determining the time when these changes occurred. Comparisons of the annual laminations of lake sediments from Finland with ^{210}Pb dates derived by using the CRS model provided information which has helped verify this interpretative method; the CIC model was found to be much less accurate than the CRS model using the data from these sediment cores with annual laminations (Appleby et al., 1979).

Anthropogenic changes in organic carbon input

One of the key concerns today is the introduction of large amounts of organic compounds into the environment, their sources and their effect on the ecosystem. The overall impact of these organic pollutants on a given region often cannot be determined by the methods of compound identification by separation and gas chromatography/mass spectrometry methods. It is not simply an identification of an increase in concentration over background as was shown previously for trace metals. The many compounds identified by the sophisticated instruments often are confused with compounds derived from natural oil seeps, coal dust, exometabolites from photosynthesis and organic degradation products from chemical and bacterial activity. Since many organic compounds are produced and decompose naturally, a means of identifying pollution chemicals from biogenic chemicals is needed.

Such a method has now been developed by Swanson in 1980 and tested on the chronology of organic pollutants in Lake Washington and Puget Sound. This carbon isotope analysis (CIA) method involves measurements of ^{12}C, ^{13}C, and ^{14}C in carbonaceous compounds found in sediment layers. The total organic carbon, total aliphatic hydrocarbons and polycyclic aromatic hydrocarbons are separated from kilogram quantities of sediments which have been dated by ^{210}Pb methods. The separation steps include freeze drying, extraction, fractionation, column chromatography, and evaporation. A large Soxhlet extraction apparatus for the sediment with ultrapure (nanograde) benzene-methanol solution is used where 97% of the total lipids and all hydrocarbons with boiling points higher than tetradecane (n-C_{14}) are extracted. Isolation of the aliphatic and aromatic hydrocarbon fractions is made by column chromatography using Sephadex and alumina/silica columns. The amount of each fraction recovered is determined using combustion and CO_2 measurements (Swanson, 1980).

The principle of the method uses measurements of the natural isotopic differences present in organic compounds derived from various sources. All modern carbon, i.e., less than 50,000 years of age, contains significant amounts of the isotope ^{14}C produced by cosmic ray interactions in the atmosphere. In old or fossil carbon, the ^{14}C has decayed ($t_{1/2}$ = 5730 years) and negligible amounts are present in organic compounds. All carbon that participates in photosynthesis now (since 1952) also contains and excessive amount of ^{14}C, i.e., greater than the cosmic-ray-produced level (13.6 dpm g^{-1}); this excessive ^{14}C has been produced by atmospheric thermonuclear explosions. The CIA method makes use of these levels of ^{14}C in organic compounds to develop a chronology and a relative anthropogenic to fossil signature of carbon compounds found in sediments. Measurements of ^{12}C and ^{13}C are used to identify sources of carbon since isotopic fractionation occurs in photosynthesis and decomposition processes of the carbon compounds (Deines, 1980). Thus, the source of the compound can often be distinguished by the

$^{13}C/^{12}C$ ratio measured. The CIA method makes use of these isotopic differences and the ^{14}C content in natural organic compounds (Swanson, 1980). By using CIA methods, additional insight into pollution can be obtained and the impact of chronic levels of toxicants can be identified.

An example of the CIA method is illustrated in Fig. 4-10 for (b) ^{14}C, and (c) for the $^{13}C/^{12}C$ ratio measurements (after Swanson, 1980). The ^{14}C profile of total organic carbon (TOC) with depth gives about 100% at the surface layers, decreases to 63% at 1930 then increases to a value of 90—95% modern at 30 cm and below. To interpret these data, three sources of ^{14}C must be imposed. Prior to 1880 the Lake Washington area was a pristine forest with no significant anthropogenic fossil carbon input. The ^{14}C in TOC of the sediments reflected the atmospheric or cosmic-ray-produced carbon in organic and inorganic detritus settling to the bottom. The 90—95% of modern illustrates that a small dilution of the contemporary produced carbon occurs, probably by a short-time storage (400—500 years) and/or by utilization of lake water CO_2 which originates partly from dissolution of carbonate rocks. At about 1905 a rapid dilution of contemporary ^{14}C in TOC occurred, reaching the greatest dilution of contemporary ^{14}C at about 1930. The source of old or fossil carbon then decreased gradually to the present as indicated by the approach to 90—95% modern ^{14}C. The minimum in the ^{14}C content of TOC at about 1930 corresponds to the greatest dilution by fossil carbon. It is expected that this time period corresponds to the maximum particulate carbon input from coal dust, coal burning or fly ash as shown in the previous section for the lead input. Subsequent to this time natural gas and/or petroleum products were burned which do not produce as many carbon particles for deposition. The increase in ^{14}C in the 1960s and 1970s is not necessarily due to the decrease in fossil carbon input but due to the excess environmental ^{14}C from atmospheric thermonuclear testing.

The upper layers contain ^{14}C in excess of the pre-industrial levels of 90—95% modern. This excess is due to the nuclear weapons tests which produced about 1 year equivalent of cosmic-ray-produced ^{14}C for each megaton of nuclear test. The first thermonuclear detonation which produced excess ^{14}C was in 1952. The increase in ^{14}C reached a maximum of about 80—100% excess ^{14}C in 1963 and 1964 (Nydal et al., 1979), and in 1980 the level has decreased to about 30% excess at this latitude. Thus, the upper layers of sediment would reflect this excess ^{14}C input to give higher ^{14}C values in organic matter deposited after 1952 as sediment. Any increase in the input of fossil carbon (dilution of biogenic carbon by pollution carbon) would not be accurately represented by the ^{14}C measurements. It would only be possible to calculate this dilution if comparable ^{14}C measurements were available, for example, on the surface sediments of a nearby lake which had not experienced pollution effects.

The extractable lipid fraction in Fig. 4-10b illustrates the dilution of biogenic carbon (with a ^{14}C content of 90—95% modern) by carbon compounds from pollution. The value at the 1915 depth is already contaminated by fossil carbon at 65% modern with the greatest contamination at the upper layers of 30—40% modern. The total aliphatic hydrocarbons and polycyclic aromatic hydrocarbons also show the contamination of biogenic carbon by fossil carbon compounds with values in the top 40 years of between 9 and 30% modern (Swanson, 1980).

The $^{13}C/^{12}C$ ratio in TOC (Fig. 4-10c) shows that values of about $-27^o/_{oo}$ are characteristic of the total organic carbon in sediment before man's impact in 1880 (Swanson, 1980). The $^{13}C/^{12}C$ ratio in organic carbon appears to have become heavier during the initial clearing of forests in 1880—1915 when large increases in the sediment deposition occurred. Above 1915 the $^{13}C/^{12}C$ ratio appears to be lighter (more depleted) up to a surface value of $-29^o/_{oo}$. These more depleted values in the upper region of the sediment layers may be due to increases in pollution chemical inputs or to processes of eutrophication in the water column. The extractable lipids measured in the upper layers also show this more depleted trend in $^{13}C/^{12}C$ ratios, although the data are limited. The trend toward lighter $^{13}C/^{12}C$ ratios with decreasing depth illustrates the fact that the lightest carbon compounds, e.g., asphalt and crankcase oil, have diluted the natural biogenic carbon isotopic compounds (Deines, 1980). The $^{13}C/^{12}C$ ratios in biogenic sources of carbon in plankton are all heavier than the surface sediment values: for example, Lake Washington plankton $-26^o/_{oo}$, terrestrial plants $-25^o/_{oo}$.

The concentration of the total aliphatic hydrocarbon (TAH) fraction extracted from the Lake Washington sediments has been measured. The present flux (surface sediment values) of 9×10^4 g m^{-2} a^{-1} is about 40 times greater than the flux in 1905, assuming no degradation (Swanson, 1980). If this flux was due only to fossil carbon, the ^{14}C content of TAH should be about 2% of modern. However, the value of 15—30% of modern ^{14}C illustrates that a significant part of the TAH comes from contemporary biogenic carbon and not from coal or oil burning.

The concentrations of the polycyclic aromatic hydrocarbon (PAH) fraction also have been measured. The present flux of 3×10^3 g m^{-2} a^{-1} is about ten times greater than the flux of PAH in 1905, assuming no degradation (Swanson, 1980). The ^{14}C content of 10% modern is about the expected level considering that all the excess PAH comes from fossil fuel (i.e., coal and oil).

The sum of the total aliphatic and polycyclic aromatic hydrocarbons comprise only about 14% of the total carbonaceous compounds in the sediment layers (Swanson, 1980). The major fraction, 86%, of the total organic compounds is non-extractable and is believed to be coal dust or fly ash. The input of coal dust has decreased from the late 1920s of 40% of the total organic carbon to between 5% and 20% at the present time. This change, which is recorded in the sediments, reflects the change in the local use pattern of energy consumption from a predominately coal to an oil-based economy. The previous atmosphere must have contained much more particulate carbon pollutants than at the present time. The quantitative amounts and the history of these contaminants may be found by further interpreting the sedimentary record.

Sediment and particulate redistribution

A comparison of ^{210}Pb, ^{137}Cs, and pollen geochronologies in Lakes Ontario and Erie has been made by Robbins et al. (1978). By using the ^{137}Cs concentration profile, the depth of surficial mixing has been identified clearly. The comparisons of mass sediment rate (g cm^{-2} a^{-1}) and linear sediment rates (cm a^{-1}) for five stations show excellent correlation for ^{210}Pb and ^{137}Cs and generally good correlations with the pollen Ambrosia counts. Anomalies of ^{210}Pb concentrations found at certain levels were correlated with some but not with all of the episodic major storm surges in the lakes. The low correlation between ^{210}Pb, ^{137}Cs and the pollen profiles for other stations in Lake Ontario would be explained by the removal of sediment at the station due to storm surges (post-depositional mobility; Edgington and Robbins, 1976a, b). It appears that small differences in the ^{210}Pb concentrations found in the individual core section of the sediment profile can give significant information on the redistribution of sediments at given stations. The reasons for the ^{210}Pb anomalies may be related to storm energy translated to the surface sediment. This energy would cause erosion in one area and deposition in another or transport of riverine material at the bottom as turbidity currents where material may settle out in regions of limited circulation. If the uppermost sediment layers were transported by these currents, then high ^{210}Pb concentrations of sediment should be advectively scoured at one area and deposited at another. The resulting ^{210}Pb profile of cores from the same lake would then include certain high and low values about a mean sedimentation rate depending on the amount of material transported. To interpret the data from these types of sediment cores, additional information would be needed on the meteorological parameters of storm surges, precipitation as well as wind speed and direction.

Mixing and transport rates in sediments

Diffusion rates for elements in sediment can be studied by using a combination of environmental radionuclides. By using the ^{210}Pb and the fallout radionuclide measurements in a core profile the diffusion rates for the fallout radionuclides can be estimated. The basic assumption used by Lerman and Lietzke (1975), is that the ^{210}Pb once scavenged by particulate matter and deposited as sediment remains fixed and does not diffuse. The time history of the profile can then be estimated. By measuring ^{137}Cs and ^{90}Sr in the sediment profiles of cores collected at Lakes Erie and Ontario, the concentration versus depth curve can be fitted by models and the diffusion rates estimated. Values of $2-4 \times 10^{-6}$ cm^2 s^{-1} (62—125 cm^2 a^{-1}) for ^{90}Sr and 2×10^{-5} cm^2 s^{-1} (628 cm^2 a^{-1}) for ^{137}Cs were found.

For ^{90}Sr the relative importance of sedimentation flux was related directly to the sedimentation rates. At Lake Erie where the sedimentation rate is high, the sedimentation flux of ^{90}Sr is more important than the diffusional flux. In Lake Ontario where the sedimentation rate is low, the diffusional flux is relatively much more important. Since ^{137}Cs is adsorbed much more strongly than ^{90}Sr, the sedimentation fluxes are important at all sites measured. In Lake Ontario the sedimentation and diffusional fluxes are comparable. In the central basin of Lake Erie, the sedimentation flux is much more important than the diffusional flux whereas in the western basin, the fluxes are in an intermediate position.

Recent measurements of the ^{239,240}Pu concentrations in sediments of Lake Washington, U.S.A., are shown in Fig. 4-10d (Schell et al., 1983); the ^{210}Pb sediment dating and accumulation rates are shown in Figs. 4-7, 4-8 and 4-9. The layers, dated by ^{210}Pb, fix the time period and the high concentrations of ^{239,240}Pu in the early 1960s confirm the chronology developed by the ^{210}Pb methods. However, the peak concentration of ^{239,240}Pu is found in the time interval 1958 to 1964 which is a few years before the maximum fallout peak for the Northern Hemisphere of 1963. This time difference may define the "error interval" for mixing of the upper sediment layers at this station. Such a mixing interval of 3—4 years may be the time the tracer spends in the upper flocule̊nt layer before final deposition as sediment particles. In addition, the ^{239,240}Pu also is found in small but significant concentrations below the 1952 layer when significant levels of radioactive fallout from nuclear weapons tests could have reached Lake Washington. The depth of penetration to the 1940 layer suggests that plutonium either may be mixed with sediment particles to a significantly deeper depth than three or four years or that part of the plutonium is not permanently fixed to the sediment particles but diffuses accross the deposited layers in the pore water. The diffusion coefficient for plutonium would necessarily be much greater than that of lead in these sediments. Plutonium can exist in four possible oxidation states in the aquatic environment: Pu (III), (IV) (V), and (VI). To explain this diffusion in the sediments, it is proposed that a portion of the plutonium initially deposited diffuses in the pore water in the form of complexes which have been produced by the organic and inorganic chemical reactions which occur during the process of sediment diagenesis.

A quantitative model for biological mixing in the upper sediment layers has been developed by Guinasso and Schink (1975) for the deep-ocean sediments. This time-dependent mixing model relates diffusion coefficients of a tracer and different mixing conditions. This model is an expansion on the work of Berger and Heath (1968), and fits observed data of sediment mixing collected in abyssal sediments. Peng et al. (1979) developed a numerical version of the Guinasso-Schink model to estimate the distribution of excess ^{210}Pb. They then apply the model to the distribution of ^{14}C in and below the mixed layer of sediment. Santschi et al. (1980) extends the multi-box sediment mixing model and applies it to the fluxes of plutonium from sediments measured in the New York Bight and Narragansett Bay waters. The sediment processes in the ocean have been reviewed by Krishnaswami and Lal (1978), Robbins (1978), Turekian et al. (1978), Berger and Johnson (1978) and Erlenkeuser (1980). These one-dimensional models were developed for mixing in deep oceans where the sedimentation rates are low. In lakes, where sedimentation rates are often high and biological mixing is minimal, the effects of both diffusion and sedimentation must still be considered. The change in radionuclide concentration, A, with time, t, at depth z, below the moving sediment interface is given by:

$$\frac{\partial}{\partial t}(\rho A) = \underbrace{\frac{\partial}{\partial z} \cdot K \frac{\partial}{\partial z}(\rho A)}_{\text{diffusive mixing}} - \underbrace{S \frac{\partial}{\partial z}(\rho A)}_{\text{sedimentation rates}} - \underbrace{\lambda(\rho A)}_{\text{radioactive decay}} \qquad (7)$$

where K is the diffusive mixing in cm^2 a^{-1}, ρ is the in-situ density in g cm^{-3} with depth and S is the sedimentation rates in cm a^{-1}. Steady-state conditions exist under boundary conditions which define the regions where the sediments consist of the upper mixed layer with constant mixing coefficient and the lower layer which is free of particle reworking processes. The equations governing the depth of distribution of a radionuclide in the two layers when K and S are constant are:

$$\frac{\partial A}{\partial t} = K \frac{\partial^2 z}{\partial z^2} - S \frac{\partial A}{\partial z} - \lambda A \qquad \text{for } 0 < z < z_m \qquad (8)$$

$$0 = -S \frac{\partial A}{\partial z} - \lambda A \qquad \text{for } z > z_m \qquad (9)$$

whose solution is equation (5). The solutions for these equations are exponential in nature and the radionuclide concentration versus depth, by itself is not a guarantee for undisturbed particle by particle accumulation (Krishanaswami and Lal, 1978). The effects of mixing, horizontal and vertical diffusion and sediment accumulation on the radionuclide distribution in the sediment column must be determined to obtain satisfactory sediment accumulation rates. If mixing and sedimentation govern the radionuclide distribution, then the sediment accumulation rate computed from the log A versus z plot, without correcting for mixing effects, will be an upper limit. The mixing effects can best be determined by measuring several short- and long-lived tracers in the sediment profile. Johansen and Robbins (1977) and Krezoski and Robbins et al. (1977) have reported evidence that biological mixing to depths of 10 cm occurs in Lake Huron using ^{210}Pb and ^{137}Cs tracers. Nozaki et al. (1977) obtained particle reworking rates in the deep seas using ^{210}Pb. Demaster (1978) utilized ^{32}Si (~100-year half-life) and ^{210}Pb to measure particle mixing rates in the Gulf of California. The use of ^{32}Si in lakes for determination of mixing has not yet been tried. For slow sedimentation rates such as those in the deep ocean, profiles of ^{14}C, ^{230}Th, ^{231}Pa, ^{10}Be could be used to define mixing, diffusion and sediment accumulation rates.

Residence times and geochemical balance

To establish a geochemical balance, the input from atmospheric deposition must be equal to the output by the sediments plus the amount lost by fluvial transport. For closed lakes and those with long water residence times, the fluvial output of ^{210}Pb is negligible. This geochemical balance for the tracer ^{210}Pb can then be utilized in studying the dynamics of nutrients and trace element cycling in lakes compared to ^{210}Pb. For any element, x, one needs to measure the average concentration in the water column, the concentration in the sediment profile and the sedimentation rate to obtain the residence time for that element in the lake. The residence time, τ, is determined by measuring the rate of input or output and the concentration in the water column as:

$$\tau = N \cdot \frac{1}{dN/dt} = \text{dpm dpm}^{-1} \text{ a} = \text{a} \tag{10}$$

It has been assumed that both ^{210}Pb and stable lead are rapidly scavenged from the water column and serve as tracers of sinking particulate material. This assumption implies that lead, ^{210}Pb and suspended particulate matter have similar residence times in the water column. The assumption may be tested if the sediment rate, water column concentrations and surface sediment concentrations of the tracers are known (Barnes et al., 1979). The mean residence time, τ, of a tracer in the water column is:

$$\tau = c_w/r \tag{11}$$

where c_w is the concentration of the tracer in the water column per unit area, dpm cm^{-2} or μg cm^{-2} defined by:

$$c_w = z \cdot A' \tag{12}$$

where z is the water column height in cm and A' is the respective tracer concentration in dpm cm^{-3} or μg cm^{-3}. The rate of deposition, r, or flux of the tracer to the sediment is given in dpm cm^{-2} a^{-1} or µg cm^{-2}. By:

$$r = c_s \cdot S' \tag{13}$$

where c_s is the concentration of the tracer in the surface sediment in dpm g^{-1} or µg g^{-1} and S' is the sedimentation rate in g cm^{-2} a^{-1}. If the sediment density profile and wet to dry weight ratio of the sediment is measured, then S' can be calculated from linear sedimentation rates.

The residence time of ^{210}Pb and stable lead in three lakes in Washington State U.S.A., may be calculated using equation (11) and the data given in Table 4-1. Because some unknown small fraction of the stable lead and ^{210}Pb may be present in the soluble form in the water column, and because the water column values are given for the total metal, the resulting residence times must be taken as a maximum estimate. By using the maximum depth of each lake for the height of the water column, z, the estimate of the residence time also tends to be maximized.

The calculated mean residence time of ^{210}Pb in the Lake Washington water column is approximate since the average water concentration of ^{210}Pb also is not accurately known over the year. ^{210}Pb measurements of water samples collected periodically in the water column over a 2-year period indicate a mean concentration of 25 dpm m^{-3} with a maximum of 40 dpm m^{-3}; thus, the residence time of 72 days is probably a more typical value. The almost identical residence time estimate for ^{210}Pb and stable lead in Lake Washington suggests that the two tracers do behave similarly in the water column and that their reactions with particulate matter are similar.

Independent estimates of particulate matter residence times can be made by considering that the Lake Washington sediment is a gyttja consisting of algal remains, clays, and fine silts. Using Stoke's Law and assuming that: (a) the clay and fine silt particles 2—5 µm in diameter, have a specific gravity of 2.65 g cm^{-3}, (b) the average water temperature is 10°C, and (c) the particles are not affected by mixing, then the residence times would be 40—260 days in the 6200-cm water column. Algal settling rates of 0.3—3.0 m per day have been measured in the nearby Lake Sammamish using sediment traps; the rates depended on season and algal species (Birch et al., 1980). On this basis algal residence times of 20—200 days are reasonable for Lake Washington. Thus, since the residence times of the clay, silt, and the algae particles have a wide range but are comparable to those estimated for ^{210}Pb and stable lead, the assumption that Pb and ^{210}Pb serve as tracers of suspended materials is valid.

With water column residence times in the smaller lakes of 100—200 days, significant amounts of ^{210}Pb, stable lead and suspended particulate matter may be lost due to flushing of lakes with hydraulic residence times of less than 1 year. The flux of ^{210}Pb to the sediments of a given lake is also dependent on the ratio of the lake surface over which the unsupported ^{210}Pb is entering, to the actual area over which sediment is being permanently accumu-

TABLE 4-1

Evaluation of the particulate flux to the bottom and mean residence time of particulate matter in Lake Washington, Lake Sammamish, and Chester Morse Reservoir (Barnes et al., 1979)

(a) Relevant characteristics of the study lakes and their sediments

Lake	Elevation[a] (m)	Mean depth, $\bar{z}$ (m)	Maximum depth (m)	Lake surface (km^2)	Catchment (km^2)	Volume (m^3)	Mean annual precipitation (cm)	Hydraulic residence time (a)
Chester Morse Reservoir	474	18.8	35.4	6.81	217	1.28×10^8	265	0.25
Sammamish	8.5	16.6	30.5	19.8	253[b]	3.28×10^8	115	1.7
Washington	4.3	32.9	65.	87.6	1560[b]	2.88×10^9	88	2.35

(b) Parameters used in estimating the sediment flux and mean water column residence times of ^{210}Pb and stable lead

Parameter	Lake Washington		Lake Sammamish	Chester Morse Reservoir
	^{210}Pb	stable lead	stable lead	stable lead
Water column height, z (cm)	6200	6200	3050	3540
Concentration in water, c_W ^{210}Pb (dpm cm^{-3})	$2.5-4.0 \times 10^{-5}$			
Pb[c] (μg cm^{-3})		$0.7-1.0 \times 10^{-3}$	0.65×10^{-3}	0.27×10^{-3}
Concentration in surface sediment, C_S	12.3 dpm g^{-1}	320 μg g^{-1}	56 μg g^{-1}	39 μg g^{-1}
Sedimentation rate, q (g cm^{-2} a^{-1})	0.0644	0.0644	0.0610	0.0360
Flux, r	0.79 dpm cm^{-2} a^{-1}	20.6 μg cm^{-2} a^{-1}	3.42 μg cm^{-2} a^{-1}	1.4 μg cm^{-2} a^{-1}
Integrated water column concentration per unit area	0.155–0.248 dpm cm^{-2}	4.34–6.2 μg cm^{-2}	1.98 μg cm^{-2}	0.96 μg cm^{-2}
Mean water column residence, τ (days)	72–114	77–110	211	250

[a] Mean sea level datum.
[b] 1073 km^2 prior to 1916; consists of 451 km^2 direct drainage, 622 km^2 of the Sammamish River drainage including 253 km^2 tributary to Lake Sammamish.
[c] Data from Barnes et al. (1979).

lated. Thus, for Lake Washington where the hydraulic residence time (τ_w = 2.35 years) is sufficiently long so that almost all the entering lead will reach the sediments, then the ratio of proportionality would be:

$$\frac{^{210}\text{Pb flux to sediments}}{^{210}\text{Pb flux to lake surface}} = \frac{\text{lake surface area}}{\text{permanent sediment area}}$$

Since the numerators of both terms are readily determined it is possible to determine either denominator if the other is known. Using the estimate of 50 km² (Spyridakis and Barnes, 1978) for the permanent sediment accumulating area of Lake Washington and the data in Table 4-1a, b, the unsupported ^{210}Pb flux to the lake surface is:

$$(50 \text{ km}^2/88 \text{ km}^2) \times (0.79 \text{ dpm cm}^{-2} \text{ a}^{-1}) = 0.45 \text{ dpm cm}^{-2} \text{ a}^{-1}$$

This estimate is in agreement with the estimate obtained from using the concentration of ^{210}Pb in marine precipitation at this latitude multiplied by the average precipitation (Fig. 4-4):

$$(4.7 \times 10^{-3} \text{ dpm cm}^{-3}) \times (88 \text{ cm a}^{-1}) = 0.41 \text{ dpm }^{210}\text{Pb cm}^{-2} \text{ a}^{-1})$$

This value can be compared with the measured value of 0.44 dpm ^{210}Pb cm^{-2} a^{-1} determined by integrating monthly bulk precipitation samples collected near the shore of Lake Washington over a 5-year period (A. Nevissi, personal communication, 1980).

Krishnaswami et al. (1980) measured the short-lived ($t_{1/2}$ = 53 days) cosmic-ray-produced ^{7}Be in sediments of Lake Whitney and found the expected steady-state inventory of 5 dpm cm^{-2}. However, the inventories of the long-lived tracers, namely, ^{210}Pb and ^{239,240}Pu were found to be increased by a factor of 2—3 in the core. They attributed these differences to lateral transport from the watershed and/or focusing of the sediment in the lake over a long time period.

Several recent studies have been made using fallout, natural radionuclides and artificial radionuclides in lakes (Lerman 1978; Hesslein et al., 1979; Stiller, 1979). As Lerman (1978) shows, a mass balance for a radionuclide tracer in a lake system can be written to relate the different rates of input, removal and the rate of change in lake water by:

change in lake water — input = outflow + decay + deposition + diffusion to sediments

Each of the removal fluxes on the right-hand side of the balance equation can be represented as a product of the radionuclide concentration c (pCi m^{-3}) and a velocity term u (m s^{-1}):

Removal flux = $c \cdot u$ pCi m^{-2} s^{-1}

For example, ^{137}Cs is strongly sorbed to sediments, especially clays, and its rate of removal through deposition is higher than the same rate for ^{90}Sr with a similar half-life. Once deposited, an element may be reintroduced in the overlying water by desorption processes, often dominated by bacterial exometabolites or other organic molecule compounds which can chelate the radionuclide tracer. Thus, the chemical speciation often dominates the concentration of radionuclides in lakes. Plutonium is removed rapidly to the sediments in the Great Lakes and its concentration in the water remains low (Edgington and Robbins, 1975). However, in Mono Lake, California, which has high carbonate and salt concentrations, the plutonium in the water column is 10^2 times higher (Simpson et al., 1980).

Labeling of a whole lake (area of 260,000 m^2) with the radionuclides ^{60}Co, ^{54}Mn, ^{59}Fe, ^{134}Cs, ^{203}Hg, ^{65}Zn, ^{51}Cr, ^{75}Se, ^{133}Ba, ^{228}Th, ^{74}As and ^{48}V has been made by Hesslein et al. (1979). The soluble and particulate fractions were measured as a function of time as was plankton, periphyton, macrophytes, crayfish, trout, white sucker, slimy sculpin and fathead minnows. The half-times for removal of the radionuclides from the water column range from 12 days for ^{59}Fe to 47 days for ^{75}Se. All the tracers in the whole lake experiment show similar short removal times but the mechanisms for removal are not the same. Some tracers are lost by settling of particles to which they have adsorbed (e.g., ^{59}Fe, ^{60}Co, ^{203}Hg) and others by direct adsorption to sediments (e.g., ^{134}Cs). The data is just now being evaluated and much more insight into the biogeochemistry of trace elements in lakes is expected (Hesslein et al., 1979).

SUMMARY AND CONCLUSIONS

Contemporary and long time-period tracers of sediment in lakes have been identified and used in studies of biogeochemical cycling of materials. The natural tracers ^{210}Pb, ^{14}C, and the bomb-produced tracers ^{137}Cs, ^{14}C, ^{239,240}Pu have been valuable in increasing our understanding of sedimentary processes in lakes. Changes in the anthropogenic sources of trace metals and trace organic compounds have been identified over the past 50 years using, in part, the ^{210}Pb and ^{137}Cs dating chronology of cores. The mixing and/or diffusion of trace constituents in the upper layers of sediment have been determined on accurately dated cores from lakes.

To evaluate the data on sedimentation rates two models are used, the constant initial concentration (CIC) and the constant rate of supply (CRS) of the ^{210}Pb. In areas where land use practices in the watershed have changed the erosion rate, the CRS model must be utilized to explain the profiles. Changes in land use practices have been demostrated by ^{210}Pb in Lake Washington and independently correlated with the times of anthropogenic events. The trace metal, namely lead, input has been correlated with coal combustion between 1890 and 1920 and with leaded gasoline combustion after 1923.

The organic compound input has been shown by ^{14}C measurements also to correlate with increasing use of coal after 1905 to a maximum use in 1930. Subsequently, the use of petroleum rather than coal as an energy source reduced the dilution of atmospheric ^{14}C after 1930. The $^{13}C/^{12}C$ ratio also indicates the fossil fuel component. In addition to the natural organic compounds transported to sediments of Lake Washington, storm-induced redistribution of shallow sediments and transfer to deeper basins for permanent deposition has been identified using the ^{210}Pb tracer. Because of a "constant input" of ^{210}Pb onto the surface of the lake, the strong sorption onto suspended sediments in the water column and the permanent deposition on the bottom, measurements of the mean residence times within the water column can be made. These residence times can be compared with elements which are reactive in the biological cycle and the cycling time for nutrients such as phosphorous can be determined. ^{14}C has been used as a tracer for long time (10,000 years) sedimentary processes. Changes in climatic conditions have occurred over this time period and has effected strongly the sediment accumulation rates in lakes using the tracer method.

REFERENCES

Appleby, P.G. and Oldfield, F., 1978. The calculation of ^{210}Pb dates assuming a constant rate of supply of unsupported ^{210}Pb to the sediment. Catena, 5: 1—8.

Appleby, P.G., Oldfield, F., Thompson, R., Huttunen, P. and Tolonen, K., 1979. ^{210}Pb dating of annually laminated lake sediments from Finland. Nature, 280: 53—55.

Ashley, G.M. and Moritz, L.E., 1979. Determination of lacustrine sedimentation by radioactive fallout (^{137}Cs), Pitt Lake, British Columbia, Canada. Can. J. Earth Sci., 16: 965—970.

Aston, S.R., 1973. Mercury in lake sediments: a possible indicator of technical growth. Nature, 241: 450—451.

Barnes, R.S. and Schell, W.R., 1973. Physical transport of trace metals in the Lake Washington watershed. In: M.G. Curry and G.M. Gigliotti (Editors), Cycling and Control of Metals. National Environmental Research Center, Cincinnati, Ohio, pp. 45—53.

Barnes, R.S., Birch, P.B., Spyradakis, D.E. and Schell, W.R., 1979. Changes in the sedimentation histories of lakes using lead-210 as a tracer of sinking particulate matter. In: Isotope Hydrology 1978. IAEA, Vienna, pp. 875—898.

Batterbee, R.W., Digerfeldt, G., Appleby, P. and Oldfield, F., 1980. Paleoecological studies of the recent development of Lake Växjösjön, III. Reassessment of recent chronology on the basis of modified ^{210}Pb dates. Arch. Hydrobiol., 89: 440—446.

Berger, W.H. and Heath, G.R., 1968. Vertical mixing in pelagic sediments. J. Mar. Res., 26: 134—143.

Berger, W.H. and Johnson, R.F., 1978. On the thickness and the ^{14}C age of the mixed layer in deep sea carbonates. Earth Planet. Sci. Lett., 41: 223—227.

Birch, P., Barnes, R.S. and Spyradakis, D.E., 1980. Recent sedimentation and its relationship with primary productivity in four western Washington lakes. Limnol. Oceanogr., 25: 240—247.

Bruland, J.W., 1974. Lead-210 geochronology in the coastal marine environment. Ph. D. Dissertation, University of California, San Diego, Calif., 106 pp.

Collinson, J.D., 1978. Lakes. In: H.G. Reading (Editor), *Sedimentary Environment and Facies*. Elsevier/North-Holland, New York, N.Y., pp. 61—79.

Crecelius, E.A., 1975. Arsenic geochemical cycle in Lake Washington. *Limnol. Oceanogr.*, 20: 441—451.

Crecelius, E.A. and Piper, D.Z., 1973. Particulate lead contamination records in sedimentary cores from Lake Washington, Seattle, WA. *Environ. Sci. Technol.*, 7: 1053—1055.

Deines, P., 1980. The isotopic composition of reduced organic carbon. In: P. Fritz and J. Ch. Fontes (Editors), *Handbook of Environmental Isotope Geochemistry, Vol 1. The Terrestrial Environmental*, A. Elsevier, Amsterdam pp. 329—406.

Demaster, D.J., 1978. ^{210}Pb, ^{32}Si and ^{14}C Chronologies in a varved Gulf of California sediment core. *Trans., Am. Geophys. Union*, 59: 1118.

Demaster, D.J. and Cochran, J.K., 1977. Rates of particle mixing in deep-sea sediments using ^{210}Pb measurements. *Trans., Am. Geophys. Union*, 58: 1154.

Donaldson, J.R., 1967. *The phosphorous budget at Illiamna Lake, Alaska, as related to the cyclic abundance of sockeye salmon*. Ph.D. Dissertation, University of Washington, Seattle, Wash., 141 pp.

Eakins, J.D. and Morrison, R.T., 1976. United Kingdom Atomic Energy Authority, Harwell, AERE-R9475.

Edgington, D.N. and Robbins, J.A., 1975. The behavior of plutonium and other long-lived radionuclides in Lake Michigan, II. Patterns of deposition in the sediments. In: *Impacts of Nuclear Releases into the Aquatic Environment*. IAEA, Vienna, pp. 245—260.

Edgington, D.N. and Robbins, J.A., 1976a. Pattern of deposition of natural and fallout radionuclides in the sediments of Lake Michigan and their relation to limnological processes. In: J.O. Nriagu (Editor), *Environmental Biogeochemistry, Vol. 2*. Ann Arbor Science Publishers, Ann. Arbor, Mich., pp. 705—728.

Edgington, D.N. and Robbins, J.A., 1976b. Records of lead deposition in Lake Michigan sediments since 1800. *Environ. Sci. Technol.*, 10: 266—274.

Edmondson, W.T. and Allison, D.E., 1970. Recording densitometry of X-radiographs for the study of cryptic laminations in the sediment of Lake Washington. *Limnol. Oceanogr.*, 15: 138—144.

El-Daoushy, M.F.A.F., 1978. The determination of ^{210}Pb and ^{226}Ra in lake sediments and dating application. *UUIP- 979:* 8.

Erlenkeuser, H., 1980. ^{14}C age and vertical mixing of deep sea sediments. *Earth Planet. Sci. Lett.*, 47: 319—326.

Förstner, U. and Muller G., 1974. *Schwermetalle in Flüssen und Seen*. Springer-Verlag.

Francis, C.W. and Brinkley, G.S., 1976. Preferential adsorption of ^{137}Cs to micaceous minerals in contaminated freshwater sediment. *Nature*, 260: 511—513.

Goldberg, E.D., 1963. Geochronology with ^{210}Pb. In: *Radioactive Dating*. IAEA, Vienna, pp. 121—131.

Gould, H.R. and Budinger, T.F., 1958. Control of sedimentations and bottom configuration in convection currents, Lake Washington. *J. Mar. Res.*, 17: 183—198.

Guinasso, N.L., Jr. and Schink, D.R., 1975. Quantitative estimates of biological mixing in abyssal sediments. *J. Geophys. Res.*, 80: 3032—3043.

Hesslein, R.H., Schindler, D.W., Broecker, W.S. and Kipput, G., 1979. Fates of metal radiotracer additions in experimental enclosures in lakes and in a whole lake. In: *Application of Nuclear Techniques to the Study of Lake Dynamics*. IAEA, Vienna, pp. 261—271.

Hutchinson, G.E., 1957. *A. Treatise on Limnology, Vols. I and II*. J. Wiley, London, 1015 pp.

Hutchinson, G.E., 1975. *A Treatise on Limnology, Vol. III*. J. Wiley, New York, N.Y., 660 pp.

Johansen, K.A. and Robbins J.A., 1977. Fallout cesium-137 in sediments of southern Lake Huron and Saginaro Bay. *Proc. 20th Conf., Great Lakes Research* (abstract)

Koide, M., Soutar, A. and Goldberg, E.D., 1971. Marine geochronology with ^{210}Pb. *Earth Planet. Sci. Lett.*, 14: 442—446.

Koide, M., Bruland, K.W. and Goldberg, E.D., 1973. ^{228}Th/^{232}Th and ^{210}Pb geochronologies in marine and lake sediments. *Geochim. Cosmochim. Acta*, 37: 1171—1184.

Kolde, E.J., 1956. From mine to market — a study of production, marketing, and consumption of coal in the Pacific Northwest. *Bur. Bus. Res., Coll. Bus. Admin., Univ. Washington, Occas. Pap.*, 3.

Krezoski, J.R. and Robbins, J.A., 1977. Radioactivity in sediments of Great Lakes: post depositional redistribution by deposit-feeding organisms. *Proc. 20th Conf., Great Lakes Research* (abstract).

Krishnaswami, S. and Lal, D., 1978. Radionuclide limnochronology. In: A. Lerman (Editor), *Lakes, Chemistry, Geology, and Physics.* Springer-Verlag, New York, N.Y., pp. 153—177.

Krishnaswami, S., Lal, D.L., Martin, J.M. and Meybeck, M., 1971. Geochronology of lake sediments. *Earth Planet. Sci. Lett.*, 11: 407—414.

Krishnaswami, S., Benninger, L.K., Aller, R.C. and Von Damm, K.L., 1980. Atmospherically-derived radionuclides as tracers of sediment mixing and accumulation in nearshore marine and lake sediments: evidence from ^{7}Be, ^{210}Pb and ^{239,240}Pu. *Earth Planet. Sci. Lett.*, 47: 307—318.

Lerman, A. (Editor), 1978. *Lakes, Chemistry, Geology, and Physics.* Springer-Verlag, New York, N.Y.

Lerman, A. and Lietzke, T.A., 1975. Uptake and migration of tracers in lakes sediments. *Limnol. Oceanogr.*, 20: 497—510.

Lewis, D.M., 1977. The use of ^{210}Pb as a heavy metal tracer in the Susquehanna River system. *Geochim. Cosmochim. Acta*, 41: 1557—1564.

Matsumoto, E., 1975a. Pb-210 geochronology of sediments from Lake Sinji. *Geochem. J.*, 9: 167—172.

Matsumoto, E., 1975b. Accumulation rate of Lake Biwa-Ko sediments by the Pb-210 method. *J. Geol. Soc. Jpn.*, 81: 301—306.

Nevissi, A., Beck, J.N. and Kuroda, P.K., 1974. Long lived radon daughters as atmospheric radioactive tracers, *Health Phys.*, 27: 181—188.

Nozaki, Y., Cochran, J.K. and Turekian, K.K., 1977. Radiocarbon and ^{210}Pb distribution in submersible-taken deep-sea cores from Project FAMOUS. *Earth Planet. Sci. Lett.*, 34: 167—173.

Nydal, R., Löuseth, K. and Gullicksen, S., 1979. A survey of radiocarbon variation in nature since the test ban treaty. In: R. Berger and H.E. Suess, (Editors), *Radiocarbon Dating.* University of California Press, Berkeley, Calif., pp. 313—323.

Oldfield, F., 1977. Lakes and their drainage basins as units of sediment based ecological study. *Progr. Phys. Geogr.*, 1: 460—504.

Oldfield, F., Appleby, P.G. and Battarbee, R.W., 1978. Alternative ^{210}Pb dating: results from the New Guinea Highlands and Lough Erne. *Nature*, 271: 339—342.

Oldfield, F., Appleby, P.G. and Petit, D., 1980. A re-evaluation of lead-210 chronology and the history of total lead influx in a small south Belgian pond. *Ambio*, 9: 97—99.

Peng, T.H., Broecker, W.S. and Berger, W.H., 1979. Rates of benthic mixing in deep-sea sediments as determined by radioactive tracers. *Quat. Res.*, 11: 141—149.

Pennington, W., Cambray, R.S. and Fisher, E.M., 1973. Observations on lake sediments using ^{137}Cs as a tracer. *Nature*, 242: 324—326.

Pennington, W., Cambray, R.S., Eakins, J.D. and Harkness, D.D., 1976. Radionuclide dating of the recent sediments of Blelham Tarn. *Freshwater Biol.*, 6: 317—333.

Petit, D., 1974. ^{210}Pb et isotopes stables du plomb dan des sédiments lacustres. *Earth Planet. Sci. Lett.*, 23: 199—205.

Poet, S.E., Moore, H.E. and Martell, E.A., 1972. Lead-210, bismuth-210, and polonium-210 in the atmosphere. *J. Geophys. Res.*, 77: 6515—6520.

Rangarajan, C.S., Gopulakvishnan, S. and Eapen, C.D., 1976. Global variation of ^{210}Pb in surface air and precipitation. *USAERDA, Health Safety Lab., Environ. Q. Rep.*, HASL-289, UC-11, 1-63-83.

Richey, J.C., McHenry, M.R. and Gill, A.C., 1973. Dating of recent reservoir sediments. *Limnol. Oceanogr.*, 18: 254—263.

Robbins, J.A., 1978. Geochemical and geophysical applications of radioactive lead. In: J.O. Nriagu (Editor), *Biogeochemistry of Lead*. Elsevier, Amsterdam.

Robbins, J.A. and Edgington, D.N., 1975. Determination of recent sedimentation rates in Lake Michigan using ^{210}Pb and ^{137}Cs. *Geochim. Cosmochim. Acta*, 39: 285—304.

Robbins, J.A., Edgington, D.N. and Kemp, A.L.U., 1978. Comparative lead-210, cesium-137 and pollen geochronologies of sediments from Lakes Ontario and Erie. *Quat. Res.*, 10: 256—258.

Robbins, J.A., Krezoski, J.R. and Mozley, S.C., 1977. Radioactivity in sediments of the Great Lakes: post-depositional redistribution by deposit feeding organisms. *Earth Planet. Sci. Lett.*, 36: 325—333.

Santschi, P.H., Li, Y.H., Bell, J.I., Trier, R. and Kawtaluk, K., 1980. Pu in coastal marine waters. *Earth Planet Sci. Lett.*, 51: 248—265.

Schell, W.E., 1977. Concentrations, physico-chemical states and mean residence times of ^{210}Pb and ^{210}Po in marine and estuarine waters. *Geochim. Cosmochim. Acta*, 41: 1019—1031.

Schell, W.R. and Barnes, R.S., 1974. Lead and mercury in the aquatic environment of Western Washington State. In: A.J. Rubin (Editor), *Aqueous-Environmental Chemistry of Metals*. Ann Arbor Science Publishers, Ann Arbor, Mich., pp. 129—165.

Schell, W.R., Jokela, T. and Eagle, R., 1973. Natural ^{210}Pb and ^{210}Po in a marine environment. In: *Radioactive Contamination of the Marine Environment*. IAEA, Vienna, pp. 701—724.

Schell, W.R., Swanson, J.R. and Currie, L.A., 1983. Anthropogenic changes in organic carbon and trace metal input to Lake Washington. *Radiocarbon*, 25: 621—628.

Shapiro, J., Edmondson, W.T. and Allison, D.E., 1971. Changes in the chemical composition of sediments of Lake Washington. *Limnol. Oceanogr.*, 16: 437—452.

Simpson, H.J., Trier, R.M., Olsen C.R., Hammond, D.E., Ege, A., Miller, L. and Melack, J.M., 1980. Fallout plutonium in an alkaline, saline lake. *Science*, 207: 1071—1073.

Spyridakis, D.E. and Barnes, R.S., 1978. Contemporary and historical trace metal loadings to the sediments of four lakes of the lake Washington drainage. Final report. *Off. Water Res. Technol., Dep. Civil Eng., Univ. Wash., OWRT Proj.*, No. A-083-WASH: 70 pp.

Stiller, M., 1979. Sedimentation patterns in Lake Kinneret (Tiberias). In: *Application of Nuclear Techniques to the Study of Lake Dynamics*. IAEA, Vienna, pp. 273—285.

Stuiver, M., 1967. Origin and extent of atmospheric ^{14}C variations during the past 10,000 years. In: *Radioactive Dating and Methods of Low Level Counting*. IAEA, Vienna, pp. 27—40.

Stumm, W. and Baccini, P., 1978. Man-made chemical purturbation of lakes. In: A. Lerman (Editor), *Lakes, Chemistry, Geology, and Physics*. Springer-Verlag, New York, N.Y., pp. 92—126.

Swanson, J.R., 1980. *Carbon isotope analysis of carbonaceous compounds in Puget Sound and Lake Washington*. Ph. D. Dissertation. University of Washington, Seattle, Wash.

Thomas, R.L., 1972. The distribution of mercury in the sediments of Lake Ontario. *Can. J. Earth Sci.*, 9: 636—651.

Tschopp, J., 1977. M.S. Thesis, Swiss Federal Institute of Technology, Zurich.

Turekian, K.K., Cochran, J.K. and Demaster, D.J., 1978. Bioturbation in deep-sea deposits; rates and consequences. *Oceanus*, 21: 34—41.

Wahlen, M. and Thompson, R.C., 1980. Pollution records from sediments of three lakes in New York State. *Geochim. Cosmochim. Acta*, 44: 333—339.

Chapter 5

STABLE ISOTOPE GEOCHEMISTRY OF TRAVERTINES

BRUNO TURI

INTRODUCTION

Travertine, in a broad sense, refers to all non-marine limestone accumulations formed in lakes, rivers, springs and caves (Sanders and Friedman, 1967). Limestones precipitated from cave waters, however, are commonly called *speleothems*. The term *travertine* or its synonym *calcareous sinter*, should be strictly used to designate a finely crystalline deposit or incrustation of white to tan colour formed by rapid inorganic precipitation of carbonate from cold or thermal waters in a surface environment. The nucleation and/or precipitation of calcium carbonate can be enhanced by the action of some algae and bacteria. Such organisms, therefore, may play a significant role in travertine formation. For example, Folk and Chafetz (1980) have observed in travertine from various occurrences bacterially controlled calcite structures. Even this mode of precipitation, however, should be considered "inorganic" because it does not involve metabolic processes (Sanders and Friedman, 1967). Travertine also occurs as incrustations on plants or vegetal remains (*phytohermal** and *stromatolitic travertines*, Fig. 5-1).

The term travertine is derived from the Latin *lapis tiburtinus*, i.e., the stone of Tibur, the former name of Tivoli, a town near Rome well known for the huge deposits (up to 100 m in thickness) of this material (see Fig. 5-6).

Stable isotope analyses proved to be very useful in investigating the genesis of travertines (Craig, 1953, 1963; Fritz, 1965, 1968; Gonfiantini et al., 1968; Savelli and Wedepohl, 1969; Friedman, 1970; O'Neil and Barnes, 1971; Demovič et al., 1972; Manfra et al., 1974, 1976; and others).

Of particular interest are the distributions of carbon and oxygen isotopes, but important information may also be obtained from the isotopic compositions of other elements such as sulphur, lead and strontium, generally present in small amounts in these carbonates (Savelli and Wedepohl, 1969: Demovič et al., 1972; Leeman et al., 1977; Barbieri et al., 1979).

---

* From the Greek words φυτόν (plant) and ἕρμα (stone)

Fig. 5-1. Travertine incrustations on aquatic plants (collection of the Institute of Geochemistry, University of Rome.)

MINERALOGY, PETROLOGY AND PHYSICAL PROPERTIES OF TRAVERTINES

The carbonate of travertines is, in most cases, simply represented by $CaCO_3$ in either of its most common natural polymorphic forms, calcite and/or aragonite. The physicochemical factors that favour the deposition of travertine as either calcite or aragonite are still under discussion. According to Malesani and Vannucci (1975), the precipitation of either of the $CaCO_3$ polymorphs from supersaturated thermal waters is mainly controlled by the absolute Sr^{2+} content: aragonite precipitates from solutions with Sr contents in the dry residue $\geq \sim 0.5$ wt.% (corresponding to about 0.95 mol.% with respect to the sum Ca + Mg + Sr), while for lower Sr contents calcite precipitates preferentially. These authors also report that, considering two solutions A and B with, respectively, low and high strontium contents and variable Mg contents, a calcite relatively rich in magnesium will precipitate from A whereas an aragonite with very low magnesium content will be deposited from B.

These observations are partially contradicted by more recent studies

(Duchi et al., 1978), which indicate that the precipitation of travertine in the aragonitic form depends on the Mg^{2+} concentration rather than on the Sr^{2+} content of the solution. Another important factor appears to be the rate of deposition of the carbonate which, in turn, is related to the temperature and to the gradient of partial pressure of CO_2 between solution and atmosphere. However, the influence of these parameters on the mineralogy of the precipitated travertine is quite variable, so that no threshold values can be given, although it would appear that from thermal, strongly degassing solutions relatively rich in magnesium, travertine precipitates preferentially in the aragonitic form.

Barnes and O'Neil (1971) have classified the Ca-Mg carbonates forming the travertines of some stream channels of the California Coast Range on the basis of their X-ray diffraction patterns. They found that calcite has compositions ranging from $CaCO_3$ to $Ca_{0.5}Mg_{0.5}CO_3$; in addition the $\{015\}$ ordering peak is lacking and the basal peak $\{006\}$ greatly weakened or totally absent in these phases. Dolomite solid solutions extend over the compositional range from $Ca_{0.6}Mg_{0.4}CO_3$ to $Ca_{0.5}Mg_{0.5}CO_3$ and yield ordering reflections.

Travertines often contain a non-carbonatic fraction, the amount of which varies within large limits (roughly, from 0.1 to 30%; Cortesi and Leoni, 1958; Cipriani et al., 1972, 1977; Ferreri and Stanzione, 1978). The mineralogical composition of this fraction is also quite variable, reflecting the geological and geochemical features of the environment of deposition. The most common constituents are sulphur-bearing minerals (mostly metallic sulphides and calcium sulphates, such as gypsum and occasionally, at least in some deposits of Central Italy, the quite rare bassanite, $2\,CaSO_4 \cdot H_2O$) and clay minerals, accompanied by a variety of detrital minerals such as quartz and feldspars. Non-mineral constituents such as remains of fossil fauna, algae and aquatic plants may also be present.

The texture of travertines varies from compact and massive to friable, porous or spongy. Spongy travertines, frequently referred to as *tufa*, exhibit porosities as high as 50%, whereas in the most compact varieties the porosity is typically lower than 10%; correspondingly, the apparent density changes from about 1.55 to 2.57 (Cipriani et al., 1977; Ferreri and Stanzione, 1978). As an example, the Tivoli travertine that is used as building stone typically has apparent density and porosity values ranging from 2.376 to 2.548 and from 13.38 to 3.99%, respectively (Penta, 1956).

GEOCHEMICAL CONDITIONS CONTROLLING TRAVERTINE DEPOSITION

Travertines, as most fresh-water carbonates, typically precipitate from Ca-bicarbonate waters in contact with the atmosphere (open-system conditions). The process involves the following reactions:

$$CO_2 \text{ (gas)} \rightleftharpoons CO_2 \text{ (aq)} \quad (1a)$$

$$CO_2 \text{ (aq)} + H_2O \rightleftharpoons H_2CO_3 \quad (1b)$$

$$CO_2 \text{ (gas)} + H_2O \rightleftharpoons H_2CO_3 \quad (1c)$$

$$H_2CO_3 \rightleftharpoons H^+ + HCO_3^- \quad (2)$$

$$HCO_3^- \rightleftharpoons CO_3^{2-} + H^+ \quad (3)$$

$$CaCO_3 \text{ (solid)} \rightleftharpoons Ca^{2+} + CO_3^{2-} \quad (4)$$

$$H_2O \rightleftharpoons H^+ + OH^- \quad (5)$$

together with the condition of electrical neutrality:

$$2(Ca^{2+}) + (H^+) = 2(CO_3^{2-}) + (HCO_3^-) + (OH^-) \quad (6)$$

where ( ) denote concentration of the enclosed species.

Equations (1) to (6) clearly show that both the deposition and the dissolution of carbonate is controlled by the physicochemical features of the solution such as pH, T, p_{CO_2} and chemical composition.

Travertine-depositing waters must have p_{CO_2} values higher than that measured in the atmosphere ($10^{-3.5}$ atm). This is a consequence of their geochemical evolution, which in most cases begins in subsurface environments where values of $p_{CO_2} \geqslant \sim 10^{-2}$ atm can be easily attained (Vogel and Ehhalt, 1963; Deines et al., 1974; Holland, 1978). As these waters reach the surface, excess CO_2 tends to escape giving rise to a metastable state characterized by a progressive supersaturation with respect to $CaCO_3$ (Mook, 1970; Usdowski et al., 1979). This supersaturation is also favoured by the fact that the Ca^{2+} concentrations are in general quite large, far exceeding the mean value of 15 ppm estimated for continental surface waters (Livingstone, 1963). Degassing of CO_2 and precipitation of travertine occur simultaneously when supersaturation is sufficiently high, i.e., for values of the saturation index (SI)* typically close to 1. For example, Jacobson and Usdowski (1975) and Usdowski et al. (1979) observed that in a calcite-precipitating spring near Westerhof, Germany, the deposition of $CaCO_3$ begins when SI $\cong$ 0.5 but becomes more rapid at values greater than 1.0.

Inasmuch as the deposition of travertine strongly depends upon the rate of escape of CO_2, it is evident that the process will be enhanced by the agitation of water which can be brought about, for instance, by falls or rapids.

---

* SI for calcite is defined by the expression (Holland, 1978):

$$SI = \log[(Ca^{2+}) \cdot (CO_3^{2-})/K_c]$$

where K_c is the solubility product of $CaCO_3$. Waters with values of SI within about ± 0.1 units of zero can be considered saturated with calcite (Deines et al., 1974).

Thus, in the above mentioned spring, the amount of carbonate precipitated was found to be the greatest where the water was the most turbulent.

A noteworthy exception to this picture is represented by the travertines associated with the calcium hydroxide ("Ca^{2+}-OH^-")-type waters described by Barnes and O'Neil (1969). The unusual chemical features of these waters, such as pH values > 11, Ca^{2+} concentrations ranging from 35 to 53 ppm and lack of carbonate, are attributable to reactions between meteoric waters and ultramafic rocks under near-surface conditions. According to O'Neil and Barnes (1971), the precipitation of travertine from such waters would be attributable to the rapid reaction of atmospheric CO_2 with thin films of water, favoured by the high pH. This mode of origin leads to travertines isotopically different from those precipitated under the conditions described above.

$^{13}C/^{12}C$ AND $^{18}O/^{16}O$ VARIATIONS IN TRAVERTINES

Carbon isotopes

Carbon species in travertine-depositing solutions may derive from various sources, namely: (1) organic CO_2; (2) atmospheric CO_2; (3) limestones; and (4) igneous sources.

(1) The principal source of organic carbon to the circulating groundwaters is the CO_2 produced in the soil by decomposition of organic material and plant-root respiration. Such CO_2 has $\delta^{13}C$ values typically ranging from -15 to $-25^0/_{00}$* (Deines, 1980, and references therein).

Travertines remarkably depleted in ^{13}C have been observed in several localities all over the world, such as the Coast Range of California (O'Neil and Barnes, 1971), the localities of West Germany studied by Savelli and Wedepohl (1969), some of the Slovakian deposits studied by Demovič et al. (1972) and a few other localities of Europe and United States (see Fig. 5-2). Excepting the very peculiar samples from California, whose extremely light carbon ($\delta^{13}C = -24$ to $-11^0/_{00}$) is thought to derive from the atmospheric CO_2, as discussed in the next section, the other travertines of this group are characterized by $\delta^{13}C$ values ranging from -10 to $-4^0/_{00}$, and for them a contribution of biogenic CO_2 seems very likely.

(2) Atmospheric CO_2 normally has a $\delta^{13}C$ close to $-7^0/_{00}$ (Keeling, 1958; Bottinga and Craig, 1969; Keeling et al., 1979). In the absence of carbon from other sources, surface waters in equilibrium with the atmosphere

---

* The carbon and oxygen isotopic compositions of the samples are expressed in the conventional δ ($^0/_{00}$) notation. The reference standards are PDB for carbon (Craig, 1957) and SMOW for oxygen (Craig, 1961).

should give origin to calcium carbonates with $^{13}C/^{12}C$ similar to marine limestone. The travertines from Stuttgart-Münster (western Germany) analyzed by Fritz (1968), for example, have $\delta^{13}C$ values compatible with such an

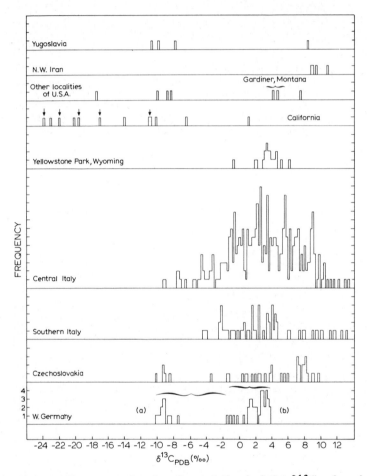

Fig. 5-2. Histograms showing the variations of the $\delta^{13}C$ values in the travertines from various localities throughout the world. Data from Craig (1953), Fritz (1965, 1968), Gonfiantini et al. (1968), Savelli and Wedepohl (1969), Friedman (1970), O'Neil and Barnes (1971), Demovič et al. (1972), Panichi and Tongiorgi (1975), Manfra et al. (1976), Leeman et al. (1977), Buccino et al. (1978) and Ferreri and Stanzione (1978). The two groups of travertines from West Germany refer to the samples studied by Savelli and Wedepohl (1969), mostly from the Westerhof, Göttingen and Iburg areas (a), and to the samples from Stuttgart-Münster analyzed by Fritz (1968) (b). For simplicity's sake, only the maximum and minimum $\delta^{13}C$ values of sets of samples from the same sampling point are reported in the central Italy histogram, which actually represents about 900 data. The analyses of the travertines from California marked by arrows refer to travertines associated with "Ca^{2+}-OH^-"-type waters.

origin of carbon. However, as a consequence of kinetic isotope effects, air-CO_2 may yield travertines with $^{13}C/^{12}C$ ratios as low as those of biogenic carbon. This occurs if the formation water has a particular chemical composition, as, for example, the calcium hydroxide-type water discussed in the previous section (see also next section).

(3) Limestone represents the largest reservoir of high-^{13}C carbon for the travertine-depositing waters. Marine limestones have $\delta^{13}C = 0 \pm 4^0/_{00}$ (Craig, 1953; Keith and Weber, 1964) so that circulating groundwaters receive an appreciable contribution of relatively heavy carbon from the dissolution of these carbonates.

In general, CO_2 with $\delta^{13}C$ values roughly within the range of the marine limestones should be regarded conservatively as being derived from these rocks (Barnes et al., 1978). High-^{13}C CO_2 can also be produced by limestone through metamorphic reactions. Shieh and Taylor (1969) report that CO_2 liberated from decarbonation of carbonate rocks during contact metamorphism is about $6^0/_{00}$ richer in ^{13}C (and about $5^0/_{00}$ richer in ^{18}O) than the original carbonate. This CO_2 dissolves in groundwaters at depth and the CO_2-charged water can, in turn, react with the carbonate rocks. A solution rich in high-^{13}C carbon species is thus obtained, which can easily give origin to "heavy" travertine under surface conditions.

On this basis, there is little doubt that a great number of the travertines so far analyzed derive from such a source. This is the case, for example, for the travertines from the majority of the deposits of Latium, central Italy, in which $\delta^{13}C$ values from about 4 to $13^0/_{00}$ were measured. The travertines from Tivoli belong to this group. Their high $\delta^{13}C$ values, mostly ranging from 8.5 to $11^0/_{00}$, seem to be due to a derivation from the Mesozoic limestones cropping out in that area, loss of CO_2 by evaporation, and precipitation at relatively low temperatures (Craig, 1963).

A similar explanation probably holds for the high-^{13}C travertines from Czechoslovakia, Iran and Yugoslavia studied by Demovič et al. (1972) and those from southern Italy reported in Buccino et al. (1978) and Ferreri and Stanzione (1978).

(4) Carbon dioxide from igneous sources has $\delta^{13}C$ values that vary over the wide range from -26 to $-3^0/_{00}$. However, the following distinction can be made (see Hoefs, 1978 and references therein): (a) CO_2 liberated in mofettes, fumaroles, etc., from volcanic areas normally has $\delta^{13}C = -8$ to $-3^0/_{00}$; (b) CO_2 samples collected directly over hot lavas has $\delta^{13}C$ ranging from -26 to $-15^0/_{00}$; and (c) mantle-derived CO_2 is thought to have $\delta^{13}C$ values in the range of -8 to $-5^0/_{00}$.

Summing up, the carbon involved in the genesis of travertines can have $\delta^{13}C$ ranging roughly from -26 to $+6^0/_{00}$. This fact, together with the isotopic effects occurring during the deposition, explains the large variability of the $^{13}C/^{12}C$ ratios observed in travertines from the various localities investigated (Fig. 5-2).

Oxygen isotopes

Water represents the main source of oxygen of travertines. A noteworthy, although rare, exception is represented by the peculiar travertines and scums formed near the surface of Ca^{2+}-OH^--type waters from the Coast Range of

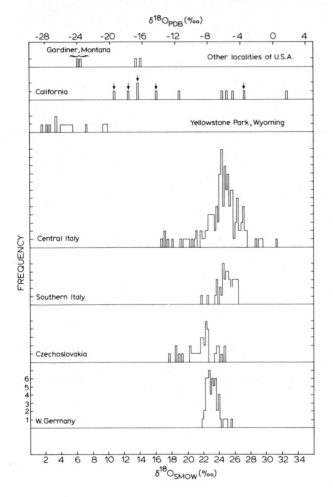

Fig. 5-3. Variations of the $\delta^{18}O$ values, expressed in both SMOW and PDB scales, in the travertines from various localities throughout the world. Data from Fritz (1965, 1968), Gonfiantini et al. (1968), Savelli and Wedepohl (1969), Friedman (1970), O'Neil and Barnes (1971), Demovič et al. (1972), Manfra et al. (1976), Leeman et al. (1977), Buccino et al. (1978) and Ferreri and Stanzione (1978). In constructing the histogram of travertines from central Italy, only the highest and lowest $\delta^{18}O$ values were used. This histogram actually summarizes about 300 $^{18}O/^{16}O$ analyses.

The analyses of the travertines from California marked by arrows refer to samples from "Ca^{2+}-OH^-"-type waters.

California studied by O'Neil and Barnes (1971). Their oxygen could derive, to a large degree, from air-CO_2. As a consequence of kinetic isotope effects accompanying their formation (see the next section), these carbonates are relatively depleted in ^{18}O, with $\delta^{18}O$ values ranging mostly from 11 to 17‰.

Except for this small group of samples, the $^{18}O/^{16}O$ ratios of travertines essentially reflect the isotopic composition of the waters from which they precipitate and the relevant physicochemical conditions. Fig. 5-3 summarizes the oxygen isotope variations in travertines throughout the world. Most travertines from Italy, West Germany and Czechoslovakia, as well as some travertines from California not directly associated with Ca^{2+}-OH^--type waters, have $\delta^{18}O$ = 20—28‰. To a first approximation, such a pattern may be ascribed to a precipitation from waters of meteoric origin with $\delta^{18}O$ ranging roughly from —9 to —4‰ at temperatures from a few tens of degrees down to normal near-surface temperatures.

The travertines from Yellowstone Park are extremely depleted in ^{18}O ($\delta^{18}O$ = 2—10‰), probably because of the very low ^{18}O contents of water ($\delta^{18}O < \sim -17$‰) and the relatively high temperatures of precipitation. A similar explanation probably also applies to the two very light travertines from Gardiner, Montana, analyzed by Friedman (1970).

ISOTOPIC FRACTIONATIONS IN TRAVERTINE FORMATION

The formation of travertines in different environments and from different source materials described in the previous sections are accompanied by fractionation of the carbon and oxygen isotopes.

As the proportions of the aqueous carbon species in groundwaters (i.e. CO_2 (aq), HCO_3^-, CO_3^{2-}) are controlled by the temperature, pH and ionic strength, I, and appreciable differences in ^{13}C contents exist among the various carbon species, the $^{13}C/^{12}C$ ratio of the precipitated carbonate will also be affected by the physicochemical conditions of the environment.

Taking CO_2(gas) as a reference under equilibrium conditions, the isotopic composition of any other carbon species i can be expressed as:

$$\delta^{13}C_i = \delta^{13}C_{CO_2 \text{ (gas)}} + \epsilon_i \quad (7)$$

where ϵ_i is the enrichment factor between the species i and CO_2(gas)*. The relationships between the enrichment factors of interest and temperature are listed in Table 5-1.

They show that in the range of temperatures at which travertine deposition occurs, the equilibrium partitioning of carbon isotopes is such that solid

---

* See footnote on next page.

TABLE 5-1

Temperature-dependence of the carbon isotopic enrichment factor ϵ_i of the various carbon species i relative to CO_2 (gas)

i	ϵ_i (‰)		
CO_2(aq)	$\epsilon_{CO_2(aq)-CO_2(gas)}$	$= -0.373 \times 10^3/T + 0.19$	(a)
HCO_3^-	$\epsilon_{HCO_3^- - CO_2(gas)}$	$= 9.483 \times 10^3/T - 23.89$	(b)
CO_3^{2-}	$\epsilon_{CO_3^{2-} - CO_2(gas)}$	$= 0.87 \times 10^6/T^2 - 3.4$	(c)
$CaCO_3$	$\epsilon_{CaCO_3 - CO_2(gas)}$	$= 1.435 \times 10^6/T^2 - 6.13$	(d)

The temperature T is in degrees Kelvin. Expressions (a) and (b) have been obtained by Mook et al. (1974) for $T = 278-298$ K; the equation for the HCO_3^--CO_2(gas) pair, however, is considered valid up to about 420 K with reasonable accuracy ($\pm$ 0.2‰). Equation (c) is from Deines et al. (1974); its accuracy, although admittedly low, is not critical to our purposes because the great majority of the travertine-depositing waters, characterized by pH values of 6—9, contain negligible amounts of CO_3^{2-} (less than about 3% of the total carbon in solution). Equation (d) is a least-squares fit of the data by Bottinga (1968) at temperatures from 273 to 373 K reported by Gonfiantini et al. (1968). This equation interpolates quite well the data acquired by Baertschi (1957), Vogel (1961) and Emrich et al. (1970).

carbonate in enriched in ^{13}C with respect to the carbon species in solution and CO_2(gas). It is also worth noting that, under the same conditions, CO_2 (aq) is slightly depleted in ^{13}C relative to CO_2(gas).

The oxygen isotope pattern is complicated by the fact that both the water and CO_2 contribute oxygen to the precipitated carbonate. At equilibrium, $CaCO_3$ is isotopically heavier than H_2O but lighter than CO_2(gas), as shown

---

* The permil isotopic fractionation between two substances, A and B, is expressed in terms of the quantity $10^3 \ln \alpha_{A-B}$, where α_{A-B} is the fractionation factor defined as:

$$\alpha_{A-B} = \frac{1000 + \delta_A}{1000 + \delta_B}$$

For δ-values $\leqslant \sim |10|$, the following approximate relationship holds:

$$10^3 \ln \alpha_{A-B} = \sim \delta_A - \delta_B$$

The enrichment factor, ϵ, is defined as:

$$\epsilon_{A-B} = 10^3 (\alpha_{A-B} - 1)$$

For small values of α, $\epsilon_{A-B} = \sim 10^3 \ln \alpha_{A-B}$ (Friedman and O'Neil, 1977).

by the following equations, valid over a wide range of temperatures (Bottinga, 1968; O'Neil et al., 1969; Friedman and O'Neil, 1977):

$$10^3 \ln \alpha_{CaCO_3-H_2O} = 2.78 \times 10^6/T^2 - 2.89 \tag{8}$$

$$10^3 \ln \alpha_{CO_2(gas)-CaCO_3} = -1.8034 \times 10^6/T^2 + 1.0611 \times 10^4/T - 2.7798 \tag{9}$$

Isotopic equilibrium, however, is seldom attained in the deposition of travertines, mainly as a consequence of kinetic effects. If the rate of escape of CO_2 from the solution is not sufficiently slow to enable chemical and isotopic equilibrium to be maintained between the carbon containing species, the isotopically lighter CO_2 molecules will be eliminated preferentially. A similar effect takes place in the evaporation of the water, resulting in an enrichment of the $H_2^{18}O$ molecules in the liquid phase. As a consequence of these effects, the travertine may acquire isotope ratios higher than those one would expect to observe in carbonates precipitated under equilibrium conditions.

Under open-system conditions, where temperature, p_{CO_2} and the ^{13}C contents of the aqueous and gaseous carbon reservoirs remain constant, the carbon isotope composition of calcium carbonate is subject to kinetic isotope effects due to variations of the rate of precipitation (Turner, 1982). At 25°C, the apparent enrichment factor $\epsilon^{app}_{HCO_3^--CaCO_3}$, that is a combination of kinetic and equilibrium enrichment factors, varies between -0.35 ± 0.23 and $-3.7 \pm 0.36°/_{00}$, depending on the calcite to aragonite ratio in the solid carbonate and the rate of precipitation, the enrichment in ^{13}C in $CaCO_3$ decreasing with increasing precipitation rate. Theoretical calculations by Bottinga (1968) yield a value of $-2.3°/_{00}$ for the equilibrium factor $\epsilon_{HCO_3^--calcite}$. Thus, under fast precipitation conditions, the ^{13}C enrichment in the carbonate relative to HCO_3^- is much lower than that at equilibrium.

These results can explain the anomalous $^{13}C/^{12}C$ fractionation between dissolved carbonate species and precipitated calcite observed by Usdowski et al. (1979) at Westerhof. Here, calcite of $\delta^{13}C = -6.9$ to $-7.9°/_{00}$ deposited under disequilibrium conditions from highly supersaturated waters of $\delta^{13}C_{TDC} = -6.4$ to $-8.0°/_{00}$, where TDC = total dissolved inorganic carbon. Theoretically, calcite was expected to be enriched in ^{13}C by about $2.4°/_{00}$. This discrepancy is attributable to the above-mentioned kinetic effects occurring at high precipitation rate, which may result in very low ^{13}C enrichments in the $CaCO_3$ relative to HCO_3^- in bulk solution.

Preferential deposition of the light isotopes in the precipitated carbonate can also occur in the case of the travertines formed by the interaction of air CO_2 with thin films of $Ca^{2+}-OH^-$-type water. Inasmuch as these waters have pH values of 11 or higher, a very rapid reaction occurs between air-CO_2 and the alkaline solution, leading to the formation of carbonate ions which will combine with Ca^{2+} and precipitate a $CaCO_3$ strongly depleted in ^{13}C and ^{18}C. Such a process is just the opposite of the one discussed above. Its influence on the isotopic composition of the resulting travertines is explained by the simple model proposed by O'Neil and Barnes (1971). The $^{13}C/^{12}C$ ratio of the CO_2 molecules incident to the surface of the water is, on the average, controlled by the relative diffusion rates of the species $^{13}C^{16}O_2$ (mass 45) and $^{12}C^{16}O_2$ (mass 44). Inasmuch as air CO_2 has $\delta^{13}C = \sim -7‰$, assuming a kinetic isotope effect of $(45/44)^{1/2}$ or about 1%, the travertines would be expected to display $\delta^{13}C$ values of at least $-17‰$, in agreement with the observed range from $\delta^{13}C = \sim -22$ to $-17‰$ obtained for most samples (Fig. 5-2).

The explanation of the marked depletion in ^{18}O noticeable in this group of travertines is more complicated, because the oxygen can be provided to the carbonate by both H_2O and CO_2. However, a kinetic model similar to that just described still applies. In fact, more $^{12}C^{16}O_2$ molecules will hit the water surface than $^{12}C^{16}O^{18}O$ molecules by a factor $(46/44)^{1/2}$ or about 2%. Assuming a $\delta^{18}O = 42‰$ for atmospheric CO_2 (Bottinga and Craig, 1969), the $\delta^{18}O$ value of the CO_2 under consideration would be about $22‰$. Unpublished experiments by O'Neil showed that the $CaCO_3$ rapidly precipitated upon addition of CO_2 to a strongly alkaline solution, derives 1/3 of its oxygen from the water and 2/3 from the CO_2. Because the Ca^{2+}-OH^--type waters from which these travertines precipitated are meteoric in origin (Barnes and O'Neil, 1969), the observed $\delta^{18}O$ values of these travertines, mostly lying between about 11 and 16 per mil (Fig. 5-3), are in agreement with the reasoning presented above.

In some areas, isotopic analyses have also been made of the travertine-depositing waters, the total inorganic carbon in solution (in general, mainly as HCO_3^-) and/or the evolved carbon dioxide. These studies represent the best approach to the understanding of the possible mechanisms of isotope fractionation in travertine formation and therefore will be considered in some detail.

Fritz (1965) and Gonfiantini et al. (1968) investigated a number of springs that are at present depositing travertine in Tuscany and Latium, central Italy. In Fig. 5-4, the measured differences in ^{18}O content between the carbonate and water are plotted as a function of temperature. For the majority of the travertine-water pairs, the values of the isotopic fractionations are larger than those corresponding to equilibrium. The deviation from the equilibrium, however, decreases as the distance from the orifice increases, as shown by the samples from Bagni S. Filippo and Bagnaccio thermal springs. The differences in ^{13}C content between calcium carbonate and carbon dioxide are also mostly higher than the equilibrium ones (Fig. 5-5). It is interesting to note that the variation in heavy isotope contents in both systems roughly tends to decrease with increasing temperature, as is the case for the equilibrium pairs. These results indicate that, in general, the travertines from these springs formed out of isotopic equilibrium with both water and carbon dioxide. The observed isotopic variations suggest that the rate of escape of CO_2 from the formation water may be the controlling factor of disequilibrium. On these grounds, the following mechanism can be proposed. The equilibrium existing in the system H_2O-CO_2-HCO_3^--CO_3^{2-} as the water issues from the spring orifice is quickly broken by the rapid outgassing of CO_2. Thereafter, the system progressively tends to re-equilibrate under the new (surface) conditions as the rate of loss of CO_2 diminishes with increasing distance from the orifice and cooling of the water.

A least-squares fit to the data of Fritz (1965) and Gonfiantini et al. (1968) gives the following relationship between the carbon isotopic compositions of CO_2 and travertine samples collected at the orifices of 11 springs from Tuscany and Latium (Panichi and Tongiorgi, 1975):

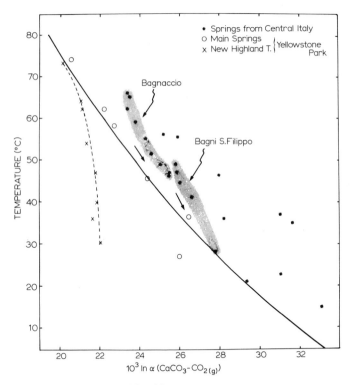

Fig. 5-4. Plot of the $^{18}O/^{16}O$ fractionations between $CaCO_3$ and H_2O as a function of temperature for the travertine-depositing springs of central Italy and Yellowstone Park studied by Gonfiantini et al. (1968) and Friedman (1970), respectively. The equilibrium curve by O'Neil et al. (1969), recalculated to the newly proposed value of $\alpha_{CO_2-H_2O}$ at 25°C (1.0412; O'Neil et al., 1975) has been drawn as a reference (solid line). The arrows indicate increasing distance of the samples from the spring orifice for Bagnaccio and Bagni S. Filippo springs.

$$\delta^{13}C_{CO_2} = 1.2\, \delta^{13}C_{travertine} - 10.5 \tag{10}$$

This equation may be used to evaluate, from the analyses of "fossil" travertines, the $^{13}C/^{12}C$ of the CO_2 originally associated with these carbonates. It was obtained from waters issuing at temperatures and pH values ranging from about 15 to 65°C and from 6.2 to 7.5, respectively (Francalanci, 1958; Fancelli and Nuti, 1975; Malesani and Vannucci, 1975). This method represents a hopeful tool which may be applied in the preliminary prospecting of geothermal areas elsewhere.

In order to avoid the uncertainty regarding the location of the spring office, several travertine samples must be collected from all over the deposit under consideration and the lightest one used for the calculations.

Substantially different isotopic patterns were observed by Friedman (1970) at Main Springs and New Highland Terrace, in the Mammoth Hot Spring area of Yellowstone Park, Wyoming. Main Springs consists of a

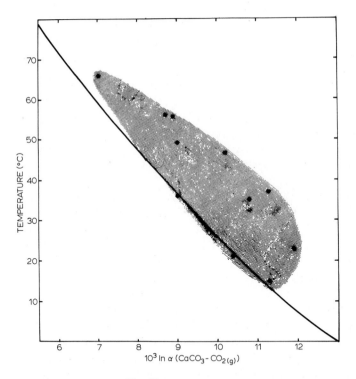

Fig. 5-5. Plot of the $^{13}C/^{12}C$ fractionations between $CaCO_3$ and CO_2 versus temperature for the travertine-depositing springs from central Italy studied by Gonfiantini et al. (1968). The equilibrium curve drawn as a reference for this system (solid line) is the least-squares fit to the low-temperature ($\leqslant 100°C$) data by Bottinga (1968) given by the aforementioned authors.

number of pools; the water flows from the higher to the progressively lower ones, depositing travertine at the bottom of them and at the lip over which it leaves the pools.

At New Highland Terrace, the water issues at the top of a mound of travertine about 18 m high, and falls rapidly over the travertine as a thin sheet. The deposition of travertine is confined to the bottom of the water. Waters issuing from these springs have the same temperature (73—75°C), pH (6.7), $\delta^{18}O$ (—18.1‰) and contain dissolved inorganic carbon with identical $^{13}C/^{12}C$ ratios ($\delta^{13}C_{TDC}$ = —2.3‰).

It is therefore not surprising that travertines being deposited at the orifice of the springs have isotopic compositions that are extremely similar ($\delta^{18}O$ = 2.1—2.5‰; $\delta^{13}C$ = 1.7—1.9‰).

The isotopic composition of the carbon in solution at the orifice of the springs closely represents the $^{13}C/^{12}C$ ratio of the spring system before the water reaches the surface. The value of —2.3‰ recorded for $\delta^{13}C_{TDC}$ suggests that the carbon is derived principally from limestone.

During their travel in the open air, the waters cool and lose CO_2; as a consequence, the pH rises progressively, attaining values as high as 8.1—8.3 at 27—30°C. This implies that the dominant dissolved carbon species are CO_2 and HCO_3^- in the warmer waters and simply HCO_3^- in the cooler waters (say, at temperatures below about 50°C, corresponding to pH = ~ 8). Such changes in the proportions of the various carbon-containing species in solution and the kinetic isotope effects accompanying the rapid escape of CO_2 may be the basis of the increase of $\delta^{13}C_{TDC}$ with decreasing temperature observed in these waters. The high rate of travertine deposition occurring at the spring orifice seems to persist to a temperature of about 50°C, indicating that a large degree of supersaturation of $CaCO_3$ must occur in waters of both springs during most of their travel in the open air.

At Main Springs, oxygen isotope equilibrium between $CaCO_3$ and H_2O is generally maintained over the entire temperature range, except for the sample collected at 27°C (Fig. 5-4). At New Highland Terrace, on the other hand, all the samples but the one corresponding to the higher temperature, are out of equilibrium, with $10^3 \ln\alpha_{CaCO_3-H_2O}$ values nearly constant below about 50°C and typically smaller than the equilibrium ones (Fig. 5-4).

Friedman (1970) basically attributes this situation to the difference in the rate of movement of the waters from the two springs. Nucleation of calcium carbonate conceivably occurs as soon as the waters issue from the orifice, and continues during most of the travel in the open air. If the water runs very rapidly, as at New Highland Terrace, it is possible that the nuclei of $CaCO_3$ deposit downstream, probably after being mixed in some proportion with those formed at lower temperature. Thus, the $^{18}O/^{16}O$ disequilibrium observed at this spring could arise, at least in part, from a delay in the deposition of solid $CaCO_3$. At Main Springs, where the water moves much less rapidly, such a process should not occur to an appreciable extent and in fact travertine appears to be in isotopic equilibrium with water at the temperature measured at the sampling sites.

DIAGENETIC EFFECTS

The lithological, mineralogical, chemical and isotopic features of travertines can be profoundly altered by postdepositional, diagenetic processes.

The effects of such processes, as well as their mechanisms, have been investigated in several large deposits of central Italy. On the basis of field and laboratory observations, these travertines can, in general, be grouped into three categories (Cipriani et al., 1972, 1977; Malesani and Vannucci, 1975): (1) *present-day travertines;* (2) *"overlying"* travertines — this category embraces the relatively younger fossil travertines of the upper layers of the deposit; and (3) *"underlying"* travertines — a category generically including the travertines of the lower (older) layers of the deposit.

The three types of travertine, which may or may not be present contemporaneously in a given deposit, display different petrographic, geochemical and physical features.

Macroscopically, the present-day and overlying travertines are in general friable and highly porous, whereas those of the third category are more compact and lithified (Fig. 5-6).

Such a pattern is thought to be due to a complex diagenetic process of the pristine carbonate material, which can be summarized as follows. The early carbonate precipitated is, as a rule, a spongy, very permeable material of calcitic or aragonitic nature, depending on the chemistry of the solution. As soon as this material is no longer exposed to its formation water, it may interact with groundwaters.

The action of these waters upon porous aragonitic travertines results in a rapid inversion of aragonite to calcite, accompanied by a loss of as much as about 50% of the strontium present in the pristine precipitate. Malesani and Vannucci (1975) observed that the present-day travertines from Bagni S. Filippo, Tuscany, which are being deposited as aragonite, completely change into calcite in 10—15 days.

In some rare cases, however, the early formed aragonitic travertine occurs as a hard, practically impermeable material through which groundwater cannot circulate to a great extent.

In such travertines, the transformation from aragonite to calcite is strongly hindered. The calcitic material resulting from the recrystallization of aragonitic travertine, as well as calcitic travertine of early precipitation, can undergo further diagenetic changes. The action of the groundwater on these highly permeable carbonates results in continuous dissolution and reprecipitation processes which lead to progressive filling of the pores and cavities by secondary calcite that is more and more depleted in strontium and magnesium.

Fig. 5-6. Section of a travertine quarry in the Tivoli area, showing the porous and compact varieties and their relative proportions. The exploitation of the raw material is restricted to the compact variety.

In a number of deposits, the observed decrease in porosity with time is in part attributable to the deposition of clay minerals along with the secondary calcite, as indicated by an increase of the amount of insoluble residue in the "underlying" travertine with respect to the "overlying" and "present-day" ones.

These processes, as a whole, account for the differences in the physical and chemical features among the various types of travertine (see Table 5-2); in addition, they strongly influence the geochemistry of strontium in travertines.

In this connection, it is also worth noting that, other things being equal, the amount of Sr which enters the structure of the solid carbonate precipitated from highly supersaturated solutions where non-equilibrium conditions prevail, as is the case for most travertine-depositing waters, strongly depends upon the precipitation rate (Jacobson and Usdowski, 1975).

Diagenesis also influences the isotopic composition of travertines. Inasmuch as most diagenetic changes basically result from the progressive action of percolating groundwaters at surface temperatures, the diagenetically altered travertines should conceivably tend to acquire, in a given area, $\delta^{18}O$ values lying within a narrow range. In general, the isotopic effects of diagenesis are expected to be particularly evident in travertines precipitated either at temperatures appreciably higher than the surface ones or/and from solutions isotopically very different from the local groundwaters.

Manfra et al. (1974, 1976) observed, in the lithified travertine layers from some of the major deposits of northwestern Latium, $\delta^{18}O$ values mostly ranging from 22 to 25‰, just the range one calculates for $CaCO_3$ in equilibrium with water isotopically similar to the local meteoric waters ($\delta^{18}O = \sim -6$ to $-8‰$; Baldi et al., 1973) at a temperature close to the mean annual temperature of the area of interest (15°C).

TABLE 5-2

Average values $(\bar{x})$ and standard deviations (σ) of strontium content and porosity calculated for the present-day, overlying and underlying travertines from 32 deposits of central Italy (data from Cipriani et al., 1977)

Type of travertine	Mineralogy	Sr (ppm)		Porosity (%)	
		$\bar{x}$	σ	$\bar{x}$	σ
Present-day	aragonite	8883	1253	42.94	16.01
	calcite	4024	1082		
Overlying	aragonite	5816	1517	14.89	7.54
	calcite	1285	884		
Underlying	calcite	681	395	6.04	3.22

The effects of diagenesis on the $^{13}C/^{12}C$ ratio of travertines are not easy to predict, because they depend on several factors such as temperature, the proportion of carbon from external sources (i.e., biogenic and atmospheric) relative to that derived from the dissolution of the carbonate and, at least in some deposits, the ^{13}C effect related to the transition aragonite → calcite (Rubinson and Clayton, 1969).

RADIOCARBON DATING OF TRAVERTINES

Recently, attempts have been made to apply the radiocarbon method to dating recent travertines. As we have just seen, carbon of travertines may have a complex origin, i.e., it actually consists of a mixture of carbon from different reservoirs in proportions that are highly variable. Two of the afore mentioned sources (limestone and igneous sources) contain no radiocarbon, so that in areas where significant contributions from such sources are possible, the ^{14}C ages of travertines may be anomalously old, unless such a "dilution effect" is not properly corrected for. Also, diagenetic effects could have some effect on the ^{14}C content of travertine.

In favourable circumstances, however, the radiocarbon dating of travertines or "calcareous tufa" is claimed to give satisfactory results, at least when the total dissolved carbon in the travertine-depositing waters is in equilibrium with the atmosphere. For example, Srdoč et al. (1979) applied this method to the dating of travertines from Plitvice National Park, central Croatia, Yugoslavia. These authors believe that these travertines, in spite of their porous nature which favours exchanges with groundwaters, retained their original ^{14}C contents. They measured ages below about 40,000 years B.P. with good reproducibility.

^{14}C measurements made in conjunction with stable isotope analyses can aid in understanding the geochemistry of carbon in hydrological systems. For example, in southern Afar, Fontes et al. (1980) showed that travertine deposition was due to the mixing of two different waters. A deep supply of hot groundwater of the calcium sulphate type came up through the bottom of a lake of sodium carbonate type. Stable isotope contents of carbon indicate that equilibrium with the atmosphere was mantained during precipitation. Radiocarbon ages ranging from about 6300 to 1600 years B.P. were interpreted as representative.

SULPHUR, STRONTIUM AND LEAD ISOTOPES

Important information regarding the pattern of underground circulation of travertine-depositing waters and the possible interactions with the rocks through which these waters have moved may be obtained from the isotopic analyses of sulphur and heavy elements such as strontium and lead, normally present in travertines as minor components.

Sulphur isotopes

The usefulness of $^{34}S/^{32}S$ ratios in deciphering some geochemical features of the travertines is demonstrated by the studies by Savelli and Wedepohl (1969) and Demovič et al. (1972) (see Fig. 5-7). Savelli and Wedepohl (1969) studied the variations of $^{13}C/^{12}C$, $^{18}O/^{16}O$ and $^{34}S/^{32}S$ ratios in travertines from a number of occurrences in the Göttingen area, West Germany. The observed $\delta^{13}C$ and $\delta^{18}O$ values, averaging —8.1 and 24.0‰, respectively (see Figs. 5-2 and 5-3), indicate an origin from fresh waters rich in ^{13}C-depleted biogenic carbon at surface temperatures. The travertines are quite rich in sulphur (on the average, 5300 ppm), mainly occurring as sulphate; in the great majority of the analyzed samples, the isotopic composition of this element range from $\delta^{34}S$ = 12.1 to 20.5‰*.

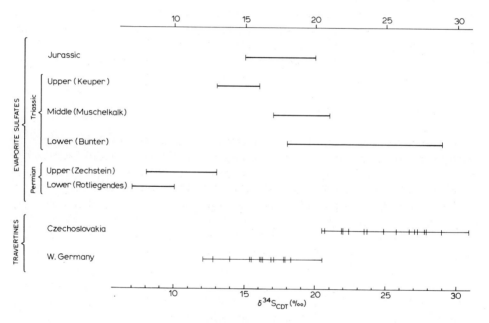

Fig. 5-7. Comparison of the $\delta^{34}S$ variations in travertines from western Germany and Czechoslovakia and in evaporite sulphates of different ages, Central Europe. Data from Savelli and Wedepohl (1969), Demovič et al. (1972) and Nielsen (1979).

---

* The $\delta^{34}S$ values are given relative to troilite sulphur from the Cañon Diablo meteorite (CDT), defined by Ault and Jensen (1962).

These data, in conjunction with the high mean strontium content (1500 ppm) strongly suggest a deep circulation of the travertine-depositing waters, which would have leached gypsum from the Zechstein (Upper Permian), Muschelkalk (Middle Triassic) and Keuper (Upper Triassic) deposits, present in the stratigraphic column. Inasmuch as marine carbonates have, on the average, S (as sulphate) and Sr contents close to 1200 and 450 ppm respectively (Clarke, 1924; Turekian and Wedepohl, 1961; Flügel and Wedepohl, 1967), it is likely that additional sources to the limestones underlying the area were involved in the travertine genesis. The low $^{13}C/^{12}C$ ratios measured in most travertine samples confirm this conclusion.

The $\delta^{34}S$ values reported by Demovič et al. (1972) indicate a contribution of evaporite sulphate of Triassic and/or Tertiary age to Slovakian travertines (Fig. 5-7). In contrast with the relative uniformity of the $^{34}S/^{32}S$ ratios ($\delta^{34}S$ = 20.5 to 30.9‰), however, other isotopic and chemical parameters, such as $\delta^{13}C$, sulphur and strontium contents, display large variations; the observed ranges are in fact $\delta^{13}C$ = −10.1 to +9.7‰ (Fig. 5-2); S = 50—7300 ppm, Sr = 25—3900 ppm. It is thus evident that the mechanism of formation is not the same for all of these travertines. For example, the fact that carbon apparently has a variety of sources (thermometamorphic reactions; organic matter; leaching of the carbonate rocks associated with the evaporite beds), suggests that the travertine-depositing waters circulated through different conduits and therefore interacted with different rock types.

Strontium and lead isotopes

Tracer studies of hydrothermal systems based on the variations in stable isotope contents of heavy elements can be particularly useful because the heavy isotopes, unlike the light ones, are not fractionated during chemical reactions of physicochemical processes. For this reason, the study of the variations of the isotopic ratios of strontium and lead in travertines can help to identify the most plausible sources of these elements, provided that representative analyses are available for the potential source rocks in the area of interest and that the differences in isotopic composition of these materials are sufficiently large compared with the analytical uncertainties. Leeman et al. (1977) utilized this technique in an attempt to elucidate the provenance of the thermal waters of Yellowstone Park. For this purpose, they analyzed a number of travertines and siliceous sinters from hot-spring deposits located both outside (Mammoth Hot Springs area, Travertine Terrace Mountain) and within (Lower and Upper Geyser Basins) the caldera rim. The data thus obtained can confidently be considered representative of the Pb and Sr isotopic compositions in the associated thermal waters. These waters are essentially of meteoric origin, and it is known that meteoric waters are in general quite low in both Pb and Sr (see, for example, Hem, 1970). Therefore, the isotopic composition of these elements in the waters associated with the hot-spring deposits are expected to be dominated by the lead and strontium in the rocks leached by such waters.

At Yellowstone, the major reservoirs of Pb and Sr are: (1) rhyolitic lavas and tuffs, and possibly a rhyolitic magma body at depth (basaltic rocks are also present, but their volume is too small to be considered as significant reservoir rocks), and (2) sedimentary (carbonate) rocks, cropping out outside the caldera. The rhyolite samples analyzed cover the entire spatial and chronological distribution of such rocks at Yellowstone. The sampling of sedimentary rocks includes a sample of Mississippian Madison limestone and a composite sample of Mesozoic sedimentary rocks from the area near Mammoth Hot Springs.

The results of the Sr and Pb isotope analyses performed on samples of travertine, siliceous sinter and source rocks are shown in Figs. 5-8 and 5-9. Also reported in Fig. 5-8 is the $^{87}Sr/^{86}Sr$ of a water sample from New Highland Spring, Mammoth Hot Springs area. The $^{87}Sr/^{86}Sr$ analyses suggest that essentially all of the potential source rocks studied could have contributed strontium to the thermal waters.

The isotope ratios of lead in the hot-spring deposits are rather uniform,

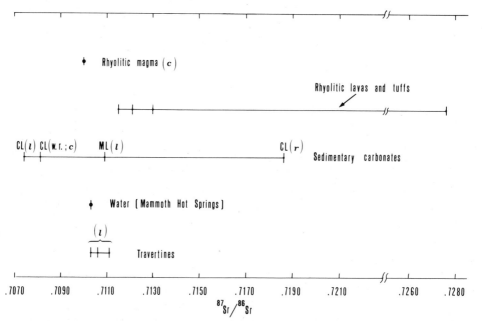

Fig. 5-8. Strontium isotopic composition of selected travertines, sedimentary carbonates, and igneous rocks from Yellowstone Park area. Also reported is the analysis of a sample of thermal water from New Highland Spring, Mammoth Hot Springs. CL = composite sample of Mesozoic limestones from the area near Mammoth Hot Springs (outside the caldera); ML = Madison Limestone (Mississippian); (c) = calculated; (l) = leach; (r) = residue after leaching; $(w.r.)$ = whole rock. Data from Leeman et al. (1977).

as those of strontium. In the diagram of Fig. 5-10, they plot between those of the limestones and of the rhyolites. Such an array conceivably represents a mixing of volcanic and sedimentary lead in the hot-spring deposits and associated waters, apparently dominated by the rhyolitic end-member.

Summing up, both sedimentary and volcanic reservoirs have contributed, to a various extent, Sr and Pb to the Yellowstone hydrothermal systems. The presence of sedimentary lead in the hot springs within the caldera indicates the occurrence of sediments in the aquifers. Inasmuch as the major rainfall in the area occurs within the caldera, it is likely that sedimentary rocks exist somewhere at depth within the caldera limits, even though they are scarce at the surface.

Strontium isotope ratios were also used, in conjunction with measure-

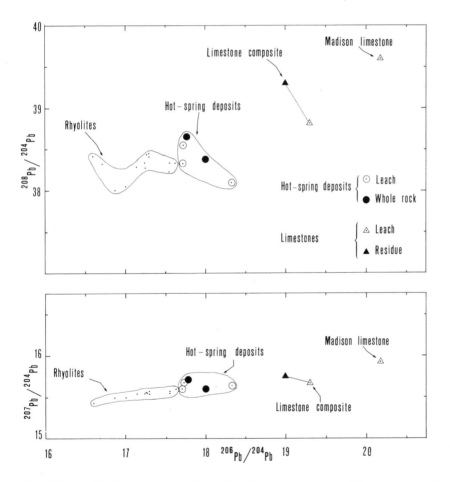

Fig. 5-9. Lead isotope contents for selected limestones, rhyolites and travertines from Yellowstone Park. Data from Leeman et al. (1977).

ments of Sr contents, to investigate the origin of this element in the travertines of Latium (Barbieri et al., 1979). In this region, strontium could have been supplied to the travertines by either sedimentary or volcanic sources, namely: (1) carbonate rocks (limestones and marly limestones of Triassic to Oligocene age) and flysch-type sediments (limestones and subordinate argillites, marls and sandstones ranging in age from Cretaceous to Miocene), present all over the region; (2) Triassic evaporites, the occurrence of which in the stratigraphic column is certain in northern Latium but not proved in southern Latium; (3) the alkali-potassic lavas, ignimbrites and pyroclastic materials of the so-called "Roman Comagmatic Region" of Pleistocene age; these rocks were erupted at a number of volcanic districts (Monti Vulsini, Vico, Monti Sabatini, Alban Hills, Monti Ernici and Roccamonfina) mostly located along the Tyrrhenian side of the region (Fig. 5-11); (4) the acidic rocks of the Monti Cimini and Tolfa-Ceriti-Manziate volcanic centers in northwestern Latium, of Pliocene and Pleistocene age.

The $^{87}Sr/^{86}Sr$ ratios of these possible source rocks are compared in Fig. 5-10 with those obtained by Barbieri et al. (1979) for eight samples from as many deposits of various localities of Latium (see Fig. 5-11).

Except for a sample distinctly enriched in ^{87}Sr, the $^{87}Sr/^{86}Sr$ ratios of the

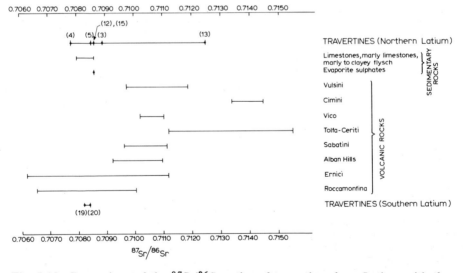

Fig. 5-10. Comparison of the $^{87}Sr/^{86}Sr$ ratios of travertines from Latium with those of the sedimentary and volcanic rocks from the same region. Data from Hurley et al. (1966), Barbieri et al. (1975, 1979), Cox et al. (1976), Vollmer (1976 and personal communication, 1978), Barbieri and Penta (1977), Carter et al. (1978), Civetta et al. (1979) and Hawkesworth and Vollmer (1979). Numbers in parentheses refer to the locations of the travertine deposits as indicated in Fig. 5-11. For the travertines and sedimentary rocks, the analytical precision of an individual run is ± 0.0007 (1σ) (Barbieri et al., 1979).

other seven are practically indistinguishable within the limits of analytical uncertainty and compatible with a provenance of strontium from sedimentary sources. This agrees with the basic geological features of the region (see, for example, Parotto and Praturlon, 1975). The ^{87}Sr-rich travertine is from the Palidoro deposit, northwestern Latium (locality *13* in Fig. 5-11). Its ^{87}Sr/^{86}Sr ratio lies within the range obtained for the nearby volcanic rocks of the Monti Ceriti district, suggesting that the bulk of strontium derives from the leaching of such rocks.

In southern Latium, the potential sedimentary reservoirs of strontium

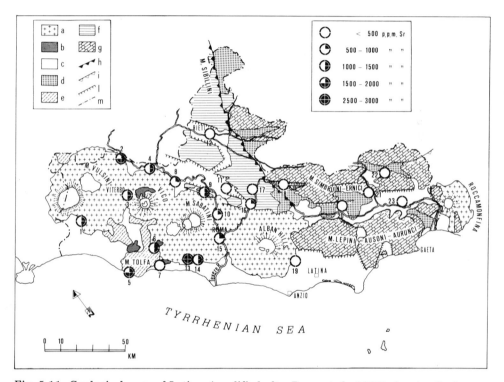

Fig. 5-11. Geological map of Latium (modified after Bono et al., 1980) showing the locations of the travertine deposits and the average Sr contents of the lithified varieties. Data from Cipriani et al. (1977) and Barbieri et al. (1979). *1* = Canino; *2* = Valteverina; *3* = Viterbo; *4* = Orte; *5* = Civitavecchia; *6* = Bagnarello; *7* = Santa Severa; *8* = Civitacastellana; *9* = Fiano Romano; *10* = Monterotondo; *11* = Poggio Moiano; *12* = Rieti; *13* = Palidoro; *14* = Torrimpietra; *15* = Roma; *16* = Tivoli; *17* = Vicovaro; *18* = Subiaco; *19* = Cisterna di Latina; *20* = Anagni-Ferentino; *21* = Isola del Liri; *22* = Casalvieri; *23* = Aquino. Legend: *a* = alkali-potassic volcanic rocks of the Roman Comagmatic Region; *b* = acidic volcanic rocks of the Tolfa-Ceriti-Manziate districts; *c* = terrigenous sediments; *d* = turbidites; *e* = allochthonous terranes ("Sicilide Complex"); *f* = carbonate pelagic sediments (Umbro-Marchigiana-Sabina Series); *g* = carbonate neritic sediments (Latium-Abruzzi Series); *h* = Sibillini-Antrodoco-Olevano line; *i* = reverse fault; *l* = normal fault; *m* = regional boundary.

have $^{87}Sr/^{86}Sr$ ratios within the range of the potassic volcanics from the Monti Ernici and Roccamonfina districts. Thus, the Sr isotope ratio does not appear to be a very promising tracer for distinguishing between volcanic and sedimentary sources of the element in the travertines from this part of Latium.

No clear-cut evidence of mixing between volcanic and sedimentary strontium is evident from the data of Fig. 5-10. If this indication proved to be true, it would have important hydrogeological implications to northwestern Latium, an area in which exploration for geothermal energy has gained increasing importance in recent years. Here, two main aquifers exist (Baldi et al., 1973): an unconfined one, essentially made up of volcanic formations covering about 3/4 of the area under consideration, and a confined one, involving carbonate and evaporite formations of the Tuscan and Umbrian series. The two aquifers are almost always separated by practically impermeable flysch-type rocks and/or clayey-sandy sediments. However, under favorable geological conditions, as in the case of the occurrence of structural "highs" of the sedimentary basement, waters from the deeper aquifer may reach the surface through fractures or faults and thus mixing with waters of the shallower aquifer. It may be significant that most travertine deposits in this part of Latium are related to structural "highs".

It is thus evident the importance of ascertaining whether mixing of waters from the two aquifers occurred in the genetic processes of travertines. Additional detailed tracer studies of the deposits of the zone are needed before significant large-scale conclusions on this subject can be drawn.

A further approach to the problem of the origin of strontium in the traver-

TABLE 5-3

Strontium contents (ppm) of the different types of travertine from various deposits of Latium

Locality	Type of travertine					
	"present day"		"overlying"		"underlying"	
Canino	—		1613 ± 631	(5)	1144 ± 267	(8)
Civitavecchia	—		2559	(2)	787 ± 265	(25)
Viterbo	8481 ± 2538	(7)	2000 ± 689	(11)	948 ± 490	(6)
Orte	2410	(2)	1558 ± 376	(9)	773 ± 329	(24)
Rieti	5376	(2)	—		403 ± 204	(7)
Tivoli	1130*	(2)	908 ± 361	(33)	856 ± 437	(69)

Classification after Cipriani et al. (1977). In parentheses is reported the number of samples analyzed. All the data are from Cipriani et al. (1977), except that noted with an asterisk which is from Barbieri et al. (1979). The locations of the deposits are shown in Fig. 5-11.

tines of Latium can be made by utilizing the broad background of knowledge available on the abundance of this element in both the travertines and the feasible source rocks of the region. As a matter of fact, the pattern of the distribution of strontium in the travertines emerging from these data is difficult to evaluate, because in the same deposit various lithofacies may occur with different strontium contents, as a consequence of the diagenetic effects discussed in the section "Diagenetic effects". This is clearly demonstrated by the data listed in Table 5-3, relative to some deposits in which at least two varieties of travertine are present.

If one is interested in the origin of strontium in travertines, the most significant data are obviously those from the materials less influenced by diagenetic processes, i.e., the present-day travertines. However, these travertines are present only in the few deposits listed in Table 5-3; therefore, the comparison of the Sr contents of travertines on a regional scale must be restricted to the older, lithified materials, arbitrarily assuming that the post-depositional processes affected the Sr content of travertines from all of the deposits in the same manner.

Fig. 5-11 was prepared by utilizing the data available for such travertines. On these grounds, it appears that the travertines from northeastern and southern Latium have relatively low Sr contents. This can be interpreted as being due to a provenance of this element from the dissolution of marine limestones (Barbieri et al., 1979). However, care must be taken in accepting such conclusion for all of the low-Sr deposits, even in the zones just mentioned. For example, in the Rieti area, northeastern Latium, the "underlying" travertines contain, on the average, about 400 ppm Sr but the "present-day" samples display Sr contents as high as about 5400 ppm (Table 5-3). This suggests that the travertine-depositing waters leached strontium from the anhydrite and/or gypsum layers present in the stratigraphic column (Parotto and Praturlon, 1975).

SUMMARY AND CONCLUSIONS

Stable-isotope studies have proved to be a suitable tool to investigate the genesis of travertines. Oxygen and carbon isotope analyses have demonstrated that the majority of travertines throughout the world deposited from waters of meteoric origin under disequilibrium conditions, mainly because of kinetic isotope effects. The deposition of these carbonates is in fact a very rapid process; deposition rates as high as $0.1-1$ mm per year have been estimated by Savelli and Wedepohl (1969). Such a process appears to be related to high concentrations of calcium- and carbon-bearing species. In fact, the deposition of travertine occurs when a sufficiently high degree of supersaturation has been achieved by the solution (i.e., for values of saturation index typically higher than about 0.5).

Travertines generally occur as calcite or aragonite. In some cases, however, Ca-Mg carbonates may form, with a spectrum of compositions from pure $CaCO_3$ to $Ca_{0.5}Mg_{0.5}CO_3$.

Carbon can be supplied to the travertine-depositing solutions by a variety of sources (organic matter, atmospheric CO_2, magmas and limestones through dissolution or decarbonation processes). The polygenic origin of this element, together with kinetic isotope effects and differences in the temperature of deposition, is at the basis of the wide range of $\delta^{13}C$ from about -24 to $+12‰$ observed in travertines.

The large range of measured $\delta^{18}O$ values from about 2 to $32‰$ vs. SMOW reflects the $^{18}O/^{16}O$ and temperature of the formation waters, as well as the equilibrium-disequilibrium conditions under which precipitation took place.

Diagenetic effects strongly influence the lithological, mineralogical, chemical and isotopic features of travertines. Dissolution-reprecipitation processes due to the action of meteoric groundwaters progressively transform the spongy, highly permeable material of early precipitation into a more compact, lithified carbonate. Aragonite, if present, rapidly changes into calcite with a loss of as much as 50% of the strontium initially present; further loss of strontium accompanies lithification of the calcitic material. Diagenesis also influences the isotopic composition of travertines. The difference in $^{18}O/^{16}O$ between the upper and lower layers observed in several travertine deposits of Latium could, at least in part, result from diagenetic effects. Studies of the contents and isotope ratios of minor components such as sulphur, strontium and lead can help to identify the types of rocks which the travertine-depositing solutions have leached. The radiogenic (Pb and Sr) isotopes are particularly useful when there is reason to believe that different rock types (e.g. sedimentary and igneous) could be involved, provided that the isotopic compositions of the potential source rocks are well established and show differences sufficiently large with respect to the analytical uncertainties.

In utilizing isotopic tracers, one should keep in mind that, in general, if the trace element selected is contributed to travertine by two (or more) reservoirs, i.e., if it has a "mixed" origin, its isotopic composition will reflect the relative proportions of the end-members. Therefore, a component of the mixing could go undetected even though isotopically quite different from the other(s), if its contribution is relatively small. Under favorable conditions, however, tracer studies can tell us a great deal about the nature of the source rocks, elemental budgets and circulation of the travertine-depositing waters and therefore have important hydrogeological implications.

ACKNOWLEDGEMENTS

The author wish to thank P. Bono, G. Cortecci, M. Fornaseri, R. Funiciello for helpful suggestions and valuable discussions. He is especially grateful to J.R. O'Neil for reviewing the manuscript and improving the English.

REFERENCES

Ault, W.U. and Jensen, M.L., 1962. Summary of sulphur isotope standards. In: M.L. Jensen (Editor), *Biogeochemistry of Sulphur Isotopes*, New Haven, Conn.

Baertschi, P., 1957. Messung und Deutung relativer Häufigkeitsvariationen von ^{18}O und ^{13}C in Karbonatgesteinen und Mineralien. *Schweiz. Mineral. Petrogr.*, 37: 73—152.

Baldi, P., Ferrara, G.C., Masselli, L. and Pieretti, G., 1973. Hydrogeochemistry of the region between Monte Amiata and Rome. *Geothermics*, 2: 124—141.

Barbieri, M. and Penta, A., 1977. La composizione isotopica dello stronzio dei complessi vulcanici Tolfetano, Cerite e Manziate (Lazio nord-occidentale). *Rend. Soc. Ital. Mineral. Petrol.*, 33: 49—53.

Barbieri, M., Penta, A. and Turi, B., 1975. Oxygen and strontium isotope ratios in some ejecta from the Alban Hills volcanic area, Roman comagmatic region. *Contrib. Mineral. Petrol.*, 51: 127—133.

Barbieri, M., Masi, U. and Tolomeo, L., 1979. Origin and distribution of strontium in the travertines of Latium (Central Italy). *Chem. Geol.*, 24: 181—188.

Barnes, I. and O'Neil, J.R., 1969. The relationship between fluids in some fresh Alpine-type ultramafics and possible modern serpentinization, western United States. *Geol. Soc. Am. Bull.*, 80: 1947—1960.

Barnes, I. and O'Neil, J.R., 1971. Calcium-magnesium carbonate solid solutions from Holocene conglomerate cements and travertines in the Coast Range of California. *Geochim. Cosmochim. Acta*, 35: 699—718.

Barnes, I., Irwin, W.P. and White, D.E., 1978. Global distribution of carbon dioxide discharges, and major zones of seismicity. *U.S. Geol. Surv., Water-Resour. Invest., Open-File Rep.*, 78—39: 12 pp.

Bono, P., Cappelli, G., Civitelli, G., Mariotti, G., Parotto, M., Fano, P. and Ventura, G., 1980. Al di là della preistoria. *E.P.T. Latina*, 43 pp.

Bottinga, Y., 1968. Calculation of fractionation factors for carbon and oxygen isotopic exchange in the system calcite—carbon dioxide—water. *J. Phys. Chem.*, 72: 800—808.

Bottinga, Y. and Craig, H., 1969. Oxygen isotope fractionation between CO_2 and water and the isotopic composition of marine atmospheric CO_2. *Earth Planet. Sci. Lett.*, 5: 285—295.

Buccino, G., D'Argenio, B., Ferreri, V., Brancaccio, L., Ferreri, M., Panichi, C. and Stanzione, D., 1978. I travertini della bassa valle del Tanagro (Campania). Studio geomorfologico, sedimentologico e geochimico. *Boll. Soc. Geol. Ital.*, 97: 1—30.

Carter, S.R., Evensen, N.M., Hamilton, P.J. and O'Nions, R.K., 1978. Continental volcanics derived from enriched and depleted source regions: Nd and Sr isotope evidence. *Earth Planet. Sci. Lett.*, 37: 401—408.

Cipriani, N., Ercoli, A., Malesani, P. and Vannucci, S., 1972. I travertini di Rapolano Terme (Siena). *Mem. Soc. Geol. Ital.*, 11: 31—46.

Cipriani, N., Malesani, P. and Vannucci, S., 1977. I travertini dell'Italia Centrale. *Boll. Serv. Geol. Ital.*, 98: 85—115.

Civetta, L., Innocenti, F., Lirer, L., Manetti, P., Munno, R., Peccerillo, A., Poli, G. and Serri, G., 1979. Serie potassica ed alta in potassio dei Monti Ernici (Lazio Meridionale):

considerazioni petrologiche e geochimiche. *Rend. Soc. Ital. Mineral. Petrol.*, 35: 227—249.

Clarke, F.W., 1924. The data of geochemistry. *U.S. Geol. Surv. Bull.*, 770: 554—580.

Cortesi, C. and Leoni M., 1958. Studio sedimentologico e geochimico del travertino di un sondaggio a Bagni di Tivoli. *Period. Mineral.*, 27: 407—458.

Cox, K.G., Hawkesworth, C.J., O'Nions, R.K. and Appleton, J.D., 1976. Isotopic evidence for the derivation of some Roman Region volcanics from anomalously enriched mantle. *Contrib. Mineral. Petrol.*, 56: 173—180.

Craig, H., 1953. The geochemistry of the stable carbon isotopes. *Geochim. Cosmochim. Acta*, 3: 53—92.

Craig, H., 1957. Isotopic standards for carbon and oxygen and correction factors for mass-spectrometric analysis of carbon dioxide. *Geochim. Cosmochim. Acta*, 12: 133—149.

Craig, H., 1961. Standard for reporting concentrations of deuterium and oxygen-18 in natural waters. *Science*, 133: 1833—1834.

Craig, H., 1963. The isotopic geochemistry of water and carbon in geothermal areas. In: E. Tongiorgi (Editor), *Nuclear Geology on Geothermal Areas — Spoleto 1963.* CNR, Pisa, pp. 17—53.

Deines, P., 1980. The isotopic composition of reduced organic carbon. In: P. Fritz and J.Ch. Fontes (Editors), *Handbook of Environmental Isotope Geochemistry, 1. The Terrestrial Environment, A.* Elsevier, Amsterdam, pp. 329—406.

Deines, P., Langmuir, D. and Harmon, R.S., 1974. Stable carbon isotope ratios and the existence of a gas phase in the evolution of carbonate ground waters. *Geochim. Cosmochim. Acta*, 38: 1147—1164.

Demovič, R., Hoefs, J. and Wedepohl, K.H., 1972. Geochemische Untersuchungen an Travertinen der Slowakei. *Contrib. Mineral. Petrol.*, 37: 15—28.

Duchi, V., Giordano, M.V. and Martini, M., 1978. Riesame del problema della precipitazione di calcite od aragonite da soluzioni naturali. *Rend. Soc. Ital. Mineral. Petrol.*, 34: 605—618.

Emrich, K., Ehhalt, D.H. and Vogel, J.C., 1970. Carbon isotope fractionation during the precipitation of calcium carbonate. *Earth Planet. Sci. Lett.*, 8: 363—371.

Fancelli, R. and Nuti, S., 1975. Studio sulle acque termali e minerali della parte orientale della Provincia di Siena. *Boll. Soc. Geol. Ital.*, 94: 135—155.

Ferreri, M. and Stanzione, D., 1978. Contributo alla conoscenza geochimica dei travertini campani: travertini di Paestum e della Valle del Tanagro (Salerno). *Rend. Accad. Sci. Fis. Mat., Soc. Naz. Sci., Lett. Arti, Napoli*, 45: 1—15.

Flügel, H.W. and Wedepohl, K.H., 1967. Die Verteilung des Strontiums in oberjurassischen Karbonatgesteinen der nördlichen Kalkalpen. *Contrib. Mineral. Petrol.*, 14: 229—249.

Folk, R.L. and Chafetz, H.S., 1980. Quaternary travertines of Tivoli (Roma), Italy: bacterially constructed carbonate rocks. *Prog., Geol. Soc. Am. Annu. Meet., Atlanta, Ga.*, p. 428 (abstract).

Fontes, J.C., Pouchan, P., Saliège, J.F. and Zuppi, G.M., 1980. Environmental isotope study of groundwater systems in the Republic of Djibouti. In: *Arid-Zone Hydrology Investigations with Isotope Techniques.* IAEA, Vienna, pp. 237—262.

Francalanci, G.P., 1958. Contributo per la conoscenza delle manifestazioni idrotermali della Toscana. *Atti Soc. Tosc. Sci. Nat.*, 45: 372—432.

Friedman, I., 1970. Some investigations of the deposition of travertine from Hot Springs, I. The isotopic chemistry of a travertine-depositing spring. *Geochim. Cosmochim. Acta*, 34: 1303—1315.

Friedman, I. and O'Neil, J.R., 1977. Compilation of stable isotope fractionation factors of geochemical interest. In: M. Fleischer (Technical Editor), *Data of Geochemistry. U.S. Geol. Surv., Prof. Pap.*, 440-KK (6th ed.).

Fritz, P., 1965. Composizione isotopica dell'ossigeno e del carbonio nei travertini della Toscana. *Boll. Geofis. Teor. Applic.*, 7: 25—30.

Fritz, P., 1968. Der Isotopengehalt der Mineralwasserquellen von Stuttgart und Umgebung und ihrer mittelpleistozänen Travertin-Ablagerungen. *Jahresber. Mitt. Oberrheinischen Geol. Ver.*, 50: 53—69.

Gonfiantini, R., Panichi, C. and Tongiorgi, E., 1968. Isotopic disequilibrium in travertine deposition. *Earth Planet. Sci. Lett.*, 5: 55—58.

Hawkesworth, C.J. and Vollmer, R., 1979. Crustal contamination versus enriched mantle: $^{143}Nd/^{144}Nd$ and $^{87}Sr/^{86}Sr$ evidence from the Italian volcanics. *Contrib. Mineral. Petrol.*, 69: 151—165.

Hem, J.D., 1970. *Study and Interpretation of the Chemical Characteristics of Natural Water. U.S. Geol. Surv., Water-Supply Pap.*, 1473: 363 pp. (2nd ed.).

Hoefs, J., 1978. Some peculiarities in the carbon isotope composition of "juvenile" carbon. In: B.W. Robinson (Editor), *Stable Isotope in the Earth Sciences. N.Z. Dep. Sci. Ind. Res., Bull.*, 220: 181—184.

Holland, H.D., 1978. *The Chemistry of the Atmosphere and Oceans.* John Wiley and Sons, New York, N.Y., 351 pp.

Hurley, P.M., Fairbairn, H.W. and Pinson, W.H., Jr., 1966. Rb-Sr isotopic evidence in the origin of potash-rich lavas of Western Italy. *Earth Planet. Sci. Lett.*, 5: 301—306.

Jacobson, R.L. and Usdowski, E., 1975. Geochemical controls on a calcite precipitating spring. *Contrib. Mineral. Petrol.*, 51: 65—74.

Keeling, C.D., 1958. The concentration and isotopic abundances of atmospheric carbon dioxide in rural areas. *Geochim. Cosmochim. Acta*, 13: 322—334.

Keeling, C.D., Mook, W.G. and Tans, P.P., 1979. Recent trends in the $^{13}C/^{12}C$ ratio of atmospheric carbon dioxide. *Nature*, 277: 121—122.

Keith, M.L. and Weber, J.N., 1964. Carbon and oxygen isotopic composition of selected limestones and fossils. *Geochim. Cosmochim. Acta*, 28: 1787—1816.

Leeman, W.P., Doe, B.R. and Whelan, J., 1977. Radiogenic and stable isotope studies of hot-spring deposits in Yellowstone National Park and their genetic implications. *Geochem. J.*, 11: 65—74.

Livingstone, D.A., 1963. Chemical composition of rivers and lakes. *U.S. Geol. Surv., Prof. Pap.*, 444-G: 1—64.

Malesani, P. and Vannucci, S., 1975. Precipitazione di calcite o di aragonite dalle acque Termominerali in relazione alla genesi e all'evoluzione dei travertini. *Atti Accad. Naz. Lincei, Rend. Sci. Fis., Mat., Nat.*, 58: 761—776.

Manfra, L., Masi, U. and Turi, B., 1974. Effetti isotopici nella diagenesi dei travertini. *Geol. Romana*, 13: 147—155.

Manfra, L., Masi, U. and Turi, B., 1976. La composizione isotopica dei travertini del Lazio. *Geol. Romana*, 15: 127—174.

Mook, W.G., 1970. Stable carbon and oxygen isotopes of natural waters in the Netherlands. In: *Isotope Hydrology 1970.* IAEA, Vienna, pp. 163—190.

Mook, W.G., Bommerson, J.C. and Staverman, W.H., 1974. Carbon isotope fractionation between dissolved bicarbonate and gaseous carbon dioxide. *Earth Planet. Sci. Lett.*, 22: 169—176.

Nielsen, H., 1979. Sulphur isotopes. In: E. Jäger and J. C. Hunziker (Editors), *Lectures in Isotope Geology*, Springer-Verlag, Berlin, pp. 283—312.

O'Neil, J.R. and Barnes, I., 1971. ^{13}C and ^{18}O compositions in some fresh-water carbonates associated with ultramafic rocks and serpentinites: western United States. *Geochim. Cosmochim. Acta*, 35: 687—697.

O'Neil, J.R., Clayton, R.N. and Mayeda, T.K., 1969. Oxygen isotope fractionation in divalent metal carbonates. *J. Chem. Phys.*, 51: 5547—5558.

O'Neil, J.R., Adami, L.H. and Epstein, S., 1975. Revised value for the ^{18}O fractionation between CO_2 and water at 25°C. *U.S. Geol. Surv., J. Res.*, 3: 623—624.

Panichi, C. and Tongiorgi, E., 1975. Carbon isotopic composition of CO_2 from springs, fumaroles, mofettes and travertines of central and southern Italy: a preliminary prospection method of geothermal area. *Proc. 2nd Symp. Development and Use of Geothermal Resources, San Francisco, Calif.*, pp. 815—825.

Parotto, M. and Praturlon, A., 1975. Geological summary of the Central Apennines. In: L. Ogniben, M. Parotto and A. Praturlon (Editors), *Structural Model of Italy. Quad. Ric. Sci. CNR*, 90: 257—311.

Penta, F., 1956. I materiali da costruzione del Lazio. *Ric. Sci CNR (Suppl.)*, 26: 1—201.

Rubinson, M. and Clayton, R.N., 1969. Carbon-13 fractionation between aragonite and calcite. *Geochim. Cosmochim. Acta*, 33: 997—1002.

Sanders, J.E. and Friedman, G.M., 1967. Origin and occurrence of limestones. In: G.V. Chilingar, H.J. Bissel and R.W. Fairbridge (Editors), *Developments in Sedimentology, 9A. Carbonate Rocks*. Elsevier, Amsterdam, pp. 169—265.

Savelli, C. and Wedepohl, K.H., 1969. Geochemische Untersuchungen an Sinterkalken (Travertinen). *Contrib. Mineral. Petrol.*, 21: 238—256.

Shieh, Y.N. and Taylor, H.P., Jr., 1969. Oxygen and carbon isotope studies of contact metamorphism of carbonate rocks. *J. Petrol.*, 10: 307—331.

Srdoč, D., Obelic, B. and Horvatinčić, 1979. Radiocarbon dating of calcareous tufa: how reliable data can we expect? *Abstr., 10th Int. Radiocarbon Conf., Berne-Heidelberg*.

Turekian, K.K. and Wedepohl, K.H., 1961. Distribution of the elements in some major units of the earth's crust. *Geol. Soc. Am. Bull.*, 72: 175—191.

Turner, J.V., 1982. Kinetic fractionation of carbon-13 during calcium carbonate precipitation. *Geochim. Cosmochim. Acta*, 46: 1183—1191.

Usdowski, E., Hoefs, J. and Menschel, G., 1979. Relationship between ^{13}C and ^{18}O fractionation and changes in major element composition in a recent calcite-depositing spring. A model of chemical variations with inorganic $CaCO_3$ precipitation. *Earth Planet. Sci. Lett.*, 42: 267—276.

Vogel, J.C., 1961. Isotope separation factors of carbon in the equilibrium system CO_2-HCO_3^--CO_3^{2-}. In: *Summer Course on Nuclear Geology, Varenna 1960*. Laboratorio di Geologia Nucleare, CNEN, Pisa, pp. 216—221.

Vogel, J.C. and Ehhalt, D., 1963. The use of the carbon isotopes in groundwater studies. In: *Radioisotopes in Hydrology*. IAEA, Vienna, pp. 385—395.

Vollmer, R., 1976. Rb-Sr and U-Th-Pb systematics of alkaline rocks: the alkaline rocks from Italy. *Geochim. Cosmochim. Acta*, 40: 283—295.

Chapter 6

ISOTOPE GEOCHEMISTRY OF CARBONATES IN THE WEATHERING ZONE

W. SALOMONS and W.G. MOOK

INTRODUCTION

Carbonates are an important constituent of soils because they act as a buffer for pH changes. The pH influences, to a large extent, the availability of macro- and micronutrients for plant growth and is a major factor in the geochemical transformations of soil constituents.

Carbonates are not a stable phase in regions with moderate to high rainfall (i.e. most temperate and tropical climates). Carbonates from soil in these regions ultimately disappear from the soil. However, the overall dissolution process may be interrupted by periods of reprecipitation in the upper part of the soil. On the other hand, carbonates which have disappeared from the upper part of the soil are (partly) reprecipitated deeper in the soil profile. It is possible to define the "efficiency" of this dissolution-reprecipitation pro-

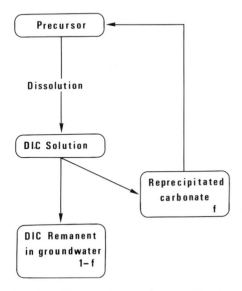

Fig. 6-1. Scheme of the processes affecting carbonates in the weathering zone.

cess as the fraction of carbonate which is reprecipitated by the initial amount of dissolved inorganic carbon (DIC) in solution (Fig. 6-1).

Apart from the transformation of pre-existing carbonates, local formation of calcium carbonate in the weathering zone may take place. It is known that earth worms excrete small globules of calcium carbonate (Wiececk and Messenger, 1972), also shells from snails may contribute to the calcium carbonate present in soils (Yapp, 1979). More important than these biogenic carbonates is the weathering of silicate rocks in which the weathered minerals are replaced by calcium carbonate (calcareous epigenesis).

The scale of carbonate accumulation in the weathering zone ranges from microscopic particles to massive developments of layers up to 30 m thick (calcrete). The latter are dominant in controlling geomorphological features and landscapes (Goudie, 1973; Fig. 6-2).

Isotope geochemistry gives unique information on the processes affecting carbonates in the weathering zone. Additionally, information on the isotopic composition of soil carbonates is a prerequisite for modelling the composition of groundwater.

Fig. 6-2. Massive calcrete crust in the landscape (Botswana; photograph by Dr. A. Goudie, University of Oxford).

ISOTOPIC COMPOSITION OF CARBON AND OXYGEN SOURCES IN THE WEATHERING ZONE

The origin of a dissolved inorganic carbon (DIC) solution in the soil can be ascribed to a number of processes (see review by Mook, 1980), such as:
— the dissolution of carbonates by the action of carbon dioxide (Münnich, 1957):

$$CaCO_3 + CO_2 + H_2O = Ca^{2+} + 2\,HCO_3^-$$

— the weathering of silicate minerals (Vogel and Ehhalt, 1963):

$$NaAlSi_3O_8(s) + CO_2 + \tfrac{11}{2} H_2O = Na^+ + HCO_3^- + 2\,H_4SiO_4 + \tfrac{1}{2} Al_2Si_2O_5(OH)_4(s)$$
albite $\qquad\qquad\qquad\qquad\qquad\qquad\qquad\qquad\qquad\qquad$ kaolinite

— the dissolution of calcium carbonate by humic acids (Vogel and Ehhalt, 1963; Mook, 1970):

$$2\,CaCO_3 + 2\,H(Hum) = Ca(Hum) + Ca^{2+} + 2\,HCO_3^-$$

The second reaction is slow but is thought to be responsible for the formation of certain calcrete formations (Millot et al., 1979) and for carbonates in soils formed on carbonate free material (Leamy and Rafter, 1972), whereas the third reaction may be important in soils with high contents of organic matter. The isotopic composition of the resulting DIC depends on the source of inorganic carbon and the oxygen isotopic composition of the water and on the processes involved.

In this section the sources for carbon and oxygen in the DIC will be discussed, the various processes leading to the formation of DIC in the soil will be discussed in the next section.

The soil CO_2 content

The atmosphere contains carbon dioxide at a partial pressure of about $10^{-3.5}$ (about 0.03%). It was shown by Keeling (1958) that the δ^{13} value of atmospheric CO_2 is closely related to the partial CO_2 pressure. Single data over a short time period obey a simple mixing equation (valid for January 1, 1956):

$$\delta^{13} = -26.97 + 6271.63/P_{CO_2}$$

where P_{CO_2} is the CO_2 concentration in ppm. The varying concentration of CO_2 was appointed to the combustion of fossil fuel and biospheric activity. A similar equation for January 1, 1978 (Keeling et al., 1979, slightly corrected):

$$\delta^{13} = -26.64 + 6346.98/P_{CO_2} \tag{0}$$

The relatively small change of δ^{13} for global average and uncontaminated (by fossil fuel CO_2) air is due to exchange with the oceans. After applying a correction for the N_2O content of air, interfering with the mass spectrometric analysis (Craig and Keeling, 1963), δ^{13} and P_{CO_2} have changed from δ^{13} = $-6.69‰$ at P_{CO_2} = 314.2 ppm (at 1-1-1956) to $\delta^{13} = -7.34‰$ at P_{CO_2} = 334.2 ppm (at 1-1-1978) (slight correction of values by Keeling et al., 1979, included). At present δ^{13} of atmospheric CO_2 is decreasing by $-0.02‰$ per ppm.

The δ^{13} value for atmospheric carbon dioxide over continents is lower due to an admixture of carbon dioxide derived from respiration, decay of plant material and combustion of fossil fuel (Keeling, 1961). During their passage through the atmosphere, rainwater droplets may take up carbon dioxide. The amount taken up is rather small (Gorham, 1955) and is in general not a major contribution to the bicarbonate in soil water.

The carbon dioxide in the soil atmosphere may be several orders of magnitude higher than in the atmosphere. For soils in temperature regions the carbon dioxide levels vary between 0.03 and 11.5% (Atkinson, 1977), with a mean value of 0.9%. CO_2 production in the soil depends on the vegetation and on the microbial degradation of organic matter. In soils from the Mediterranean, the production is low in the dry summer periods, increases with the autumn rains and decreases again during the cold season (Billès et al., 1971). In temperate climates there is in general a net flux of carbon dioxide from the soil to the atmosphere. Vogel (1970) observed values for the carbon isotopic composition of carbon dioxide above the soil in dense forests in western Germany of $-23.4‰$, Broecker and Olsson (1961) gave a value of $-24.9‰$ for carbon dioxide emanating from soils in forests. Also, in some deserts a net efflux of carbon dioxide from the soil is found during the growing season (Caldwell et al., 1977).

Direct determination of the carbon dioxide in the soil atmosphere shows a range of values from -10 to $-28‰$ (see also Deines, 1980, fig. 9-9). The observed values clearly show two maxima, one at -23 and the other at $-16‰$. The isotopic composition appears to be related to the photosynthetic cycle used by the vegetation.

Three photosynthetic pathways have been recognized with differences in the isotopic composition of the plant material (Smith and Epstein, 1971; Bender, 1971; Lerman, 1972; Throughton, 1972; Allaway et al., 1974; for a review see Deines, 1980):

— the C_3 or Calvin cycle. The isotopic composition of the various plants ranges from -22 to $-34‰$ at with a maximum at $-27 \pm 2‰$.

— the C_4 or Hatch-Slack cycle. The isotopic composition of the various plants varies between -9 and $-19‰$ with a maximum at $12 \pm 2‰$.

— the Crassulacean acid metabolism (CAM) cycle. The carbon isotopic composition is intermediate between C_3 and C_4 averaging around $-17‰$.

In areas covered by vegetation consisting of an isotopically light plant material, the carbon isotopic composition of the soil-CO_2 also has low values (range between -20 and $-25‰$). Values higher than $-20‰$ are observed for areas with an isotopically heavy plant cover (C_4 dominated). In areas from which the composition of the soil-CO_2 and plant cover was determined, it was found that the soil-CO_2 is always slightly less negative. Rightmire and Hanshaw (1973) observed an enrichment of $2.6-9.1‰$ for samples from central Florida, while Fritz et al. (1978b) found an enrichment of $5-8‰$ for samples from Canada. The enrichment has been explained by the preferential uptake of ^{12}C during the formation of plant material and the release of ^{13}C enriched respiration CO_2 (Park and Epstein, 1960). Fritz et al. (1978a) observed no fractionation during decay of organic matter into CO_2. However, a microbiological fractionation cannot be fully excluded.

Rightmire (1978) found distinct seasonal cycles in the composition of soil-CO_2. During the winter period, with low P_{CO_2}, the isotopic composition of the soil-CO_2 is close to atmospheric values (about $-9.6‰$). Between May and June with the onset of vegetation growth a rapid decrease to $-21.5‰$ parallel to an increase in P_{CO_2} was observed. A similar correlation between low P_{CO_2} and high δ^{13} on the one hand, high P_{CO_2} and low δ^{13} on the other has also been reported by Fritz et al. (1978b) for forest soils in Canada. The high values for soil-CO_2 during the periods of low plant activity were believed to be caused by an admixture of atmospheric carbon dioxide.

A micro-oscillation of the atmospheric pressure might produce a slight (apparent) eddy diffusion in the uppermost soil layers (Kraner et al., 1964); however, not at a depth of several centimeters.

Dörr and Münnich (1980) recently explained a ^{13}C enrichment in soil-CO_2 between the root zone and the soil surface by the process of diffusional fractionation. This effect is similar to that discussed by Craig (1954). During diffusion of CO_2 through air, the fractionation is given by:

$$\alpha = \left(\frac{m^*_{CO_2} \cdot m_A}{m^*_{CO_2} + m_A} \cdot \frac{m_{CO_2} + m_A}{m_{CO_2} \cdot m_A} \right)^{1/2}$$

where the asterisk refers to $^{13}CO_2$ and m_A stands for the average molecular mass of air ($\simeq 29$). The resulting fractionation $\alpha = 1.0044$, which means that the average ^{13}C content of the soil-CO_2 between root zone and soil surface is $4.4‰$ higher than the produced CO_2, for instance $-21.6‰$ versus $-25‰$.

The soil carbonate content

The precursors for the carbonates in the soil (Figs. 6-1 and 6-3) are mainly ancient marine carbonate rocks. The carbon isotopic composition of carbonate rocks of various ages has been reviewed by Keith and Weber (1964) and by Veizer and Hoefs (1976). The majority of the ancient marine carbonates has a carbon isotopic composition between +2 and $-2^0/_{00}$.

The isotopic composition of biogenic carbonates (shells from snails) has been discussed by Manze et al. (1974), Yapp (1979) and Magaritz et al. (1981). In arid environments bacteria seem to be involved in the formation of carbonates (Krumbein, 1968. 1972); unfortunately, information on the isotopic composition of these carbonates is lacking.

The soil water

The oxygen isotopic composition of the soil water is initially and primarily determined by that of the precipitation and, therefore, shows regional variations. It decreases from the equator to the poles and from the coast to

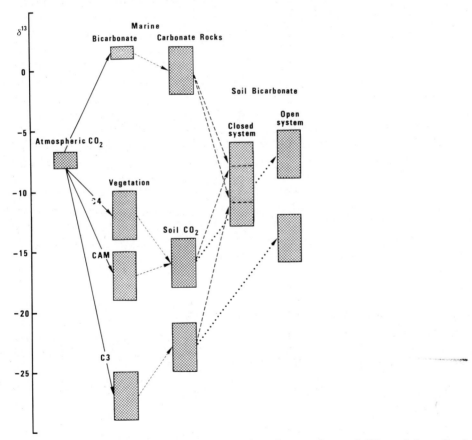

Fig. 6-3. Scheme of carbon isotopic composition of vegetation, soil-CO_2 and the soil solution in the weathering zone.

TABLE 6-1

Reactions, equilibrium constants and isotope geochemical definitions for the system CO_2-H_2O-$CaCO_3$[a]

Reactions

$$CO_{2\,g} + H_2O \rightleftarrows H_2CO_3 \tag{1}$$

$$H_2CO_3 \rightleftarrows H^+ + HCO_3^- \tag{2}$$

$$HCO_3^- \rightleftarrows H^+ + CO_3^{2-} \tag{3}$$

$$CO_3^{2-} + Ca^{2+} \rightleftarrows CaCO_3 \tag{4}$$

Equilibrium constants

$$K_W = [H^+][OH^-] \tag{5}$$

$$K_H = [H_2CO_3]/P_{CO_2} \tag{6}$$

$$K_1 = [H^+][HCO_3^-]/[H_2CO_3] \tag{7}$$

$$K_2 = [H^+][CO_3^{2-}]/[HCO_3^-] \tag{8}$$

$$K_c = [Ca^{2+}][CO_3^{2-}] \tag{9}$$

Concentration condition and fractional carbon species

$$C_T = [H_2CO_3] + [HCO_3^-] + [CO_3^{2-}] \tag{10}$$

$$[H_2CO_3] = f_1 C_T \tag{11}$$

$$[HCO_3^-] = f_2 C_T \tag{12}$$

$$[CO_3^{2-}] = f_3 C_T \tag{13}$$

Definitions isotope geochemistry

Isotopic ratio:

$$R = \frac{\text{abundance of the rare isotope}}{\text{abundance of the abundant isotope}} \quad \text{where } R = {}^{13}C/{}^{12}C,\ {}^{18}O/{}^{16}O \tag{14}$$

Isotopic fractionation factor between substances A and B:

$$\alpha_A(B) = R_B/R_A \tag{15}$$

Fractionation:

$$\epsilon = (\alpha - 1) \times 10^3\ \permil \tag{16}$$

Isotopic ratio of sample A relative to that of standard S:

$$\delta_S(A) = (R_A - R_S)/R_S \times 10^3\ \permil \tag{17}$$

Differences in isotopic ratio between two substances:

$$\delta_S(B) - \delta_S(A) = \epsilon_A(B)[1 + \delta_S(A) \times 10^{-3}] \simeq \epsilon_A(B) \tag{18}$$

$$\delta_S(B) \simeq \delta_S(A) + \epsilon_A(B) \tag{19}$$

Carbon isotopic composition in the CO_2-$CaCO_3$ system

$$\delta^{13}(C_T) = f_1\,\delta^{13}(CO_{2\,aq}) + f_2\,\delta^{13}(HCO_3^-) + f_3\,\delta^{13}(CO_3^{2-}) \tag{20}$$

$$\delta^{13}(HCO_3^-) = \delta^{13}(CO_{2\,g}) + \epsilon^{13}_{CO_{2\,g}}(HCO_3^-) \tag{21}$$

$$\delta^{13}(HCO_3^-) = \delta^{13}(C_T) - f_1\,\epsilon^{13}_{HCO_3^-}(CO_{2\,aq}) - f_3\,\epsilon^{13}_{HCO_3^-}(CO_3^{2-}) \tag{22}$$

$$\delta^{13}(CaCO_3) \simeq \delta^{13}(HCO_3^-) + \epsilon^{13}_{HCO_3^-}(CaCO_3) \tag{23}$$

$$\delta^{18}(CaCO_3) \simeq \delta^{18}(H_2O) + \epsilon^{18}_{H_2O}(CaCO_3) \tag{24}$$

[a] Values for the equilibrium constants have been given by Harned and Davis, 1943; Harned and Scholes, 1941; Langmuir, 1971; Jacobson and Langmuir, 1974; Picknett, 1964, 1973; Mook and Koene, 1975. Data on the isotope fractionations are given in Table 6-3.

the interior of the continents. A discussion of the processes affecting the isotopic composition of precipitation has been given by Gat (1980).

Changes in the isotopic composition are induced during infiltration in the upper part of the soil. From data obtained in Israël, it appears that one has to allow for an enrichment (compared to δ^{18} of the rainfall) of up to $0.5^o/_{oo}$ by evaporation (Gat and Tzur, 1967). The evapotranspiration process as such does not introduce a fractionation. As the water is taken up by the roots in large quantities relative to the volume of the plant veins, the δ^{18} of the water cannot be affected by diffusional fractionation.

Large fractionation for ^{18}O is to be expected, however, if the surface water or shallow soil water is subjected to evaporation. This effect has been observed in many groundwaters in semi-arid regions (Fontes, 1980).

THE FORMATION OF DISSOLVED INORGANIC CARBON

In discussing the dissolution of carbonates a distinction has to be made between open and closed systems (Carrels and Christ, 1965; Deines et al., 1974; Mook, 1980; Stumm and Morgan, 1981). In an open system, calcium carbonate reacts with water in contact with a gas phase of fixed P_{CO_2}. In a closed system the water first equilibrates with a carbon dioxide reservoir, is subsequently isolated from the reservoir and finally reacts with the carbonates (Fig. 6-3).

Open-system dissolution of calcium carbonates

During the dissolution of the calcium carbonate, the system remains open to the soil-CO_2 reservoir. The concentration of the dissolved CO_2 is given by Henry's law (Stumm and Morgan, 1981):

$$K_H = [CO_{2\ aq}]/P_{CO_2} \tag{6}$$

Combining this with equation (11) from Table 6-1 gives:

$$C_T = K_H P_{CO_2}/f_1 \tag{25}$$

and the dissolved bicarbonate concentration (using equation (12)):

$$[HCO_3^-] = \frac{f_2}{f_3} K_H P_{CO_2} \tag{26}$$

The logarithm of the dissolved bicarbonate concentration is a linear function of the pH (Fig. 6-4).

The extent to which calcium carbonate is dissolved is equivalent to the increase in the calcium concentration:

$$\Delta[Ca^{2+}] \simeq \tfrac{1}{2}\Delta[HCO_3^-] \tag{27}$$

This means that the dissolution of calcium carbonate follows the line given by equation (26).

The solubility of calcium carbonate is given by:

$$CaCO_{3\,s} + H^+ = Ca^{2+} + HCO_3^- \tag{28}$$

The equilibrium constant for this reaction is:

$$K_S = [Ca^{2+}][HCO_3^-]/[H^+] \tag{29}$$

If it is assumed that the alkalinity in the soil solution originates from the calcium carbonate dissolution, then:

$$2[Ca^{2+}] = [HCO_3^-] \tag{30}$$

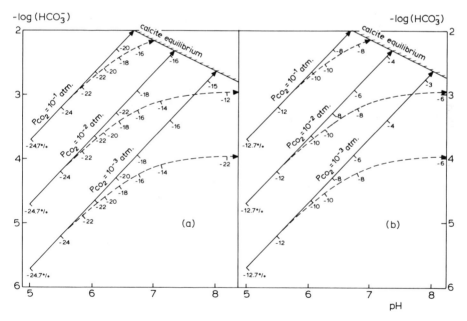

Fig. 6-4. Dissolution paths of calcium carbonate for open (diagonal lines) and closed (curved lines) systems: (a) For soil dominated by C_3 vegetation; (b) for soils dominated by C_4 vegetation.

and equation (29) can be rewritten as:

$$[HCO_3^-] = (K_S \times 2[H^+])^{1/2} \tag{31}$$

In an open system complete isotopic equilibrium may ultimately occur between the soil-CO_2 and the inorganic carbon species in solution. The composition then is not influenced by that of the dissolving calcium carbonate, but is determined by the δ^{13} of the soil-CO_2, the relative proportions of the inorganic species (f_1 to f_3 in equations (11) to (13)) and the fractionation factors involved between $CO_{2\,g}$ and the inorganic species (Table 6-2A):

$$\delta^{13}(C_T) = f_1 \delta^{13}(CO_{2\,aq}) + f_2 \delta^{13}(HCO_3^-) + f_3 \delta^{13}(CO_3^{2-}) \tag{32}$$

TABLE 6-2A

Isotope fractionation (in ⁰/₀₀ for carbon in the system CO_2 g, CO_2 aq, HCO_3^-, CO_3^{2-}, $CaCO_3$

t (°C)	$\epsilon^{13}_{HCO_3^-}(CO_{2\,g})$ [a]	$\epsilon^{13}_{CO_{2\,g}}(CO_{2\,aq})$ [b]	$\epsilon^{13}_{HCO_3^-}(CO_{2\,aq})$ [c]	$\epsilon^{13}_{HCO_3^-}(CO_3^{2-})$ [d]	$\epsilon^{13}_{HCO_3^-}(CaCO_3)$ [e]
5	−10.20	−1.15	−11.35	−0.60	−0.11
15	−9.02	−1.10	−10.12	−0.49	+0.41
25	−7.92	−1.06	−8.97	−0.39	+0.91
35	−6.88	−1.02	−7.90	−0.29	+1.37

[a] Mook et al. (1974).
[b] Vogel et al. (1970).
[c] From Mook et al. (1974) and Vogel et al. (1970).
[d] Thode et al. (1965) and Mook et al. (1974); values for CO_3^{2-} are derived from theory should be considered with caution.
[e] According to our evaluation of the original data by Emrich et al. (1970) and Rubinson Clayton (1969).

TABLE 6-2B

Oxygen isotope fractionations (in ⁰/₀₀) in the system H_2O-$CaCO_3$

t (°C)	$\epsilon^{18}_{H_2O\,liq}(H_2O\,vap)$ [a]	$\epsilon^{18}_{H_2O}(CaCO_3)$ [b]
5	−11.09	+2.67
15	−10.17	+0.33
25	−9.32	−1.85
35	−8.51	−3.89

[a] Majoube (1971).
[b] Epstein (1976) (equilibrium refers to CO_2 from $CaCO_3$ by H_3PO_4 and CO_2 in equilibrium with H_2O at 25°C).

The fractional amounts f_1 to f_3 can be calculated from the equilibrium constants and are a function of the pH. At normal pH values for calcite saturation of soils (around 7), the dominant species is HCO_3^- and the isotopic composition of the soil solution is approximately given by:

$$\delta^{13}(C_T) \simeq \delta^{13}(HCO_3^-) = \delta^{13}(CO_{2\,g}) + \epsilon^{13}_{CO_{2\,g}}(HCO_3^-) \tag{33}$$

For C_4-dominated vegetation with an isotopic composition of soil-CO_2 of about $-16‰$, this results in a $\delta^{13}(HCO_3^-)$ of about $-7‰$ at $15°C$, and for C_3-dominated vegetation (with an isotopic composition of soil-CO_2 of about $-23‰$) this results in a $\delta^{13}(HCO_3^-)$ of about $-14‰$ at $15°C$ (Fig. 6-3).

Closed-system dissolution of calcium carbonate

In a closed system, the water equilibrates first with the gas phase and is then brought into contact with the calcium carbonate; no further exchange with the gas reservoir takes place. The dissolved carbon dioxide (H_2CO_3) reacts with the $CaCO_3$ according to:

$$H_2O + CO_2 + CaCO_{3\,s} = Ca^{2+} + 2\,HCO_3^- \tag{34}$$

The extent of the calcium carbonate dissolution is given by equation (29). The acidity of the system:

$$Acy = 2[H_2CO_3] + [HCO_3^-] + [H^+] - [OH^-] \tag{35}$$

does not change with the dissolution of calcium carbonate (see also Deffeyes, 1965):

$$Acy \simeq C_T(2f_1 + f_2) \simeq \text{constant} \tag{36}$$

Acy can be calculated for the same initial conditions as for the open system using equation (25):

$$Acy \simeq \frac{2f_1 + f_2}{f_1} K_H P_{CO_2} \simeq \text{constant} \tag{37}$$

During the dissolution of calcium carbonate C_T increases, however Acy has to remain constant; this can only be accomplished by changes in f_1 and f_2. Above pH values of 7.5, f_1 becomes very low and C_T virtually only consists of HCO_3^-, because the acidity has to remain constant. As a consequence little calcium carbonate will dissolve, ultimately for partial pressures of 10^{-2} and 10^{-3} leading to soil solutions undersaturated with respect to calcite.

In the closed system, the water first equilibrates with the gas and is then brought into contact with the calcium carbonate; no further exchange with the gas reservoir takes place. Therefore, the $\delta^{13}(C_T)$ depends on the original composition $(\delta^{13}(C_{T,i}))$ and the amount of dissolved calcium carbonate, which in its turn is determined by P_{CO_2}. The $\delta^{13}(C_T)$ is given by:

$$\delta^{13}(C_T) = \{\delta^{13}(C_{T,i})[C_{T,i}] + \delta^{13}(CaCO_3)[Ca^{2+} - Ca_i^{2+}]\}/[C_T] \tag{38}$$

Selected $\delta^{13}(C_T)$ values for the dissolution paths are given in Fig. 6-4.

At normal pH values for calcareous soils, the reactions (1) to (4) reduce to the reaction (Münnich, 1957; Vogel, 1959):

$$n\,H_2O + m\,CO_2 + n\,CaCO_3 = n\,Ca^{2+} + 2n\,HCO_3^- + (m-n)\,CO_2 \tag{39}$$

Half of the bicarbonate in the soil solution is derived from the dissolved carbonate and the other half from soil-CO_2. The isotopic composition of the soil solution is then given by:

$$\delta^{13}(C_T) = n\,\delta^{13}(CaCO_3) + m\,\delta^{13}(CO_2)/(n+m) \tag{40}$$

where $n + m = [C_T]$ and $n = [Ca^{2+}] = \tfrac{1}{2}[HCO_3^-]$.
On the other hand:

$$\delta^{13}(C_T) = \tfrac{1}{2} f_2 [\delta^{13}(CaCO_3) + \delta^{13}(CO_2)] + f_1\,\delta^{13}(CO_{2\,aq}) =$$
$$= (f_1 + f_2)\delta^{13}(HCO_3^-) + f_1\,\epsilon^{13}_{HCO_3^-}(CO_{2\,aq}) \tag{41}$$

The latter relation includes the fact that the HCO_3^- and $CO_{2\,aq}$ fractions are in isotopic equilibrium. If calcium carbonate with a δ^{13} value of $+1‰$ is dissolved by soil-CO_2 of $-23‰$, the resulting soil solution will have an isotopic composition of about $-11‰$, and for soil-CO_2 with an isotopic composition of $-16‰$, the resulting $\delta^{13}(HCO_3^-)$ under closed conditions is about $-8‰$ (Fig. 6-4).

This simplified treatment of the DIC soil solution only holds, however, if the additional calcium concentration $[Ca^{2+} - Ca_i^{2+}]$ is derived from dissolution of calcium carbonate. In case of the additional dissolution of gypsum, the $[Ca^{2+}]$ increase has to be corrected, for instance, by means of the associated increase in $[SO_4^{2-}]$ (Reardon and Fritz, 1978; Fontes and Garnier, 1979). These authors similarly allow for the water-clay interaction resulting in exchange of Ca^{2+} for Na^+.

Another point of interest discussed by Reardon and Fritz (1978) and by Wigley et al. (1978) is the phenomenon of *incongruent dissolution*, instead of the *congruent dissolution* according to equation (39). In this process (secondary) calcite precipitates from the groundwater after calcite saturation has been reached, while dissolution of, for instance, dolomite occurs. The extent of this process is indicated by an increase in the [Mg^{2+}] concentration, which should be accounted for in the chemical groundwater modelling.

REPRECIPITATION PROCESSES IN THE WEATHERING ZONE

Two processes can be responsible for the reprecipitation of dissolved calcium carbonate from the soil solution: (1) *degassing*, i.e. the loss of CO_2 from the soil solution after movement towards an environment with a lower partial CO_2 pressure, and (2) *evaporation*, i.e. the loss of H_2O vapour from the soil solution.

Again we have to distinguish between an *open system*, where CO_2 as well as H_2O can be transferred out of as well as into the soil solution and a *semi-open system* (also referred to as a closed system; Hendy, 1971), where CO_2 and H_2O transport is only out of the solution.

The semi-open (or closed) system

In this idealized case the operating process is simply of a Rayleigh distillation type (Gat, 1980, p. 34). The solution becomes enriched in ^{12}C or in ^{18}O by the preferential removal of the light isotopic molecules ($^{12}CO_2$ or $H_2^{16}O$). The resulting isotope enrichment depends on the fractionation (ϵ) and the remaining fraction of the DIC or H_2O. The respective changes in δ are given by:

$$\ln [R^{13}(HCO_3^-)/R^{13}(HCO_{3,i}^-)] \simeq \Delta\delta^{13}(HCO_3^-) = \epsilon^{13}_{HCO_3^-}(X) \ln [C_T/C_{T,i}] \quad (42)$$

where X stands for CO_2 in the case of CO_2 degassing, while in the case of evaporation:

$$\Delta\delta^{18}(H_2O) = \epsilon^{18}_{H_2O}(Y) \ln m_{H_2O}/m_{H_2O_i} \quad (43)$$

where m is the amount of water and Y refers to the escaping water (vapour). In a semi-open system the removal of water from the soil solution is generally by (evapo)transpiration, which process does not cause an isotope effect (p. 246), i.e. $\epsilon^{18} = 0$, and, consequently, $\Delta\delta^{18} = 0$. Precipitating $CaCO_3$ is simply in isotopic equilibrium with the water at the prevailing temperature (Table 6-2B). An example of the carbon isotopic shift resulting from CO_2 degassing and the simultaneous reprecipitation of calcium carbonate has been discussed by Hendy (1971). In this latter case the ϵ^{13} value in equation (42) reades (Mook, 1976):

$$\epsilon^{13}_{HCO_3^-}(X) = \tfrac{1}{2}[\epsilon^{13}_{HCO_3^-}(CO_2) + \epsilon^{13}_{HCO_3^-}(CaCO_3)] \quad (44)$$

The δ^{13} value of the reprecipitating calcium carbonate increases from $-11 \pm 1‰$ and that of the reprecipitated calcium carbonate as a whole varies between $-10 \pm 1‰$ and -8 to $-5‰$. The actual δ^{13} values observed depend on the efficiency of the reprecipitation process (Fig. 6-1). This efficiency is visualized schematically in Fig. 6-5. The rare isotopic mass balance gives:

$$RN = (R - dR)(N - dN) + \alpha_1 R(1-f)dN + \alpha_2 RfdN$$

Here the basic point is, that $CaCO_3$ precipitates by degassing of an equal molar amount of CO_2. Solving this differential equation gives:

$$R = R_0 (N/N_0)^{(1-f)\epsilon_1 + f\epsilon_2} \tag{45}$$

The isotopic composition of $CaCO_3$ precipitating after the DIC concentration has been reduced from $C_{T,i}$ to C_T then is:

$$\delta^{13}_{calc} = \alpha^2 (1 + \delta^{13}_i)(C_T/C_{T,i})^{\bar{\epsilon}} - 1 \tag{46}$$

where $\bar{\epsilon} = (1-f)\epsilon_1 + f\epsilon_2$ and δ^{13}_i is the original δ^{13} value of the dissolved bicarbonate fraction. The average isotopic composition $\bar{R}$ of the integrated secondary carbonate at a reprecipitation efficiency $(1 - C_T/C_{T,i})$ is:

$$\bar{R} = \frac{\alpha_2}{f(N_0 - N)} \int_{N_0}^{N} -R \cdot f dN \tag{47}$$

where R is given by equation (45). The average δ^{13} value of the integrated amount of reprecipitated $CaCO_3$ then is:

$$\overline{\delta^{13}_{calc}} = \frac{\alpha_2}{\bar{\alpha}} (1 + \delta^{13}_i) \left[\frac{1 - (C_T/C_{T,i})^{\bar{\alpha}}}{1 - (C_T/C_{T,i})} \right] - 1 \tag{48}$$

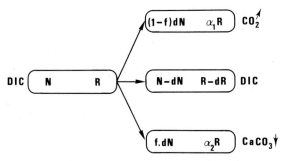

Fig. 6-5. Schematic presentation of the Rayleigh model applied to the precipitation of $CaCO_3$ from a bicarbonate solution by CO_2 degassing.

where $\bar{\alpha} = \bar{\epsilon} + 1$. The fraction f of the decrease in C_T, precipitating as $CaCO_3$, is obtained by an iterative calculation where P_{CO_2} is allowed to decrease:

$$\overline{\delta^{13}_{calc}} = \frac{\alpha_2}{\bar{\alpha}} (1 + \delta_i^{13}) \frac{1 - [C_T(P_{CO_2})/C_{T,i}]^{\bar{\alpha}}}{1 - [C_T(P_{CO_2})/C_{T,i}]} - 1 \qquad (49)$$

Using this procedure we obtain a set of graphs as presented by Hendy (1971) (Fig. 6-6a). This model specifically applies to $CaCO_3$ deposition in caves.

The open system

In the *open degassing system* it is necessary to distinguish between the diffusion of carbon dioxide from the soil solution to the atmosphere and that from the soil atmosphere into the soil solution (Hendy, 1971). The net rate of transport of carbon dioxide is:

$$\frac{dCO_2}{dt} = K \left(P_{CO_2 \text{ solution}} - P_{CO_2 \text{ soil air}} \right)$$

where dCO_2 is the amount of carbon dioxide evolved as given by the Rayleigh distillation iteration, while K is a transport constant. The amounts of carbon dioxide transported in either direction are given by:

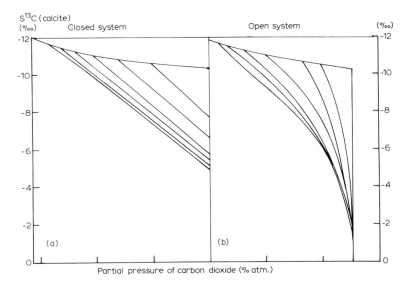

Fig. 6-6. The isotopic composition of precipitated calciumcarbonate for (a) closed and (b) open systems (redrawn after Hendy, 1971).

$$dCO_{2\,F} = \frac{P_{CO_2\,solution} \cdot dCO_2}{P_{CO_2\,solution} - P_{CO_2\,soil\,air}} \qquad (51)$$

$$dCO_{2\,R} = \frac{P_{CO_2\,soil\,air} \cdot dCO_2}{P_{CO_2\,solution} - P_{CO_2\,soil\,air}} \qquad (52)$$

in which the subscript "F" denotes the transport from solution to the atmosphere and "R" for the transport from the atmosphere to the solution. The $\delta^{13}(C_T)$ values during an iterative Rayleigh process are given by:

$$\delta^{13}(C_T) = \frac{\delta^{13}(C_T)_x \cdot C_{T,x} - \delta^{13}_F \cdot dCO_{2F} + \delta^{13}_R \cdot dCO_{2R} - \delta^{13}(CaCO_3) \cdot dCaCO_3}{C_{T,x} - dCO_2 - dCaCO_3} \qquad (53)$$

where x refers to the conditions prior to the iterative step. From this equation and the pH the isotopic composition of the various species in solution and the isotopic composition of the reprecipitating calcium carbonate can be calculated. These calculations have shown (Hendy, 1971) that at partial pressures exceeding 0.1% atm the effects of isotopic exchange between atmosphere and solution are small compared to those of the Rayleigh distillation and the isotopic composition of the reprecipitating calcium carbonate is similar to the one given in Fig. 6-6a. However, at lower partial pressures the effects of isotopic exchange predominate, so that as the partial pressure approaches 0.03% atm the calcite precipitated approaches isotopic equilibrium with the atmospheric carbon dioxide (Fig. 6-6b).

During *open-system evaporation* both the carbon (as a consequence of CO_2 loss) as well as the oxygen isotopic composition will change. The fact is that the water becomes enriched in ^{18}O during evaporation, according to the simple Rayleigh process (equation (45)), where the exponent is replaced by $\alpha_{H_2O\,liq}(H_2O_{vap}) - 1$. Also δ^{13} will be shifted towards higher values (Salomons et al., 1978).

Although it is conceivable that $CaCO_3$ reprecipitation is caused by merely CO_2 degassing, the reprecipitation under natural conditions probably does not take place according to one single process. It is more likely that combined processes act during soil evolution.

Furthermore, the processes of dissolution and partial reprecipitation of carbonates are not isolated events, but are repeated many times in the course of the soil evolution. The question arises as to what changes are to be expected due to such a multistage process. The isotopic composition at each stage depends on the relative amount of newly formed carbonates present, which again depends on the efficiency of the reprecipitation process. Zero efficiency means that no reprecipitation takes place: the calcium carbonate content of a soil decreases with time and its isotopic composition remains equal to that of the parent material. A high efficiency causes the addition of carbonates with an isotopic composition different from that of the parent material, already in the early stages of the soil evolution.

A simple model of the dissolution-reprecipitation process is shown in Fig. 6-7. Calculated curves for the expected isotopic composition of the soil carbonates for different efficiencies are given in Fig. 6-10 (Salomons and Mook, 1976). The isotopic composition of the carbonates reaches low values when the carbonate content decreases. From each dissolved carbonate dN a fraction f_r reprecipitates. The rare isotope mass balance gives:

$$(R - dR)[N - (1-f)dN] = R(N-dN) + \alpha R f dN$$
$$\quad\quad\quad\quad\quad\quad\quad\quad\quad\quad\quad\text{(parent}\quad\quad\text{(secondary}$$
$$\quad\quad\quad\quad\quad\quad\quad\quad\quad\quad\quad\text{CaCO}_3)\quad\quad\text{CaCO}_3)$$

Integration, and applying the boundary condition that R_o is the isotopic ratio of the parent rock (N_o) results in:

$$R/R_o = (N/N_o)^{-f(\alpha-1)} \tag{54}$$

or if we take $(N_o-N)/N_o = f_d$, the fraction of parent rock which has been dissolved:

$$\delta^{13}_{calc} = (1 + \delta^{13}_0)(1-f_d)^{-f_r \epsilon^{13}} - 1 \tag{55}$$

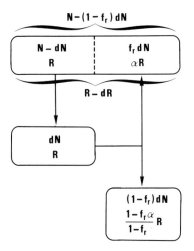

Fig. 6-7. Schematic presentation of the isotopic changes induced by the dissolution-reprecipitation process.

where δ_0^{13} refers to the primary (parent) carbonate rock and $\epsilon^{13} = \epsilon^{13}_{HCO_3^-}$ (CaCO$_3$).

ISOTOPIC COMPOSITION OF CARBONATES IN THE WEATHERING ZONE

Authigenous carbonates occur in different forms in the soil. Gile et al. (1966), Ruellan (1967) and Millot et al. (1979) have given a classification which will be used in this chapter. It is based on three different morphological forms of secondary calcium carbonate: (1) *diffuse accumulations* — this type is difficult to observe with the naked eye; (2) *discontinuous concentrations* — calcium carbonate becomes visible with the naked eye and accumulates in a discontinuous way; and (3) *continuous concentrations* — an almost pervasive accumulation of calcium carbonate.

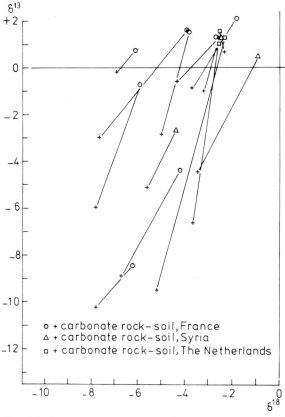

Fig. 6-8. Isotopic composition of carbonates in the soil compared with that of the underlying parent material (data from Salomons et al., 1977, Salomons and Mook, 1976, and unpublished data).

TABLE 6-3

Stable isotopic composition and apparent ^{14}C age of the carbonates in profiles from Dolni Vestonice and Oberfella Brunn[a]

Depth (cm)	Carbonates (%)	δ^{13}_{PDB} (°/₀₀)	δ^{18}_{PDB} (°/₀₀)	Apparent ^{14}C age, carbonates (years)	^{14}C age, organic matter (years)
Profile Dolni Vestonice (48°53'N, 16°40'E)					
420	8.0	−3.58	−5.53	25,790 ± 320	15,350 ± 1000
560	—	—	—	—	18,400 ± 700
680	0.15	−13.20	−11.18	10,000 ± 500	28,300 ± 300
920	10.1	−5.66	−6.61	32,850 ± 660	> 34,000
1150	—	—	—	—	> 51,800
1200	—	−8.13	−8.13	27,790 ± 370	—
1390	2.2	−9.86	−10.70	29,180 ± 460	> 50,000
Profile Oberfella Brunn (48°34'N, 16°00'E)					
120	17.1	−6.73	−6.82	15,680 ± 190	16,700 ± 800
230	14.0	−6.26	−6.38	24,275 ± 410	31,800 ± 500
300	7.9	−8.26	−7.78	22,425 ± 350	43,000 ± 700
350	—	—	—	—	42,300 ± 800
400	27.9	−3.02	−5.07	—	42,100 ± 800

[a] The samples were provided by B. Klima and by F. Felgenhauer and J. Flink, respectively. The ^{14}C ages of the organic matter are from Vogel and Zagwijn (1967). All ^{14}C ages are conventional B.P. (Salomons and Mook, 1976).

Diffuse accumulation of calcium carbonate

Diffuse accumulation of calcium carbonate has been studied in soils developed on carbonate rocks (Salomons, 1973; Salomons and Mook, 1976) in carbonate-rich marine clay soils and salt marshes (Salomons and Mook, 1976), in loess deposits (Vidal et al., 1966; Manze et al., 1974; Salomons and Mook, 1976) and in brown-grey earth soils developed in carbonate free parent material (Leamy and Rafter, 1972).

The isotopic composition of carbonates in the A-horizon of soils developed on carbonate rock differs from that of the parent material. As can be seen from Fig. 6-8, δ is shifted to lower values. Such shifts are to be expected if the carbonates in the soil are subjected to dissolution-reprecipitation processes. If it is assumed that the carbonates are precipitated at a low efficiency, the carbon isotopic composition will be about −10°/₀₀, $\epsilon^{13}_{HCO_3^-}(CaCO_3)$ higher than the δ^{13} value of the dissolved bicarbonate (Table 6-3). Using this value, δ^{13}(new), and the value of the parent material δ^{13}(pm), the amount of newly formed carbonates can be calculated from:

$$\text{\% newly formed carbonates} = \frac{\delta^{13}(\text{soil}) - \delta^{13}(\text{pm})}{\delta^{13}(\text{new}) - \delta^{13}(\text{pm})} \times 100 \tag{56}$$

According to this equation the amount of newly formed carbonates for the soils shown in Fig. 6-9 varies between 10 and 50% of the total carbonate content.

The isotopic composition of soil carbonates changes vertically along profiles. Two profiles, one from an ancient loess deposit in the Netherlands and one from a carbonate-rich marine clay soil, are presented in Fig. 6-9. Loess in The Netherlands was deposited in several stages. After each depositional stage soil formation took place resulting in carbonate losses and in the accumulation of organic matter. The uppermost part of the profile given in Fig. 6-10 was decalcified. In the preserved soil profile of one depositional stage the carbonate content increases with depth from about 8 to 16%. The δ^{13} values in the upper part of the profile are shifted towards values lower than those in the overlying profile as well as at higher depth. The isotopic composition in the lower part of the profile is within the range observed for Mesozoic carbonate rocks in Europe, which is one of the carbonate sources in European Loess (Manze et al., 1974). In loess profiles from Austria and Czechoslovakia (Table 6-3), the ^{14}C age of the organic fraction showed a chronological sequence whereas the apparent ^{14}C ages of the carbonates showed age reversals (Salomons and Mook, 1976). This shows that carbonate depositions took place from infiltrating groundwater after renewed loess deposition.

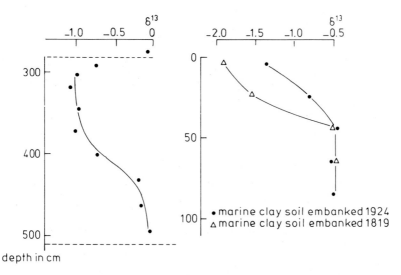

Fig. 6-9. The carbon isotopic composition of carbonates in a profile from a loess deposit and in two marine clay soils from The Netherlands.

In a study of the isotopic composition of loess deposits from Europe and the Mediterranean, Manze and Brunnacker (1977) observed δ^{13} values covering the whole range from soil carbonates to Mesozoic carbonate rocks. However, regional differences in carbon and oxygen isotopic composition has enabled a classification of these loess deposits.

As we mentioned before, the process of dissolution followed by a partial reprecipitation is repeated many times in the course of the soil evolution. Consequently, the carbonate content decreases and the isotopic composition of the carbonate is changed progressively to lower values. In Fig. 6-10 the isotopic composition of a large number of soils is plotted against the remaining carbonates in the soil (initial carbonate content is 100%). With decreasing carbonate content and increasing age of the soil, the δ^{13} shifts towards lower values, whereas the relative amount of newly formed carbonates in the soil increases. A comparison with the calculated curves for various efficiencies of the dissolution-reprecipitation process shows that the efficiency in the soils studied (France and The Netherlands) is about 10—30%.

The reprecipitation of calcium carbonate results into incorporation of ^{14}C in the carbonate fraction of the soil. Studies of soils from Germany (Geyh, 1970), France (Salomons, 1973) and Texas (Wallick, 1976) showed that the

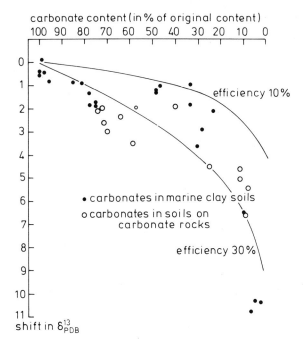

Fig. 6-10. The isotopic composition of soil carbonates as a function of the percentage of remaining carbonate content of the soil, compared with theoretical calculations for the efficiency of the dissolution-reprecipitation process.

^{14}C content varies considerably. Mean values for the ^{14}C content of soil carbonates are presented in Table 6-4. Both Geyh (1970) and Wallick (1976) observed a negative correlation between the δ^{13} and the ^{14}C content, the latter increasing with lower δ^{13}.

Discontinuous concentrations of calcium carbonate

Common discontinuous concentrations are nodules which are found in a variety of soils. Less common are secondary carbonates like lublinite or soil base (a thin layer of compact finely crystalline limestone which forms a transition zone between the soil and the underlying limestone). Data on the isotopic composition of discontinuous accumulations are presented in Table 6-5.

TABLE 6-4

The ^{14}C content (in % of the NBS standard activity) of soil carbonates

	^{14}C content	Reference
Germany	< 0.6– 79 (14)[a]	Geyh, 1970
France	15 (1)	Salomons, 1973
Texas	< 0.7–152 (15)	Wallick, 1976

[a] Number of samples analyzed.

TABLE 6-5

Isotopic composition of discontinuous accumulations of calcium carbonate in soils

Type of accumulation	δ^{13} (‰)	δ^{18} (‰)	Reference
Concretion in loess, Germany	−8.1		Vogel, 1959
Soil base from Bermuda	−7.0	−3.0	Gross, 1964
Nodules from South Australia	−2.25 ± 1.2 (14)[a]		Williams and Polach, 1971
Concretions, Queensland Australia	−4.8 (21)		Hendy et al., 1972
Lublinite, The Netherlands	−8.8 ± 1.2	−4.6 ± 0.6 (5)	Salomons et al., 1977
Concretion in loess, The Netherlands	−9.6 ± 0.4	−4.5 ± 0.5 (6)	Salomons et al., 1977
Concretion in fluvial clay, The Netherlands	−10.9	−5.4	Salomons et al., 1977

[a] Number of samples analyzed.

The secondary carbonates from The Netherlands have δ^{13} values indicating isotopic equilibrium with the groundwater bicarbonate in that region (δ^{13} (HCO_3^-) $\simeq$ −10 to −11‰). The isotopic composition of the concretions from the Talgai Region (Hendy et al., 1972) and the factors leading to their formation were studied in detail by analyzing both the plant cover and the groundwater. The area is dominated by plants with the C_4-photosynthetic pathway, as is reflected by the isotopic composition of the dissolved carbonate of the groundwater. Theorectical calculations showed that the nodules were reprecipitated in equilibrium with CO_2 of equal isotopic composition as that of the vegetation. The high values for the nodules from South Australia (Williams and Polach, 1971) are probably caused by an exchange with atmospheric carbon dioxide or a high efficiency of the reprecipitation process.

Continuous concentrations of calcium carbonate

Well known continuous concentrations of calcium carbonate are calcrete formations covering a large part of continental surfaces (Netterberg, 1969; Goudie, 1973; Reeves, 1976). Examples of continuous concentrations of calcium carbonate, formed in a region with a temperate climate, are also the Alm or Wiesenkalk deposits in southern Germany and the Alpine regions (Scheffer and Schachtshabel, 1976). The Alm from southern Germany consists of finely-grained calcium carbonate (95—99%) and occurs in layers up to 3 m thick (Vidal et al., 1966). The carbonate appears to be deposited from rising groundwater rich in dissolved carbonates. The isotopic composition of Alm deposits and the related Wiesenkalk and carbonate tufa is given in Table 6-6. The carbon isotopic composition of these accumulations of carbonate are close to values expected for C_3-dominated vegetation. The somewhat high values for the Alm deposits may point to an exchange with atmo-

TABLE 6-6

Isotopic composition of carbonates from continuous accumulations in Germany

	δ^{13} (‰)	δ^{18} (‰)
Erdinger Moos, southern Germany:[a]		
alm	−6.7 ± 1.0	−8.6 ± 0.3
carbonate tufa	−8.7	−7.9
Carbonate tufa, Schwäbische Alp:[a]		
Geilinger Steige (6 samples)	−9.6 to −10.1	
Blautops, Blaubeiern	−9.8	
Wiesenkalk from Neuwieder Becken, Rheintal[b]	−11 to 13	

[a] Data from Vidal et al. (1966).
[b] Data from Manze et al. (1974), fig. 3).

spheric carbon dioxide, whereas the low values of the Wiesenkalk may indicate exchange with soil-CO_2.

Several data are available on the isotopic composition of calcrete. Histograms showing the distribution of δ^{13} and δ^{18} of about 250 and 165 analyses, respectively, are presented in Fig. 6-11.

The *carbon isotopic composition* shows three more or less regionally significant modes. One mode around —8 to —7, the second and largest mode at —4 to —3 and a third mode around —1‰:

(1) High δ^{13} values for calcrete samples from India and part of the samples from the Sahara and South Africa.

(2) Intermediate values (around —3 to —4) in samples from South Africa, Texas, Spain and part of the samples from the Sahara.

(3) Low values (—7 to —9) for countries around the Mediterranean. The lowest values in this group are found for Italy.

The three modes agree with the values expected for carbonates precipitating from soil solutions which derive their carbon from soil-CO_2 dominated by C_4 and C_3 vegetation and from atmospheric carbon dioxide, respectively (Fig. 6-3). Variations within these modes are probably mainly due to the precipitation conditions with regard to the carbonate chemistry. Furthermore, the intermediate range is the same as found for the nodules from the Talgai region (Table 6-5) for which strong proof has been given that they were formed in equilibrium with CO_2 produced from C_4 vegetation (Hendy et al., 1972). The values for samples from Italy are very close to those found for carbonate nodules in soils dominated by C_4 vegetation (Table 6-5). The differences between the samples from Italy and the higher values for Algeria, Tunisia and Cyprus are probably caused by the temperature dependence of the fractionation between soil-CO_2 and the precipitated $CaCO_3$ (e.g. a 10°C increase in temperature causes a shift in $\epsilon^{13}_{HCO_3^-}(CaCO_3)$ of +1‰).

Additional differences, other than the temperature of formation, may be the efficiency of the reprecipitation process (Salomons et al., 1978). The isotopic composition of calcrete along profiles was found to be more or less constant for profiles (up to 1 m length) from India, Cyprus and Libya (Salomons et al., 1978). Apparently the upper part of these profiles, consisting of an indurated crust, was formed by isotopically similar processes as the more friable form of calcrete deeper in the profile. However, in a more extended calcrete profile from the Kalahari, high values for the carbon isotopic composition (about —3‰) were observed for the first 10 m and low values (about —5‰) at greater depth. In a 15-cm-thick crust from Greece, Manze and Brunnacker (1977) also observed that the isotopic composition increases progressively towards the surface which was interpreted as an exchange with atmospheric carbon dioxide during the formation.

The *oxygen isotopic composition* shows only one mode at about —5 to —4‰ (Fig. 6-11). δ^{18} is determined by that of the soil water, the kind of process responsible for the precipitation, and the fractionation factors in-

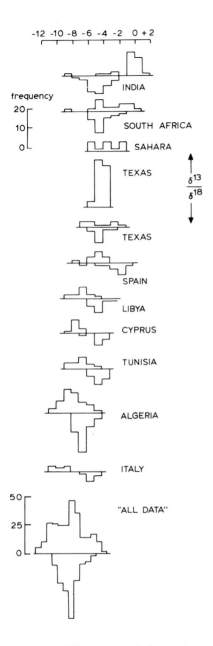

Fig. 6-11. Histograms of the carbon and oxygen isotopic composition of calcrete. The data used for the construction of the individual histograms are from Manze and Brunnacker (1977) (Algeria), Rightmire in Valastro et al. (1968) (U.S.A), Polach et al (1978) (Sahara), all other data from Salomons et al. (1978). For the histograms presenting all available data (often only carbon isotopic composition are available), data from Manze and Brunnacker (1977), Williams and Polach (1971), Wallick (1976) and Verhagen et al. (1978) have been used.

volved. The single mode is partly due to the fact that the influence of temperature is two-fold:

(1) In areas with a high average temperature (low latitude) δ^{18} of the rainfall generally is relatively high, and conversely. According to Van der Straaten and Mook (1982) the latitude effect in the average temperature range of 10—20°C is about +0.6 to +0.5‰ °C^{-1}, respectively.

(2) On the other hand, at high temperatures in the soil the oxygen isotope fractionation between carbonate and water is such that the δ^{18} values for the carbonate are lower. This effect amounts to —0.24‰ °C^{-1}. These two effects will partially cancel each other.

In Table 6-7 the isotopic composition of calcrete for a number of individual areas is compared with the equilibrium values for present day rainfall and temperature. For calcrete samples from South Africa, Libya and Italy the agreement between calculated and measured values is quite reasonable. The samples from Spain, India and Cyprus differ significantly from the calculated semi-open system values. This positive deviation suggests that for these areas evaporation of the soil solution played an important role in the formation of calcrete. The samples from Spain show the expected correlation between carbon and oxygen isotopic composition for evaporation-controlled precipitation (Salomons et al., 1978)

The results of the stable isotopic composition show that both degassing and evaporation have caused the formation of calcrete. It should be noted, however, that evaporation may initiate degassing in the manner described by Netterberg (1969).

TABLE 6-7

The isotope composition of calcium carbonate in equilibrium with present-day rainfall at the prevailing temperatures (data from IAEA precipitation network; IAEA Technical Reports Series), compared with the measured values for calcrete (Salomons et al., 1978)

Locality		δ^{18} equilibrium (‰)	δ^{18} measured (‰)
France	(3)[a]	—4.6	—6.0 ± 1.4
Italy	(8)	—5.8	—5.5 ± 0.6
Spain	(13)	—4.5	—2.3 ± 1.5
Cyprus	(6)	—7.3	—4.5 ± 0.7
Libya	(9)	—5.4	—4.7 ± 1.0
South Africa	(21)	—4.6	—4.2 ± 1.0
India	(19)	—8.1	—4.7 ± 1.1

[a] Number of sites studied.

SUMMARY AND CONCLUSIONS

The simple model originally put forward by Münnich (1957) and by Vogel (1959) for the dissolution of calcium carbonate in calcareous soils:

$$CaCO_3 + CO_2 + H_2O \leftrightarrows Ca^{2+} + 2\,HCO_3^-$$

which shows that half the carbon in the soil solution is derived from "light" biogenic carbon dioxide and the other half from "heavy" carbonate rocks, is able to explain in a first approximation the isotopic composition of carbonates in the soil. However, other factors have to be taken into account:

(1) The isotopic composition of the biogenic carbon dioxide in the soil (which ranges from -12 to -25, depending on the photosynthetic cycle used by the vegetation).

(2) The exchange with atmospheric carbon dioxide or evaporation (open or semi-open system).

(3) The mode of reprecipitation, degassing or evaporation.

(4) The efficiency of the reprecipitation process.

For an understanding of the isotopic composition of carbonates in calcareous soils an analysis of the carbonates alone is generally not sufficient; additional analyses of the vegetation and/or soil solution have to be carried out. This approach has been used by Hendy et al. (1972).

For an understanding of the more complicated processes involved in the formation of calcrete, the stable isotope investigations should be combined with the mineralogy of the formations as well theoretical models for explaining the mineralogical and the isotopic composition. Here not only the transformation of pre-existing carbonates has to be taken into account but also the additional calcareous epigenesis.

REFERENCES

Allaway, W.G., Osmond, C.B. and Throughton, J.H., 1974. Environmental regulation of growth, photosynthetic pathway and carbon isotope discrimination ratio in plants capable of Crassulancean acid metabolism. In: R.L. Bieleski, A.R. Ferguson and M.M. Cresswell (Editors), *Mechanisms of Regulation of Plant Growth. R. Soc. N.Z., Bull.*, 12: 195—202.

Atkinson, T.C., 1977. Carbon dioxide in the atmosphere of the unsaturated zone: an important control of ground water hardness in limestones. *J. Hydrol.*, 35: 111—123.

Bender, M.M., 1971. Variation in the $^{13}C/^{12}C$ ratio of plants in relation to the pathway of photosynthetic carbon dioxide fixation. *Phytochemistry*, 10: 1239—1244.

Billès, G., Cortez J. and Lossaint, P., 1971. L'activité biologique des sols dans les écosystèmes méditerranéens, I. Minéralisation du carbone. *Rev. Ecol. Biol. Sol*, 8: 375—395.

Broecker, W.S. and Olson, E.A., 1961. Lamont radiocarbon measurements, VIII. *Radiocarbon*, 3: 176—204.

Caldwell, M.M., White, R.S., Moore, R.T. and Camp, L.B., 1977. Carbon balance, productivity and water use of cold-winter desert shrub communities dominated by C_3 and C_4 species. *Oecologia*, 29: 275—300.

Craig, H., 1954. Carbon-13 in plants and the relationships between carbon-13 and carbon-14 variations in nature. *J. Geol.*, 62: 115—149.

Craig, H. and Keeling, C.D., 1963. The effects of atmospheric N_2O on the measured isotopic composition of atmospheric CO_2. *Geochim. Cosmochim. Acta*, 27: 549—551.

Deffeyes, K.S., 1965. Carbonate equilibria: a graphic and algebraic approach. *Limnol. Oceanogr.*, 10: 412—426.

Degens, E.T. and Epstein, S., 1964. Oxygen and carbon isotope ratios in coexisting calcites and dolomites from recent and ancient sediments. *Geochim. Cosmochim. Acta*, 28: 23—44.

Deines, P., 1980. The isotopic composition of reduced organic carbon. In: P. Fritz and J. Ch. Fontes (Editors), *Handbook of Environmental Isotope Geochemistry 1. The Terrestrial Environment, A.* Elsevier, Amsterdam, pp. 329—406.

Deines, P., Langmuir, D. and Harmon, R.S., 1974. Stable carbon isotope ratios and the existence of a gas phase in the evolution of carbonate ground-waters. *Geochim. Cosmochim. Acta*, 38: 1147—1164.

Dörr, H. and Münnich, K.O., 1980. Carbon-14 and carbon-13 in soil-CO_2. *Radiocarbon*, 22: 909—918.

Emrich, K., Ehhalt, D. and Vogel, J.C., 1970. Carbon isotope fractionation during the precipitation of calcium carbonate. *Earth Planet. Sci. Lett.*, 8: 363—371.

Epstein, S., 1976. A revised oxygen paleotemperature scale (personal communication).

Fontes, J.Ch., 1980. Environmental isotopes in groundwater hydrology. In: P. Fritz and J.Ch. Fontes (Editors), *Handbook of Environmental Isotope Geochemistry, 1. The Terrestrial Environment, A.* Elsevier, Amsterdam, pp. 75—140.

Fontes, J.Ch. and Garnier, J.M., 1979. Determination of the initial activity of the total dissolved carbon. A review of the existing models and a new approach. *Water Resour. Res.*, 12: 399—413.

Fritz, P., Hennings, C.S., Suzuki, O. and Salati, E., 1978a Isotope hydrology in northern Chile. In: *Isotope Hydrology*. IAEA, Vienna, pp. 525—544.

Fritz, P., Reardon, E.J., Barker, J., Brown, R.M., Cherry, J.A., Killey, R.W.D. and McNaughton, D., 1978b. The carbon isotope geochemistry of a small groundwater system in northeastern Ontario. *Water Resour. Res.*, 14: 1059—1067.

Garrels, R.M. and Christ, C.L., 1965. *Solutions, Minerals and Equilibria*. Harper and Row, New York, N.Y., 450 pp.

Gat, J.R., 1980. The isotopes of hydrogen and oxygen in precipitation. In: P. Fritz and J. Ch. Fontes (Editors), *Handbook of Environmental Isotope Geochemistry, 1. The Terrestrial Environment, A.* Elsevier, Amsterdam, pp. 21—48.

Gat, J.R. and Tzur, Y., 1967. Modification of isotopic composition of rainwater by processes which occur before groundwater recharge. In: *Isotopes in Hydrology*, IAEA, Vienna, pp. 49—60.

Geyh, M.A., 1970. Carbon-14 concentration of lime in soils and aspects of the carbon-14 dating of groundwater. In: *Isotope Hydrology*. IAEA, Vienna, pp. 215—223.

Gile, L.H., Peterson, F.F. and Grossman, R.B., 1966. Morphological and genetic sequences of carbonate accumulation in desert soils. *Soil Sci.*, 101: 347—360.

Gorham, E., 1955. On the acidity and salinity of rain. *Geochim. Cosmochim. Acta*, 7: 231—239.

Goudie, A., 1973. *Duricrusts in Tropical and Subtropical Landscapes*. University Press/Clarendon Press, Oxford, 174 pp.

Gross, M.G., 1964. Variations in the $^{18}O/^{16}O$ and $^{13}C/^{12}C$ ratios of diagenetically altered limestones in the Bermuda Islands. *J. Geol.*, 72: 170—194.

Harned, H.S. and Davis, R., 1943. The ionisation constant of carbonic acid in water and the solubility of carbon dioxide in water and aqueous salt solutions from 0 to 50°C. *J. Am. Chem. Soc.*, 65: 2030—2037.

Harned, H.S. and Scholes, R.S., 1941. The ionisation constant of HCO_3^- from 0 to 50°C. *J. Am. Chem. Soc.*, 63: 1706—1709.

Hendy, C.H., 1971. The isotopic geochemistry of speleothems, I. The calculation of the effects of different modes of formation on the isotopic composition of speleothems and their applicability as palaeoclimatic indicators. *Geochim. Cosmochim. Acta*, 35: 801—824.

Hendy, C.H., Rafter, T.A. and MacIntosh, N.W.G., 1972. The formation of carbonate nodules in the soils of the Darling Downs, Queensland Australia, and the dating of the Talgai Cranium. *Proc. 8th Int. Conf. Radiocarbon Dating, Wellington, 1972*, 1: D106—D126.

Jacobson, R.L. and Langmuir, D., 1974. Dissociation constants of calcite and $CaHCO_3^+$ from 0 to 50°C. *Geochim. Cosmochim. Acta*, 38: 301—318.

Keeling, C.D., 1958. The concentration and isotopic abundance of atmospheric carbon dioxide in rural areas. *Geochim. Cosmochim. Acta*, 13: 322—334.

Keeling, C.D., 1961. The concentration and isotopic abundances of atmospheric carbon dioxide in rural and marine air. *Geochim. Cosmochim. Acta*, 24: 277—298.

Keeling, C.D., Mook, W.G. and Tans, P.P., 1979. Recent trend in the $^{13}C/^{12}C$ ratio of atmospheric carbon dioxide. *Nature*, 277: 121—123.

Keith, M.L. and Weber, J.N., 1964. Carbon and oxygen isotopic composition of selected limestones and fossils. *Geochim. Cosmochim. Acta*, 28: 1787—1816.

Kraner, H.W., Schroeder, G.L. and Evans, R.D., 1964. Measurement of the effects of atmospheric variables on Ru-222 flux and soil-gas concentrations. In: *The Natural Radiation Environments*. Houston Rice Unversity, pp. 191—215.

Krumbein, W.E., 1968. Geomicrobiology and geochemistry of the "Narilimecrust" (Israel). In: G. Müller and G.M. Friedman (Editors), *Recent Developments in Carbonate Sedimentology in Central Europe*. Springer, Berlin, pp. 138—147.

Krumbein, W.E., 1972. Rôle des micro-organisms dans la génèse et la dégradation des roches en place. *Rev. Ecol. Biol. Sol*, 9: 283—319.

Langmuir, F., 1971. The geochemistry of some carbonate groundwaters in Central Pennsylvania. *Geochim. Cosmochim. Acta*, 35: 1023—1045.

Leamy, M.L. and Rafter, T.A., 1972. Isotope ratios preserved in pedogenic carbonate and their applications in paleopedology. *Proc. 8th Int. Conf. Radiocarbon Dating, Wellington, 1972*, 1: 353—368.

Lerman, J.C., 1972. Soil-CO_2 and groundwater: carbon isotope compositions. *Proc. 8th Int. Conf. Radiocarbon Dating, Wellington, 1972*, 1: D93—D105.

Magaritz, M., Heller, J. and Volokita, M., 1981. Land-air boundary environment as recorded by the $^{18}O/^{16}O$ and $^{13}C/^{12}C$ isotope ratios in the shells of land snails. *Earth Planet. Sci. Lett.*, 52: 101—106.

Majoube, M., 1971. Fractionnement en oxygen-18 et deuterium entre l'eau et sa vapeur. *J. Chim. Phys.*, 68: 1423—1436.

Manze, U. and Brunnacker, K., 1977. Über das Verhalten der Sauerstoff- und Kohlenstoff-Isotope in Kalkrusten und Kalktuffen des mediterranen Raumes und der Sahara. *Z. Geomorphol.*, 21: 343—353.

Manze, U., Vogel, J.C., Streit, R. and Brunnacker, K., 1974. Isotopenuntersuchungen zum Kalkumsatz im Löss. *Geol. Rundsch.*, 63: 885—897.

Millot, G., Ruellan, A., Nahon, D., Paquet, H. and Fritz, B., 1979. Geochemistry of calcareous epigenesis in calcretes (unpublished).

Mook, W.G., 1970. Stable carbon and oxygen isotopes of natural waters in the Netherlands. In: *Isotope Hydrology*. IAEA, Vienna, pp. 163—189.

Mook, W.G., 1976. The dissolution-exchange model for dating groundwater with ^{14}C. In:

Interpretation of Environmental Isotope and Hydrochemical Data in Groundwater Hydrology. IAEA, Vienna, pp. 213—225.

Mook, W.G., 1979. Evaluation of literature data (unpublished).

Mook W.G., 1980. Carbon-14 in hydrogeological studies. In: P. Fritz and J.Ch. Fontes (Editors), *Handbook of Environmental Isotope Geochemistry 1. The Terrestrial Environment, A*, Elsevier, Amsterdam, pp. 49—47.

Mook, W.G. and Koene, B.K.S., 1975. Chemistry of dissolved inorganic carbon in estuarine and coastal brackish waters. *Estuarine Coastal Mar. Sci.*, 3: 325—336.

Mook, W.G., Bommerson, J.C. and Staverman, W.H., 1974. Carbon isotope fractionation between dissolved bicarbonate and gaseous carbon dioxide. *Earth Planet. Sci. Lett.*, 22: 169—176.

Münnich, K.O., 1957. Messung des ^{14}C-Gehaltes vom hartem Grundwasser. *Naturwissenschaften*, 44: 32—39.

Netterberg, F., 1969. *The geology and engineering properties of South African calcretes.* Ph.D. Dissertation, University of the Witwaters Rand, Johannesburg.

Park, R. and Epstein, S., 1960. Carbon isotope fractionation during photosynthesis. *Geochim. Cosmochim. Acta*, 21: 110—126.

Picknett, R.G., 1964. A study of calcite solutions at 10°C. *Trans. Cave. Res. Group. G.B.*, 7: 39—62.

Picknett, R.G., 1973. Saturated calcite solutions from 10° to 40°: a theoretical study evaluating the solubility product and other constants. *Trans. Cave. Res. Group. G.B.*, 15: 67—80.

Polach, H.A., Head, M.J. and Gower, J.D., 1978. ANU Radiocarbon date list, VI. *Radiocarbon*, 20: 360—385.

Reardon, E.J. and Fritz, P., 1978. Computer modelling of groundwater ^{13}C and ^{14}C isotope compositions. *J. Hydrol.*, 36: 201—224.

Reeves, C.C., Jr., 1976. *Caliche Origin, Classification, Morphology and Uses.* Carftsman Printers, Inc., Estacado Book, Lubbock, Texas, 233 pp.

Rightmire, C.T., 1978. Seasonal variation in P_{CO_2} and ^{13}C content of soil atmosphere. *Water Resour. Res.*, 14: 691—692.

Rightmire, C.T. and Hanshaw, B.B., 1973. Relationship between the carbon isotope composition of soil CO_2 and dissolved carbonate. *Water Resour. Res.*, 9: 958—966.

Rubinson, M. and Clayton, R.N., 1969. Carbon-13 fractionation between aragonite and calcite. *Geochim. Cosmochim. Acta*, 33: 997—1002.

Ruellan, A., 1967. Individualisation et accumulation du calcaire dans les sols et les dépôts quaternaires du Maroc. *Cah. ORSTOM, Sér. Pédol.*, 5: 421—462.

Salomons, W., 1973. *Chemical and isotopic compositions of carbonates during an erosion-sedimentation cycle.* Dissertation, University of Groningen, Groningen, 118 pp. (Xerox Univ. Micro film, Ann Arbor, Order No. 74 20094).

Salomons, W. and Mook, W.G., 1976. Isotope geochemistry of carbonate dissolution and reprecipitation in soils. *Soil Sci.*, 122: 15—24.

Salomons, W., Mook, W.G. and Poelman, J.N.B., 1977. Toepassing van stabiele isotopen bij bestudering van kalk in bodems. *Landbouwkd. Tijdschr.*, 89: 5—9.

Salomons, W., Goudie, A. and Mook, W.G., 1978. Isotopic composition of calcrete deposits from Europe, Africa and India. *Earth Surf. Proces.*, 3: 43—57.

Scheffer, F. and Schachtshabel, P., 1976. *Lehrbuch der Bodenkunde.* Ferdinent Enke Verlag, Stuttgart, 394 pp.

Smith, B.N. and Epstein, S., 1971. Two categories of $^{13}C/^{12}C$ ratios for higher plants. *Plant Physiol.*, 47: 380—384.

Stumm, W. and Morgan, J.J., 1981. *Aquatic Chemistry.* J. Wiley and Sons, New York, N.Y., 2nd ed.

Thode, H.G., Shima, M., Rees, C.E. and Krishnamurthy, K.V., 1965. Carbon-13 isotope effects in systems containing carbon dioxide, bicarbonate, carbonate and metal ions. *Can. J. Chem.*, 43: 582—595.

Throughton, J.H., 1972. Carbon isotope fraction by plants. *Proc. 8th Int. Conf. Radiocarbon Dating, Wellington, 1972*, 2: 421—437.

Valastro, S., Jr., Davis, E.M. and Rightmire, C.T., 1968. University of Texas at Austin radiocarbon dates, VI. *Radiocarbon*, 10: 384—401.

Van der Straaten, C.M. and Mook, W.G., 1982. The stable isotopic composition of precipitation and climatic variability. In: *The Variations of the Isotopic Composition of Precipitation and Groundwater during the Quaternary as a Consequence of Climatic Changes*, IAEA, Vienna.

Veizer, J. and Hoefs, J., 1976. The nature of $^{18}O/^{16}O$ and $^{13}C/^{12}C$ secular trends in sedimentary carbonate rocks. *Geochim. Cosmochim. Acta*, 40: 1387—1395.

Verhagen, B.Th., Smith, P.E., McGeorge, J and Dziembowski, Z., 1978. Tritium profiles in Kalahari sands as a measure of rainwater recharge. In: *Isotope Hydrology 1978, Vol. 2*. IAEA, Vienna, pp. 733—751.

Vidal, H., Brunnacker, K., Brunnacker, M., Körner, H., Hartel, F., Schuch, M. and Vogel, J.C., 1966. Der Alm im Erdinger Moos. *Geol. Bavarica*, 56: 177—200.

Vogel, J.C., 1959. Über den Isotopengehalt des Kohlenstoffs in Süsswasser-Kalkablagerungen. *Geochim. Cosmochim. Acta*, 16: 236—242.

Vogel, J.C., 1970. Carbon-14 dating of groundwater. In: *Isotope Hydrology*. IAEA, Vienna, pp. 225—239.

Vogel, J.C. and Ehhalt, D., 1963. The use of the carbon isotopes in groundwater studies. In: Radioisotopes in Hydrology. IAEA, Vienna, pp. 383—396.

Vogel, J.C. and Zagwijn, 1967. Groningen radiocarbon dates, VI. *Radiocarbon*, 9: 63—106.

Vogel, J.C., Grootes, P.M. and Mook, W.G., 1970. Isotope fractionation between gaseous and dissolved carbon dioxide. *Z. Phys.*, 230: 255—258.

Wallick, E.J., 1976. Isotopic and chemical considerations in radiocarbon dating of groundwater within the semi-arid Tucson Basin, Arizona. In: *Interpretation of Environmental Isotope and Hydrochemical Data in Groundwater Hydrology*. IAEA, Vienna, pp. 195—212.

Wiececk, C.S. and Messenger, A.S., 1972. Calcite contributions by earthworms to forest soils in northern Illinois. *Soil Sci. Soc. Am., Proc.*, 36: 478—480.

Wigley, T.M.L., Plummer, L.N. and Pearson, F.J., Jr., 1978. Mass transfer and carbon isotope evolution in natural water systems. *Geochim. Cosmochim. Acta*, 42: 1117—1139.

Williams, G.E. and Polach, H.A., 1971. Radiocarbon Dating of Aridzone calcareous paleosols. *Geol. Soc. Am. Bull.*, 82: 3069—3086.

Yapp, C.J., 1979. Oxygen and carbon isotope measurements of land snail shell carbonate. *Geochim. Cosmochim. Acta*, 43: 629—635.

Chapter 7

GEOCHRONOLOGY AND ISOTOPIC GEOCHEMISTRY OF SPELEOTHEMS

H.P. SCHWARCZ

THE CAVE ENVIRONMENT

Carbonate sedimentary rocks, which underlie about 20% of the continental surface of the earth, are commonly subject to dissolution by ground water. As a result, cavities are formed which range in size from microscopic vugs to large rooms and passages up to tens or hundreds of kilometres in length. Such large solutional features are collectively known as karst. Karstic caves are commonly partially filled by three types of materials: (a) detritus from outside the cave, carried in by water or air currents and other means; (b) collapsed blocks of the limestone walls of the cave; and (c) chemically deposited calcium carbonate, collectively known as *speleothem*. Minor amounts of other chemical precipitates such as gypsum and ferric hydroxides may also be deposited. This chapter is concerned with the methods of analysis of the age and isotopic composition of such speleothems, and the paleoenvironmental interpretation that can be obtained from such analyses.

The advantage of caves as repositories of paleoenvironmental information, is the great stability of climatic conditions within the cave. Whereas the entrance regions of caves typically undergo climatic fluctuations in sympathy with day-to-day meteorological fluctuations, the inner passages of some caves have air temperatures which do not vary by more than a few tenths of a degree from winter to summer and which are very close approximations to the mean temperature on the ground above the cave (Wigley and Brown, 1976). This condition is observed in passages and rooms through which relatively little air or water is circulating, such as higher, older sections of caves which have been deepened by lowering of the base level. In such passages, relative humidity is maintained at 100% as long as there is vigorous recharge of meteoric precipitation from the surface; the concentration of dissolved species in seepage waters entering these spaces may vary, however, due to changes in organic activity in the soil above the cave through the year. It has been found that the speleothems which are deposited in such chambers preserve a record of long-term fluctuations in the isotopic and chemical composition of the recharge waters. Furthermore, the isotopic record can be interpreted as a record of past temperature variations in the cave. By dating these speleothems with one of the methods described below, these paleoenvironmental records may be set in a temporal framework, and related to other climatic records.

Not only speleothems record past environmental conditions in the cave. Detrital sediment, which is intermittently washed into the cave by streams, or deposited from stagnant waters ponded in the cave, can yield information about conditions of erosion and past water flow conditions. Such detritus can be dated through its stratigraphic relationship with sheets of speleothem (flowstone). Also, the very erosion and dissolution of the cave walls records stages in the evolution of the groundwater system of which the cave is a part. Therefore, dates obtained for various levels of the cave system are valuable in reconstruction of this evolution, and in particular the rate of lowering of local base level. Pollen has been recovered from speleothem (Bastin, 1978) and can possibly preserve a datable record of climatic variations on the surface above the cave. And of course, some parts of cave systems have at various times been occupied by humans and other vertebrates. The skeletal remains of the latter and the artifacts of the former are a valuable record of the past history of the cave and its environs, that can in fortuitous instances be dated by the presence of radiometrically datable travertine sheets. Thus we see that the karstic cave provides a unique and potentially valuable resource for reconstruction of environmental conditions on the continent over the last few hundreds of thousands to millions of years of earth history. One problem which must be recognized by anyone seeking to use this record, is that, like all stratigraphic sequences, the older parts of the record tend to be buried by the younger parts. Whereas stratigraphically older deposits on the surface of the earth, are revealed by down-cutting erosion, erosive processes in caves tend to be less effective or absent. Furthermore, caves are rare and beautiful locales, and amateur as well as professional students of caves are enjoined to disturb these settings as little as possible in the course of their studies. Therefore, it is often impossible to penetrate to the oldest parts of the record present in a cave.

The nature of speleothems

Calcium carbonate or dolomite in the bedrock above a cave is dissolved by water containing carbonic acid obtained from atmospheric CO_2 and through the decomposition and respiration of plants. Dissolution occurs via the reaction:

$$CaCO_3 + CO_2 + H_2O \rightleftharpoons Ca^{2+} + 2HCO_3^- \qquad (1)$$

The water penetrates through the soil and the underlying, fissured and jointed limestone bedrock, and finally emerges through a stalactite tip or other opening, into a cave, containing an atmosphere with a lower P_{CO_2} than that in the solution. Therefore, calcite precipitates from the solution through the reversal of reaction (1). The precipitate forms either at the point of emergence of the water on the roof (forming a *stalactite, curtain* or other deposit); or on the floor where it may build up a mound or upward growing column called a *stalagmite*, or simply form a planar deposit on the cave floor, called *flowstone*. The precipitate is usually calcite, though rarely aragonite or other polymorphs of $CaCO_3$ have been observed. Dolomite is also rarely formed (White, 1976). The most common objects for isotopic study are stalagmites; a typical one is shown in Fig. 7-1. It consists of a series of stacked, inverted, cup-like deposits, representing successive generations of precipitate which have maintained an essentially uniform morphology. The apex of each inverted cup is more or less planar, and an axial section through the stalagmite intersects these planar zones and provides a continuous stratigraphic sequence through the stalagmite. By contrast, a section normal to the growth axis intersects only a small portion of the total temporal record, in which the various layers are not equally represented. Each cup-like deposit, called a "growth

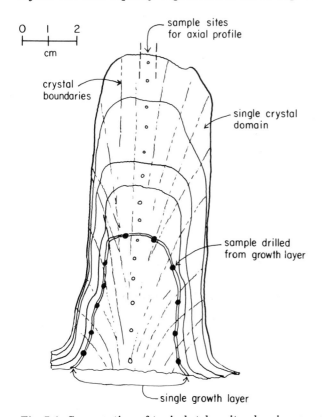

Fig. 7-1. Cross-section of typical stalagmite, showing growth layers.

layer", can be recognized by colour bands or other textural features which mark it off from older and younger deposits. The colour is believed to be largely due to the presence of organic matter bound in some way to the calcite crystals; traces of Fe and Mn may also be partly responsible (Gascoyne, 1977). Thus it is possible to extract, with a fine drill, samples from a single growth layer, as exposed in an axial section, such as shown in Fig. 7-1. The texture of the growth layers also allows us to recognize hiatuses in growth, if present, usually marked by dissolution and corrosion of older layers, and accumulation of some insoluble residue consisting of clay minerals and hydrous oxides of Fe and Al.

Flowstone deposits representing tens of thousands of years of growth are also found on the floors of caves. However, it is difficult to obtain samples because flowstone generally forms continuous, uninterrupted sheets, seldom exposed in eroded sections or fractured blocks; however, flowstone is in some ways preferable as sample material because it is easier to obtain replicate samples from a single growth layer for multiple isotopic analyses. Stalactites, hanging from cave roofs, are the poorest samples for studies of secular variation in speleothem deposits. They consist of stacks of conical deposits; growth layers are very thin and there is an axial hole through which water continuously flows, possibly causing dissolution and reprecipitation. Small, tubular stalactites actively growing on the cave roof are however ideal for obtaining samples of modern deposits, to test whether isotopic equilibrium currently exists between seepage waters and calcite.

Fine lamination occurs in many speleothems; Broecker and Olson (1960) concluded that these were annual layers, but other radiometric studies (e.g., Harmon et al., 1978d) show that such layers may represent tens or hundreds of years' growth. While some speleothems display very regularly and finely spaced growth layers (with spacings of 0.1 mm or less), discrete layers are much more commonly irregularly spaced and too thick (> 1 mm) to represent annual deposits. Typically growth rates in speleothems from temperate regions range from 0.05 to 0.1 mm/y (Harmon et al., 1975). Growth layers do not represent hiatuses in growth but only changes in composition of the continuously deposited calcite. Single calcite crystals transect these growth layers and persist for distances of up to several tens of centimetres along the axis of a stalagmite, giving it the appearance of a recrystallized limestone. However, Kendall and Broughton (1978) have shown that this very coarsely crystalline texture is a primary feature of the speleothem. Evidently, once preferred directions of crystal orientation have been established, further growth is persistently homoaxial with these few crystals, regardless of variations in growth rate or chemical composition of the calcite (as marked by growth layers). Nevertheless, recrystallization fabrics have been observed in speleothems (Folk and Asseretto, 1976), and in addition, isotopic evidence has occasionally been found that speleothems have been partially or wholly recrystallized. Considering that they are perpetually subjected to a shower of seepage water which may periodically vary from undersaturation with $CaCO_3$ (during times of very low or very high organic activity in the overlying soil) to supersaturation, it is surprising that we do not see such recrystallization effects more often.

Although the relative ages of successive layers within a single speleothem are obvious, it is impossible to estimate the relative ages of two speleothems from a given cave. Radiometric dating has shown that two stalagmites of approximately identical appearance, grown side by side, can differ in age by at least several tens of thousands of years (Thompson et al., 1976). Most samples collected for study are broken fragments of stalagmites or columns (stalagmites connected to stalactites). Therefore it is generally impossible to collect a systematic sequence of samples spanning a previously specified part of the depositional history of a cave. The exception to this is in the fortuitous case where development of a cave for public viewing has required cutting through thick flowstone layers. Although such caves are seldom isotopically ideal, they may provide the best samples for some purposes.

GEOCHRONOLOGY OF SPELEOTHEMS: METHODS

Radiocarbon dating

Dissolved inorganic carbon in surface water is derived partly from atmospheric CO_2, but principally from organic matter, either through bacterial decomposition or respiration. Both sources contribute a component of H_2CO_3 which is labelled with the atmospheric level of ^{14}C activity. This carbonic acid will react with calcite in the soil to produce HCO_3^- and Ca^{2+} ions as shown in reaction (1). From that reaction we would expect the ^{14}C activity of dissolved HCO_3^- ions in the soil to be 50 pmc (percent modern ^{14}C activity). Generally it will be even higher than this due to additional exchange with soil CO_2 and presence of excess H_2CO_3, sufficient to bring it to chemical equilibrium with the prevailing partial pressure of CO_2 (Mook, 1980). Subsequently, the water percolates through limestone to reach the cave. On its way, dissolved inorganic carbon can exchange with the limestone, lowering the ^{14}C activity of the water. In the case of incongruent dissolution of dolomite (Wigley et al., 1978) the activity can fall to less than 1% of the initial activity, although passage through calcite bedrock with no precipitation of carbonates or sulphates should result in very small changes. It is possible in principle to estimate the age of a water from its ^{14}C activity and the stable carbon isotopic composition where the water has interacted with a limestone bedrock and a soil-generated CO_2, both of known $\delta^{13}C$. Conversely, given the $\delta^{13}C$ of rock and soil gas, it ought to be possible to infer what the ^{14}C activity of the inorganic carbon of a drip water source in a cave was at any time in the past. Thus we might be able to use $\delta^{13}C$ values of speleothem to obtain

TABLE 7-1
Methods of dating speleothems

Method	Half-life of isotope or reaction time (years)	Range of applicability (years)	Size of sample required (g)	Comments
^{230}Th/^{234}U	^{230}Th: 75×10^3	$1-400 \times 10^3$	5–50	initial ^{234}U/^{238}U ratio must be known
^{234}U/^{238}U	^{234}U: 248×10^3	$50-2500 \times 10^3$	5–50	concentration of U must be greater than 1 ppm
^{231}Pa/^{235}U	^{231}Pa: 34×10^3	$5-200 \times 10^3$	10–100	initial ^{226}Ra activity must be known
^{226}Ra	^{226}Ra: 1.6×10^3	100–10000	10–50	initial ^{14}C activity must be known
^{14}C	^{14}C: 5730	100–40000	10–20	
Thermoluminescence	—	10^2 to 10^6	0.1	dose rate must be known
Electron spin resonance	—	10^2 to 10^6	1	dose rate must be known

their initial ^{14}C activities and thus be able to use current ^{14}C values to determine their ages. There are at least two problems with this theory. First, the δ^{13}C of carbonate precipitated in a cave differs from that of the feed water by an amount that depends not only on the temperature of precipitation, but on the extent of precipitation, that is, the fraction of initial carbonate that was left at the time the analysed sample was precipitated (Wigley et al., 1978). Also, the isotopic composition of the soil-gas component might have changed considerably due to a climate-controlled change in the proportion of C_3 and C_4 plants in the soil. In practice it appears to be more reliable to measure ^{14}C in an actively growing deposit and to assume that this value has persisted throughout the history of the site, to the limit of ^{14}C dating. If the initial activity had in fact changed by, e.g., 25%, at some time in the past, this would result in an error of 1.4 ky which is small compared with errors in U-series dating.

Dates obtained in this manner have been reported by Geyh (1970), Hendy (1970), Franke and Geyh (1972) and Talma et al. (1974). The limit of this dating method is about 40 ky (Table 7-1).

Uranium-series dating

Uranium possesses two isotopes, ^{235}U and ^{238}U, which decay to isotopes of Pb through a number of intermediate, radioactive isotopes, some of whose half-lives are of geological magnitude. Therefore, some geological deposits can be dated by measuring the degree to which these intermediate isotopes have come into secular, radioactive equilibrium with their parents. The principal decay schemes which can be used in this way are shown in Fig. 7-2. The application of these methods has been described by Ku (1976). In general the principal requirement is that the parent isotope (uranium) be deposited in the presence of a known deficiency or excess of one or more of its daughters. The optimal case is that of a parent deposited in the complete absence of its daughter; the ratio of daughter to parent then increases with time to a limit of unit activity ratio (secular equilibrium).

As has been noted, speleothem is precipitated from recharge water that contain dissolved components from the bedrock. Uranium and thorium are always present in the carbonate rock and also any other noncarbonate rocks that the water may have contacted. Recharge waters commonly contain from 0.1 to 3 ppb of U, carried as a complex of uranyl (UO_2^{2+}) ions with either carbonate or sulphate ions, or dissolved organic species (Langmuir, 1978). In contrast, both thorium and protactinium are commonly undetectable in such waters and are presumably present at levels of less than 0.1 ppb. This is due to their very low solubility in neutral waters (Langmuir and Herman, 1980). At the time of its deposition, speleothem takes up U and Th, possibly by substitution of the 4^+ cations at Ca-sites in the crystal lattice; fission track maps show a homogeneous distribution of U, unrelated to grain boundaries

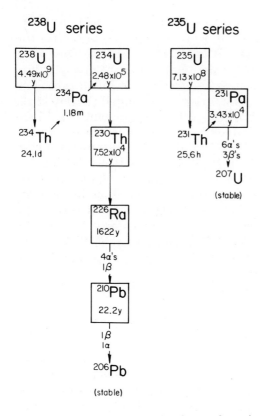

Fig. 7-2. Decay scheme for daughters of uranium isotopes.

or growth layers (Truscott and Schwarcz, unpublished data). The low content of Th in recharge waters leads to correspondingly low Th contents of freshly deposited speleothem. The isotope ^{230}Th, the daughter of ^{234}U, is essentially absent from modern speleothem; ^{230}Th/^{234}U activity ratios are less than 0.01. In some cases significant amounts of "common" thorium are present in speleothem, always associated with detrital matter, or other impurities in the speleothem, revealed by the presence of non-radiogenic ^{232}Th; this problem is discussed below. Concentrations of U in speleothem range from about 0.01 to several hundred parts per million. The average value is around 0.5 ppm. In view of its moderate U content and its low initial Th/U ratio, speleothem normally satisfies the principal criterion for the applicability of uranium-series dating, namely that parent ^{234}U is initially deposited in the absence of daughter ^{230}Th, which then grows into equilibrium with the uranium, with a half-life of 75 ky. At any time t after deposition of the speleothem, its ^{230}Th/^{234}U ratio is given by the relationship:

$$\frac{^{230}\text{Th}}{^{234}\text{U}} = \frac{^{238}\text{U}}{^{234}\text{U}} (1 - e^{-\lambda_{230}t}) + \frac{\lambda_{230}}{\lambda_{230}-\lambda_{234}} \cdot \left(1 - \frac{^{238}\text{U}}{^{234}\text{U}}\right)(1 - e^{-(\lambda_{230}-\lambda_{234})t})$$

(2)

where λ_{234} and λ_{230} are the decay constants of ^{234}U and ^{230}Th, respectively.

Note that to obtain the age, both the ^{230}Th/^{234}U and ^{234}U/^{238}U ratios must be determined. The latter ratio is unity in very old deposits such as the parent limestones in which caves form. However, during the decay of ^{238}U to ^{234}U through two intermediate isotopes, the uranium atom may undergo a recoil process or be altered in its oxidation state, resulting in the ^{234}U atoms becoming more readily leachable from the rock during weathering. Therefore, surface and ground waters commonly have ^{234}U/^{238}U ratios greater than unity, ranging up to 20 (see Osmond, 1980, for a fuller discussion of uranium isotope ratios in natural waters). Speleothem precipitated from surface-derived waters, acquires in turn ^{234}U/^{238}U ratios greater than (or rarely, less than) unity. In rare cases it is possible to demonstrate that a chemically or biologically precipitated deposit was precipitated with a *constant* initial ^{234}U/^{238}U ratio, B_0. Then the age of the deposit at time t can by given by the relation:

$$(B_t - 1)/(B_0 - 1) = e^{-\lambda_{234}t}$$

(3)

This method cannot, however be generally applied to speleothems because studies have shown (P. Thompson et al., 1975; Harmon et al., 1978e) that in many caves B_0 varies with time; furthermore, it was impossible in the caves studied by these workers to estimate B_0 by analysis of dissolved U in modern seepage water, because the latter is isotopically different from U extracted from adjacent speleothems. They speculated that different chemical species of U present in the drip waters were isotopically inhomogeneous, and unequally represented in speleothem. G. Thompson et al. (1975) attempted to date speleothems by this method but their results are in conflict with ^{230}Th/^{234}U ages from the same speleothems (Harmon et al., 1978f). However, Gascoyne and Schwarcz (1981) have found that ^{234}U/^{238}U ratios in drip water averaged over sufficiently long periods of collection are indistinguishable from the isotope ratio of the speleothems growing from them, in caves in North America and England.

The joint variation in ^{234}U/^{238}U and ^{230}Th/^{234}U ratios is given in Fig. 7-3. Note that the maximum age to which a speleothem can be dated increases somewhat with ^{234}U/^{238}U ratio. However, the upper limit for most speleothems is about 350 ky. The lower limit is determined by counter background and reagent blanks during speleothem analysis, but samples as young as 1 ky have been dated with a precision of ± 100 years (Gascoyne et al., 1983a).

The procedures for isotopic analysis of Th and U have been described by

Gascoyne et al. (1978). In brief, the speleothem sample is dissolved, Th and U are coprecipitated on Fe(OH)$_3$, and are separated from one another by ion exchange. They are plated onto steel discs and their alpha-particle activities are determined with solid-state detectors.

^{235}U decays through a number of short-lived intermediate isotopes to ^{231}Pa, with a half-life of 34 ky (Fig. 7-2). Like thorium, protactinium is excluded from surface waters, and therefore speleothem is initially deposited free of this daughter. Therefore ^{231}Pa/^{235}U ratios can be used to determine the age of a speleothem. However, the absolute activity of ^{235}U is only 1/21.7 that of its associated ^{238}U. Therefore, only samples with U concentrations greater than about 1 ppm can be dated by this means.

The possibility of using ^{226}Ra ($t_{1/2}$ = 1640 y) to measure the age of very young speleothems has not yet been experimentally investigated. Radium, unsupported by ^{230}Th, is presumably coprecipitated with Ca in young speleothems. If the initial specific activity of ^{226}Ra were known, ages up to about 10 ky could be determined from the residual activity of ^{226}Ra.

We have assumed up to now that the speleothem is free of detrital contamination. If any such detritus or bone, guano, etc., is present, these contaminants may contribute excess Th, Pa or U to the speleothem. During analysis of the speleothem the detrital component is generally left behind as a residue insoluble in the weak acid used to dissolve the CaCO$_3$. However, some Th, U

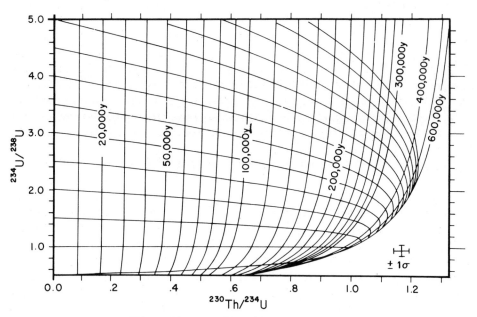

Fig. 7-3. Variation of ^{230}Th/^{234}U and ^{234}U/^{238}U ratios in closed systems. Near-vertical curves are isochrons; near horizontal curves are growth curves for systems of constant initial ^{234}U/^{238}U.

or Pa may be leached from the detritus. The presence of contaminant Th can be recognized in the Th alpha spectrum by the presence of a peak for ^{232}Th, which is not a daughter of U. Detrital U cannot be so recognized, but we may assume that detritus typically has a chemical Th/U ratio of about 4 (as is commonly observed in sedimentary rocks) and that these two elements are leachable by cold, $2N$ HNO$_3$ in approximately equal amounts (Schwarcz, 1980). Various schemes for correcting for this detrital contaminant have been devised (Kaufman, 1971; Turekian and Nelson, 1976; Ku et al., 1979; Schwarcz, 1980).

Electron spin resonance and thermoluminescence

Electrons which are formed by ionizing radiation striking calcite, can be trapped at defects in the crystal lattice structure (Aitken, 1974). The number of these trapped electrons increases steadily with time, in a crystal subjected to a constant radiation environment. Therefore the age of the crystal can be determined by obtaining an inventory of the trapped electrons. Two methods to do this have been devised:

(1) *Thermoluminescence:* The sample is heated before a photomultiplier tube, the trapped electrons are excited to the point where they are able to recombine with "holes" (positive charges in the lattice) and in so doing energy is emitted in the form of visible light. The intensity of the light is proportional to the number of trapped electrons (Aitken, 1974, p. 85 et seq.).

(2) *Electron spin resonance:* The sample is placed in a strong magnetic field within a resonant microwave cavity. A modulating magnetic field sweeps through field values corresponding to the conditions of frequency and field at which the trapped electrons resonate (through precession of their spins). The intensity of the resonance (observed as an absorption of the microwave radiation) is proportional to the number of electrons (Ikeya, 1978; Hennig et al., 1981).

In both methods, the number of electrons trapped per unit time is calibrated by exposing the sample to a radiation source of known intensity and determining, by extrapolation, the "equivalent" initial dose that the sample was subjected to in nature. Then the radiation dose rate at the site is evaluated either by radiation dosimetry or by measurement of the concentration of radionuclides in the sample. The thermoluminescence method has the advantage of small sample size (1—10 mg). However, Wintle (1976) has found that small variations in the procedure of preparation of calcite samples for analysis result in significant changes in the character of the thermoluminescent signal. For relatively young speleothem samples (age < 400 ky) the internal dose rate due to U and its daughters in the calcite must be corrected for the disequilibrium of the radioisotopes (Wintle, 1978). While neither of these dating methods can be very precise due to the numerous corrections that are required, they provide the possibility of obtaining dates beyond the 350-ky limit of U-series dating.

Paleomagnetic stratigraphy

Speleothems containing only a few parts per million of iron (presumably as very small crystals of an iron oxide) exhibit paleomagnetic signals with intensities ranging from 1 to 100×10^{-9} oersteds per gram. These paleomagnetic signals can be detected on superconducting Josephson-effect magnetometers, and the paleomagnetic orientation of the sample can be determined. Latham et al. (1979) have shown that curves of secular variation in declination and inclination can thus be obtained. Some samples, too old to be dated by U series, have displayed reversed magnetic polarity with respect to the present geomagnetic field at their sites of deposition. They were therefore deposited at least 720 ky ago, the time of the Brunhes/Matuyama magnetic reversal. Younger samples could in principle be assigned dates by identifying characteristic "signature" events on their secular D, I curves, through matching with a master curve for the same region, obtained on a dated speleothem or from cores taken from nearby lakes.

APPLICATIONS OF GEOCHRONOLOGY OF SPELEOTHEMS

The principal use of absolute ages obtained from speleothems is to provide a time base for geochemical and paleomagnetic data which have been obtained on the speleothem. However, other important applications have been made as follows.

Paleosea-level determination. Oceanic islands and coastal regions in the tropics and subtropics tend to be underlain by extensive, young carbonate deposits. During glacial periods of low sea-stands, karst features may develop in such terranes, including speleothem-decorated caves. When sea levels rise during interglacials such as the present, the "drowned" karst remains as a marker of past low sea-stands.

In Bahamas Islands, there are numerous caves of this type, some of which are accessible through submarine sinkholes known as "blue holes". Spalding and Matthews, (1972) dated a stalagmite from 37 feet below sea level at 22 ky B.P., showing that sea level stood below that level late in the Wisconsin glacial stage. Gascoyne et al. (1979) analysed stalagmites from 30 m depth in a blue hole off Andros Islands, and obtained ages of 130 ky; they showed that the age-depth relation for these samples is consistent with the sea-level curve of Shackleton and Opdyke (1976). On Bermuda, Harmon et al. (1978a, 1981) have used the ages of submarine stalagmites to derive a sea-level/age curve. The growth of at least one stalagmite was shown to have been interrupted at 114 ky by the presence of a coating of marine calcite overlying stalagmitic growth layers of that age.

The advantage of these studies over other estimates of paleosea-level is that it permits direct dating of low sea-stands, whereas conventional studies of raised reefs and fossil beaches are indicators of high sea-stands only. The method suffers from two disadvantages: (a) only highly trained diving teams can recover speleothems from such drowned caves; (b) speleothems in tropical seawater are heavily eroded through infestation by various boring organisms (sponges, brachiopods, etc.), many of which deposit calcite or aragonite which is younger than that of the host stalagmite. Gascoyne et al. (1979b) have used various chemical and physical methods to separate these two carbonate components in order to obtain the true age of the stalagmite.

Rates of geomorphic evolution. As noted above, the downcutting of hilly or mountainous karstic terrain results in the stranding of high, vadose passages which were originally phreatic, that is, filled with groundwater. The age of speleothem decorating such features then provides an index of the rate of downcutting of the valley floors. This technique has been applied in the Canadian Rocky Mountains by Ford et al. (1972, 1981) and Ford (1976). Ford et al. were able to demonstrate that the minimum possible age of the present relief of 1.3 km was 1.2—6 My while its maximum age was approximately 12 My. The oldest cave deposit used in this study, at Eagle's Nest Cave, was assigned a minimum age of 710 ky on the basis of its reversed paleomagnetic remanence. This flowstone was assigned to the Matuyama chron rather than the older Gilbert reversed chron, because of its relatively high $^{234}U/^{238}U$ ratio of 1.26 ± 0.05 showing that its age was probably not more than 1 My. Its position high on the side of a mountain ridge, was consistent with its great age. Atkinson et al. (1978) have studied the development of the Mendip Hills, England, in this fashion, while Gascoyne (1983b) have made a similar study of cave development in Lancashire.

Climatic studies. The rate of deposition of spleothem as determined for example by the increase in height per year of a stalagmite, is determined in part by climatic influences. The rate of production of CO_2 in the soil above the cave increases with temperature and humidity of the soil. The freezing of water in the soil and in the cave stops all precipitation of speleothem. Marked decrease in humidity of the cave due to lowering of surface precipitation may slow down precipitation of calcite in the cave. The most intense effects, however, should be expected in association with glaciation above the cave. With the loss of a soil zone, stripped away by a flowing ice sheet, CO_2 production stops. Under subglacial and periglacial (permafrost) conditions, speleothem deposition should be severely slowed or stopped. Therefore a record of the frequency of speleothem deposition versus time should reflect the advance and retreat of glaciers in regions close to or under ice sheets. This effect has been observed in alpine terranes of the Rocky Mountains of North America (Harmon et al., 1977). They observe four distinct periods of high deposition rates which they correlate with periods of low continental ice volume as inferred from deep-sea isotopic records. A similar frequency distribution was obtained for the ages of speleothems from caves in northern England (Gascoyne et al., 1983a; Fig. 7-4), an area that was ice-covered only in the severest part of the last glaciation, around 20 ky B.P. Their data also show distinct peaks of speleothem frequency during interglacials, in addition to an overall exponential decrease in age frequency due to the burial of older speleothems by younger deposits, and the loss of old deposits by erosion.

Archaeometry. During the Middle Pleistocene the hominid ancestors of man made use of cave mouths as living areas or as hunting camps. In many

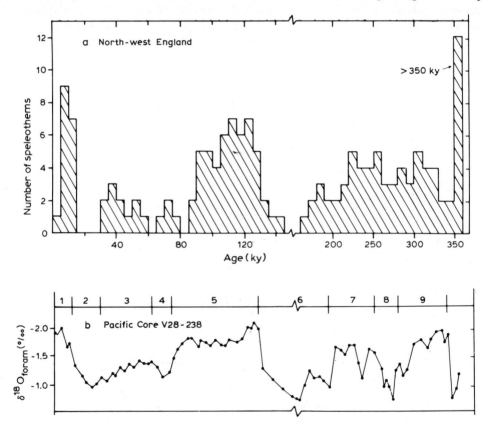

Fig. 7-4. (a) Histogram of speleothem ages for caves in Yorkshire and Lancashire, northern England (Gascoyne et al., 1983a). (b) $\delta^{18}O$ of planktonic foraminifera (*Globigerinoides sacculifer*) from deep-sea core V28-238 (Shackleton and Opdyke, 1973). Note coincidence of isotopic peaks in two records; $\delta^{18}O$ minima in deep-sea record correspond to interglacials. Isotope stages are numbered.

instances archaeologists have subsequently found artefacts and skeletal remains of these early humans interstratified with speleothems, or in layers which enclose stalagmitic or stalactitic deposits. By dating these deposits it has been possible to estimate the age of occupation of some of the sites, over a time range from 350 ky B.P. to near the present (Turekian and Nelson, 1976; Schwarcz, 1980).

One example of such a study is that of Blackwell et al. (1983) at the Lower and Middle Paleolithic site of La Chaise in the Charente district of France. Here, clastic, detrital deposits fill two adjacent chambers, but are interrupted by discrete layers of travertine up to 0.5 m thick. The lowest layer dates from 240 ky and underlies Acheulian layers containing a primitive Neanderthaloid hominid. A higher travertine layer, resting on an early Neanderthal mandible associated with late Acheulian artefacts yields a date of 151 ± 15 ky. Still higher, resting on Middle Paleolithic artefacts, a calcite layer with stalagmites dated to about 70 ky; Upper Paleolithic (Aurignacian) tools rěst just above this layer. These deposits are some of the richest and best dated in France (Fig. 7-5).

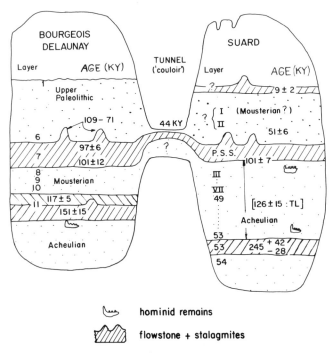

Fig. 7-5. Schematic cross-section of cave deposits at La Chaise-de-Vouthon, Charente, France, showing stratigraphic location of dated stalagmite "floors" (from Blackwell et al., 1983).

STABLE ISOTOPE GEOCHEMISTRY OF SPELEOTHEMS

Principles of isotopic exchange between $CaCO_3$ and water

Speleothems are inorganic precipitates of calcium carbonate, generally calcite, formed from dilute aqueous solutions. They are capable of being formed in a state of oxygen isotopic equilibrium with the water from which they are precipitated. This can be represented by the exchange equation:

$$\tfrac{1}{3} CaC^{16}O_3 + H_2{}^{18}O \leftrightarrows \tfrac{1}{3} CaC^{18}O_3 + H_2{}^{16}O \tag{4}$$

The equilibrium constant for this reaction α_{cw} is equal to:

$$\alpha_{cw} = \frac{(^{18}O/^{16}O)_{ct}}{(^{18}O/^{16}O)_{H_2O}} \tag{5}$$

This equilibrium constant has been shown by Epstein et al. (1953) to be dependent on temperature and independent of all other environmental variables. Therefore, if the calcite is deposited at equilibrium, its isotopic composition can yield a record of the temperature of deposition. The temperature dependence has been determined by O'Neil et al. (1969) to be as follows:

$$\Delta_{cw}\ 10^3 \ln \alpha_{cw} = 2.78 \times 10^6\ T^{-2} - 2.89 \tag{6}$$

(this is modified to take into account that $\alpha_{CO_2-H_2O}$ at 25°C is 1.04012, Friedman and O'Neil, 1977).

The temperature of deposition of a speleothem can only be obtained if the following conditions are met:

(1) The speleothem was deposited in isotopic equilibrium with its parent water; that is, it did not undergo some kinetic isotope effect during deposition, which would result in isotopic fractionation that was not strictly temperature-dependent.

(2) The speleothem was deposited from a water of known isotopic composition.

(3) The speleothem has not been isotopically altered (e.g., recrystallized) subsequent to its deposition.

Furthermore, if the speleothem was deposited in a region of the cave where temperature is stable year-round, then we can use the speleothem's temperature record as an estimate of surface temperatures above the cave.

Equilibrium and disequilibrium deposition of speleothems

Early studies of the oxygen isotopic composition of speleothems by Fornaca-Rinaldi et al. (1968) suggested that these deposits were not formed in isotopic equilibrium with water. This was further shown by Fantidis and Ehhalt (1970), who also grew artificial stalagmitic deposits and observed kinetic isotopic effects in these deposits. However, Hendy (1969, 1971), in a thorough study of the conditions of formation of speleothems, showed that some speleothems are deposited in oxygen isotopic equilibrium, and that they can be recognized by the fact that the oxygen isotopic composition measured along a single growth layer must be constant; any slight changes which are observed must not be correlated with changes in the isotopic composition of carbon. The locality in the cave where the sample is collected can be shown to be suitable for equilibrium deposition if a speleothem, forming at present at the site, is in isotopic equilibrium with the water from which it is being deposited. Typically such sites are far from the cave mouth, in passages which are blocked at one end, and exhibit relative humidities of 100%.

The criterion of uniform $\delta^{18}O$ within a growth layer eliminates from consideration speleothems which have experienced non-equilibrium, kinetic isotope effects in the fractionation between dissolved bicarbonate ions and the precipitated calcite. Such kinetic isotope effects invariably arise from the preferential loss of light isotopes of both C and O from solution during rapid, irreversible loss of CO_2 from solution and therefore one can commonly observe a correlation between the oxygen and carbon isotopic composition of such disequilibrium deposits *within a single growth layer*. Examples of $\delta^{18}O$ and $\delta^{13}C$ plots for equilibrium and disequilibrium deposits are shown in Fig. 7-6*.

Hendy furthermore observed that such equilibrium deposits are invariably found in poorly ventilated sections of caves where humidity is close to 100%. If the humidity is below this value then water passing over the surface of the growing speleothem will be enriched in ^{18}O due to evaporation of ^{16}O-enriched water vapour. Where the cave is greatly undersaturated in CO_2 with respect to chemical equilibrium between calcite and water, the rate of precipitation of the calcite will be so rapid as to prevent isotopic equilibrium between calcite and the water. This will also result in a fractionated speleothem. Note that isotopic equilibrium between calcite and water takes place via the equilibrium between H_2O and HCO_3^- ions through the reaction

$$CO_2 + H_2O \rightleftarrows HCO_3^- + H^+ \tag{7}$$

---

* Isotopic compositions are normally represented in the δ-notation

$$\delta^{18}O_x = [(R_x/R_s) - 1] \times 1000$$

where x = sample; s = standard and $R = {}^{18}O/{}^{16}O$. In this chapter the standards are as follows: PDB for ^{13}C and ^{18}O content of calcite; SMOW for ^{18}O and ^2H contents of water. We shall use the words "heavy" and "light" to refer to materials enriched or depleted in a heavy isotope (e.g., ^{18}O), respectively.

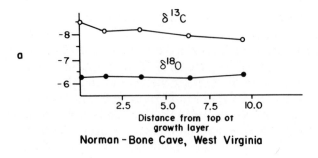

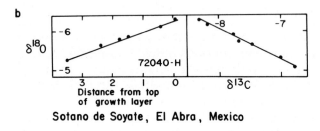

Fig. 7-6. Covariation of $\delta^{13}C$ and $\delta^{18}O$ in speleothems formed in isotopic equilibrium with seepage water (a); and at disequilibrium (b). Samples from Norman-Bone Cave, West Virginia (Thompson et al., 1976), and Sotano de Soyate, Mexico (Harmon, 1976).

Because this reaction is slow at and below 25°C, the rate of equilibration between calcite and water is effectively limited by this reaction. Therefore only relatively slow-growing speleothems will be formed in oxygen isotopic equilibrium.

In addition to Hendy's isotopic criteria we may also require that the speleothem should show no signs of recrystallization or inversion from aragonite to calcite (Folk and Asseretto, 1976).

Speleothems as paleotemperature indicators

Having determined that some speleothems are formed in isotopic equilibrium with their parent seepage waters, it remains to interpret the isotopic composition of ancient speleothems in terms of paleotemperatures of the caves in which they grew. Hendy and Wilson (1968) suggested that the isotopic composition of the seepage water was determined by the temperature of the surface precipitation according to the relation observed for maritime continental rain by Dansgaard (1964): $d(\delta^{18}O)/dT = 0.69‰$ $°C^{-1}$. It can be shown that seepage water in caves resembles isotopically the average meteoric water falling on the soil above the cave (see, e.g., Harmon and Schwarcz, 1981) and the temperature in the cave is equal to the average surface temperature. We would therefore expect $d(\delta^{18}O)/dT$ within the cave to equal the value of this derivative for local meteoric water. Hendy and Wilson also noted that the isotopic composition of sea water has changed during the Pleistocene due to growth of continental ice. Therefore, to simplify their analysis, they considered the change $\Delta^{18}O$ in the oxygen isotopic composition of speleothem deposited at a site at which temperature has decreased by $T_x\,°C$, while, concurrently the isotopic composition of tropical seawater (the source of meteoric precipitation at x) is changed by $Y°/_{oo}$ and its temperature drops by $T\,°C$. Then:

$$\Delta^{18}O = Y - 0.69(T_x - T) + 0.24 T_x = Y + 0.24T - 0.45(T_x - T) \qquad (8)$$

The coefficient 0.24 derives from the value of $d(\Delta_{cw})/dT$ at 25°C. For a tropical site, $T_x = T$ and the temperature response of speleothem is the same as for planktonic foraminifera observed in deep-sea cores.

Speleothems formed in non-tropical but maritime climates would show an isotopic response to the temperature gradient between the tropics and their locale, as indicated by equation (8). Hendy and Wilson obtained a sequence of $\delta^{18}O$ values for two stalagmites and Hendy (1970) added a third record, all of which showed that modern calcite was up to $1°/_{oo}$ lighter than calcite deposited during the last glacial epoch; ages of the stalagmites were determined by ^{14}C analysis (Hendy, 1970). Hendy and Wilson interpreted their record in terms of a variation of about $10°C$ in the variable $T_x - T$. However, we shall see that this interpretation is subject to serious reservations.

Various interpretations of stable isotopic records of speleothems have been suggested. Talma et al. (1974) assumed that in the region of the Transvaal which they studied, the isotopic composition of Ice-age groundwater was $0.8°/_{oo}$ lighter than at present, and used this to interpret a record of $\delta^{18}O$ dated by ^{14}C, from a stalagmite from a cave in Transvaal. They also observed that modern calcite was lighter than that of glacial periods. Duplessy et al. (1970), having demonstrated that the seepage waters in caves are isotopically identical to meteoric waters, argued that $\delta^{18}O$ of the speleothem should vary with temperature in response to: (1) the assumed $+ 0.69°/_{oo}\,°C^{-1}$ temperature dependence of the isotopic composition of rain observed by Dansgaard; (2) the $- 0.24°/_{oo}\,°C^{-1}$ dependence of Δ_{cw} at about 25°C. The sum of these effects yields $d(\delta^{18}O)/dT = + 0.45°/_{oo}\,°C^{-1}$. In the French cave of Orgnac, where modern speleothem has a $\delta^{18}O = - 5.9$, a stalagmite was found by them to have deposited calcite between 130 and 90 ky B.P., with an isotopic composition that varied from $- 6.3$ to $- 4.4°/_{oo}$. The heaviest speleothem was deposited between 120 and 98 ky B.P., a period which is believed to represent the last interglacial (Imbrie et al., 1976). Duplessy et al. argued that the shift to deposition of lighter speleothems at about 98 ky B.P. represented a shift to lower temperatures in the cave, and that the last interglacial was consequently much warmer than the present. They unfortunately did not measure the variation in $\delta^{18}O$ along single growth layers to test whether the stalagmite had been deposited in equilibrium. Their plot of $\delta^{18}O$ vs. $\delta^{13}C$ for *axial* samples do not test this question.

Emiliani (1971) reinterpreted the data of Duplessy et al., noting that the lightest calcites measured in this speleothem had $\delta^{18}O$ values similar to those observed in a modern stalagmite from the same cave. He therefore proposed that light calcite had been deposited in interglacials (stage 7 and 5) at 122 ky and 95 ky B.P., respectively; in this writer's view it is more likely that these periods of deposition of light calcite correspond to isotope stages 5e and 5c, respectively, although the intervening 1.3‰ increase in $\delta^{18}O$ is surprisingly large.

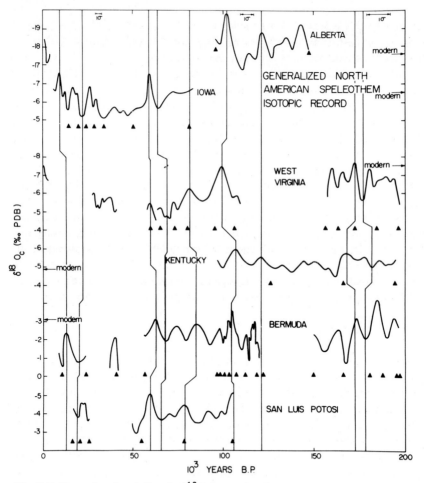

Fig. 7-7. Records of variation in $\delta^{18}O$ for speleothems from North America and Bermuda, (from Harmon et al., 1978b). Vertical lines are suggested correlations between isotopic maxima and minima corresponding to climatic events.

It thus appears that at two localities (New Zealand and South Africa) $\delta^{18}O$ has increased with decreasing temperature, while in France the opposite may have occurred. Further studies by Thompson et al. (1976), Harmon et al. (1978b) and M. Gascoyne (unpublished data) show that both situations do arise in differing environments. Typical records of $\delta^{18}O$ in speleothems from North America are given in Fig. 7-7.

The interpretation of paleotemperatures from isotopic analyses of speleothems seems fraught with difficulty. Indeed, we have now obtained records from several sites in North America at which essentially no change in $\delta^{18}O$ of calcite occurs through the last glacial epoch. This can be understood as follows. The temperature dependence of the isotopic composition of precipitation, $\delta^{18}O_p$, is not a constant value of $0.69‰\,°C^{-1}$, as suggested by Dansgaard. Rather, this coefficient varies greatly from place to place, and in general it is much less than this value in mid-continental regions. Values of $0.18-0.3‰\,°C^{-1}$ have been observed in these regions (based on data obtained by IAEA, 1981). Consequently the local effect of change in temperature on the isotopic composition of speleothems will vary depending on the relative magnitudes of $d(\delta^{18}O_p)/dT$ and $d(\Delta_{cw})/dT$. Where the sum of these two coefficients is approximately zero, there will be no net effect of temperature on speleothem isotopic composition; the only effects which will remain will be changes in $\delta^{18}O_p$ due to: (a) growth and melting of continental ice, (b) changes in storm paths and in the average condensation process from clouds reaching the site (Gat, 1980). The temperature coefficient of $\delta^{18}O_p$ as determined from records of modern precipitation is a measure of the effect of seasonal (winter to summer) changes in temperature on $\delta^{18}O$ and may not necessarily equal the *secular* dependence of $\delta^{18}O_p$ on long-term changes in temperature; it is this secular coefficient which should be considered in obtaining paleotemperature data from isotopic compositions of speleothems. Unfortunately neither this secular temperature coefficient nor the above mentioned factors (a) and (b) can be calculated from other isotopic data.

Fluid inclusions in speleothems

Fortunately there appears to be a solution to the dilemma posed in the previous section. Many speleothems, and in particular many of those deposited at isotopic equilibrium, are found to contain varying amounts of water trapped inside them as inclusions (Fig. 7-8; Schwarcz et al., 1976; Kendall and Broughton, 1978). These inclusions appear to be trapped samples of the water from which the speleothem grew, as can be demonstrated in various ways (Schwarcz et al., 1976; Harmon et al., 1978c). These water samples have been trapped in the speleothem, enclosed in cavities lined with calcite; it is therefore possible that the oxygen of the water has continued to exchange with the walls of the cavity since the time of entrapment. To avoid this problem, we have made use of the fact that, in a given region, the hydrogen isotopic composition of the water, δ^2H, is known to have a well-defined relation to $\delta^{18}O_p$ for all modern precipitation and also for Pleistocene glacial ice (Johnsen et al., 1972). Since there is no hydrogen in calcite with which to exchange, δ^2H of a trapped inclusion should be unchanged from its initial value. We have analysed only inclusions from speleothems showing no change in $\delta^{18}O$ of calcite along a growth layer. The seepage water of such equilibrium deposits cannot have been fractionated by evaporation as this would result in a progressive increase in $\delta^{18}O$ of the calcite precipitated from it. The lack of evaporation is due to the 100% humidity which occurs in cave chambers where equilibrium deposits are typically found. We have extracted fluid inclusions and used their δ^2H values to calculate $\delta^{18}O$ of the seepage water from the relationship:

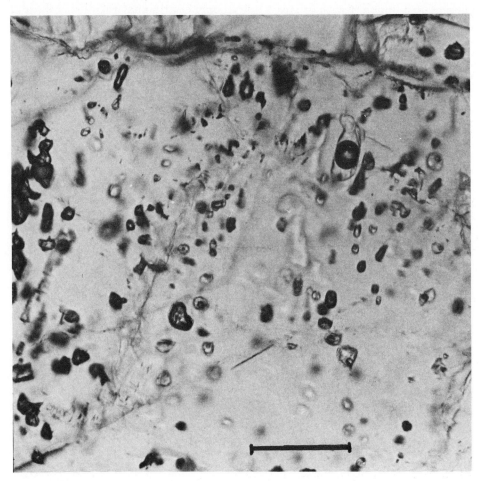

Fig. 7-8. Photograph of fluid inclusions in stalagmites. Scale bar = 0.1 mm.

$$\delta^2H = 8\,\delta^{18}O + 10 \tag{9}$$

as obtained by Dansgaard (1964) and Craig (1961). Paleotemperatures calculated using this relationship generally lie in the range from $-2°C$ to $+25°C$ depending on the age and geographic location of the samples. The principal application of this method was initially to establish the sign of $d(\delta^{18}O_c)/dT$ where $\delta^{18}O_c$ is the isotopic composition of calcite in a speleothem. Schwarcz et al. (1976) found that in 5 out of 6 sites studied this coefficient was negative, that is, that the dominant control on the isotopic composition of speleothem was the temperature dependence of the fractionation factor Δ_{cw}, together with the shift toward higher $\delta^{18}O$ values in seawater during glacial epochs.

Fluid inclusions can be extracted for analysis of either hydrogen or oxygen isotopes by crushing the sample in an evacuated steel tube attached to a cold finger. The water vapour must be frozen off immediately or it will exchange oxygen isotopes with the crushed calcite.

Applications of oxygen and hydrogen isotope geochemistry of speleothems

Harmon et al. (1978b) described records of secular variation over the past 200 ky of $\delta^{18}O_c$ of speleothems from North America and Bermuda. The speleothems had been dated by the $^{230}Th/^{234}U$ method. They made use of the previously noted estimates of the sign of $d(\delta^{18}O_c)/dT$ to assign a paleoclimatic significance to the variations in $\delta^{18}O_c$. They noted correlations between "cold" and "warm" events on these speleothem records and corresponding features on various other dated paleoclimate indicators including $\delta^{18}O$ of planktonic foraminifera from deep-sea sediments. Thompson et al. (1976) and Harmon et al. (1978d) demonstrated in caves of West Virginia and Kentucky that speleothem records were correlated with global and regional paleoclimate trends.

More recently, Harmon et al. (1979) have studied a detailed fluid-inclusion profile from a stalagmite in Coldwater Cave, northeastern Iowa. They found that the temperatures calculated from δ^2H of the included waters using equation (9) are unacceptably low, extending well below $0°C$. They interpret this record as indicating that during the last glacial epoch, the relation between δ^2H and $\delta^{18}O$ must have differed from that observed for modern rain and snow. Support for this is obtained from studies of Antarctic ice by Epstein et al. (1970) who observed the relation:

$$\delta^2H = 7.9\,\delta^{18}O \tag{10}$$

These data, if fitted to a line of slope 8, would have a δ^2H intercept of $4^0/_{00}$, rather than $10^0/_{00}$ as in equation (9). Using this revised relationship, values for the temperature of formation of speleothem in Coldwater Cave appear to have dropped to a minimum of 1.6°C at about 10 ky B.P.,* , which coincides with the period of furthest southerly ice advance of the Wisconsin glaciation, as inferred from ^{14}C dating of glacial moraines.

Harmon and Schwarcz (1981) in a study of fluid inclusions from a number of localities in North America, conclude that the intercept of equation (9) must have been lower than 10°/$_{00}$ in general during glacial periods. We can use data from Antarctic and Greenland ice cores to estimate shifts in the intercept of equation (9) for the last 100 ky. However, these data may not be strictly applicable to meteoric precipitation in lower latitudes; this fact can partly be checked by analysis of dated groundwaters, although the dating methods are so far not very precise. Therefore, it is not yet possible to attribute exact temperatures to fractionations between speleothem calcite and fluid inclusions during transitions between glacial and interglacial periods when the intercept is expected to vary between 10 and about 4°/$_{00}$. This will result in an uncertainty in calculated temperatures of about 5°C. Nevertheless the "paleotemperature" record calculated from δ^2H values will reflect approximate fluctuations in cave temperature, while the exact temperatures may not be known. The resolution of this dilemma may rest in the direct analysis of the δ^{18}O of the fluid inclusions. Some preliminary data are presented in Fig. 7-9, obtained by crushing 1- to 5-g samples of speleothems in vacuum, and reacting the liberated water with BrF_5. The data scatter rather widely around the meteoric water line and lie beyond the limits expected for shifts due to change in the δ^{18}O-δ^2H intercept. Temperatures calculated using these data are generally too low, being either negative (which is impossible, since the speleothems were deposited from liquid water) or too low, compared with paleoclimate estimates for the locality where the speleothem was collected, at the date indicated by U-series analysis of the sample. These results suggest that some sort of exchange process, not yet well-understood, is altering the oxygen isotopic composition, other than simple isotopic exchange between calcite and water (which would preserve the relationship of equation (9)).

While earlier studies of water bound in calcite were carried out by crushing the calcite, this procedure was found to result in rather poor precision of isotopic analyses, and required large samples (2—5 g) to outweigh traces of water present as background in the apparatus. More recently we have been obtaining water by heating pieces of speleothem weighing about 100 mg (cut with a very thin, diamond-impregnated band saw blade to minimize disturbance of the calcite crystals) (Schwarcz and Yonge, 1983). The samples must

---

* The estimates of the age of this event as originally presented by Harmon et al. (1979) appear now to have been in error and the profile of δ^2H for stalagmite 74014 presented in that paper more likely corresponds to a time span from 12 to 6 ky, that is, beginning at the very end of the Wisconsin glacial stage and extending into the early Holocene.

be heated to 700°C to liberate all their water, at which temperature they also break down to CaO and CO_2. The amount of water liberated is typically about 0.1—0.2 wt% of the calcite. The δ^2H values of water from modern, equilibrium deposits of stalagmite is about 20‰ lighter than associated drip waters, indicating that the water has been fractionated in some way when bound to the calcite; the fractionation factor does not appear to depend on the temperature of deposition (as inferred from studies of caves at various latitudes). Infrared absorption studies of oriented single crystals of calcite from a speleothem show no preferred orientation to the water molecules, indicating that they are present as liquid water droplets rather than as a water of hydration. These waters, when corrected for the 20‰ fractionation observed in modern deposits (assumed to be constant through time) yield paleotemperature estimates for the speleothem that are consistent with the expected trend as inferred from U-series ages and comparison with the foraminiferal isotope record of global ice volume (see below for example).

Besides allowing us to infer cave paleotemperatures, fluid inclusion data on speleothems of known age are interesting in themselves as a record of

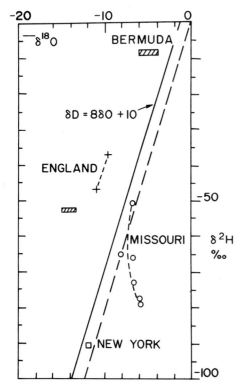

Fig. 7-9. Oxygen and hydrogen isotopic composition of fluid inclusions in speleothems, obtained by crushing (Yamamoto and Schwarcz, unpublished data).

variation in the isotopic composition of meteoric precipitation through the past. Data from studies of fossil groundwater from Africa, Europe and the Middle East suggest that during the last ice age the deuterium (^{2}H) and ^{18}O content of precipitation was less than or equal to that observed in the same areas today (Münnich and Vogel, 1962; Gat and Issar, 1974). Yapp and Epstein (1977) have, however, found evidence that ^{2}H contents of rainwater in parts of North America were higher than at present during the last ice age. Harmon et al. (1978c) have studied the variations of δ^2H of fluid inclusions over the past 250 ky at sites in east central North America and found that there is a tendency for glacial-age waters to be lighter than interglacial waters. The average shift in δ^2H at a given site from glacial to interglacial times is 12‰. By sampling speleothems of approximately the same age it should be possible to plot contours of δ^2H for a continent at any time in the

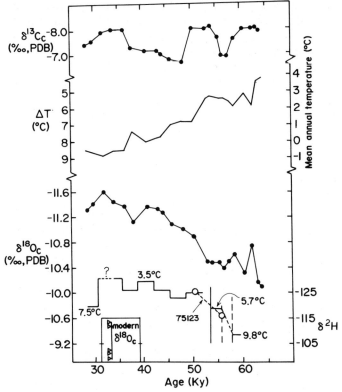

Fig. 7-10. Isotopic composition of calcite speleothems from Cascade Cave, British Columbia. Temperatures are shown as inferred from variation in δ^{18}O of seawater, temperature dependences of Δ_{cw} (−0.24‰ °C^{-1}) and meteoric precipitation (0.70‰ °C^{-1}) and modern temperature of cave (Gascoyne et al., 1979). Also shown are δ^2H values of water extracted by decrepitation at 700°C from calcite samples, and temperatures calculated from them assuming δ^2H = 8 δ^{18}O + 10. Note also δ^{18}O$_c$ of modern calcite from cave.

past. Harmon et al. (1978c) show that for eastern North America, the contours are approximately parallel to the present-day contours of δ^2H of meteoric precipitation plotted by Taylor (1974), but shifted southwards.

Although speleothem depositional sites in the mid-continent have been shown by Schwarcz et al. (1976) to have negative temperature coefficients of $\delta^{18}O_c$, positive values of d $\delta^{18}O_c/dT$ are theoretically possible if the local temperature coefficient of $\delta^{18}O_p$ is sufficiently large. For maritime regions, Dansgaard (1964) found a value of 0.70‰ $°C^{-1}$. We have obtained data for speleothems which grew in a cave in British Columbia, during the first half of the Wisconsin stage (Gascoyne et al., 1980). The $\delta^{18}O_c$ of speleothem decreased by 1.3‰ from 63 to 30 ky B.P., presumably in response to cooling of the cave (Fig. 7-10). If one takes into account the increase in $\delta^{18}O$ of seawater over that period by about 0.7‰ as estimated from the isotopic record of benthic foraminifera in core V19-29 (Shackleton, 1977) and one accepts Dansgaard's value for the temperature coefficient of meteoric precipitation, then the observed shift in $\delta^{18}O_c$ corresponds to a cooling of the cave by about 4.5°C. Comparison of the glacial-age speleothems with modern deposits suggests that deposition began at a temperature of + 4°C, and the cave's temperature then fell almost to the freezing point at about 30 ky B.P. when speleothem deposition ceased. We expect similar isotopic behaviour in other maritime caves. For comparison, paleotemperatures estimated from δ^2H values obtained by heating of samples of this speleothem are also shown; $\delta^{18}O$ of drip waters were calculated using the correlation between δ^2H and $\delta^{18}O$ as described above.

Carbon isotope variations in speleothems

Although in theory it is possible to use fractionation of carbon isotopes between dissolved HCO_3^- ions and calcite as a means of determining paleotemperature, this proves to be impossible in practice. First of all, there is commonly a progressive fractionation of carbon isotopes in successive deposits along a single growth layer, even where oxygen- isotope equilibrium is observed. Secondly, the isotopic fractionation between calcite and solution depends on P_{CO_2} and pH of the solution as well as temperature. Nevertheless we may expect $\delta^{13}C$ values to yield some useful information about the history of the cave and the climate above it.

Dissolved carbon in the seepage water is derived from three sources: atmospheric CO_2; decomposing organic matter and respiring plants in the soil; and the limestone which is being dissolved to form the cave. The net isotopic composition of the dissolved bicarbonate or the carbonate which is precipitated from it will vary in a complex fashion in response to changes in the degree of saturation of the solution with respect to $CaCO_3$ (or dolomite), the presence of a gas phase, the rate of precipitation, and other variables. Wigley et al. (1978) have discussed the chemical evolution of this system in great detail. In the discussion at the end of their paper they consider the case of a speleothem growing duw to degassing of CO_2 from water seeping into a cave. They found that for seepage water saturated with respect to calcite and gypsum, starting at a P_{CO_2} of 10^{-1} and outgassing in a cave at $P_{CO_2} = 10^{-3.3}$, and containing dissolved carbon with a total $\delta^{13}C$ of $-13‰$, that the isotopic composition of precipitated calcite increases from an initial value of $-7.8‰$ to a final value of $-2.1‰$. This variation is larger than is typically observed in single growth layers of speleothems. Evidently most of the variation that is observed can be attributed to variations in the chemical path taken by the water to reach the point of precipitation of calcite. However, axial samples of speleothems fed by seepages directly from a small opening in the roof of the cave, such as a joint or microfissure, should preserve a $\delta^{13}C$ record which is largely controlled by changes in $\delta^{13}C$ of total dissolved carbon.

To a large extent this is in turn controlled by the relative contribution from the three sources previously mentioned, which will in turn be a function of: (1) the rate of production of CO_2 in the soil above the cave, itself a function of climate; (2) the thickness of the soil zone above the cave, a function of relative rates of soil formation and erosion, and subject to alteration by such climate-controlled processes as deflation and glaciation; (3) the nature of the vegetation above the cave. The degree of ^{13}C enrichment in the organic matter produced by that vegetation depends on the photosynthetic mechanism used by the plant. Plants using the C_4 (Hatch-Slack) cycle of photosynthesis live largely in semi-arid environments and generate cellulose that averages around $5‰$ lighter than atmospheric CO_2, whereas C_3 (Calvin cycle) plants of more humid habitats produce carbon compounds up to $19‰$ depleted in ^{13}C with respect to the atmosphere. This isotopic label is transmitted in turn to organic matter in the soil and to CO_2 formed by its oxidation. Thus shifts in $\delta^{13}C$ of speleothems in a cave under a soil transitional between these climatic regimes could reflect shifts through time in the position of the vegetational boundary. Incongruent dissolution of dolomite in the limestone above the cave can also strongly affect $\delta^{13}C$ (Wigley et al., 1978). When studying $\delta^{13}C$ variations in speleothems as a possible climatic indicator, care must be taken to analyse speleothems precipitated from water that has undergone as little prior CO_2 loss as possible (e.g., stalactites).

Some marked fluctuations in $\delta^{13}C$ have been observed in speleothems but at present these have not been adequately interpreted in terms of possible climatic or biotic effects. Fig. 7-11 shows a record from Victoria Cave, northwestern England (Gascoyne, 1979). In this site $\delta^{13}C$ does not appear to have been strongly influenced by transitions from glacial to interglacial climate inferred to have occurred at the times of the growth hiatuses. In other caves, however, $\delta^{13}C$ fluctuations have been found to be more pronounced than those in $\delta^{18}O$, presumably reflecting climatic influence on the biota.

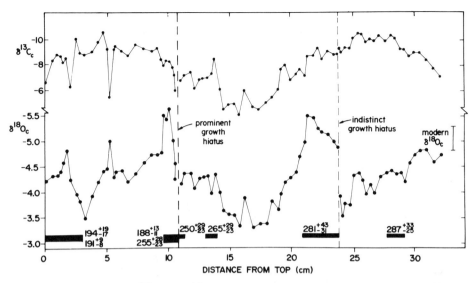

Fig. 7-11. Variation of $\delta^{13}C$ and $\delta^{18}O$ along the axis of a stalagmite from Victoria Cave, northwestern England (from Gascoyne, 1979). Stalagmite grew in oxygen isotopic equilibrium with its parent water.

CONCLUSIONS

Speleothems form a remarkably well-preserved record of past climatic and environmental changes at and near the site of their deposition. Their purity and chemical and physical stability make them exceptionally well suited to detailed geochemical and especially isotopic analysis. Unfortunately, they are also valued as objects of natural beauty, so that extreme circumspection must be employed in obtaining samples. We have so far exclusively used samples which had already fallen from their original growth position or were in such an inaccessible and unobtrusive locale that no aesthetic detriment resulted from their removal. Is is hoped that improved field sampling techniques will allow us some day to obtain cores or other in-situ samples of large speleothems while leaving them for others to admire.

The possibility of obtaining continuous, dated records of paleotemperatures for the continents and for many oceanic islands is tantalizingly close. With further improvements in our ability to analyse the fluid inclusions in these objects and to obtain fine-scale records of variations in their isotopic composition, we should be able to construct paleoclimate records which allow greater resolution than obtainable by any other means, including deep-sea isotopic records of foraminifera. This greater resolution is made possible by the presence of a continuous isotopic record, undisturbed by bioturbation, which can be sampled on a scale such that individual stable isotope analyses correspond to only a few years of growth. These isotopic records must, however, be matched by higher-resolution geochronology. It is expected that some of the techniques described above (paleomagnetic correlation, ESR dating, etc.) will with further improvement allow more precise time-correlation.

REFERENCES

Aitken, M., 1974, Physics and Archaeology. Clarendon Press, Oxford, 291 pp.
Atkinson, T., Smart, P.L., Harmon, R. and Waltham, A.C., 1978. Paleoclimatic and geomorphic implications of ^{230}Th/^{234}U dated speleothems from Britain. Nature, 272: 24—28.
Bastin, B., 1978. L'analyse pollinique des stalagmites: une nouvelle possibilité d'approche des fluctuations climatiques du Quaternaire. Ann. Soc. Geol. Belg., 101: 13—19.
Blackwell, B., Schwarcz, H.P. and Debénath, A., 1983. Absolute dating of homonids and Paleolithic artifacts of the cave of La Chaise-de-Vouthon (Charente), France. J. Archeol. Sci., 10: 493—513.
Broecker, W. and Olson, E.A., 1960. Radiocarbon measurements and annual rings in cave formations. Nature, 185: 93—94.
Craig, H., 1961. Isotopic variations in meteoric waters. Science, 133: 1702—1703.
Dansgaard, W., 1964. Stable isotopes in precipitation. Tellus, 16: 438—468.
Duplessy, J-C., Labeyrie, J., Lalou, C. and Ngyuen, H.V., 1970. Continental climatic variations between 130,000 and 90,000 years B.P. Nature, 226: 631—633.
Emiliani, C., 1971. The last interglacial: paleotemperatures and chronology. Science, 171: 571—573.
Epstein, S., Lowenstam. H., Buchsbaum, R. and Urey, H.C., 1953. A revised carbonate-water isotopic temperature scale. Geol. Soc. Am. Bull., 64: 1315—1326.
Epstein, S., Sharp, R.P. and Gow, A.J., 1970. Antarctic ice sheet: stable-isotope analyses of Byrd Station cores and interhemispheric climatic implications. Science, 168: 1570—1572.
Fantidis, J. and Ehhalt, D.H., 1970. Variations of the carbon and oxygen isotopic composition in stalagmites and stalactites: evidence of non-equilibrium isotopic fractionation. Earth Planet. Sci. Lett., 10: 136—144.
Folk, R. and Asseretto, R., 1976. Comparative fabrics of length-slow and length-fast calcite and calcitized aragonite, in a Holocene speleothem, Carlsbad Caverns, New Mexico. J. Sediment. Petrol., 46: 486—496.
Ford, D.C., 1976. Evidence of multiple glaciation in South Nahanni National Park, Northwest Territories. Can. J. Earth Sci., 13: 1433—1445.
Ford, D.C., Thompson, P. and Schwarcz, H., 1972. Dating cave calcite deposits by the uranium disequilibrium method: some preliminary results from Crowsnest Pass, Alberta. 2nd Guelph Symp. Geomorphol., Guelph, Ont., pp. 247—255.
Ford, D.C., Schwarcz, H.P., Drake, J., Gascoyne, M., Harmon, R. and Latham, A., 1981. On the age of the extant relief in the southern Rocky Mountains of Canada. Arct. Alpine Res., 13: 1—10.
Fornaca-Rinaldi, G., Panichi, C. and Tongiorgi, E., 1968. Some causes of the variations of the isotopic composition of carbon and oxygen on cave concretions. Earth Planet. Sci. Lett., 4: 321—324.
Franke, H. and Geyh, M., 1972. Tropfsteinwachstum und Datierung. Mitt. Dtsch. Höhlen Karstforsch., 18: 59—60.
Friedman, I. and O.Neil, J.R., 1977. Compilation of stable isotope fractionation factors of geochemical interest. U.S. Geol. Surv., Prof. Pap., 440-KK: 3.

Gascoyne, M., 1977. Trace elements in calcite — the only cause of speleothem color? *Natl. Speleol. Soc. Annu. Conv., Proc.*, pp. 39—42.

Gascoyne, M., 1979. *Isotopic and geochronological studies of speleothems.* Ph.D. Thesis, McMaster University, Hamilton, Ont. (unpublished).

Gascoyne, M. and Schwarcz, H.P., 1981. Carbonate and sulfate precipitates. In: M. Ivanovich and R.S. Harmon, (Editors), *Uranium Series Disequilibrium: Applications to Environmental Problems in the Earth Sciences.* Oxford University Press, Oxford, pp. 268—301.

Gascoyne, M., Schwarcz, H.P. and Ford, D., 1978. Uranium series dating and stable isotope studies of speleothem: I. Theory and techniques. *Brit. Cave Res. Assoc. Trans.*, 5: 91—111.

Gascoyne, M., Schwarcz, H.P. and Ford, D., 1980. A paleotemperature record for the mid-Wisconsin in Vancouver Island. *Nature*, 285: 474—476.

Gascoyne, M., Benjamin, G., Schwarcz, H. and Ford, D.C., 1979. Sea level lowering during Illinoian glaciation: evidence from a Bahama bluehole. *Science*, 205: 806—807.

Gascoyne, M., Schwarcz, H.P. and Ford, D.C., 1983a. Uranium series ages of speleothem from northwest England: correlation with Quaternary climate. *Philos. Trans. R. Soc. London, Ser. B*, 301: 143—164.

Gascoyne, M., Ford, D.C. and Schwarcz, H.P., 1983b. Rates of cave and landform development in the Yorkshire Dales from speleothem age data. *Earth Surf. Proc. Landf.*, 8: 557—568.

Gat, J.R., 1980. The isotopes of hydrogen and oxygen in precipitation. In: P. Fritz and J.Ch. Fontes (Editors), *Handbook of Environmental Isotope Geochemistry, 1. The Terrestrial Environment, A.* Elsevier, Amsterdam, pp. 21—47.

Gat, J. and Issar, A., 1974. Desert isotope hydrology: water sources of the Sinai Desert. *Geochim. Cosmochim. Acta.* 36: 1117—1131.

Geyh, M., 1970. Isotopenphysikalische Untersuchungen an Kalksinter, ihre Bedeutung für die ^{14}C-Alterbestimmung von Grundwasser und die Erforschung des Paläoklimas. *Geol. Jahrb.*, 88: 149—158.

Harmon, R.S., 1976. *Late Pleistocene paleoclimates in North America as inferred from isotopic variations in speleothems.* Ph.D. Thesis, McMaster University, Hamilton, Ont. (unpublished).

Harmon, R.S. and Schwarcz, H.P., 1981. Changes of ^{2}H and ^{18}O enrichment of meteoric water and Pleistocene glaciation. *Nature*, 290: 125—128.

Harmon, R.S., Thompson, P., Schwarcz, H.P. and Ford, D.C., 1975. Uranium series dating of speleothems. *Natl. Speleol. Soc. Bull.*, 37: 21—33.

Harmon, R.S., Ford, D.C. and Schwarcz, H.P., 1977. Interglacial chronology of the Rocky and Mackenzie Mountains based on ^{230}Th/^{234}U dating of calcite speleothems. *Can. J. Earth Sci.*, 14: 1543—1552.

Harmon, R.S., Schwarcz, H.P. and Ford, D.C., 1978a. Late Pleistocene sea level history of Bermuda. *Quat. Res.*, 9: 205—218.

Harmon, R.S., Thompson, P., Schwarcz, H.P. and Ford, D.C., 1978b. Late Pleistocene paleoclimates of North America as inferred from stable isotope studies of speleothems. *Quat. Res.*, 9; 54—70.

Harmon, R.S., Schwarcz, H.P. and O'Neil, J.R., 1978c. D/H ratios in speleothem fluid inclusions: a guide to variations in the isotopic composition of meteoric precipitation? *Earth Planet. Sci. Lett.*, 42: 254—266.

Harmon, R.S., Schwarcz, H.P. and Ford, D.C., 1978d. Stable isotope geochemistry of speleothems and cave-waters from the Flint Ridge—Mammoth cave system, Kentucky: implications for terrestrial climate change during the period 230,000 to 100,000 yr B.P. *J. Geol.*, 86: 373—384.

Harmon, R.S., Schwarcz, H.P. and Aley, T., 1978e. Isotopic studies of speleothem from a cave in southern Missouri, U.S.A. In: R. Zartman (Editor), *Short Papers of the Fourth*

International Conference on Geochronology, Cosmochronology and Isotope Geology. U.S. Geol. Surv. Open File Rep., 78-701: 165—166.
Harmon, R.S., Schwarcz, H., Thompson, P. and Ford, D.C., 1978f. Critical comments on "Uranium series dating of stalagmites from Blanchard Springs Caverns, Arkansas, U.S.A", Geochim. Cosmochim. Acta, 42: 433—437.
Harmon, R.S., Schwarcz, H.P., Ford, D.C. and Koch, D.L., 1979. An isotope paleotemperature curve for Late Wisconsinan time. Geology, 7: 430—433.
Harmon, R.S., Land, L.S., Mitterer, R.M., Garrett, P., Schwarcz, H.P. and Larson, G., 1981, Bermuda sea level during the last interglacial. Nature, 289: 481—483.
Hendy, C.H., 1969. *The Isotopic Geochemistry of Speleothems and its Application to the Study of Past Climates.* Ph.D. Thesis, Victoria University, Wellington, 425 pp. (unpublished).
Hendy, C.H., 1970. The use of C-14 in the study of cave processes In: I. Olsen (Editor), *Radiocarbon Variations and Absolute Chronology. Proceedings 12th Nobel Symposium, Uppsala, 1969.* Wiley, New York, N.Y., pp. 419—443.
Hendy, C.H., 1971. The isotopic geochemistry of speleothems, 1. The calculation of the effects of different modes of formation on the isotopic composition of speleothems and their applicability as paleoclimatic indicators. Geochim. Cosmochim. Acta, 35: 801—824.
Hendy, C.H. and Wilson, A., 1968. Paleoclimatic data from speleothems. Nature, 219: 48—51.
Hennig, G., Herr, W., Weber, E. and Xirotiris, N.I., 1981. ESR-dating of the fossil hominid cranium from Petralona Cave, Greece. Nature, 292: 533—536.
IAEA, 1981. Statistical treatment of environmental isotope data in precipitation. Int. At. Energy Agency Tech. Rep. Ser., 206: 256pp.
Ikeya, M., 1978. Electron spin resonance as a method of dating. Archaeometry, 20: 147—158.
Imbrie, J., Hays, R. and Shackleton, N., 1976. Variations in the Earth's orbit: pacemaker of the Ice Ages. Science, 194: 1121—1132.
Johnsen, S., Dansgaard, W., Clausen, H. and Langway, C.C., 1972. Oxygen isotope profiles through the Antarctic and Greenland ice sheets. Nature, 235: 429—434.
Kaufman, A., 1971. U-series dating of Dead Sea basin carbonates. Geochim. Cosmochim. Acta, 35: 1269—1281.
Kendall, A. and Broughton, P., 1978. Origin of fabrics in speleothems composed of columnar calcite crystals. J. Sediment. Petrol., 48: 519—538.
Ku, T.L., 1976. The uranium series methods of age determination. Annu. Rev. Earth Planet Sci., 4: 347—379.
Ku, T.L., Bull, W.G., Freeman, S. and Knauss, K.G., 1979. ^{230}Th/^{234}U dating of pedogenic carbonates in gravelly desert soils of Vidal Valley, southeastern California. Geol. Soc. Am. Bull., 90: 1063—1073.
Langmuir, D., 1978. Uranium solution-mineral equilibria at low temperatures with applications to sedimentary ore deposits. Geochim. Cosmochim. Acta, 42: 547—569.
Langmuir, D. and Herman, J., 1980. The mobility of thorium in natural waters at low temperatures. Geochim. Cosmochim. Acta, 44: 1753—1766.
Latham, A., Schwarcz, H.P., Ford, D.C. and Pearce, G.W., 1979. The paleomagnetism of stalagmite deposits. Nature, 280: 282—284.
Mook, W.G., 1980. Carbon-14 in hydrogeological studies. In: P. Fritz and J.Ch. Fontes (Editors), *Handbook of Environmental Isotope Geochemistry, 1. The Terrestrial Environment,* A. Elsevier, Amsterdam, pp. 49—74.
Münnich, K. and Vogel, J.C., 1962. ^{14}C Alterbestimmung von Süsswasser-Kalkablagerungen. Naturwissenschaften, 46: 168—169.
O'Neil, J.R., Clayton, R.N. and Mayeda, T., 1969. Oxygen isotope fractionation in divalent metal carbonates. J. Chem. Phys., 30: 5547—5558.

Osmond, J.K., 1980. Uranium disequilibrium in hydrologic studies. In: P. Fritz and J.Ch. Fontes (Editors), *Handbook of Environmental Isotope Geochemistry, 1. The Terrestrial Environment, A*. Elsevier, Amsterdam, pp. 259—282.
Schwarcz, H.P., 1980. Uranium series dating of speleothems from archaeological sites. *Archaeometry*, 22: 3—24.
Schwarcz, H.P. and Yonge, C., 1983. Isotopic composition of paleowaters as inferred from speleothem and its fluid inclusions. In: R. Gonfiantini (Editor), *Paleoclimates and Paleowaters: A Collection of Environmental Isotope Studies*. IAEA, Vienna, pp. 115—133.
Schwarcz, H.P., Harmon, R.S., Thompson, P. and Ford, D.C., 1976. Stable isotope studies of fluid inclusions in speleothems and their paleoclimatic significance. *Geochim. Cosmochim. Acta*, 40: 657—665.
Shackleton, N.J., 1977. The oxygen isotope stratigraphic record of the Late Pleistocene. *Philos. Trans. R. Soc. London, Ser. B*, 280: 169—182.
Shackleton, N.J. and Opdyke, N.J., 1973. Oxygen isotope and paleomagnetic stratigraphy of equatorial Pacific core V28-238: oxygen isotope temperatures and ice volumes on a 10^5 year and 10^6 year time scale. *Quat. Res.*, 3: 39—55.
Shackleton, N. and Opdyke, N., 1976. Oxygen-isotope paleomagnetic stratigraphy of Pacific core V28-239, Late Pliocene to latest Pleistocene, In: R. Cline and J.D. Hays (Editors). *Investigation of Late Quaternary Paleoceanography and Paleoclimatology. Geol. Soc. Am. Mem.*, 145: 449—464.
Spalding, R.F. and Mathews, T.D., 1972. Stalagmites from caves in the Bahamas; indicators of low sea-level stand. *Quat. Res.*, 2: 470—472.
Talma, A.S., Vogel, J.C. and Partridge, T.C., 1974. Isotopic contents of some Transvaal speleothems and their paleoclimatic significance. *S. Afr. J. Sci.*, 70: 135—140.
Taylor, H.P., Jr., 1974. The application of oxygen and hydrogen isotope studies to problems of hydrothermal alteration and ore deposition. *Econ. Geol.*, 69: 843—883.
Thompson, G.M., Lumsden, D., Walker, R.L. and Carter, J.A., 1975. Uranium series dating of stalagmites from Blanchard Springs Caverns, Arkansas, U.S.A. *Geochim. Cosmochim. Acta*, 39: 1211—1218.
Thompson, P., Ford, D.C. and Schwarcz, H.P., 1975. $^{234}U/^{238}U$ ratios in limestone cave seepage waters, and speleothem from West Virginia. *Geochim. Cosmochim. Acta*, 39: 661—669.
Thompson, P., Schwarcz, H.P. and Ford, D.C., 1976. Stable isotope geochemistry, geothermometry and geochronology of speleothems from West Virginia. *Geol. Soc. Am. Bull.*, 87: 1730—1738.
Turekian, K. and Nelson, E., 1976. Uranium series dating of the travertines from Caune de l'Arago (France). *Union Int. Sci. Prehist., Protohist., IX Congr., Nice.* pp. 172—179.
White, W., 1976. Cave minerals and speleothems, In: T.D. Ford and C.M.D. Cullingford (Editors), *The Science of Speleology*. Academic Press, New York, N.Y., pp. 267—327.
Wigley, T.M.L. and Brown, M.C., 1976. Cave physics. In: T.D. Ford and C.M.D. Cullingford (Editors), *The Science of Speleology*. Academic Press, New York, N.Y., pp. 329—358.
Wigley, T.M.L., Plummer, L.N. and Pearson, F.J., 1978. Mass transfer and carbon isotope evolution in natural water systems. *Geochim. Cosmochim. Acta*, 42: 1117—1140.
Wintle, A., 1976. Effects of sample preparation on the thermoluminescence characteristics of calcite. *Modern Geol.*, 5: 165—167.
Wintle, A., 1978. A thermoluminescence dating study of some Quaternary calcite: potential and problems. *Can. J. Earth Sci.*, 15: 1977—1986.
Yapp, D. and Epstein, S., 1977. Climatic implications of D/H ratios of meteoric water over North America (9500—22,000 B.P.) as inferred from ancient wood cellulose C-H hydrogen. *Earth Planet. Sci. Lett.*, 34: 333—350.

Chapter 8

OXYGEN AND HYDROGEN ISOTOPE GEOCHEMISTRY OF DEEP BASIN BRINES

YOUSIF K. KHARAKA and WILLIAM W. CAROTHERS

INTRODUCTION

Oxygen and hydrogen isotope analyses have proven extremely important in the study of the origin of deep basin brines. Prior to these isotopic studies, it was generally assumed that most saline waters of marine sedimentary basins were original ocean water trapped with the sediments at the time of deposition (White, 1965). Clayton et al. (1966) were the first to show that waters from many petroleum fields were predominately of local meteoric origin, the original connate water of deposition being lost by compaction and subsequent flushing. Craig and coworkers (Craig et al., 1956; Craig, 1963) had shown earlier that geothermal fluids were meteoric in origin as their deuterium content was similar to that of local precipitation.

Determinations of stable isotope ratios and fluid potentials, on the other hand, were used by Kharaka et al. (1978a) to show that formation waters in Tertiary sandstones from the northern Gulf of Mexico basin were connate waters representing the original marine water of deposition. White et al. (1973) showed that some thermal spring waters from the northern California Coast Ranges are predominantly metamorphic in origin.

This paper reviews the application of stable isotope analyses of oxygen and hydrogen to the study of the origin of formation waters in sedimentary basins. Stable isotope variations of oxygen and hydrogen in present and old meteoric and ocean waters are reviewed and used in conjunction with the isotopic composition of deep basin waters to indicate the origins of formation waters, especially those associated with petroleum. The processes responsible for modification of the original isotopic composition of these waters including evaporation, mixing, exchange with rocks and membrane behavior of shales are discussed in some detail. Finally, we discuss the application of these analyses to geothermometry and mineral diagenesis.

A number of excellent reviews on stable isotope analyses and their use in geology and hydrology have appeared in the last decade. Volume 1 of the present Handbook (Fritz and Fontes, 1980) contains many excellent chapters with information pertinent to this chapter. Volume II/1 of the Handbook of Geochemistry (Wedepohl, 1978) also contains excellent chapters on the isotopes of hydrogen, oxygen and carbon. Other recent books and symposium volumes on this subject include those by Faure (1977), Gonfiantini (1977), O'Neil (1977), IAEA (1978), Jäger and Hunziker (1979), and Hoefs (1980).

TERMS APPLIED TO SUBSURFACE WATERS

A number of descriptive terms have been used to classify subsurface waters. No overall satisfactory classification system exists, due to the fact that subsurface waters can be assessed by several different criteria, such as: the salinity of the water, the concentrations of the dissolved constituents, the origin of the water, and the origin of the solutes which commonly are different from that of the transporting water. Our views on the various terms used in this paper are summarized below. The interested reader should consult White (1957, 1965), and White et al. (1963), for a more complete review of water terms.

Brine. The definition preferred for this term is "Water with salinity higher than that of average sea water, i.e., more than 35,000 mg l^{-1} dissolved solids" (Hem, 1970). The majority of oil field waters are brines according to this definition; whereas, only a small fraction of these waters could be classified as brines if we accept the definition of brine given by Davis (1964) and Carpenter et al. (1974) which is "Water containing more than 100,000 mg l^{-1} dissolved solids".

Formation water. The accepted definition of this term is "Water present in the rocks immediately before drilling" (Case, 1955). This term is used extensively in the petroleum industry, but has no genetic or age significance.

Meteoric water. The generally accepted definition of this term is that given by White (1957) as "Water that was recently involved in atmospheric circulation. The age of meteoric ground waters is slight compared with the age of the enclosing rocks, and is not more than a small part of a geologic period." Recent evidence from stable isotope ratios of water indicates that meteoric water can percolate to great depths (up to 10 km) in geothermal systems (Truesdell and Hulston, 1980), hydrothermal systems (White, 1974; Hattori and Sakai, 1979), batholiths (Taylor and Margaritz, 1978; Norton and Taylor 1979), and deep sedimentary basins (Clayton et al., 1966). The time required for percolation to these depths and total distances from recharge areas that may exceed 100 km (Manheim and Sayles, 1974) will not be "slight", but may require millions of years. Kharaka and Carothers (1982) showed that oil field waters from North Slope, Alaska, originated as surface waters, but that the recharge took place in Miocene or earlier times.

In this report we redefine subsurface water of meteoric origin as "water derived from rain, snow, water courses, and other bodies of surface water that percolates in rocks and displaces their interstitial water that may have been connate, meteoric, or of any other origin." The majority of meteoric waters in sedimentary basins are nonmarine and are generally recharged at high elevations in the margins of the basins. The time of last contact with the atmosphere was intentionally omitted from the definition but may be specified to further define meteoric water. Thus, "Recent meteoric water", "Pleistocene meteoric water", or "Tertiary meteoric water", would indicate the time of last contact with the atmosphere.

Connate water. Is here defined as "water that was deposited with sediments or other rocks in the basin and which has been out of contact with the atmosphere since its deposition."

This definition is similar to that given by White (1965) in that the water need not be present in the same rocks with which it was deposited. Formation water in sandstone beds, present within or above thick shale and siltstone sequences, is almost always water squeezed from the underlying shales and siltstones (Berry, 1973; Kharaka et al., 1979). Connate waters, as here defined, are generally marine in origin and are similar in age to their associated rocks in thick shale and siltstone sequences, but are generally older than their associated rocks when present in sandstone and limestone aquifers within these sequences; it is also possible, under the right combination of lithology and fluid potentials, for connate waters to be younger than the enclosed rocks.

Metamorphic water. Defined by White (1957) as "water that is, or that has been, associated with rocks during their metamorphism." This water is probably not important quantitatively in deep basin brines associated with petroleum because the temperatures reached are generally lower than that expected in metamorphic reactions.

Diagenetic water. Defined here as "water released from solid phases as a result of mineral reactions and transformations in the zone of diagenesis." Water released from the transformation of gypsum to anhydrite and smectite to illite are examples of this type of water. "Absorbed water" on clay minerals, released as a result of compaction, is excluded from this definition of diagenetic water.

ISOTOPIC COMPOSITION OF SURFACE WATERS

Application of stable isotope analysis has shown that subsurface waters are mainly related to local meteoric water of different ages and marine connate water deposited with shales and siltstones. The isotopic compositions of present-day and old surface waters (including ocean water) are discussed below even though they are covered in more detail in Volume 1 of this series (Fritz and Fontes, 1980) because an understanding of these is needed in interpreting the origin of deep basin brines.

Stable isotope compositions of recent meteoric water

The relations governing the distribution of stable isotopes of oxygen and hydrogen in present-day surface waters are reasonably well known. Craig (1961) showed that surface waters, other than those from closed basins undergoing extensive non-equilibrium evaporation and from hot springs, have isotopic compositions that lie on a "global meteoric water line" that can be described by the equation:

$$\delta^2 H = 8\,\delta^{18}O + 10 \tag{1}$$

The δ values are given in the permil (parts per thousand, $^0/_{00}$) notation. The equation obtained by Gat (1980) from a plot of the mean $\delta^2 H$ and $\delta^{18}O$ values of all precipitation samples from the International Atomic Energy Agency (IAEA) network is:

$$\delta^2 H = 8.17\,\delta^{18}O + 10.56 \tag{2}$$

which is essentially that of the "global meteoric water line".

A more general equation describing the relationship between $\delta^2 H$ and $\delta^{18}O$ values of surface waters is of the form;

$$\delta^2 H = x\,\delta^{18}O + d \tag{3}$$

where the value of x is generally equal to 8, but is about 6 for precipitation in tropical island stations (Yurtsever, 1975) and is 2—5 in closed basin-lakes undergoing non-equilibrium fractionation during evaporation. Dansgaard (1964) refers to the "d-parameter" in equation (3) as the "deuterium excess". The value of d is generally equal to 10, but may be less than 10 over many areas including wet coastal areas and tropical islands; its value is 22 in the dry eastern Mediterranean area (Gat and Carmi, 1970). The "global meteoric water line", however, should be used for surface waters in any area where no specific line has been established.

The distribution of the mean annual $\delta^2 H$ values of surface and groundwaters in North America (Taylor and Margaritz, 1978) is shown in Fig. 8-1. The distribution of the $\delta^{18}O$ values for these waters may be approximated, if not already available, from equation (1) describing the "global meteoric water line".

The precipitation temperature, which can be correlated with latitude, altitude and distance from the ocean, was shown to be the most important parameter controlling the δ^2H (Friedman et al., 1964) and $\delta^{18}O$ (Dansgaard, 1954, 1961) values of precipitation. Dansgaard (1964) gave the following equation to describe the relationship between temperature and $\delta^{18}O$ values for Northern Atlantic precipitation:

$$\delta^{18}O = 0.695t - 13.6 \tag{4}$$

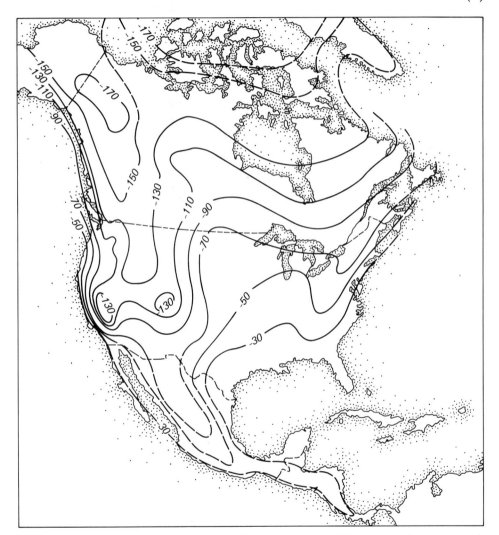

Fig. 8-1. Map of North America showing the approximate distribution of mean annual δ^2H values of modern meteoric water (modified from Taylor and Margaritz, 1978). Data for Arctic water from Kharaka and Carothers (1984) and Yurtsever (1975).

where t is the mean annual temperature in degrees Celsius. Evans et al. (1978), however, obtained a different slope and intercept than those in equation (4) when he plotted the $\delta^{18}O$ and temperatures from stations in northwest Europe. On a worldwide basis, the relationship between temperature and isotope values is not linear, but is represented by a set of curves which all approach values close to $\delta = 0$ at the higher temperatures. The latitude effect over the North American Continent was shown to be about $0.5^{0}/_{00}$ $\delta^{18}O$ per degree latitude (Yurtsever, 1975). The altitude effect is variable depending on local topography and climate, but depletions of $12-40^{0}/_{00}$ in δ^2H and $1.5-5^{0}/_{00}$ in $\delta^{18}O$ per kilometer are typical (Gat, 1980).

Stable isotope compositions of old meteoric water

Knowledge of the variations in the isotopic composition of surface waters through geologic time is necessary in order to study the origin of some deep basin brines which may have been recharged millions of years ago. These variations result from two main causes: (1) changes in the isotopic composition of the oceans through geologic time (see next section) and (2) changes in the climatic and topographic conditions at the area of recharge. These climatic changes in turn result from two factors: (1) continental drift and the phenomena connected with it, including latitude changes, mountain building, and the ensuing changes in oceanic currents and wind circulation patterns; and (2) astronomical factors not related to continental drift that include changes in the sun's radiation and changes in the composition of the earth's atmosphere.

It is generally recognized that "the present is not the key to the past" as far as the climate is concerned (Nairn, 1964; Lamb, 1977; Habicht, 1979; and others). This is illustrated in the general distribution of temperatures, as a function of latitude for the present, glacial, interglacial and non-glacial periods (Fig. 8-2). More detailed studies reported in Lamb (1977) and Gerasimov (1979) show temperature fluctuations of the order of $\pm 2°C$ in the middle latitudes within the Holocene Epoch alone. It should be noted that Dansgaard (1964) has shown that present-day precipitation changes in δ^2H and $\delta^{18}O$ values by 5.6 and $0.7^{0}/_{00}$ respectively, for every $1°C$ change in mean annual temperature (see equation (4) for $\delta^{18}O$ change). This shows clearly that paleoclimatic and paleogeographic reconstructions (see Habicht, 1979, for worldwide general reconstructions) should be attempted to aid in interpreting the isotopic data (see also our discussion of deep basin brines for North Slope, Alaska).

Isotopic composition of ocean water through time

The δ^2H and $\delta^{18}O$ values of present-day seawater are close to zero and vary only within narrow limits. Redfield and Friedman (1965) and Craig and

Gordon (1965) showed that the δ^2H values of deep water masses had a narrow range from —1.4 to +2.2. The variations in the δ^2H and δ^{18}O values of the surface waters in the ocean, on the other hand, are much larger and can be related to evaporation, precipitation, and the freezing and melting of ice. The processes modifying the isotopic composition of surface waters also affect the salinity of these waters, and the relationships between salinity and isotope values can be used to infer the origin of the modifications (Craig and Gordon, 1965; Redfield and Friedman, 1965).

The interstitial water of recent sediments deposited in the oceans will have δ^2H and δ^{18}O values close to zero because these values will be controlled by the deeper parts of the water column and not by surface waters. Sediments deposited in coastal areas may acquire interstitial water that is a mixture of fresh, surface and sea waters. The isotopic composition and the salinity of these waters will reflect the amount of mixing between these waters.

Unanimity is lacking in estimates of the isotopic composition of oceanic water in the past. The importance of continental glaciation on the isotopic composition of the ocean is widely recognized, although the magnitude of this effect is uncertain. Craig (1965) determined that the mean δ^{18}O value of the ocean could be —0.6, instead of its present value of —0.1‰ if all the continental glaciers were to melt; the δ^2H value would be about —5‰. The mean δ^{18}O value of the ocean during the time of maximum glaciation during

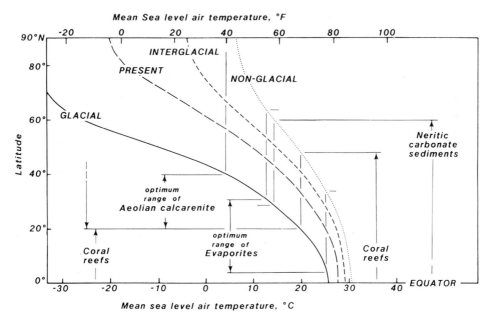

Fig. 8-2. Surface mean temperatures for the present, glacial, non-glacial and interglacial periods at different latitudes (modified from Fairbridge, 1964). Note the large differences in temperatures between different periods at high latitudes.

the Pleistocene epoch has been estimated to vary from 0.4 to 1.6‰ (see Williams et al., 1981, and Emiliani and Shackleton, 1974, for discussion of this problem).

There are two contrasting views on the isotopic composition of ocean water through geologic time. The first view is based on analyses of marine cherts and carbonates (Weber, 1965; Perry, 1967; Perry and Tan, 1972; Brand and Veizer, 1981) that suggest (Fig. 8-3) that the $\delta^{18}O$ value of sea water has increased monotonically at a rate of about 5‰ per billion years. Taylor (1977), on the other hand, concluded that ocean water must have been relatively uniform in δ^2H and $\delta^{18}O$ values during the past 2.5 billion years. Taylor's conclusion is based on the fact granites affected by meteoric-hydrothermal waters have similar isotopic characteristics over this time span.

Studies of the $\delta^{18}O$ values of the oceanic crust by Muehlenbachs and Clayton (1976) support the conclusion that the $\delta^{18}O$ values of ancient oceans have been relatively constant. Material balance calculations based on the processes that deplete seawater in ^{18}O (low-temperature submarine weathering

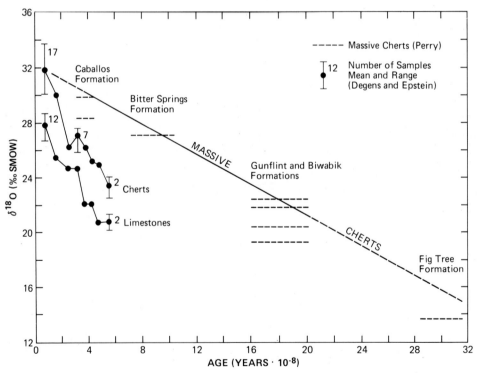

Fig. 8-3. Isotopic composition of limestones, cherts and massive cherts as a function of age (modified from Wedepohl, 1978). Data from Degens and Epstein (1962) and Perry (1967). All the data show a general decrease in the $\delta^{18}O$ values with increasing age of the rocks.

and sedimentation) and enrich it in ^{18}O (high-temperature hydrothermal alteration of the oceanic crust and cycling of chemically bound water from the mantle) showed that the $\delta^{18}O$ of the ocean water probably has been held constant at about zero (Fig. 8-4) at least since the beginning of sea-floor spreadings. $\delta^{18}O$-buffering of the ocean by seawater-hydrothermal circulation at mid-oceanic ridges was also invoked by Gregory and Taylor (1981) to explain the $\delta^{18}O$ values of the Cretaceous oceanic crust, Samail Ophiolite, Oman. Isotopic composition of connate waters from oil and gas fields in the Gulf Coast (to be discussed later) shows that the δ^2H and $\delta^{18}O$ values of ocean water, during the Tertiary period, were close to that of the average ocean water today. Also, it is now believed (Knauth and Epstein, 1976; and others) that the lower $\delta^{18}O$ values of cherts and carbonates of older rocks result from post-depositional exchange with meteoric waters combined, perhaps, with warmer temperatures during parts of the Precambrian. The important question of whether or not the isotopic composition of ocean water has changed significantly with time, however, remains unresolved.

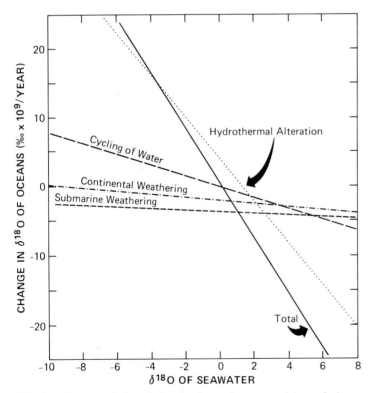

Fig. 8-4. Changes induced in the isotopic composition of the ocean by major oceanic processes versus $\delta^{18}O$ of the ocean (Muehlenbachs and Clayton, 1972). Note that at the present time when $\delta^{18}O = 0$ for seawater, there is no tendency for the isotopic composition of the ocean to change.

ORIGIN OF WATER IN DEEP BASIN BRINES

Application of stable isotope measurements to the study of subsurface waters has shown that locally derived meteoric water is the dominant component of most of these fluids. Stable isotope analyses have also shown that magmatic and juvenile waters (as defined by White, 1957) constitute a negligible portion of subsurface waters even in hydrothermal or deep geothermal systems (White, 1974; Truesdell and Hulston, 1980; and many others). The role of magmatic and juvenile waters to deep basin brines is so negligible, in our opinion, that it will receive no more mention in this paper. Deep basin waters are mainly related to local meteoric water of different ages and marine connate water deposited with shales and siltstones in a number of deep sedimentary basins; diagenetic water, as defined in this report, may contribute an important portion especially for the higher temperature brines.

Deep basin brines derived from Holocene meteoric water

Craig and his coworkers (Craig et al., 1956; Craig, 1963) were the first to show that geothermal waters were related to local precipitation. These and other authors (see Truesdell and Hulston, 1980, for many references) have

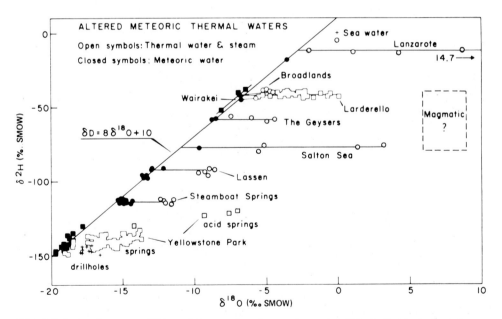

Fig. 8-5. Isotopic compositions of hot-spring, fumarole, and drill-hole fluids derived from meteoric water (open symbols) and of meteoric water local to each system (closed symbols) (Truesdell and Hulston, 1980).

shown that most geothermal waters have δ^2H values similar to local precipitation, but that the $\delta^{18}O$ values are generally displaced towards higher ^{18}O content (Fig. 8-5). The "oxygen isotope shift" to higher ^{18}O content was shown (Craig et al., 1956) to result from water/rock exchange that increases table ^{18}O content in water and lowers it in minerals. The absence of a comparable "hydrogen isotope shift" in most geothermal systems is believed to be due to the fact that minerals in these systems contain little hydrogen and that the water/rock ratios are very high so that the hydrogen in the minerals is a small fraction of the total hydrogen.

The isotopic composition of formation waters associated with petroleum from many sedimentary basins was also shown to be related principally to recent local meteoric water (Clayton et al., 1966; Hitchon and Friedman, 1969; Kharaka et al., 1973; Gutsalo et al., 1978; and others). The δ^2H values in these waters (Fig. 8-6) show a "hydrogen isotope shift" in addition to the generally observed "oxygen isotope shift". Combined hydrogen- and oxygen-isotope shifts have also been reported for a number of geothermal waters from sedimentary basins (Coplen et al., 1975; Coplen, 1976; Vetshteyn et al., 1981; Dowgiallo and Lesniak, 1980; Truesdell and Hulston, 1980; and many others). Combined hydrogen- and oxygen-isotope shifts have also been

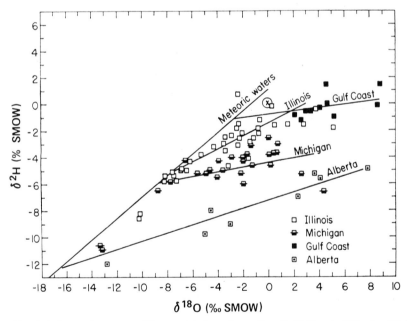

Fig. 8-6. Isotopic compositions of brines from the Gulf Coast, Illinois, Michigan and Alberta sedimentary basins reported by Clayton et al. (1966). Cross represents SMOW and "meteoric water line" is after Craig (1961). Note that in each basin the isotope line intersects the "global meteoric water line" at a point with δ^2H and $\delta^{18}O$ values of present-day meteoric water.

TABLE 8-1

Chemical analyses (mg l^{-1}), well data and stable isotopes ($^o/_{oo}$ relative to SMOW) for oil field waters, North Slope Alaska

	Barrow			Prudhoe Bay			Pitt Point			Simpson Bay
Sample No.:	78-AK-51	78-AK-52	78-AK-53	78-AK-54[a]	78-AK-55[b]	78-AK-56[c]	79-AK-1	79-AK-5	79-AK-8	80-AK-53
Well name	So. Barrow No. 11	So. Barrow No. 5	So. Barrow No. 7	Arco 13 D.S. #1	Arco 8 D.S. #3	Webster Lake	Dalton No. 1	Dalton No. 1	Dalton No. 1	E. Simpson No. 2
Location	S14T22N R18W			S8T10N R15E	S11T10N R15E			S14T18N R5W		S23T19N R11W
Reservoir	Barrow Sand	Barrow Sand	Barrow Sand	Sadlerochit	Sadlerochit	surface	Sadlerochit	Lisburne	Sadlerochit	Sadlerochit
Depth (m)	66–718	701–749	503–683	2804–2835	2682–2782	surface	2381–2481	2012–2641	2430–2436	2181–2195
Subsurface temperature (°C)	17	16	13	94	90	4	–	–	–	–
TDS (mg l^{-1})	20,600	22,100	19,000	21,900	17,000	730	23,000	21,800	21,900	24,200
Na	7260	7980	7070	7600	5720	6.4	8720	8360	8130	8700
Li	1.3	2.1	1.5	4.0	3.3	0.02	6.1	2.6	4.3	1.5
K	3.3	3.0	19	86	68	1.1	57	43	49	35
Rb	0.2	0.3	0.1	0.5	0.5	<0.01	0.5	0.23	0.5	0.2
Cs	0.8	1.6	0.8	1.1	0.9	<0.3	3.3	1.6	3.0	0.9
Mg	66	67	57	20	16	8.3	22	2.7	22	54
Ca	224	119	166	182	151	71	140	31	185	280
Sr	134	16.1	12.5	20.2	13.1	0.28	27	1.6	26	30
Ba	53	175	51	3.8	3.1	<0.5	3.1	42	34	6.3
Fe	138	5.5	4.6	63	57	1.9	<0.1	2.3	0.6	<0.1
Cl	10,700	11,800	10,400	10,600	7940	14	12,600	12,100	12,100	13,900
F	0.7	1.6	1.1	1.5	0.2	<0.1	1.2	0.1	0.7	1.5
Br	49	62	36	54	40	1.2	61	–	59	53
I	11	28	11	19	14	0.1	30	–	31	29
B	20	42	19	158	115	–	130	–	120	112
NH$_3$	9	19	9	17	5	<0.3	15	–	–	18
H$_2$S	<0.1	<0.1	<0.1	<0.1	<0.1	<0.1	–	–	–	–
HCO$_3^-$	320	1480	720	2680	2560	205	1100[d]	1280[d]	1150[d]	870[d]
CH$_3$COO$^-$	1600	230	290	250	240	–	–	–	–	–
SO$_4^{2-}$	3.1	–	2.2	69	33	18	31	–	–	34
SiO$_2$	9	11	16	62	23	0.8	42	–	–	26
pH	6.8	7.2	7.3	6.5	6.3	–	8.6[d]	8.1[d]	8.0[d]	6.7[d]
δ^{18}O	–	−1.66	−0.94	+6.06	+2.61	−18.54	+2.78	+1.85	+1.55	+2.57
δ^{2}H	–	−44.8	−50.5	−34.4	−38.9	−149.6	−35.5	−40.8	−40.2	−34.9

[a] Sample taken below oil-water contact; [b] water sample diluted by 30% condensed water vapor; [c] water sample collected from lake containing 40% ice. [d] Measured in the laboratory.

reported (Fritz and Frape, 1982) in some highly saline brines in the Canadian Shield with somewhat unusual isotope values (the trend line is above instead of below the global meteoric water line). The origins of the shifts (see the section on modification of waters, for discussion) are different from basin to basin, but the line fitted to the δ^2H and $\delta^{18}O$ plots (Fig. 8-6) in each case intersects the meteoric water line (Craig, 1961) at a point with δ^2H and $\delta^{18}O$ values that equal those of present-day meteoric water in the area of recharge.

Deep basin brines derived from "old" meteoric water

Stable isotopes of water in addition to age determinations, using ^{14}C and other isotopes and flow rates, have shown that the waters in many groundwater aquifers are "old" meteoric water, i.e., older than Holocene (> 10,000 years) (Degens, 1962; Pearson and White, 1967; Gat and Issar, 1974; Downing et al., 1977; Bath et al., 1978; Schwartz and Muehlenbachs, 1979; Fontes, 1980; and many others). Clayton et al. (1966) showed that a number of formation water samples from the Michigan and Illinois basins were probably recharged during the Pleistocene Epoch because they had $\delta^{18}O$ values that were much lower than the values for present-day meteoric water in the respective areas.

Kharaka and Carothers (1984) are probably the first to give evidence of meteoric water older than the Pleistocene Epoch. These authors presented isotopic and chemical data for formation waters from exploration and producing oil and gas wells from the North Slope of Alaska. The water samples were obtained from reservoir rocks varying in depth from 700 to 2800 m and in age from Mississippian to Triassic. The chemical data (Table 8-1) show that the formation waters, regardless of the wells sampled and the age, depth, or lithology of the reservoir rocks, are remarkably similar in total dissolved solids (19,000—24,000 mg l^{-1}) and distribution of major cations and anions. The δ^2H and $\delta^{18}O$ values for these waters with the least-squares line drawn through them are shown in Fig. 8-7. This least-squares line intersects the meteoric water line at δ^2H and $\delta^{18}O$ values of —65 and —7‰ respectively. This line for formation waters, it should be noted, does not pass through the values for standard mean ocean water (SMOW) or the present-day meteoric water of the region.

The conclusion drawn by Kharaka and Carothers (1984) was that the formation waters are meteoric in origin, but that recharge took place when North Slope had an entirely different climate from that of today. This climate may be inferred from the data of Dansgaard (1964), who showed that present-day meteoric waters change in δ^2H and $\delta^{18}O$ by 5.6 and 0.7‰, respectively, for every 1°C change in mean annual temperatures. Using the above relationship, which is rather approximate, suggests that the mean annual temperatures at Brooks Range (the most likely recharge area) were 15—20°C higher at the time of recharge than today.

318

Two possibilities exist that would result in meteoric waters at North Slope having isotopic compositions equal to those of the presumed recharge waters. The first assumes topographic conditions at the time of recharge were similar to those of the present but that the North Slope was at a different latitude at the time of recharge. The map of North America showing the distribution of δ^2H values for present-day meteoric water (Fig. 8-1) indicates a location for North Slope at a latitude of 48°N or further south. At about this latitude, the contour line with δ^2H value equal to that of the presumed

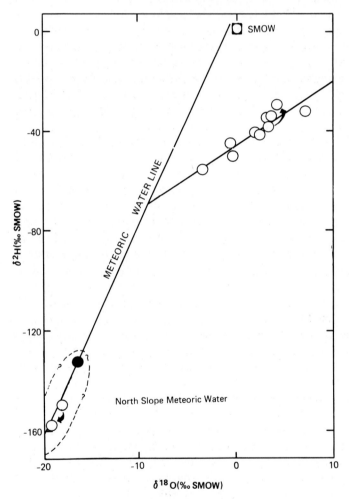

Fig. 8-7. Isotopic compositions of formation waters from North Slope, Alaska. The solid line is the least-squares line drawn through the data. Also shown are values for SMOW, meteoric water in the area and the global "meteoric water line". Arrows connect corrected and uncorrected values (Kharaka and Carothers, 1984). The line through the data, it should be noted, does not pass through SMOW or the meteoric water in the area.

recharge water, crosses from the continent to the Pacific Ocean. Placing northern Alaska at a more southerly latitude from its present position may be in agreement with current theories on the tectonic framework of Alaska (Tailleur and Brosgé, 1970; Grantz and Eittreim, 1979; Coney et al. 1980). These theories indicate a very complex tectonic framework for Alaska that involves rifting in the Arctic basin, and the building of Alaska from numerous allochthonous blocks that collided and accreted the North American Cratonic Margin mostly during the Mesozoic and early Cenozoic.

A more plausible explanation, however, would keep northern Alaska in its present location, but requires changing the climatic conditions so that the mean annual temperatures were significantly higher, perhaps 15—20°C higher, than their present values. Mean annual temperatures during parts of interglacial periods may have been higher in northern Alaska than their present values, but the increases were far less than those required by the waters of this study (Habicht, 1979; D.M. Hopkins, oral communication, 1981). However, paleoclimatic indicators show that North America (including northern Alaska) had warmer climates than that of the present through most of Mesozoic and Tertiary (Hopkins et al., 1971; Irving, 1978; Habicht, 1979; Wolfe, 1980). These studies show that the mean annual temperatures in northern Alaska required by the recharge waters of this study (15—20°C higher than at present) were reached as recently as Miocene and then throughout most of early Cenozoic and Mesozoic. The actual time of recharge cannot be determined but it is clear that the formation waters of this study, even though meteoric, are extremely old. The presence of permafrost in North Slope probably helps to preserve the ancient water in the sediments by preventing new recharge.

Deep basin brines of connate origin

There are a number of sedimentary basins in the world where geological and hydrodynamic considerations indicate that the formation waters in the basins are connate in origin. The Gulf of Mexico is a prime example of such basins. The extensive drilling for petroleum, especially in the northern half of this basin, has shown a very thick (up to 15,000 m) terrigenous sequence of shales, siltstones and sandstones of Cenozoic age that is essentially devoid of major unconformities. The presence of a thick sequence of fine-grained sediments in the section is important because these sediments contain the largest portion of connate water at the time of deposition (initial porosity up to 80% for clays), and that most of this water is squeezed out into the standstones in the section by subsequent compaction.

Abnormally high fluid pressures (geopressured zones) are another important characteristic of these basins. In the northern Gulf of Mexico basin, geopressured zones with hydraulic pressure gradients higher than hydrostatic (10.5 kPa m^{-1} ≈ 0.465 psi ft^{-1}) are encountered at depths that vary from

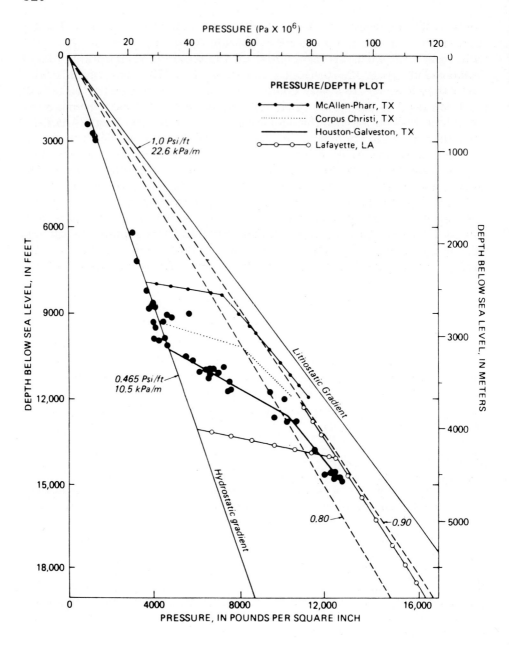

Fig. 8-8. Original bottom-hole pressures from oil and gas wells in coastal Texas and Louisiana plotted against depths below sea level. Note the different depths at which the geopressured zone is encountered in different areas of the Gulf Coast, U.S.A. (Kharaka et al., 1978).

about 2500 m in south Texas to more than 4000 m in parts of coastal Louisiana (Fig. 8-8). Fluid potentials calculated from the pressure data (Fig. 8-8) give water heads that are higher than any recharge areas in the basin, indicating water flow upward and updip in this section. The geological and paleogeographic histories of the northern Gulf of Mexico basin indicate that the flow has always been updip and toward the recharge areas to the northwest indicating that the water, at least in the geopressured zones, is connate water squeezed from the marine shales and siltstones in the basin (Hardin, 1962; Jones, 1975; Kharaka et al., 1979).

The isotopic compositions of formation waters from the geopressured and normally pressured zones from the northern Gulf of Mexico basin are shown in Fig. 8-9. Also shown in this figure are the isotopic values of local groundwaters for comparison. The data for formation waters from the geopressured zones and from the normally pressured zone, fall on a general trend that passes through SMOW and away from the meteoric water of the area. The isotopic data show that the formation waters in the geopressured and nor-

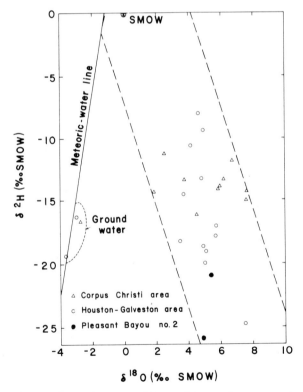

Fig. 8-9. Isotopic compositions of formation waters from northern Gulf of Mexico basin. Note that the trend shows decreased $\delta^2 H$ values with increasing $\delta^{18}O$ values and that the trend goes through SMOW and away from meteoric water of the area (Kharaka et al., 1979).

TABLE 8-2

Chemical composition (mg l^{-1}) and production data of formation waters in the geopressured zones from coastal Texas and Louisiana

Field:	Lafayette, La.		Houston-Galveston, Tx.		Corpus Christi, Tx.		McAllen-Pharr, Tx.	
	Weeks Island	Tigre Lagoon	Chocolate Bayou	Halls Bayou	Portland	East Midway	Pharr	La Blanca
Sample number	77-GG-19	77-GG-55	76-GG-7	76-GG-24	76-GG-63	76-GG-73	77-GG-107	77-GG-117
Well name	St. Un. A #9	Edna Delcombre #1	Angle #3	Houston "FF" #1	Portland A-3	Taylor E-2	Kelly A-1	La Blanca #12
Production zone	S-Sand	Sand #3	Upper Weiting	Schenck	Morris	Lower Frio	Marks	7150 Sand
Depth (m)	4275	3928	3444	4161	3514	3622	3018	2903
Temperature (°C)	117	114	118	150	123	128	127	148
Pressure	43.1	75.8	52.4	80.0	58.0	66.2	52.4	56.6
Fluid production:								
oil/condensate	21.9	0	0.5	3.8	4.8	2.7	0	0.3
water	56.0	633	6.7	57.9	7.5	0.2	7.1	51.0
gas	6.1	7.9	5.1	70.8	25.1	4.9	3.2	17.1
TDS	235,700	112,200	73,300	58,100	17,800	36,000	36,600	7500
Na	78,000	40,000	26,500	20,500	6500	13,250	9420	2680
Li	16	7.1	9.9	15	3.6	4.2	7.5	1.2
K	1065	265	400	180	68	72	240	46
Rb	3.4	0.8	0.4	0.9	0.3	0.5	0.8	0.1
Cs	11.8	3.5	—	—	—	—	2.9	0.3
Mg	1140	270	220	170	15	48	18	3.3
Ca	10,250	1860	2000	800	89	330	4225	150
Si	920	320	365	170	7.0	23	256	9.6
Ba	185	8.2	290	59	1.4	13	27	1.5
Fe	84	0.4	10.2	22	2.3	1.6	4.1	< 0.1

	1	2	3	4	5	6	7	8
Cl	143,000	67,900	42,700	34,500	9270	21,000	22,000	3950
F	0.8	0.8	0.8	—	1.5	7.3	3.9	5.7
Br	419	63	52	32	19	45	78	15
I	18	26	16	11	25	45	22	16
B	44	57	35	91	62	35	105	117
NH_3	100	69	29	13	5.8	13.5	21.5	4.2
H_2S	0.4	0.5	1.2	1.4	<0.1	0.04	<0.1	<0.1
HCO_3^-	450	1050	455	409	1600	1180	114	400
SO_4^{2-}	6.4	220	2.7	16	110	42	7.0	57
SiO_2	48	57	87	110	93	132	90	88
pH	6.2	6.3	5.9	6.8	6.8	6.4	6.8	7.3

Note: Depth is depth below sea level of midpoint of perforation. Temperature is measured subsurface temperature. Pressure is original bottom-hole pressure to MPa (1 psi = 6.9 kPa). Liquid production is in m^3 day^{-1}; gas in 1000 m^3 day^{-1}. Production zones are names used by oil companies. TDS is calculated total dissolved solids. HCO_3 is the field titrated alkalinity.

mally pressured zones are not meteoric in origin or the result of mixing of meteoric and ocean waters. These isotopic data indicate that the formation waters are connate representing the original marine water of deposition of shales and siltstones.

The general trend observed in Fig. 8-9 is unusual in that the δ^2H values decrease (are lighter) instead of increasing with increasing $\delta^{18}O$ values which in turn increase with increasing subsurface temperatures. The origin of this unusual "hydrogen isotope shift" is discussed in detail below in the section on, " 'Hydrogen isotope shift' from exchange with clays". In that section we show that isotopic exchange between ocean water and clay minerals having very light original δ^2H value of $-70‰$ (Yeh, 1980) are responsible for this unusual shift.

The chemical composition and the origin of solutes in formation waters from the northern Gulf of Mexico basin have been studied extensively (Dickey et al. 1972; Kharaka et al., 1979; and many others). Chemical anal-

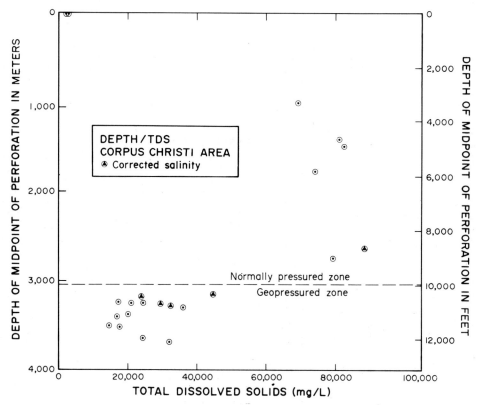

Fig. 8-10. Distribution of salinity of formation waters from oil and gas fields in the Corpus Christi area, Texas (Kharaka et al., 1977). The salinity of formation waters in the geopressured zone is much lower than that in the overlying normally pressured zone.

yses of eight formation water samples selected as typical of the geopressured zone from three areas in Coastal Texas and one area in Coastal Louisiana are shown in Table 8-2. Table 8-2 also gives the locations of the sampled wells subsurface temperatures, pressures and other pertinent data. More complete chemical analyses, as well as the field and laboratory procedures used for collection and analysis, are given in Kharaka et al. (1977, 1978b) and Lico et al. (1982).

The salinity of formation waters from the geopressured zones ranges from about 10,000 to about 275,000 mg l^{-1} dissolved solids. The salinity of waters in the Lafayette area, Louisiana, is generally the highest, ranging from 20,000 to 275,000 mg l^{-1} dissolved solids; salinity is generally the lowest in the McAllen-Pharr area, Texas, where the range is from about 10,000 to about 40,000 mg l^{-1}. The salinity ranges from about 55,000 to about 130,000 mg l^{-1} in the Huston-Galveston area and from about 15,000 to about 45,000 mg l^{-1} in the Corpus Christi area.

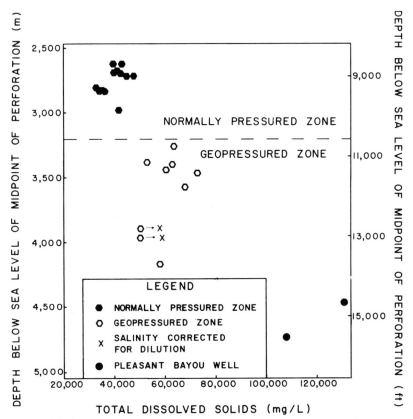

Fig. 8-11. Distribution of salinity of formation waters from Pleasant Bayou No. 2 well and adjacent areas in Brazoria and Galveston Counties, Texas (Kharaka et al., 1979). Note the general increase of salinity with increasing depth.

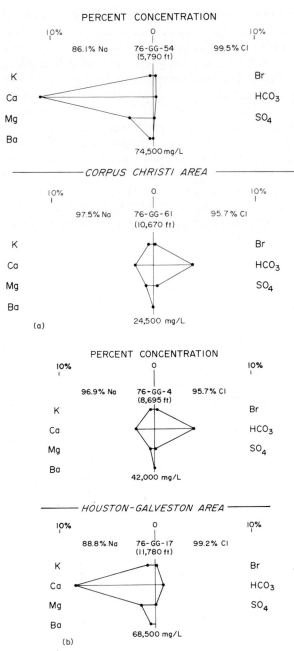

Fig. 8-12. Modified Stiff diagrams showing the percent concentrations of the major cations and anions in formation waters from the normally pressured (upper) and geopressured (lower) zones from (a) Corpus Christi area, Texas, and (b) Houston-Galveston area, Texas (Kharaka et al., 1979).

The salinity of formation waters from the geopressured zones may be higher or lower than the salinity from the normally pressured zones. In the Corpus Christi area (Fig. 8-10) the salinity in the geopressured zone is lower, showing the trend reported by previous investigators (Dickey et al., 1972; Schmidt, 1973; Jones, 1975). In the Huston-Galveston area (Fig. 8-11) the salinity in the geopressured zone is higher than that in the normally pressured zone. In the Lafayette area, different fault blocks have different salinity variations and both salinity trends are observed.

All the formation water are Na-Cl-type waters. Na generally constitutes more than 90% of the total cations and Cl constitutes more than 90% and up to 99.8% of the total anions. The proportions of other major cations and anions in samples typical of geopressured and normally pressured zones from two areas are shown in Fig. 8-12. This figure and the data of Table 8-2 show that Ca concentrations increase with increasing salinity and bicarbonate al-

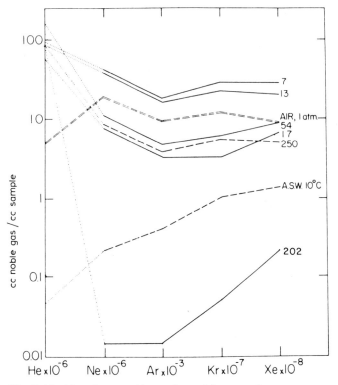

Fig. 8-13. Abundance patterns for noble gases from geopressured-geothermal and natural gas wells from coastal Texas and Louisiana, and for air and seawater saturated at 10°C. Note that for Ar, only the atmospheric component is plotted and that the concentration units are at standard conditions (Mazor and Kharaka, 1981).

kalinity increases with decreasing Ca concentrations. Sulfate and magnesium concentrations are generally low.

It is clear from Table 8-2 and the above discussion that the chemical composition of the formation waters is markedly different from that of ocean water. Modification of the original marine water of deposition to the present formation water composition were shown (Kharaka et al., 1978a, 1979) to result from a combination of the following: (1) interaction of the water with salt beds, (2) interaction of the waters with minerals and organic matter present in the enclosing sedimentary rocks, and (3) membrane-squeezing and membrane-filtration properties of shales.

The concentrations and isotopic compositions of noble gases in natural gas produced from geopressured geothermal and natural gas wells in the northern Gulf of Mexico Basin, also support the conclusion that the formation waters from this basin are marine connate in origin (Mazor and Kharaka, 1981). These authors showed (Fig. 8-13) that the abundance pattern from sample 202 (where all the gas produced is solution gas as no separate gas phase is present at depth) is essentially similar to that of air-saturated seawater. The abundance patterns for the other samples where separate gas phases are present, are similar to that of air indicating degassing of a small fraction (up to about 1%) of dissolved noble gasses from a large volume of connate water.

Deep basin brines from mixing of different waters

The distribution of oxygen isotopes, water salinities, and the concentrations of amino acids in 45 formation water samples in reservoir rocks of Paleozoic, Mesozoic, and Cenozoic age from the United States of America was interpreted by Degens et al. (1964) to indicate no alteration of the oxygen isotope ratios of formation waters by diagenetic processes but modification due to mixing of marine (connate) and meteoric waters. Clayton et al. (1966) (see later section for many other references on this subject) presented convincing evidence showing extensive oxygen isotope exchange between the water and the rock matrix and rejected the hypothesis of mixing presented by Degens and Epstein (1964) which was based on the oxygen isotopes alone. As was noted earlier in this paper, Clayton et al. place considerable emphasis on the trends obtained for different basins when both the $\delta^2 H$ and $\delta^{18} O$ values (Fig. 8-6) indicated a meteoric origin for the waters.

Data on the distribution of oxygen and hydrogen isotopes in formation waters, we believe, are not sufficient by themselves to indicate the origin of these waters. Hitchon and Friedman (1969) made a very detailed isotopic and chemical study of the surface and formation waters of the western Canada sedimentary basin. Using mass balance calculations of deuterium and total dissolved solids, they concluded that the observed distribution of deuterium in the formation waters could best be derived through mixing of the diagenetically modified seawater with not more than 2.9 times as much fresh meteoric water. They attributed the ^{18}O enrichment of formation waters to extensive exchange with the carbonate rocks.

Kharaka et al. (1973) showed on the basis of stable isotopes, water salinities and fluid potentials, that at Kettleman North Dome, California, formation waters from Miocene reservoir rocks are meteoric in origin, whereas those from Eocene reservoir rocks are marine connate squeezed from the underlying shales and siltstones. The isotopic composition of the waters alone (Fig. 8-14) would have indicated a meteoric origin for all the waters at Kettleman North Dome. Combining the data on the isotopic and chemical

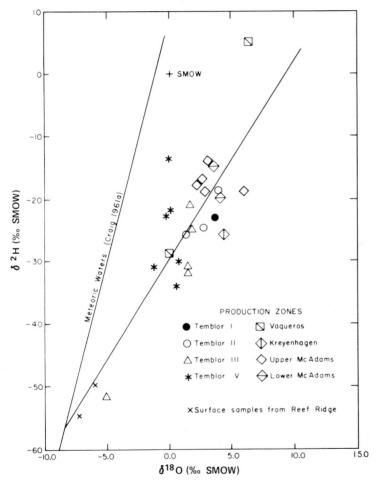

Fig. 8-14. Isotopic compositions of formation- and groundwaters from Kettleman North Dome oil field, California. The solid line is the least-squares line drawn through the data for the formation water samples (Kharaka et al., 1973).

compositon of the waters, on the other hand, indicated (Fig. 8-15) separate origin for the waters from the Eocene reservoir rocks. Data on fluid potentials and geology of the area supported the hypothesis of different origins for the waters.

The ambiguity in the interpretation of the trends from $\delta^2 H$ versus $\delta^{18}O$ plots are depicted in Fig. 8-16. All the formation waters resulting from simple mixing of meteoric water of composition (A) and unmodified marine connate water (SMOW) would be on the line A-SMOW. The δ values of formation waters will plot below the proportional mixing line, if mixing is accompanied by a process that differentially increases the ^{18}O concentration in the waters, but does not affect the $\delta^2 H$ values; the waters will plot in this region, also, if affected by a process that depletes ^2H concentration in the waters and either increases or does not affect the concentration of ^{18}O. The

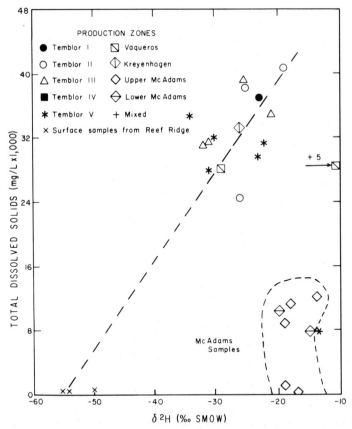

Fig. 8-15. Variations of the $\delta^2 H$ values of formation- and groundwater samples of Fig. 8-14 with the salinity of the waters. Note that the data for the McAdams samples fall on a separate grouping from the remainder of the samples (Kharaka et al. 1973).

waters will plot above this line when mixing is accompanied by a fractionation process that increases the δ^2H, but does not affect the $\delta^{18}O$ values and/or by a process that depletes the ^{18}O and either increases or does not affect the ^2H concentrations in the waters. A number of processes other than mixing, to be discussed in the next section, may affect both the ^{18}O and ^2H content of the formation waters and interpretation of the origin of the waters may require considerations of geology, geochemistry and hydrodynamics in addition to the isotopes.

Examples of deep basin waters that can be related to simple proportional mixing between meteoric water and marine connate waters are shown in (Fig. 8-17). Vetshteyn et al. (1981) showed that the waters from the Dnepr-Donets basin and Ukrainian Shield (line A-SMOW, Fig. 8-17) are essentially a mixture of meteoric and connate waters. Rózkowski and Przwłowcki (1974) reported the isotopic composition of brines from the Upper Silesian Coal basin, Poland, which lie on a simple mixing line between local meteoric water and SMOW. The isotopic data for Sacramento Valley, California, also shown in Fig. 8-17, again indicate mixing of meteoric and marine connate waters, accompanied by some ^{18}O enrichment. Mixing of meteoric and ocean water in the Sacramento Valley is supported by geological and hydrologic evidence as well as isotope evidence. Berry (1973) and Kharaka et al. (1981) used evidence from the chemical composition of formation waters, fluid hydrodynamics and regional geology to show that connate water was being

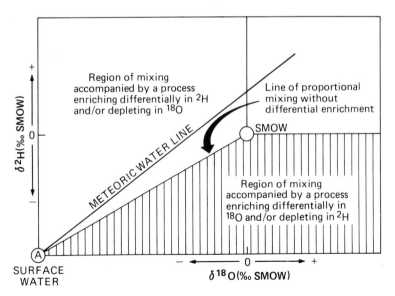

Fig. 8-16. Schematic plot of δ^2H and $\delta^{18}O$ values indicating the effects of various mixing and fractionating processes on the isotopic composition of water from any basin (after Hitchon et al., 1969).

squeezed from shales and siltstones in this basin. This water moved upward in the section and mixed with meteoric water that was present in the sandstones from which the samples were obtained.

MODIFICATION OF STABLE ISOTOPES IN DEEP BASIN WATERS

Reactions between water and minerals, dissolved species, associated gases, and liquids with which they come into contact, modify the isotopic composition of water and the other reactants. Mixing between waters of different origins or between waters affected by the different processes discussed above, can also complicate the interpretation of the isotopic data. All these processes also cause scatter in the isotopic data, but in most situations, as we will discuss below, the complicating effects of these reactions and mixing can be identified and the origin of the water determined. The following are

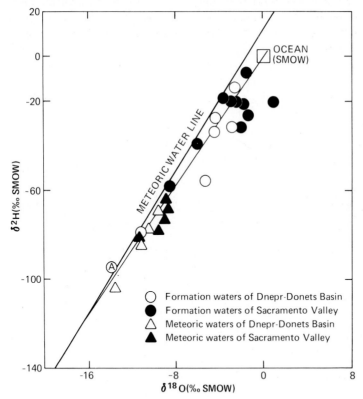

Fig. 8-17. Isotopic composition of meteoric water (triangles) and formation waters (circles) from Dnepr-Donets basins, U.S.S.R. (Vetshteyn et al., 1981), and Sacramento Valley, California, U.S.A. Note that in both basins the formation waters are probably mixtures of meteoric and connate (marine) waters.

the main processes that modify the isotopic composition of formation waters in deep sedimentary basins: (1) isotopic exchange between water and rocks; (2) evaporation and condensation; (3) fractionations caused by membrane properties of rocks; and (4) isotopic exchange between water and other fluids, especially petroleum.

Isotopic exchange between water and rocks

Isotopic exchange between water and minerals with which they come into contact and formation of authigenic minerals are probably the most important processes responsible for modifying the original $\delta^2 H$ and $\delta^{18} O$ values of deep basin brines. The extent of this modification depends on a number of factors which include:

(1) The original isotopic compositions of the water and the various minerals with which the water may have come into contact.

(2) The relative number of atoms of oxygen and hydrogen in the water and the various minerals (the water/rock ratio) that the water contacts throughout its history. In general, water is the main reservoir for the hydrogen

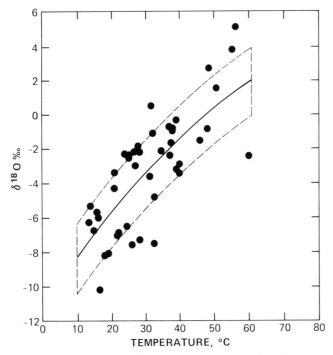

Fig. 8-18. Variations of $\delta^{18} O$ values of brines with well temperature for the Illinois basin. The equilibrium curve is drawn for calcite of $\delta^{18} O = 24.2^0/_{00}$. Note that almost all the data can be explained by isotopic exchange with calcite (Clayton et al., 1966).

atoms in sedimentary basins especially at shallow depths where the porosity and permeability of the rocks are high, whereas the minerals are normally the main reservoir for the oxygen atoms. This is the reason why extensive "oxygen isotope shifts" are common and "hydrogen isotope shifts" are relatively rare.

(3) The subsurface temperatures at which isotopic exchange between water and minerals occurred. Temperatures are an important variable because the isotopic exchange rates between water and minerals increase with increasing temperature, and the magnitude of fractionation factors between water and minerals are strongly temperature-dependent.

Isotopic equilibrium between water and the important rock-forming minerals in sedimentary basins (calcite being the exception) is generally not attained at sedimentary temperatures which are generally less than 150°C,

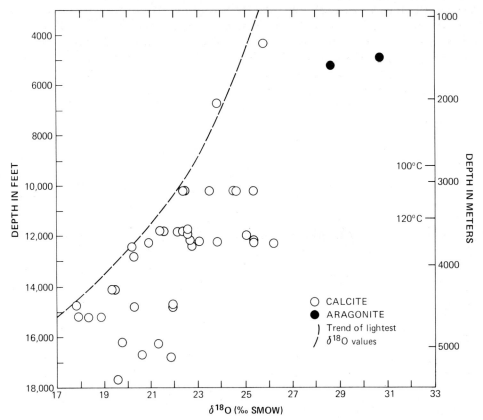

Fig. 8-19. $\delta^{18}O$ values of calcium carbonate versus depth of the samples from wells in Brazoria County, Texas. Dashed line indicates the trend for the lightest $\delta^{18}O$ values at any given depth. Note the general decrease of $\delta^{18}O$ values of carbonate with increasing depth (modified from Milliken et al., 1981).

but may be as high as 200°C. This lack of equilibrium is caused by the fact that the rates of isotopic exchange reactions between water and most minerals are low at these temperatures except when mineralogical or chemical reactions between water and minerals occur. Studies of minerals in cores and cuttings from geothermal areas and laboratory experiments have shown that temperatures much higher than 200°C are required for the establishment of isotopic equilibrium between formation water and detrital quartz, feldspars and other aluminosilicates (O'Neil and Taylor, 1967, 1969; Clayton et al., 1968; Eslinger and Savin, 1973; Yund and Anderson, 1974; Kendall, 1976; Coplen, 1976; Stewart, 1978; and many others).

Isotopic equilibrium between water and calcite and other carbonate minerals and various polymorphs of chert is probably attained at sedimentary temperatures (Clayton et al., 1968; Knauth and Epstein, 1976; Murata et al., 1977; and see Savin, 1980, for an excellent review). Clayton et al. (1966) showed that the extensive "oxygen isotope shift" observed in oil field waters could be related to isotopic equilibrium with calcite at subsurface temperatures as low as 10°C (Fig. 8-18). Milliken et al. (1981) showed that the $\delta^{18}O$ values of carbonate cements from the northern Gulf of Mexico basin (Fig. 8-19) decreased with increasing depth of burial (increasing temperatures) which are mirrored by the general increase of the $\delta^{18}O$ values of oil field waters from the same area reported by Kharaka et al. (1979).

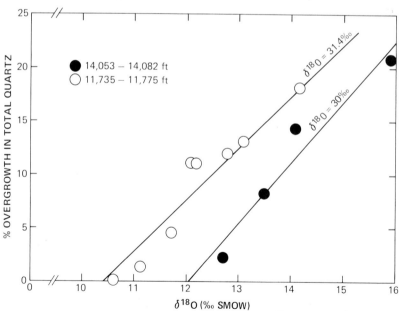

Fig. 8-20. $\delta^{18}O$ values for quartz, corrected for rock fragments, versus percent overgrowth in total quartz. Linear regression lines extrapolate to $\delta^{18}O$ value for quartz overgrowths of about 30‰ (modified from Milliken et al., 1981).

Authigenic minerals in addition to carbonates, including quartz, feldspars and clays approach isotopic equilibrium with water at the temperature of their formation (see Savin, 1980, for many references). In many sedimentary sequences detrital and authigenic portions of the same mineral are present. The δ^2H and $\delta^{18}O$ values of the different portions will have to be computed from plots similar to that in Fig. 8-20, or measured if possible in order to study isotopic exchange between water and such minerals.

The isotopic shifts resulting from reactions between water and minerals in sedimentary basins can be calculated using the law of conservation of mass. In a closed system and for water reacting with a single mineral only, the equation is:

$$W \delta^i_{H_2O} + M \delta^i_{mineral} = W \delta^f_{H_2O} + M \delta^f_{mineral} \tag{5}$$

where the superscripts "i" and "f" are the initial and final δ values for water and minerals, respectively, W and M are the atom percent of water and mineral oxygen or hydrogen in the total system. Equation (5) may be rearranged to give:

$$\delta^f_{H_2O} - \delta^i_{H_2O} = \frac{M}{W} (\delta^i_{mineral} - \delta^f_{mineral}) \tag{6}$$

Assuming isotopic equilibrium between water and this mineral results in:

$$\delta^f_{mineral} - \delta^f_{H_2O} \simeq 1000 \ln \alpha_{mineral-H_2O} \tag{7}$$

where $\alpha_{mineral-H_2O}$ is the fractionation factor between water and mineral at the temperature of this system. Combining equations (6) and (7) allows us to calculate the original isotopic value of water $\delta^i_{H_2O}$ in situations where either $\delta^f_{H_2O}$ or $\delta^f_{mineral}$ is known. For example, where $\delta^f_{H_2O}$ is not known, the equation is:

$$\delta^i_{H_2O} = \delta^f_{mineral} - 1000 \ln \alpha_{mineral-H_2O} - \frac{M}{W} (\delta^i_{mineral} - \delta^f_{mineral}) \tag{8}$$

The importance of water/rock ratios (W/M) in modifying the isotopic compositions of biotite and feldspar (quartz was assumed not to exchange) in a typical grandiorite subjected to meteoric-hydrothermal alteration was investigated by Taylor (1977). Taylor showed (Fig. 8-21) that the δ^2H of biotite changes rapidly even at low water/rock ratios while the $\delta^{18}O$ of feldspar remains essentially constant and only when the water/rock ratios are greater than 0.1 will a change in $\delta^{18}O$ begin to be detected and then increase as the ratios increase. The changes in the δ^2H and $\delta^{18}O$ values of water will parallel those of the minerals but with shifts equal to the fractionation factors between water and minerals considered. These isotopic changes result from the fact that in granodiorite and most sedimentary rocks the mineral fraction is the main reservoir for the oxygen, and the water fraction the main reservoir for the hydrogen.

In deep sedimentary basins a number of minerals including calcite, authigenic feldspars, quartz, and various clays are involved in the oxygen isotope exchange with the formation waters. The $\delta^{18}O$ values of these minerals are useful in studies of geothermometry and mineral diagenesis (see a later section) but they cannot be used in material balance calculations such as those in equations (5) to (8) or in plots similar to that shown in Fig. 8-21 because combined and detailed mineral and isotopic analyses are simply not carried out on basin-wide scales. Also, the permeabilities and thus the flow rates in the sedimentary section are often not available, but these determine the effective water/rock ratios and thus the final isotopic composition of

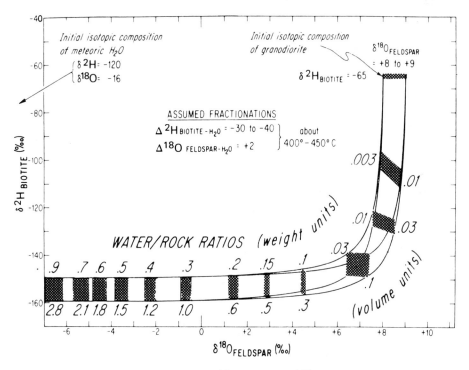

Fig. 8-21. Plot of calculated values of $\delta^2H_{biotite}$ and $\delta^{18}O_{feldspar}$ that would be obtained during the hypothetical meteoric-hydrothermal alteration of a typical granodiorite (25% quartz, 50% plagioclase, 15% K-feldspar, 10% biotite) at varying water/rock ratios and specified isotopic values and fractionation factors (Taylor, 1977).

water and minerals (Coplen, 1976). Qualitatively, however, the trends observed in the $\delta^{18}O$ values of formation waters in deep basins can be explained by isotopic exchange between water and minerals. Such exchange almost always results in an increase in the $\delta^{18}O$ values of the water and a decrease in the $\delta^{18}O$ values of the minerals involved. The "oxygen isotope shift" in the case of water almost always increases as the subsurface temperature and depth of the reservoir rocks increase. This increased "oxygen isotope shift" with increasing temperatures results from an enhanced degree of exchange in the low water/rock ratio environment and the fact that the frac-

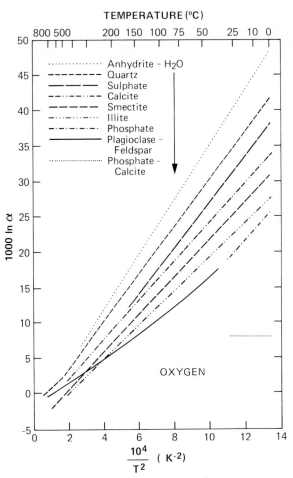

Fig. 8-22. Oxygen isotope fractionations between water and minerals plotted as a function of temperature showing the general decreases in the fractionation factors with increasing temperatures. Data for feldspars from O'Neil and Taylor (1967) are added to a modified plot from Savin (1980).

TABLE 8-3

Hydrogen and oxygen fractionation factors between water and clay minerals at earth surface temperatures (after Savin, 1980, p. 288)

Mineral	Type of occurrence	Oxygen $\alpha_{mineral\text{-}H_2O}$	Hydrogen $\alpha_{mineral\text{-}H_2O}$	Source
Glauconite	glauconite dredged from Blake Plateau	1.026	0.93	Savin and Epstein (1970)
Illite	laboratory experiments (partial equilibration)	1.024		James and Baker (1976)
Smectite	marine bentonites; authigenic marine smectites	1.027	0.94	Savin and Epstein (1970)
	soils from continental U.S.A., Hawaii, Central America	1.027	0.970	Lawrence and Taylor (1971)
	authigenic smectite from sea floor		0.985	Yeh and Epstein (1978)
	authigenic smectite from sea floor	1.0308		Yeh and Savin (1976)
Kaolinite	southeastern U.S.A. kaolinite deposits	1.027	0.97	Savin and Epstein (1970)
Gibbsite	gibbsite in soils	1.018	0.985	Lawrence and Taylor (1971)

tionation factors for oxygen between water and all the important minerals with which the water may exchange in sedimentary basins (Fig. 8-22) decrease with increasing temperatures.

The $\delta^{18}O$ values of interstitial water obtained from cores during the Deep Sea Drilling Program (DSDP), however, generally show decreased $\delta^{18}O$ values with increasing depth of the profile (Lawrence et al., 1975; Gieskes et al., 1978; Randall et al., 1979). In some sites this decrease can be attributed to the influx of water of lower ^{18}O concentrations, but in other sites it can only be attributed to the formation of clays from reaction of seawater with basalt (Randall et al., 1979). Formation of clay enriched in $\delta^{18}O$ results in residual interstitial water that is depleted in $\delta^{18}O$ values.

Hydrogen isotope exchange between water and clay minerals

Clay minerals are the most important reservoirs of hydrogen in sedimentary rocks. In many sedimentary basins (northern Gulf of Mexico, Central Valley, California, and others), shales contribute the bulk of the rocks in the basin, and the hydrogen in clay minerals comprises an important portion of the total hydrogen in these basins. In a shale sequence, consisting of illite for example, the proportion of hydrogen in illite will be about equal to that in the water (M/W = 1 in equation (6)) at a porosity of about 10% which under normally compacted shale sequences is reached at a depth of about 2000 m. The ratio M/W will be equal to 2 and 0.5 if the porosity is 5%, and 20%, respectively. Also, in many shales, quartz, feldspars and other minerals are important constituents of shales and the M/W ratios should be adjusted accordingly. On the other hand, water flow in the section may be important which leads to decreased effective M/W ratios.

There are two important differences between hydrogen isotope exchange between water and clay minerals and oxygen isotope exchange between water and clay and other sedimentary minerals. First, the hydrogen isotope fractionation factors between clay minerals and water (Table 8-3) are less than 1.0 indicating that at equilibrium the clay minerals will be depleted in deuterium with respect to the water. The reported hydrogen fractionation factors between clay minerals, especially smectites, and water show large variation at earth surface conditions ($\alpha_{\text{smectite-water}}$ = 0.94—0.98), but they increase (approaching unity) with increasing temperature (Fig. 8-23). The second difference is that the $\delta^2 H$ values of clays and other minerals in igneous, metamorphic and sedimentary rocks vary between about −80 to −30‰ and may be lower than that of the water with which they may exchange in sedimentary basins.

Experimental and field evidence leads us to believe that significant hydrogen isotope exchange takes place between water and clays with which they come into contact at all sedimentary temperatures, and that at temperatures higher than about 100°C equilibrium is generally reached in sedimentary basins. Yeh and Epstein (1978) showed 8—28% hydrogen isotope exchange between seawater and clay minerals < 0.1-μm size fraction at oceanic temperatures over periods of 2—3 million years. The clay minerals from three deep wells in the northern Gulf of Mexico basin reported by Yeh (1980) are in apparent isotopic equilibrium with the formation waters in the wells at temperatures higher than about 100°C. Isotopic equilibrium between waters and clays in this basin is inferred from the decrease in the range of the

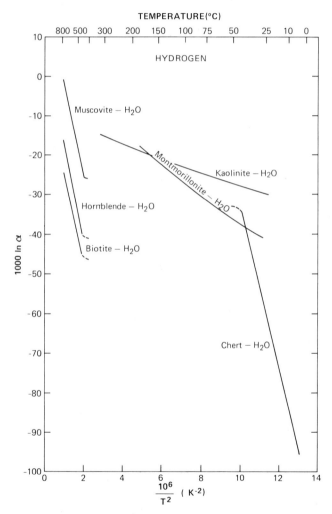

Fig. 8-23. Hydrogen isotope fractionations between water and minerals plotted as a function of temperature. The fractionation factors, as in the case of oxygen isotopes, decrease with increasing temperatures. The semectite-water curve from Yeh (1980) is added to a modified plot from Savin (1980).

δ^2H values of clays in the different size fractions from the same depth (Fig. 8-24) at temperatures higher than about 100°C and from the fact that the increase of δ^2H values of clays with increasing temperatures are mirrored by the decrease in the δ^2H value of formation waters (Fig. 8-9) from the same area (Kharaka et al., 1979).

Experiments on hydrogen and oxygen isotopic exchange of illite, smectite, and kaolinite with water conducted by O'Neil and Kharaka (1976) in the temperature range 100—350°C showed significant (up to 28% for smectite) hydrogen isotope exchange even at 100°C. The exchange increased with increased temperatures and duration of the runs (up to about 2 years) and were much higher for smectite than the other minerals at all temperatures. At 200°C the maximum hydrogen isotope exchange for kaolinite, illite and smectite were 29, 28, and 82%, respectively. These extensive exchanges took place with no significant changes in the mineral composition of the minerals. Oxygen isotope exchange in the same systems was minor to negligible at all temperatures indicating that hydrogen isotope exchange occurred by a

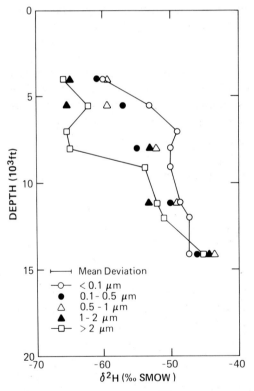

Fig. 8-24. Variations in the δ^2H values of shale with burial depth from CWRU Gulf Coast Well No. 10 (Yeh, 1980). Note the general increase in the δ^2H values of shale with increasing temperatures as a result of exchange with formation waters.

mechanism of proton exchange. These experiments, however, showed near complete oxygen isotope exchange (86%) at 350°C between kaolinite and water but this resulted from the complete transformation of kaolinite to pyrophyllite and diaspore in that experiment.

"Hydrogen isotope shift" from exchange with clays

The northern Gulf of Mexico basin is an ideal place to study isotopic exchange between water and clay minerals. The Cenozoic section in this basin consists of a thick sequence (up to 15,000 m) of shales, siltstones and sandstones. The gross mineral compositions of shales (Burst, 1969; Hower et al., 1976) and sandstones (Loucks et al., 1980) have been studied. Data have also been reported on the oxygen isotopic composition of minerals in shales (Yeh and Savin, 1977) and sandstones (Milliken et al., 1981), hydrogen isotopic composition of minerals in shales (Yeh, 1980), and oxygen and hydrogen isotope composition of formation waters (Kharaka et al., 1979).

Material balance calculations were made for hydrogen isotopes using equations (6) and (7) in order to investigate the origin of formation waters and the "hydrogen isotope shift" in this basin. The $\delta^2 H$ values of water were calculated assuming a closed shale/water system with reasonable porosities decreasing from 20% to 5% at temperatures ranging from 30°C to 150°C. The shale was assumed to consist of 50% illite $KAl_5 Si_7 O_{20}(OH)_4$ with an initial (unreacted) $\delta^2 H$ values of $-70‰$ and $1000 \ln \alpha_{clay-H_2O} = -19.6 (10^3/T) + 25$ (Yeh, 1980). Three possible $\delta^2 H$ values for the initial water were investigated. These were $\delta^2 H^i_{H_2O} = 0$ (ocean water), $\delta^2 H^i_{H_2O} = -20‰$ (present-day meteoric waters in the area; Kharaka et al., 1979) and $\delta^2 H^i_{H_2O} = -70‰$ (Mississippi river water near the coast; Knauth et al., 1980).

The material balance equation at 30°C (hydrogen in water = 81% of total) for this system with $\delta^2 H^i_{H_2O} = 0$ is:

$$\delta^2 H^{30}_{H_2O} \times 0.81 + (\delta^2 H^{30}_{H_2O} - 40) \times 0.19 = -70 \times 0.19 + 0 \times 0.81$$

which results in $\delta^2 H^{30}_{H_2O} = -5.7‰$.

Calculations at higher temperatures for this system result in $\delta^2 H$ values of water decreasing (becoming more negative) with increasing temperatures (higher $\delta^{18}O$ values) in a manner similar to that observed in the formation waters of this basin (Fig. 8-9). On the other hand, we were unable to reproduce the $\delta^2 H$ values of formation waters in this basin, especially at the shallower depths, when we made the calculation with meteoric water or Mississippi river water indicating that the water in this basin is connate marine in origin.

We believe this material balance approach is useful in explaining the trends obtained in the δ^2H values and investigating the origin of formation waters. Unfortunately, we are not aware of data on δ^2H values of minerals in other basins. Material balance calculations using δ^2H values of clays from the literature and different and hypothetical initial values for water gave somewhat unexpected results. The majority of these results showed that the δ^2H values of water decrease with increasing temperature. Increasing δ^2H values of formation waters with increasing temperature (the general trend in most sedimentary basins) resulted only when we made the calculations using relatively light water reacting with very heavy clay (e.g., $\delta^2 H^i_{H_2O} = -100$, $\delta^2 H^i_{clay} = -30^0/_{00}$). The results of these calculations, though not specific to real basins, may indicate that isotopic exchange between water and detrital clays is not responsible for the general trend of increased δ^2H values with depth that is generally observed in sedimentary basins.

Clays formed by reaction between feldspars and formation water inherit all their hydrogen from the water phase. The deuterium content of water after clay formation will be higher than that of the initial water because the δ^2H value of the clay formed will be lower than that of the water, assuming isotopic equilibrium between water and clay. As the reaction progresses with increasing time and subsurface temperatures the δ^2H values of the water will continue to increase as more clay minerals are formed. This process leads to increased δ^2H values of water with depth, but this increase probably will not be quantitatively significant in most sedimentary basins except, perhaps, in a few basins dominated by arkosic sandstones of low ($< 5\%$) porosity where the feldspars have been diagenetically altered to clay minerals.

Evaporation, boiling and condensation

Evaporation and boiling concentrate the light isotopes of oxygen (at all temperatures) and hydrogen (at temperatures less than about 240°C) in the vapor phase, and result in higher δ^2H and δ^{18}O values of the remaining water with increasing evaporation and boiling. Evaporation from surface waters, especially from lakes in closed basins and in the unsaturated zone, is very important in determining the isotopic composition of these waters. δ^2H and δ^{18}O values from these evaporating waters generally lie on straight lines with slopes that vary (depending on humidity and local conditions of evaporation) from 2 to 5, and that intersect the "meteoric water line" at the initial (unevaporated) δ values characteristic of the area. Boiling and evaporation modify the isotopic composition of water in many geothermal systems. The degree of modification depends on the amount of steam lost, whether steam separation occurred in a single stage or was continuous, the temperature of steam separation and the chemical composition of the water (see Truesdell, 1980, for references and an excellent treatment of this topic).

Evaporation is probably not an important process that modifies the isotopic composition of formation waters in deep sedimentary basins except in special situations. The isotopic and chemical composition of formation waters produced from many gas fields, for example, are strongly affected by mixing with condensed water which is present as water vapor in the gas phase at subsurface conditions but condenses to liquid water as the gas cools on entering the well. Kharaka et al. (1977) have shown that the amount of this condensed water which can be calculated per unit volume of gas produced (Fig. 8-25), increases as the temperatures increase and subsurface pressures and salinities decrease. They also showed that this condensed water may comprise a small fraction or almost all the formation water produced from gas wells.

The isotopic composition of formation waters from gas wells diluted by condensed water should not be used uncorrected in isotope studies because the effects are variable and may be very large (Kharaka et al., 1977). The isotopic composition of undiluted formation water from these gas wells, however, can be calculated using the fractionation factors between water/water vapor (Fig. 8-26) and the amount of condensed water produced with the natural gas (Fig. 8-25). The fractionation factors between water and water vapor, it should be noted, are modified (Fig. 8-27) by changes in the chemi-

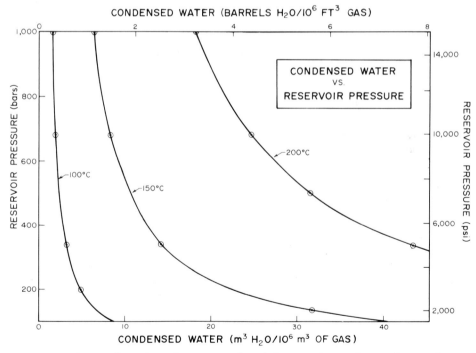

Fig. 8-25. Volume of condensed water produced from a given volume of natural gas as a function of reservoir pressure and temperature (Kharaka et al., 1977).

cal composition of the water (Sofer and Gat, 1972; Truesdell, 1974; Stewart and Friedman, 1975).

Membrane properties of rocks

The ability of soils, clays, and shales (the geological membranes) to serve as semipermeable membranes has been demonstrated by field data and laboratory experiments (McKelvey and Milne, 1962; White, 1965; Berry, 1973; Hanshaw and Coplen, 1973; Kharaka and Berry, 1973; and many others). Geologic membranes retard or restrict, by varying degrees, the flow of dissolved species through them. The concentrations of solutes are lower in the effluent (throughput) solution compared to their concentrations in the hyperfiltrated solutions remaining on the input side of the membrane. The membrane behavior of clays and shales has been invoked by many investigations to explain the salinities and chemical compositions of formation waters in deep sedimentary basins (De Sitter, 1947; Bredehoeft et al., 1963; Graf et al., 1966; Hanshaw and Hill, 1969; Hitchon et al., 1971; Kharaka and Berry, 1974; 1976; and many others).

There is field and laboratory evidence suggesting that shales act as membranes fractionating the stable isotopes of water. The concentrations of ^2H

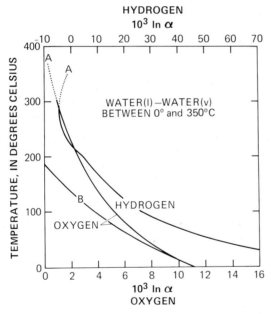

Fig. 8-26. Oxygen and hydrogen fractionation factors between water and water vapor. Oxygen line A from Bottinga (1968a) and Bottinga and Craig (1968); line B from Uvarov et al. (1962). Hydrogen line A from Merlivat, Botter and Nief (1963), Bottinga (1968b) and Majzoub (1971) (modified from Friedman and O'Neil, 1977).

and ^{18}O are lower in the effluent water compared to their concentrations in the hyperfiltrated water. Graf et al. (1965) showed (Fig. 8-28) that there is isotopic fractionation resulting from the passage of water through micropores of shales in the Illinois, Michigan, Gulf Coast and Alberta basins. The $\delta^2 H$ and $\delta^{18} O$ values shown in Fig. 8-28 are residual values as the data from different basins were normalized to the value for meteoric precipitation at Chicago and to the value at 25°C for equilibrium with pre-tertiary marine limestone. This normalization was carried out in order to separate the membrane from other effects. Hitchon et al. (1971) and Kharaka and Berry (1973) reported isotopic compositions of waters from the western Canada and California basins, respectively, that suggest that isotopic fractionations arise from the membrane properties of shales.

Laboratory experiments were carried out by Coplen and Hanshaw (1973) to determine the isotopic fractionation factors when distilled water and dilute NaCl solution were forced through compacted montmorillonite at ambient temperatures. Results show that the effluent waters were depleted by 2.5°/00 and 0.8°/00 in ^2H and ^{18}O, respectively relative to the hyperfiltrated (residual) solutions. A plot of $\delta^2 H$ versus $\delta^{18} O$ values from these experiments

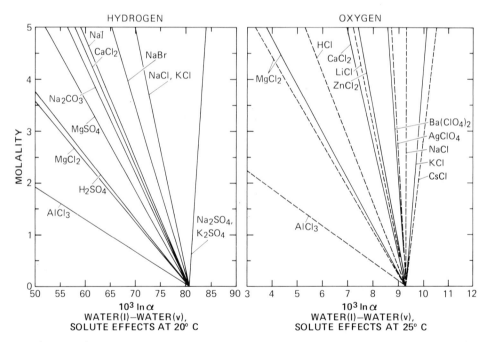

Fig. 8-27. Isotopic fractionation factors between water of different chemical composition and water vapor (modified from Friedman and O'Neil, 1977). For hydrogen, data from Stewart and Friedman (1975); for oxygen, data for solid curves from Sofer and Gat (1972) and for dashed curves from Taube (1954).

(Fig. 8-29) falls on a line with a slope of 3.1. The slope of this line, it should be noted, is exactly the same as that given by Graf et al. (1965) for the residual $\delta^2 H$ and $\delta^{18}O$ values for formation waters (Fig. 8-28) that they attributed to membrane effects. We believe, however, that more work is needed to determine the isotopic fractionation factors of clays and shales as functions of temperature, different compaction and hydraulic pressures and water composition. Experiments (Hanshaw and Coplen, 1973; Kharaka and Smalley, 1976) have shown that these parameters affect strongly the selectivity of geological membranes to different dissolved species and likely will affect isotopic fractionations.

Isotope exchange between water and other fluids

Hydrogen isotope exchange between H_2S and water and oxygen isotope exchange between CO_2 and water will occur in nature where water comes in contact with H_2S- or CO_2-rich gases. Isotopic exchange with these gases, however, will not in general be quantitatively important in modifying the isotopic composition of formation waters from deep sedimentary basins. On

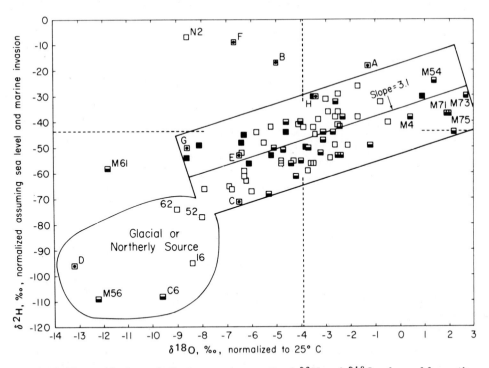

Fig. 8-28. The residual regularity between normalized $\delta^2 H$ and $\delta^{18}O$ values of formation waters from Illinois, Michigan, Gulf Coast and Alberta basins showing the membrane effects of shales (Graf et al., 1965).

a local scale, isotopic exchange between water and these gases may be significant expecially in the case of H_2S in sour gas fields (Hitchon and Friedman, 1969) because the fractionation factor between H_2S and water is very large. Isotopic exchange between water and gaseous or liquid hydrocarbons is probably not quantitatively important because exchange rates are extremely low at sedimentary temperatures and these effects can not be properly evaluated at present (Alexeev and Kontsova, 1962). Finally, isotopic exchange between water and hydrogen gas may be very important in sedimentary basins because the fractionation factors are very large (Bottinga, 1969) ($\alpha_{H_2O(v)-H_2(g)}$ = 1400—640 at 0—200°C) and hydrogen gas may be an important component of the basins that are rich in organic matter, but the importance of this reaction needs investigation.

OTHER APPLICATIONS OF STABLE ISOTOPES OF WATER

Geothermometry

A number of geothermometers, based on the chemical composition of water and the isotopic composition of water and dissolved species and asso-

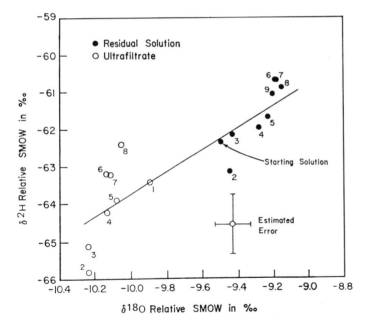

Fig. 8-29. Isotopic compositions of residual (hyperfiltrated) and ultrafiltrate (effluent) solutions of $0.01N$ NaCl forced through compacted clay (Coplen and Hanshaw, 1973). Note that the slope of the line in this figure is equal to that obtained (Fig. 8-28) from natural systems.

ciated gases, have been used successfully to estimate the subsurface temperatures or the temperatures of last equilibration between water/solutes and gases in geothermal systems. The principles and equations used in these estimations together with the limitations of each geothermometer have been covered extensively in Volume 1 of this series (Truesdell and Hulston, 1980). It suffices to mention here, that the isotopic geothermometers have not been used extensively in deep sedimentary basins. There are three main reasons for this discrepancy. First, knowledge of the subsurface temperatures are not as critical to exploration for oil and gas as they are for exploration for geothermal energy. Second, there are conventional methods for estimating these temperatures including measurement with maximum reading thermometers. And third, the majority of these isotopic geothermometers can not be used in sedimentary environments as they give erroneous results. The erroneous temperatures result mainly from the fact that isotopic equilibrium is not normally reached at the lower temperatures encountered in sedimentary basins. However, the sulfate-water isotopic geothermometer should be investigated as it may prove applicable to these waters.

The application of geothermometers, based on the isotopic composition of water and minerals, to sedimentary environments has had mixed results. Hydrogen isotopic fractionations factors between clay minerals and waters of different compositions are not known accurately enough to be of use in hydrogen isotope geothermometry. Yeh and Savin (1977) investigated the use of oxygen isotope geothermometry by comparing measured temperatures with those calculated assuming oxygen isotopic equilibrium between authigenic quartz, clay and calcite in three deep wells from the northern Gulf of Mexico basin. They report that isotopic temperature calculated from the $\delta^{18}O$ values of the finest fractions of coexisting clay and quartz, approach measured subsurface temperatures with increasing temperatures, and at temperatures higher than about 100°C there is reasonable agreement between measured and calculated temperatures (Fig. 8-30). Coplen et al. (1975) also reported that the quartz-water and alkali feldspar-water isotope geothermometers agree well with in-situ temperatures of 104°C at Dunes geothermal anomaly, Imperial Valley, California. Yeh and Savin (1977), on the other hand, reported that temperatures based on $\delta^{18}O$ values of calcite were unreasonable and indicate that calcite continued to undergo isotopic exchange with the pore waters.

Mineral diagenesis

Authigenic minerals including quartz, feldspars, clay minerals and carbonates most likely are in isotopic equilibrium with water at the temperatures of their formations (see the section "Isotopic exchange between water and rock" for discussion). In the case of non-carbonate minerals this "isotopic signature" is then preserved at sedimentary temperatures and can be used to distinguish authigenic from detrital portions of the same mineral and in the study of the "history of diagenesis" in sedimentary basins. This approach, which has been used extensively in the study of oceanic sediments (Savin, 1980) is only recently (Milliken et al., 1981) being applied to diagenetic minerals in sedimentary basins.

The history of burial diagenesis, using measurements of the stable isotope ratios of oxygen and carbon in authigenic minerals in standstone cores from two deep wells in the Gulf Coast, U.S.A., was investigated in detail by Milliken et al. (1981). They assumed that the authigenic minerals were in isotopic equilibrium with the formation waters from the same depth at the time of their formation. They then calculated the range of temperatures for each authigenic mineral studied from the $\delta^{18}O$ values of the minerals, the $\delta^{18}O$ values of the formation water reported for waters of the same area by Kharaka et al. (1979), and the equations giving the fractionation factors of minerals with water. These authors were able to show (Fig. 8-31) that quartz cement formed at temperatures between 75 and 80° and kaolinite at about 100°C. Albitization of plagioclase feldspar was most rapid near 150°C, whereas authigenic carbonates formed over a wide range of temperatures. They also showed that the relative sequence of events determined from isotopic data largely corroborate paragenetic sequences determined for the cements by petrographic studies conducted by Loucks et al. (1979, 1980).

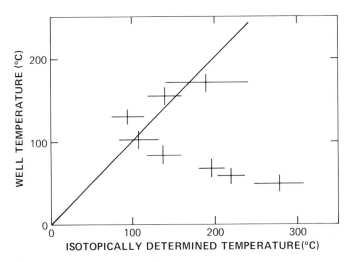

Fig. 8-30. Temperatures measured and calculated from oxygen isotope fractionations between coexisting clay (0.1 μm) and quartz (0.1—0.5 μm) from CWRU Gulf Coast Well No. 6 (Yeh and Savin, 1977). Note that the agreement is reasonable only at well temperatures higher than about 100°C.

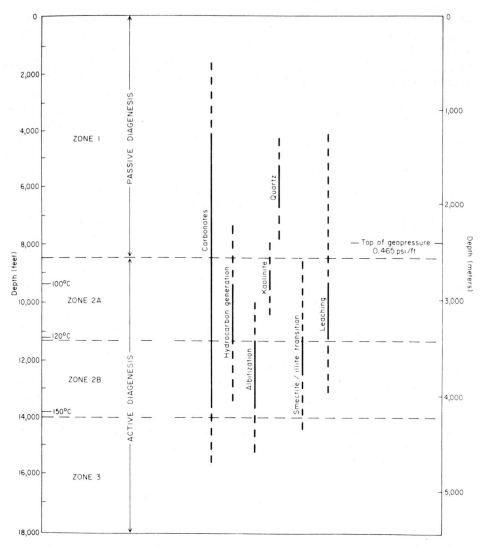

Fig. 8-31. Occurrence of diagenetic reactions, based on isotopic and petrographic data, during burial of Frio Formation, Brazoria County, Texas, U.S.A. (modified from Milliken et al., 1981).

SUMMARY

Oxygen and hydrogen isotope analyses are extremely useful in the study of the origin of the waters in sedimentary basins. Application of such measurements together with regional geology, hydrology, and chemical composition of the water have shown that the waters in the majority of the basins are predominantly of local meteoric origin. In the North Slope, Alaska, the evidence suggests that the formation water, even though meteoric, was recharged in Miocene or earlier times when the climate was warmer than at present. Marine connate water is encountered in the northern Gulf of Mexico and other deep sedimentary basins where a thick sequence of conformable rocks is dominated by marine shales and siltstones.

Studies of the isotopic composition of water and minerals in sedimentary basins have also proven very useful in determining the provenance of detrital minerals and the "history of diagenesis" in the case of authigenic minerals. The use of these and other isotopic ratios in geothermometry has been successful in geothermal systems, but less so in sedimentary basins where isotopic equilibrium may not be attained because of the lower temperatures.

A number of reactions and processes modify the original isotopic composition of water and other reactants in sedimentary basins. Isotopic exchange between water and minerals is by far the most important reaction and always results in "oxygen isotope shifts" to higher ^{18}O content of water, but "hydrogen isotope shifts" may result in higher or lower $\delta^2 H$ values of water. Determinations of accurate fractionation factors and rates of exchange between water and minerals at sedimentary temperatures are needed for a better interpretation of the isotopic data. Fractionations caused by evaporation and condensation may not be important except in interpreting the isotopic composition of water samples from natural gas wells. Field and laboratory experiments indicate that fractionations caused by membrane properties of shales may be very important but more laboratory experiments simulating the field environments are needed to evaluate this process. More studies are also needed to evaluate the importance of hydrogen isotope exchange between water and liquid, and gaseous hydrocarbons and hydrogen gas.

REFERENCES

Alekseev, F.A. and Kontsova, V.V., 1962. Distribution of deuterium in interstitial waters of oil pools. *Yad. Geofiz. Sb. Statei*, 1961, 196—201.
Bath, A.H., Edmunds, W.M. and Andrews, J.N., 1978. Paleoclimatic trends deduced from the hydrochemistry of Triassic sandstone aquifer, United Kingdom. In: *Isotope Hydrology 1978*, 2. IAEA, Vienna, pp. 545—568.
Berry, F.A.F., 1973. High fluid potentials in the California coast ranges and their tectonic significance. *Bull. Am. Assoc. Pet. Geol.*, 57: 1219—1249.
Bottinga, Y., 1968a. Calculation of fractionation factors for carbon and oxygen exchange in the system calcite—carbon dioxide—water. *J. Phys. Chem.*, 72: 800—808.
Bottinga, Y., 1968b. Hydrogen isotope equilibria in the system hydrogen-water. *J. Phys. Chem.*, 72: 4338—4340.
Bottinga, Y., 1969. Calculated fractionation factors for carbon and hydrogen isotope ex-

change in the system calcite—CO_2—graphite—methane—hydrogen and water vapor. *Geochim. Cosmochim. Acta*, 33: 49—64.

Bottinga, Y. and Craig, H., 1968. High temperature liquid-vapor fractionation factors for H_2O-HDO-$H_2^{18}O$. *EOS*, 49: 356.

Brand, U. and Veizer, J., 1981. Chemical diagenesis of multicomponent carbonate system-2: stable isotopes. *J. Sediment. Petrol.*, 51: 987—997.

Bredehoeft, J.D., Blyth, C.R., White, W.A. and Maxey, G.B., 1963. Possible mechanism for concentration of brines in subsurface formations. *Bull. Am. Assoc. Pet. Geol.*, 47: 257—269.

Burst, J.F., 1969. Diagenesis of Gulf Coast clay sediments and its possible relation to petroleum migration. *Bull. Am. Assoc. Pet. Geol.*, 53: 73—93.

Carpenter, A.B., Trout, M.L. and Pickett, E.E., 1974. Preliminary report on the origin and chemical evolution of lead- and zinc-rich brines in central Mississippi. *Econ. Geol.*, 69: 1191—1206.

Case, L.C., 1955. Origin and current usage of the term, "connate water". *Bull. Am. Assoc. Pet. Geol.*, 39: 1879—1882.

Clayton, R.N., Friedman, I., Graf, D.L., Mayeda, T.K., Meents, W.F. and Shimp, N.F., 1966. The origin of saline formation waters, I. Isotopic composition. *J. Geophys. Res.*, 71: 3869—82.

Clayton, R.N., Muffler, L.J.P. and White, D.E., 1968. Oxygen isotope study of calcite and silicates of the River Ranch No. 1 well, Salton Sea Geothermal field, California. *Am. J. Sci.*, 266: 968—979.

Coney, P.J., Jones, D.L. and Monger, J.W.H., 1980. Cordilleran suspect terranes. *Nature*, 288: 329—333.

Coplen, T.B., 1976. Cooperative geochemical resource assessment of the Mesa geothermal system. *Univ. Calif., Riverside, Rep.*, IGPP-UCR-76-1, 97 pp.

Coplen, T.B. and Hanshaw, B.B., 1973. Ultrafiltration by a compacted clay membrane, I. Oxygen and hydrogen fractionation. *Geochim. Cosmochim. Acta*, 37: 2295—2310.

Coplen, T.B., Kolesar, P., Taylor, R.E., Kendall, C. and Mooser, C., 1975. Investigations of the Dunes geothermal anomaly, Imperial Valley, California, IV. Geochemical studies of water, calcite, and silicates. *Univ. Calif., Riverside, Rep.*, ICPP-UCR-75, 42 pp.

Craig, H., 1961. Isotopic variations in meteoric waters. *Science*, 133: 1702.

Craig, H., 1963. The isotopic geochemistry of water and carbon in geothermal areas. In: E. Tongiorgi (Editor), *Nuclear Geology on Geothermal Areas — Spoleto 1963*. CNR, Pisa, pp. 17—53.

Craig, H., 1965. The measurement of oxygen isotope paleotemperatures. In: E. Tongiorgi (Editor), *Stable Isotopes in Oceanographic Studies and Paleotemperatures — Spoleto 1965*. CNR, Pisa, pp. 161—182.

Craig, H. and Gordon, L., 1965. Deuterium and oxygen-18 variation in the ocean and the marine atmospheres. In: E. Tongiorgi (Editor), *Stable isotopes in Oceanographic Studies and Paleotemperatures — Spoleto 1965*. CNR, Pisa, pp. 9—13.

Craig, H., Boato, G. and White, D.E., 1956. Isotopic geochemistry of thermal waters. *U.S. Natl. Acad. Sci.-Natl. Res. Counc., Publ.*, 400: 29—38.

Dansgaard, W., 1954. The ^{18}O abundance in fresh water. *Geochim. Cosmochim. Acta*, 6: 241—260.

Dansgaard, W., 1961. The isotopic composition of natural waters with special reference to the Greenland ice cap. *Medd. Groenl.*, 165: 1—120.

Dansgaard, W., 1964. Stable isotopes in precipitation. *Tellus*, 16: 436—468.

Davis, S.N., 1964. "The chemistry of saline waters" by Kreiger, R.A. — discussion. *Groundwater*, 2: 51.

Degens, E.T., 1962. Geochemische Untersuchungen von Wasser aus der Ägyptischen Sahara. *Geol. Rundsch.*, 52: 625—639.

Degens, E.T., Hunt, J.M., Reuther, J.H. and Reed, W.E., 1964. Data on distribution of amino acids and oxygen isotops in petroleum brine waters of various geologic ages. *Sedimentology*, 3: 199—225.

De Sitter, L.U., 1947. Diagenesis of oil-field brines. *Bull. Am. Assoc. Pet. Geol.*, 31: 2030—2040.

Dickey, P.A., Collins, A.G. and Fajardo, I.M., 1972. Chemical composition of deep formation waters in southwestern Louisiana. *Bull. Am. Assoc. Pet. Geol.*, 56: 1530—1570.

Dowgiallo, J. and Lesniak, P., 1980. The origin of chloride waters in the Polish Flysch Carpathians. *Proc. 3rd Int. Symp., Water-Rock Interaction, Edmonton, Alta.*, pp. 20—24.

Downing, R.A., Smith, D.B., Pearson, F.J., Monkhouse, R.A. and Otlet, R.L., 1977. The age of groundwater in the Lincolnshire Limestone, England, and its relevance to the flow mechanism. *J. Hydrol.*, 33: 201—216.

Emiliani, C. and Shackleton, N.J., 1974. The Brunhes epoch: isotopic paleotemperatures and geochronology. *Science*, 183: 511—514.

Eslinger, E.V. and Savin, S.M., 1973. Mineralogy and oxygen isotope geochemistry of the hydrothermally altered rocks of the Ohaki-Broadlands, New Zealand geothermal area. *Am. J. Sci.*, 273: 240—267.

Evans, G.V., Otlet, R.L., Downing, R.A., Monkhouse, R.A. and Rae, G., 1978. Some problems in the interpretation of isotope measurements in United Kingdom aquifers. In: *Isotope Hydrology 1978, 2*. IAEA, Vienna, pp. 679—708.

Fairbridge, R.W., 1964. The importance of limestone and its Ca/Mg content to paleoclimatology, In: A.E.M. Nairn (Editor), *Problems in Paleoclimatology*. Interscience, New York, N.Y. pp. 431—530.

Faure, G., 1977. *Principles of Isotope Geology*. John Wiley and Sons, New York, N.Y., 464 pp.

Fontes, J.Ch., 1980. Environmental isotopes in groundwater hydrology. In: P. Fritz and J.Ch. Fontes (Editors), *Handbook of Environmental Isotope Geochemistry, 1. The Terrestrial Environment A*. Elsevier, Amsterdam, pp. 75—134.

Friedman, I. and O'Neil, J.R., 1977. Compilation of stable isotope fractionation factors of geochemical interest. In: M. Fleischer (Editor), *Data of Geochemistry. U.S. Geol. Surv., Prof. Pap.*, 440-KK: 12 pp.

Friedman, I., Redfield, A.C., Schoen, B. and Harris, J., 1964. The variation in the deuterium content of natural waters in the hydrologic cycle. *Rev. Geophys.*, 2: 177—224.

Fritz, P. and Fontes, J.Ch. (Editors), 1980. *Handbook of Environmental Isotope Geochemistry, 1. The Terrestrial Environment, A*. Elsevier, Amsterdam, 545 pp.

Fritz, P. and Frape, S.K., 1982. Saline groundwaters in the Canadian Shield, a first review. *Chem. Geol.* (in press).

Gat, J.R., 1980. The isotopes of hydrogen and oxygen in precipitation. In: P. Fritz and J.Ch. Fontes (Editors), *Handbook of Environmental Isotope Geochemistry, 1. The Terrestrial Environment, A*. Elsevier, Amsterdam, pp. 21—48.

Gat, J.R. and Carmi, I., 1970. Evolution of the isotopic composition of atmospheric waters in the Mediterranean Sea area. *J. Geophys. Res.*, 75: 3039—3048.

Gat, J.R. and Issar, A., 1974. Desert isotope hydrology: water sources of the Sinai Desert. *Geochim. Cosmochim. Acta*, 38: 1117—1131.

Gerasimov, I.P., 1979. Climates of past geological epochs. *Proc. World Climate Conf., Geneva, WMO*, 537: 88—111.

Gieskes, J.M., Lawrence, J.R. and Galleisky, G., 1978. Interstitial water studies. In: *Initial Reports of the Deep Sea Drilling Project, 38*. U.S. Government Printing Office, Washington, D.C., 121 pp.

Gonfiantini, R. (Editor), 1977. The application of nuclear techniques to geothermal studies; report of the Advisory Group Meeting, Pisa (Italy), 1975. *Geothermics*, 5: 1—184.

Graf, D.L., Friedman, I. and Meents, W.F., 1965. The origin of saline formation waters, II. Isotopic fractionation by shale micropore systems. *Ill. State Geol. Surv., Circ.*, 393: 32 pp.

Graf, D.L., Meents, W.F., Friedman, I. and Shimp, N.F., 1966. The origin of saline formation waters, III. Calcium chloride waters. *Ill. State Geol. Surv., Circ.*, 397: 60 pp.

Grantz, A. and Eittreim, S., 1979. Geology and physiography of the continental margin north of Alaska and implications for the origin of the Canada basin. *U.S. Geol. Surv., Open-File Rep.*, 79—288; 61 pp.

Gregory, R.T. and Taylor, H.P., Jr., 1981. An oxygen isotope profile in a section of Cretaceous oceanic crust, Samail Ophiolite, Oman: evidence for $\delta^{18}O$-buffering of the oceans by deep (> 5 km) seawater-hydrothermal circulation at mid-ocean ridges. *J. Geophys. Res.*, 86: 2737—2755.

Gutsalo, L.K., Vetshteyn, V.V. and Lebedev, L.M., 1978. Recent metalliferous ore-formation in hydrothermal solutions of the Cheleken region characterized by isotopic compositions of hydrogen and oxygen. *Geochem. Int.*, 7: 116—122.

Habicht, J.K.A., 1979. Paleoclimate, paleomagnetism and continental drift. *Am. Assoc. Pet. Geol., Stud. Geol.*, 9: 29 pp.

Hanshaw, B.B. and Coplen, T.B., 1973. Ultrafiltration by a compacted clay membrane, II. Sodium ion exclusion at various ionic strengths. *Geochim. Cosmochim. Acta*, 37: 2311—2327.

Hanshaw, B.B. and Hill, G.A., 1969. Geochemistry and hydrodynamics of the Paradox basin region, Utah, Colorado and New Mexico. *Chem. Geol.*, 4: 264—294.

Hardin, G.C., 1962. Notes on Cenozoic sedimentation in the Gulf Coast geosyncline, U.S.A. In: *Geology of the Gulf Coast and Central Texas. Houston Geol. Soc. Guidebk.*, 1: 1—15.

Hattori, K. and Sakai, H., 1979. D/H ratios, origins, and evaluation of the ore-forming fluids for the Neogene veins and Kuroko deposits of Japan. *Econ. Geol.*, 74: 535—555.

Hem, J.D., 1970. Study and interpretation of the chemical characteristics of natural water. *U.S. Geol. Surv., Water Supply Pap.*, 1473: 363 pp.

Hitchon, B. and Friedman, I., 1969. Geochemistry and origin of formation waters in the western Canada sedimentary basin, I. Stable isotopes of hydrogen and oxygen. *Geochim, Cosmochim. Acta*, 33: 1321—1349.

Hitchon, B., Billings, G.K. and Klovan, J.E., 1971. Geochemistry and origin of formation waters in the western Canada sedimentary basin, 3. Factors controlling chemical composition. *Geochim. Cosmochim. Acta*, 35: 567—598.

Hoefs, J., 1980. *Stable Isotope Geochemistry*. Springer-Verlag, New York, N.Y., 208 pp.

Hopkins, D.M., Matthews, J.V., Wolfe, J.A. and Silberman, M.L., 1971. A Pliocene flora and insect fauna from the Bering Strait region. *Palaeogeogr., Palaeoclimatol., Palaeoecol.*, 9: 211—231.

Hower, J., Eslinger, E.V., Hower, M.E. and Perry, E.A., 1976. Mechanism of burial metamorphism of argillaceous sediment; I. Mineralogical and chemical evidence. *Geol. Soc. Am. Bull.*, 87: 725—737.

International Atomic Energy Agency, 1978. *Isotope Hydrology 1978. 1 and 2. Proc. Int. Symp. on Isotope Hydrology, Neuherberg, 1978*. IAEA, Vienna, 984 pp.

Irving, E., 1978. Paleopoles and paleolatitudes of North America and speculations about displaced terrains. *Can. J. Earth Sci.*, 16: 669—694.

Jäger, E. and Hunziker, J.C., 1979. *Lectures in Isotope Geology*. Springer-Verlag, New York, N.Y., 329 pp.

James, A.T. and Baker, D.R., 1976. Oxygen isotope exchange between illite and water at 22°C. *Geochim. Cosmochim. Acta*, 40: 235—239.
Jones, P.H., 1975. Geothermal and hydrocarbon regimes, northern Gulf of Mexico basin. *Proc. 1st Conf., Geopressured-Geothermal Energy, Austin, Texas*, pp. 15—90.
Kendall, C., 1976. *Petrology and stable isotope geochemistry of three wells in the Buttes area of the Salton Sea geothermal field, California, U.S.A.* M.Sc. Thesis, University of California, Riverside, Calif., 211 pp.
Kharaka, Y.K. and Berry, F.A.F., 1973. Simultaneous flow of water and solutes through geological membranes, I. Experimental investigation. *Geochim. Cosmochim. Acta*, 37: 2577—2603.
Kharaka, Y.K. and Berry, F.A.F., 1974. The influence of geological membranes on the geochemistry of subsurface waters from Miocene sediments at Kettleman North Dome, California. *Water Resour. Res.*, 10: 313—327.
Kharaka, Y.K. and Berry, F.A.F., 1976. The influence of geological membranes on the geochemistry of subsurface waters from Eocene sediments at Kettleman North Dome, California—an example of effluent-type waters. *Proc. Int. Symp. Water-Rock Interaction, Prague, September 1974*, pp. 268—277.
Kharaka, Y.K. and Carothers, W.W., 1984. Geochemistry of oil-field waters from North Slope, Alaska. *U.S. Geol. Surv., Prof. Pap.* (in press).
Kharaka, Y.K. and Smalley, W.C., 1976. Flow of water and solutes through compacted clays. *Bull. Am. Assoc. Pet. Geol.*, 60: 973—980.
Kharaka, Y.K., Berry, F.A.F. and Friedman, I., 1973. Isotopic composition of oil-field brines from Kettleman North Dome oil field, California, and their geologic implications. *Geochim. Cosmochim. Acta*, 37: 1899—1908.
Kharaka, Y.K., Callender, E. and Carothers, W.W., 1977. Geochemistry of geopressured geothermal waters from the Texas Gulf Coast. *Proc. 3rd Conf., Geopressured-Geothermal Energy, Univ. Southwestern Louisiana, La.*, pp. GI121—GI165.
Kharaka, Y.K., Carothers, W.W. and Brown, P.M., 1978a. Origins of water and solutes in the geopressured zones of the northern Gulf of Mexico Basin. *Proc. 53rd Annu. Conf., Soc. Petrol. Eng. AIME*, SPE 7505: 8 pp.
Kharaka, Y.K., Brown, P.M. and Carothers, W.W., 1978b. Chemistry of waters in the geopressured zone from coastal Louisiana—implications for the geothermal development. *Geotherm. Res. Counc. Trans.*, 2: 371—374.
Kharaka, Y.K., Lico, M.S., Wright, V.A. and Carothers, W.W., 1979. Geochemistry of formation waters from Pleasant Bayou No. 2 well and adjacent areas in Coastal Texas. *Proc. 4th Conf., Geopressured-Geothermal Resources Austin, Texas*, pp. 168—193.
Kharaka, Y.K., Lico, M.S., Law, L.M. and Carothers, W.W., 1981. Geopressured-geothermal resources in California. *Geotherm. Res. Counc. Trans.*, 5: 721—724.
Knauth, L.P. and Epstein, S., 1976. Hydrogen and oxygen isotope ratios in nodular and bedded cherts. *Geochim. Cosmochim. Acta*, 40: 1095—1108.
Knauth, L.P., Kumar, M.B. and Martinez, J.D., 1980. Isotope geochemistry of water in Gulf Coast salt domes. *J. Geophys. Res.*, 85: 4863—4871.
Lamb, H.H., 1977. *Climate, Present, Past and Future*, 2. Methuen, London, 835 pp.
Lawrence, J.R. and Taylor, H.P., Jr., 1971. Deuterium and oxygen-18 correlation: clay minerals and hydroxides in Quaternary soils compared to meteoric waters. *Geochim. Cosmochim. Acta*, 35: 993—1003.
Lawrence, J.R., Gieskes, J.M. and Broecker, W.S., 1975. Oxygen isotope and cation composition of DSDP pore waters and alteration of layer II basalts. *Earth Planet. Sci. Lett.*, 27: 1—10.
Lico, M.S., Kharaka, Y.K., Carothers, W.W. and Wright, V.A., 1982. Methods for collection and analysis of geopressured geothermal and oil field waters. *U.S. Geol. Surv., Water-Supply Pap.*, 2194: 21 pp.

Loucks, R.G., Dodge, M.M. and Galloway, W.E., 1979. Sandstone consolidation analysis to delineate areas of high-quality reservoirs suitable for production of geopressured-geothermal energy along the Texas Gulf Coast. *Texas Univ. Bur. Econ. Geol., Contr. Rep. for U.S. Dep. Energy*, EG-77-05-554: 98 pp.

Loucks, R.G., Richmann, D.L. and Milliken, K.L., 1980. Factors controling porosity and permeability in geopressured Frio sandstone reservoirs, General Crude Oil/Department of Energy Pleasant Bayou test wells, Brazoria County, Texas. *Proc. 4th Conf., Geopressured-Geothermal Energy, Univ. Texas, Austin, Texas*, pp. 46—82.

Majzoub, M., 1971. Fractionnement en oxygen-18 et en deuterium entre l'eau et sa vapeur. *J. Chem. Phys.*, 68: 1423—1436.

Manheim, F.T. and Sayles, F.L., 1974. Composition and origin of intersitial waters of marine sediments, based on deep-sea drill cores. In: E.D. Golberg (Editor), *The Sea, 5*. John Wiley and Sons, New York, N.Y., pp. 527—568.

Mazor, E. and Kharaka, Y.K., 1981. Atmospheric and radiogenic noble gases in geopressured-geothermal fluids: northern Gulf of Mexico basin. *Proc. 5th Conf., Geopressured-Geothermal Energy, Louisiana State Univ., Baton Rouge, La.*, pp. 197—200.

McKelvey, J.G. and Milne, I.H., 1962. The flow of salt solutions through compacted clay. *Clays Clay Miner.*, 9: 248—259.

Merlivat, L., Botter, R. and Nief, G., 1963. Fractionation isotopique au cours de la distillation de l'eau. *J. Chem. Phys.*, 60: 56—59.

Milliken, K.L., Land, L.S. and Loucks, R.G., 1981. History of burial diagenesis determined from isotopic geochemistry, Frio Formation, Brazoria County, Texas. *Bull. Am. Assoc. Pet. Geol.*, 65: 1397—1413.

Muehlenbachs, K. and Clayton, R.N., 1976. Oxygen isotope studies of fresh and weathered submarine basalts. *Can. J. Earth Sci.*, 9: 172—184.

Murata, K.J., Friedman, I. and Gleason, J.D., 1977. Oxygen isotope relations between diagenetic silica minerals in Monterey Shale, Temblor Range, California. *Am. J. Sci.*, 277: 259—272.

Nairn, A.E.M., 1964. *Problems in Paleoclimatology*. Interscience, New York, N.Y., 705 pp.

Norton, D. and Taylor, H.P., Jr., 1979. Quantitative simulation of the hydrothermal systems of crystallizing magmas on the basis of transport theory and oxygen isotope data: an analysis of the Skaergaard intrusion. *J. Petrol.*, 20: 421—486.

O'Neil, J.R., 1977. Stable isotopes in mineralogy. *Phys. Chem. Miner.*, 2: 105—123.

O'Neil, J.R. and Kharaka, Y.K., 1976. Hydrogen and oxygen isotope exchange reactions between clay minerals and water. *Geochim. Cosmochim. Acta*, 40: 241—246.

O'Neil, J.R. and Taylor, H.P., 1967. The oxygen isotope cation exchange chemistry of feldspars. *Am. Mineral.*, 52: 1414—1437.

O'Neil, J.R. and Taylor, H.P., 1969. Oxygen isotope equilibrium between muscovite and water. *J. Geophys. Res.*, 74: 6012—6022.

Pearson, F.J., Jr. and White, D.E., 1967. Carbon-14 ages and flow rates of water in Carrizo Sand, Atascosa County, Texas. *Water Resour. Res.*, 3: 251—261.

Perry, E.C., Jr., 1967. The oxygen isotope chemistry of ancient cherts. *Earth Planet. Sci. Lett.*, 3: 62—66.

Perry, E.C., Jr. and Tan, F.C., 1972. Significance of oxygen and carbon isotope variations in early Precambrian cherts and carbonate rocks of South Africa. *Geol. Soc. Am. Bull.*, 83: 647—664.

Randall, S.M., Lawrence, J.R. and Gieskes, J.M., 1979. Interstitial water studies, sites 386 and 387. In: *Initial Reports of the Deep Sea Drilling Project, 43*. U.S. Government Printing Office, Washington, D.C., 669 pp.

Redfield, A.C. and Friedman, I., 1965. Factors affecting the distribution of deuterium in the ocean. In: *Symposium on Marine Geochemistry. Narraganset Mar. Lab., Univ. R.I., Occas. Publ.*, 3: 149 pp.

Rózkowski, A. and Przwłocki, K., 1974. Application of stable environmental isotopes in mine hydrogeology taking Polish coal basins as an example. In: *Isotope Techniques in Groundwater Hydrology, 1.* IAEA, Vienna, pp. 481—488.

Savin, S.M., 1980. Oxygen and hydrogen isotope effects in low-temperature mineral-water interactions. In: P. Fritz and J.Ch. Fontes (Editors), *Handbook of Environmental Isotope Geochemistry, 1. The Terrestrial Environment, A.* Elsevier, Amsterdam, pp. 283—328.

Savin, S.M. and Epstein, S., 1970. The oxygen and hydrogen isotope geochemistry of clay minerals. *Geochim. Cosmochim. Acta*, 34: 25—42.

Schmidt, G.W., 1973. Interstitial water composition and geochemistry of deep Gulf Coast shales and sandstones. *Bull. Am. Assoc. Pet. Geol.*, 57: 321—337.

Schwartz, F.W. and Muehlenbachs, K., 1979. Isotope and ion geochemistry of groundwaters in the Milk River aquifer, Alberta. *Water Resour. Res.*, 15: 259—268.

Sofer, Z. and Gat, J., 1972. Activities and concentrations of ^{18}O in concentrated aqueous salt solutions: analytical and geophysical implications. *Earth Planet. Sci. Lett.*, 15: 232—238.

Stewart, M.K., 1978. Stable isotopes in waters of the Wairakei geothermal area, New Zealand. In: *Stable Isotopes in the Earth Sciences. N.Z. Dep. Sci. Ind. Res., Bull.*, 220: 113—119.

Stewart, M.K. and Friedman, I., 1975. Deuterium fractionation between aqueous salt solutions and water vapor. *J. Geophys. Res.*, 80: 3812—3818.

Tailleur, I.L. and Brosgé, W.P., 1970. Tectonic history of northern Alaska. In: W.L. Adkison and M.M. Brosgé (Editors), *Geological Seminar on the North Slope of Alaska, Palo Alto, Calif., 1970. Proc. Am. Assoc. Pet. Geol., Pacif. Sect. Meet.*, pp. E1—E18.

Taube, H., 1954. Use of oxygen isotope effects in the study of hydration of ions. *J. Phys. Chem.*, 58: 523—528.

Taylor, H.P., Jr., 1977. Water/rock interactions and the origin of H_2O in granitic batholiths. *J. Geol. Soc., London*, 133: 509—558.

Taylor, H.P., Jr. and Magaritz, M., 1978. Oxygen and hydrogen isotope studies of the Cordilleran batholiths of western North America; *N.Z. Dept. Sci., Ind. Res., Bull.*, 220: 151—173.

Truesdell, A.H., 1974. Oxygen isotope activities and concentrations in aqueous salt solutions at elevated temperature-consequences for isotope geochemistry. *Earth Planet Sci. Lett.*, 23: 387—396.

Truesdell, A.H. and Hulston, J.R., 1980. Isotopic evidence in environments of geothermal systems. In: P. Fritz and J.Ch. Fontes (Editors), *Handbook of Environmental Isotope Geochemistry, 1. The Terrestrial Environment, A.* Elsevier, Amsterdam, pp. 179—226.

Uyarov, D.V., Sokolov, N.M. and Zavoronkov, N.M., 1962. Physikalisch-chemische konstanten von $H_2{}^{18}O$. *Kernenergie*, 5: 323—329.

Vetshteyn, V.V., Gavish, V.K. and Gutsalo, L.K., 1981. Hydrogen and oxygen isotope composition of waters in deep seated fault zones. *Int. Geol. Rev.*, 23: 302—310.

Weber, J.N., 1965. Changes in the oxygen isotopic composition of sea water during the Phanerozoic evolution of the oceans. *Geol. Soc. Am., Spec. Pap.*, 82: 218—219.

Wedepohl, K.H. (Executive Editor), *Handbook of Geochemistry, II/1.* Springer-Verlag, New York, N.Y. (Elements H (1) to Al (12)).

White, D.E., 1957. Thermal waters of volcanic origin. *Geol. Soc. Am. Bull.*, 68: 1637—1658.

White, D.E., 1965. Saline waters of sedimentary rocks. In: A. Yound and G.E. Galley (Editors), *Fluids in Subsurface Environments. Am. Assoc. Pet. Geol., Mem.*, 4; 342—366.

White, D.E., 1974. Diverse origins of hydrothermal ore fluids. *Econ. Geol.*, 69: 954—973.
White, D.E., Barnes, I. and O'Neil, J.R., 1973. Thermal and mineral waters of nonmeteoric origin, California Coast Ranges. *Geol. Soc. Am. Bull.*, 84: 547—560.
White, D.E., Hem, J.D. and Waring, G.A., 1963. Chemical composition of subsurface waters. In: *Data of Geochemistry. U.S. Geol. Surv., Prof. Pap.*, 440F: 67 pp.
Williams, F.D., Moore, W.S. and Fillon, R.H., 1981. Role of glacial Arctic Ocean ice sheets in Pleistocene oxygen isotope and sea-level records. *Earth Planet Sci. Lett.*, 56: 157—166.
Wolfe, J.A., 1980. Tertiary climates and floristic relationships at high latitudes in the northern hemisphere. *Palaeogeogr., Palaeoclimatol., Palaeoecol.*, 30: 313—323.
Yeh, H.W., 1980. D/H ratios and late-stage dehydration of shales during burial. *Geochim. Cosmochim. Acta*, 44: 341—352.
Yeh, H.W. and Epstein, S., 1978. The hydrogen isotope exchange between clay minerals and sea water. *Geochim. Cosmochim. Acta*, 42: 140—143.
Yeh, H.W. and Savin, S.M., 1976. The extent of oxygen isotope exchange between clay minerals and sea water. *Geochim. Cosmochim. Acta*, 40: 743—748.
Yeh, H.W. and Savin, S.M., 1977. The mechanism of burial metamorphism of argillaceous sediments, 3. Oxygen isotope evidence. *Geol. Soc. Am. Bull.*, 88: 1321—1330.
Yund, R.A. and Anderson, R.F., 1974. Oxygen isotope exchange between potassium feldspar and KCl solutions. In: A.W. Hofman, B.J. Giletti, H.S. Yoder and R.A. Yund (Editors), *Geochemical Transport and Kinetics.* Carnegie Institution of Washington, Washington, D.C., pp. 99—105.
Yurtsever, Y., 1975. Worldwide survey of stable isotopes in precipitation. *Rep. Sect. Isot. Hydrol., IAEA*, 40 pp.

Chapter 9

ISOTOPE EFFECTS OF NITROGEN IN THE SOIL AND BIOSPHERE

H. HÜBNER

INTRODUCTION

The chemical element nitrogen is one of the principal components of proteins and nucleic acids which represent the hereditary material of the animate world. It is attributable to the history of its discovery that it was originally called azote (from Greek words meaning "no life" and still applied in the French and in the Russian languages and in accordance with this meaning in German Stickstoff), a name given by A.L. Lavoisier because the gas did not sustain life. Eighteenth century chemists and physicists, who could avail themselves of no other apparatus than what is known as the pneumatic trough, were able to distinguish only between an essential gas contained in the air, i.e., dephlogisticated air as O_2, and phlogisticated or mephitic air, i.e., N_2. The latter was found to be stuffy and fuggy and incapable of supporting life, and Lavoisier was the first to show that this component of the air is a chemical element. H. Cavendish performed a very large number of tests to determine the composition of the air (making more than 400 measurements in 60 days at different places in England and under different weather conditions), and he found the proportion of oxygen to be 20.84% by volume, a value that differs from the present value of 20.946 ± 0.0017% by volume (Machta and Hughes, 1970) by only about 0.5%.

With few exceptions, nitrogen is present in all natural products in its most strongly reduced form, the oxidation number being —3. The nitro group with +5 charged nitrogen is contained in heptagenic acid (β-nitro-propionic acid), in chloramphenicol, and in aristolochic acid in which the nitro groups are attached to aromatic rings. The azoxy group (elaiomycin, cycasine, macrozamine) is found somewhat more frequently. On the other hand, a multitude of nitrogen compounds can be, and are, taken up by living beings, although it is not possible to make as clear a distinction between N-autotrophic and N-heterotrophic organisms as in the case of carbon. N-heterotrophic organisms are capable of taking up only organically fixed nitrogen in the form of proteins (amino acids). In the case of N-autotrophic organisms, numerous transitions are possible, e.g., from purely inorganic compounds (NH_4^+, NO_3^-, NO_2^-, N_2) with transitions (urea, carbamic acid) to organic compounds (amino acids) including even those which are today the focal point of environmental research, and more particularly the nitrosamines of which not inconsiderable amounts are invariably taken up through the roots of autotrophic lettuce and spinach plants (Dean-Raymond and Alexander, 1976). Many details of the nitrogen balance of plants are as yet imperfectly understood, and it always amazes us to learn of the discovery, e.g., in lentils (Sulser and Sager, 1976), of unusual amino acids and even toxic amino acids (for instance, β-cyano-L-alanine) that may perhaps be considered to be responsible for lathyrism in Mediterranean countries. Also astonishing and improperly understood at present are nitrogen food chains in biomes and even in tall individual trees, which involves not only the nitrogen input from

TABLE 9-1

Global inventories of nitrogen (after Söderlund and Svensson, 1976; Winteringham, 1980)

	Nitrogen (Tg)[a]	Nitrogen (%)
Terrestrial	1.9×10^{11}	97.98
Plant biomass	$1.1–1.4 \times 10^4$	6.4×10^{-6}
Animal biomass	2×10^2	1.0×10^{-7}
Litter	$1.9–3.3 \times 10^3$	1.3×10^{-6}
Soil	3.17×10^5	1.6×10^{-4}
Soil organic matter	3×10^5	1.5×10^{-4}
Soil, insoluble inorganic	1.6×10^4	8.2×10^{-6}
Soil, soluble inorganic	3×10^3	1.5×10^{-6}
Soil, micro-organisms included in total soil organic matter	5×10^2	2.6×10^{-7}
Rocks	1.9×10^{11}	99.775
Sediments	4×10^8	0.206
Coal	1.2×10^5	6.2×10^{-5}
Oceanic	2.3×10^7	1.19×10^{-2}
Plant biomass	3×10^2	1.5×10^{-7}
Animal biomass	1.7×10^2	8.6×10^{-8}
Organic matter	5.43×10^5	2.8×10^{-4}
Organic matter dissolved	5.3×10^5	2.7×10^{-4}
Organic matter particulate	$0.3–2.4 \times 10^4$	6.9×10^{-6}
N_2 dissolved	2.2×10^7	0.0113
NO_3^- dissolved	5.7×10^5	2.9×10^{-4}
NH_4^+ dissolved	7×10^3	3.6×10^{-6}
Atmospheric	3.9×10^9	2.007
N_2	3.9×10^9	2.007
N_2O	1.3×10^3	6.7×10^{-7}
Trace gases (NH_3, NO_x, etc.)	6.7	3.4×10^{-9}
Total	194.3×10^9	100

[a] 1 Tg (teragram) = 10^{12} g.

precipitations but nitrogen-fixing bacteria, lichens, epiphytes, and the soil flora as well (Denison, 1973).

Geochemically, nitrogen is counted among the atmophile elements, on the one hand, and the biophile elements, on the other, because of its involvement in the constitution of the atmosphere and biosphere respectively. Accounting for 0.03%, nitrogen, as far as its terrestrial abundance is concerned (Remy, 1973), takes the sixteenth place, thus neighbouring elements such as Rb (0.029%), F (0.028%), Ba (0.026%), and Zr (0.021%). The total quantity of nitrogen is calculated as roughly 2×10^{11} Tg N ($\equiv 2 \times 10^{23}$ g N $\equiv 2 \times 10^{17}$ t N) (Söderlund and Svensson, 1976; Winteringham, 1980), of which 97.98% are fixed in the lithosphere. The atmosphere contains 2.007% of the total quantity, and the biosphere contains as little as 0.001%. The remaining shares are summarized in Tables 9-1 and 9-2 and are also shown in Fig. 9-1.

In the literature on isotope geology and isotope geochemistry, isotopic variations are stated in δ values, defined as:

$$\delta^{15}N \text{ \textperthousand} = \left(\frac{R_{sample}}{R_{standard}} - 1 \right) \times 10^3$$

TABLE 9-2

Global inventaries of nitrogen in the biosphere (after Söderlund and Svensson, 1976; Winteringham, 1980)

	Nitrogen (Tg)	%	Living biomass
Terrestrial	3.32×10^5	37.9	
Plant biomass	$1.1-1.4 \times 10^4$		$1.32 \times 10^4 \hat{=} 96.56\%$
Animal biomass	2×10^2		
Litter	$1.9-3.3 \times 10^3$		
Soil	3.17×10^5		
Soil organic matter	3×10^5		
Soil, insoluble inorganic	1.6×10^4		
Soil, soluble inorganic	3×10^3		
Soil micro-organisms included in total soil organic matter	5×10^2		
Oceanic	5.44×10^5	62.1	
Plant biomass	3×10^2		$4.7 \times 10^2 \hat{=} 3.44\%$
Animal biomass	1.7×10^2		
Organic matter	5.43×10^5		
Organic matter dissolved	5.3×10^5		
Organic matter particulate	$0.3-2.4 \times 10^4$		
	8.76×10^5	100	1.37×10^4
	$\hat{=} 4.5 \times 10^{-4}\%$ from total N		$\hat{=} 7.0 \times 10^{-6}\%$ from total N

where R is the isotope ratio of $^{15}N/^{14}N$ in the case of nitrogen which has only two stable isotopes. In the literature on agricultural chemistry, isotopic variations are not infrequently expressed by $\delta_a^{15}N$ or $\Delta^{15}N$, defined as:

$$\delta_a^{15}N \text{ \textperthousand} = \left(\frac{\text{at.\% sample}}{\text{at.\% standard}} - 1\right) \times 10^3$$

However, since at.% $^{15}N \times 100/(^{14}N + {}^{15}N) \neq$ isotope ratio $^{15}N/^{14}N$, there is a difference (though only a slight one) between these statements. A conversion formula for this is given by:

$$\delta^{15}N = \left(\frac{99.6336 (1 + 10^{-3} \delta_a^{15}N)}{100 - 0.3663 (1 + 10^{-3} \delta_a^{15}N)} - 1\right) \times 10^3$$

with $R_{standard} = (^{15}N/^{14}N)_{air} = 1/272$ which corresponds to 0.3663 at.% ^{15}N.

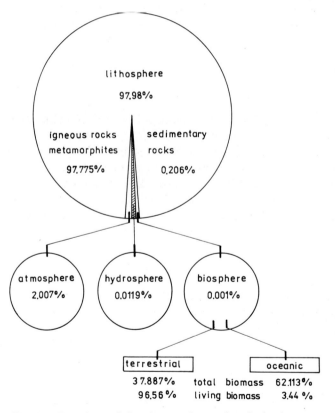

Fig. 9-1. Fractions of the nitrogen in geochemical spheres.

A numerical example can be used to show that there is only a relatively slight difference:

$\delta_a{}^{15}N = +10‰;$ $\delta^{15}N = +10.03_6‰$
$\delta_a{}^{15}N = -10‰;$ $\delta^{15}N = -10.03_7‰.$

^{15}N IN THE PEDOSPHERE

Isotopic variations

Isotopic variations of nitrogen in the pedosphere have been observed for many years; one of the first studies was done in 1964 (Cheng et al., 1964), in which samples were collected from different depths of soil in fifteen different locations (Iowa, U.S.A.) and in which the $\delta^{15}N$ values were determined. The mean total nitrogen was $\delta_a{}^{15}N = 6.28 \pm 5.2‰$, with extreme values from -1 to $+17‰$. Despite wide ranges of variation, even then it had been observed that cultivated land shows a somewhat smaller $\delta^{15}N$ value than uncultivated land, the numerical expressions (for five soils at depths of from 0 to 15 cm: A_p horizon) being as follows:

$\delta^{15}N$ (cultivated) $= 5.0 \pm 3.5‰$

$\delta^{15}N$ (uncultivated) $= 6.8 \pm 6.4‰.$

This illustrates that cultivated land has, in general, a lower $\delta^{15}N$ value than uncultivated land, since the ^{15}N content of nitrogenous fertilizers is smaller as compared with nitrogen in the soil. However, this does not hold for all cases. Studies of variations of $\delta^{15}N$ in dependence upon location, time of year, depth, compounds in the soil (total nitrogen, NH_4^+, NO_3^-, amino acids, hexosamines, hydrolyzable and non-hydrolyzable fractions, etc.), utilization of land (cultivated and uncultivated land), and other factors have been, and are being, done on a wide scale. For example, Broadbent et al. (1980) made a total of 1713 isotopic analyses of corresponding soil samples collected from three cultivated and three uncultivated locations, with an overall mean of 1.69 ± 2.9 being determined at depths of soil from 0 to 120 cm. The overall mean obtained for the uppermost layer of soil (0—15 cm) is 0.53 ± 3.5. Separation of overall means into cultivated and uncultivated soil samples again shows that the ^{15}N content of cultivated land is smaller than that of uncultivated land, the numerical expressions being respectively:

$\delta^{15}N$ (cultivated) $= 0.65 \pm 2.6‰$ and

$\delta^{15}N$ (uncultivated) $= 2.73 \pm 3.4‰.$

In some soil profiles one observes a significant dependence of $\delta^{15}N$ values upon soil depth, whereas other profiles do not show such a dependence. By correlating depth (cm) = $a\, \delta^{15}N + b$, Table 9-3 is obtained (from values given by Broadbent et al., 1980; Riga et al., 1971; Shearer et al., 1978). As seen in this table four soil profiles, three of which are forest soils, show significant correlation coefficients. However, this statement cannot be considered as reflecting a generally valid law. A similar observation has been made in another extensive study where 139 soil samples from twenty states of the U.S.A. were analyzed. Of those samples, the majority had been collected from the uppermost horizon (0—15 cm) the following relationships were observed:

$\delta^{15}N = -0.032$ (annual precipitation in cm) $- 0.022$ (% sand) $+ 11.1$

or:

$\delta^{15}N = 9.41$ (% nitrogen) $- 0.032$ (annual precipitation in cm) $+ 9.1$.

Since total nitrogen generally decreases as depth of soil increases, it is recommended to correlate the total nitrogen content with the corresponding $\delta^{15}N$ value. Thus, forest soil (Broadbent et al., 1980) yields:

$\delta^{15}N = -0.0061$ (total nitrogen, ppm) $+ 10.606$
$r = -0.8771$,

since the relation between depth of soil (cm) and total nitrogen (ppm) holding for this case is:

TABLE 9-3

Dependence of $\delta^{15}N$ values upon soil depth

	a	b	r
Uncultivated			
Forest soil:			
Tahoe, U.S.A.	+11.13	−6.237	0.8975
Maine, U.S.A.	−0.052	+8.102	−0.7332
Saint-Hubert, Ardennes, Belgium	+6.32	+29.35	0.8146
Croix-Scaille, Ardennes, Belgium	+8.93	+34.47	0.8837
Stony loam	−3.67	+68.876	−0.3137
Loam	+4.79	+60.31	0.1560
Cultivated			
Loamy sand	+5.65	+143.1	0.1767
Sand	−19.43	+120.89	−0.9295
Loam	+6.66	+26.64	0.2225

depth (cm) = -0.08 (total nitrogen, ppm) $+ 119.7$
$r = -0.9274$.

To date, only a small number of investigators have presented $\delta^{15}N$ values separated into total nitrogen and NO_3-N (nitrogen nitrate). On the contrary, it is sometimes plainly expressed that the $\delta^{15}N$ values, because of the very small contents of NO_3^-, yield quite erratic results and therefore are not included (Broadbent et al., 1980). Although a comparison of both values for nine different Canadian soils (A_p horizon) (Rennie et al., 1976) yields nearly identical mean values, namely:

$\delta^{15}N_{N\text{-total}} = 8.2 \pm 1.73$ and $\delta^{15}N_{NO_3^-} = 8.83 \pm 2.94$.

There is a substiantially greater correlation between NO_3^- content and $\delta^{15}N_{NO_3^-}$ than between total nitrogen content and the corresponding $\delta^{15}N$ value:

N_{total} (ppm) $= 1772.2 + 46.57\ \delta^{15}N \quad r = 0.1385$

NO_3-N (ppm) $= 45.91 - 2.656\ \delta^{15}N \quad r = -0.7507$.

This finding suggests that the NO_3-N abundance in soil is more or less controlled by the same processes which control the $\delta^{15}N_{NO_3^-}$ values as well. On the other hand the amount of NO_3-N has the smallest share of the amounts of nitrogen which is bound in other chemical forms. According to Table 9-1 the amount of soluble inorganic N (mainly NO_3-N) comes to about 1% compared with the other nitrogen compounds in soil. Such small pools (compartments) are much more sensitive in changing their isotope ratio than the greater ones if a permanent turn over with high frequency takes place.

Causes of isotopic variations in the pedosphere

Two factors will control the $\delta^{15}N$ values of a particular chemical compound in a soil horizon, namely: (1) variations in (or constancy of) the $\delta^{15}N$ values of the input of the chemical compound being examined, and (2) processes of transformation of materials that lead the particular chemical compound being studied and which are accompanied by isotopic effects.

The input variables include the following:
— precipitations, dusts, aerosols, etc.
— fertilizers
— loadings exerted by NO_x from areas extending around chemical plants or power plants using coal as fuel for power generation (the compounds involved including amines, nitriles, and nitrogen-containing intermediate chemicals)

— contents of NO_3^- in the groundwater, which may enter the respective soil horizon through capillary forces or groundwater table fluctuations (gley, pseudogley).

As far as processes of transformation are concerned, a distinction can initially be made between biological and abiological transformation processes.

Additional processes of transformation include:
— phase changes, e.g. NH_3 evaporation
— transport processes, e.g. N_2 diffusion accompanying N_2 fixation
— ion exchange processes, especially those of NH_4^+ ions
— chemical conversions during protein hydrolyses and chemical denitrification
— biological transformation during assimilations, and aerobic and anaerobic dissimilations such as nitrification and denitrification.

A schematic representation is given in Fig. 9-2. As far as the input/output ratio is concerned, it is of interest to note that an input or output of nitrogen can take place through both the atmosphere and the groundwater and that in the case of farmland, harvest represents a special case in which no isotopic effects whatsoever are involved.

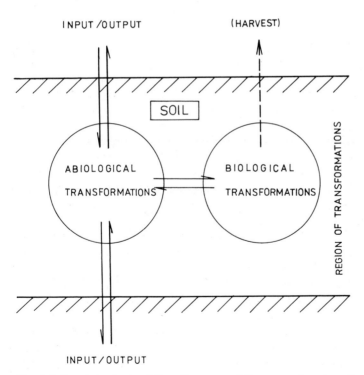

Fig. 9-2. Inputs/outputs of the nitrogen and its conversions in soils.

ISOTOPE INPUTS IN THE PEDOSPHERE

Isotopic variations of nitrogen in precipitation

For the forests in Europe and North America it can be assumed that approximately one third of the annual input of nitrogen from precipitations accumulates in drainage water, with roughly two thirds thus being temporarily retained by the pedosphere (Abrahamsen, 1980). Data (Heinrichs and Mayer, 1977) collected for Sollingen (Federal Republic of Germany), for example, shows 6 kg ha^{-1} y^{-1} in drainage water of the 23 kg ha^{-1} y^{-1} in precipitations [NH_4^+, NO_3^-]. As far as the fate of nitrogen compounds in wet and dry fall out is concerned, one can observe that at relative air humidity below 81% $(NH_4)_2SO_4$ will precipitate in crystalline form, whereas droplets of solution will be formed when the relative air humidity is greater than 81% (Healy et al., 1970). Values ranging from 0.43 to 1.0 have been reported (Orel et al., 1977) for the [NH_4^+]/([NO_3^-] + 2[SO_4^{2-}]) ratio in precipitation. Oxides of nitrogen are produced mainly during combustion processes as well as in the chemical industry (NH_3 production, nitration, etc.). The following percentage emissions have been reported (Laue et al., 1980):

Coal- and oil-fired boilers in power plants	24.7%
Oil-, coal-, and gas-fired industrial boilers	11.1%
Industrial internal combustion engines	9.5%
Industrial gas turbines	1.1%
Other industrial emissions	4.5%
Sum of industrial emissions	50.9%
Automobile engines, domestic fuels, etc.	49.1%

In the air of big cities, for example, 0.01—0.03 ppm NO_x is recorded, which for an output rate of 4% h^{-1} during the daytime almost equals the input rate and leads to a retention period of about 2 days (Chang et al., 1979). Some attempts are made to understand globally the atmospheric cycles of nitrogenous compounds, e.g. by Böttger et al. (1978), or in the special case of ammonia by McConnell, 1973. As expected, complex aerochemical reactions will result in quite different $\delta^{15}N$ values depending upon the particular chemical compound involved, the time of day and year, and the form of compound (Table 9-4).

Simple isotopic exchange reactions are not sufficient to explain the different $\delta^{15}N$ values of NH_4^+ and NH_3 (Moore, 1977); kinetic and absorptive effects are superimposed on these exchange reactions. No significant relations exist between the NH_3 content of the air and its $\delta^{15}N$ value. However, it is quite possible that $\delta^{15}N_{NH_3}$ values over the oceans show a far more significant dependence upon NH_3 concentrations, since regular relations of the kind:

$$p\,NH_3 = \frac{K_H \cdot K_w\,[NH_4^+]}{K_b\,[H^+]}$$

could be recognized where $K_H = 8.26 \times 10^{-3}$ atm mole^{-1}, $K_w = 2.92 \times 10^{-15}$ mole^2 l^{-2}, and $K_b = 1.57 \times 10^{-5}$ mole l^{-1}. This leads to a concentration of 0.06 μg NH$_3$ m^{-3} under normal conditions (Ayers and Gras, 1980).

Some idea of seasonal variations of δ^{15}N values of NH$_4^+$ and NO$_3^-$ in precipitations may be obtained from Fig. 9-3; they were measured at Jülich, Federal Republic of Germany, in 1975 and 1976 (Freyer, 1978a). While individual measured values were not given in that paper, it was still possible to make a few correlations, for example:

$$\delta^{15}N_{NH_4^+} = -2.894 \cdot \frac{C_{NH_4\text{-}N}}{C_{NO_3\text{-}N}} - 5.874 \qquad r = -0.5529$$

$$\delta^{15}N_{NO_3^-} = 4.916 \cdot \frac{C_{NH_4\text{-}N}}{C_{NO_3\text{-}N}} - 12.872 \qquad r = 0.6016$$

where $C_{NH_4\text{-}N}$ and $C_{NO_3\text{-}N}$ are the concentrations of NH$_4$-N and NO$_3$-N, respectively, expressed in mg l^{-1}. That is, the ratio of NH$_4$-N to NO$_3$-N in the precipitation has a distinct influence upon the δ^{15}N value of both NH$_4$-N and NO$_3$-N. The relationship will become even more conspicuous when $\Delta\delta^{14}N = \delta^{15}N_{NH_4^+} - \delta^{15}N_{NO_3^-}$ is correlated, which leads to:

$$\Delta\delta^{15}N = -7.995 \cdot \frac{C_{NH_4\text{-}N}}{C_{NO_3\text{-}N}} + 7.476 \qquad r = -0.7834$$

TABLE 9-4

Isotopic composition of NH$_3$, NO$_x$ and NO$_3^-$ in the atmosphere

		δ^{15}N (‰)
NH$_3$	pure air	-10.0 ± 2.6
NH$_3$	barnyard	$+24.9 \pm 3.4$
NO$_2$	pure air	-9.3 ± 3.5
NO$_x$	automobile exhaust gases	$+3.7 \pm 0.3$
NH$_4^+$	particulates	$+5.6 \pm 5.5$
NH$_4^+$	in rain	-1.4 ± 3.5
NO$_3^-$	particulates	$+5.0 \pm 5.7$
NO$_3^-$	in rain	-6.6 ± 3.9

with Student's t-test giving a confidence probability of 95% and more ($\nu = n - 2 = 6$; $t = 3.09$). Correlation of the mean values of the two years yields:

$$\Delta \delta^{15}N = -12.742 \cdot \frac{C_{NH_4-N}}{C_{NO_3-N}} + 17.558 \qquad r = -0.9677$$

with $t = 5.41$ for $\nu = n - 2 = 2$. The extent to which these relations can be considered to be of a general nature will have to be elucidated by further studies. Meteorological conditions (thunderstorms, etc.), snow, and the near-shore location should have a major influence. In another paper (Freyer, 1978b) it is reported that $\delta^{15}N$ values recorded in Brittany, France, were respectively $\delta^{15}N_{NH_4^+} = +2.6 \ldots +3.0$ and $\delta^{15}N_{NO_3^-} = +6.2 \ldots +6.9$, thus representing values that were not reached in Jülich, Federal Republic of Germany. Also given in Fig. 9-3 are $\delta^{15}N_{total}$ values which were calculated from the isotope balance equation:

$$\delta^{15}N_{total} = \frac{\delta^{15}N_{NH_4^+} \cdot C_{NH_4-N} + \delta^{15}N_{NO_3^-} \cdot C_{NO_3-N}}{C_{NH_4-N} + C_{NO_3-N}}$$

and which, when correlated, give:

$$\delta^{15}N_{total} = -0.996 (N_{total}) - 6.862$$

with $r = -0.4740$.

Additional values of precipitations are shown in Table 9-5 for Japan (Wada et al., 1975) and for the German Democratic Republic (Maass and Weise, 1981; Weise and Maass, 1981).

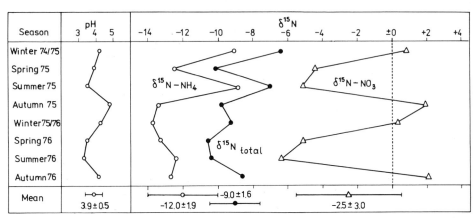

Fig. 9-3. Measured $\delta^{15}N$ values in precipitations at Jülich (F.R.G.) after Freyer (1978a).

Extreme values of NO_2^-/NO_3^- in precipitations must be expected to occur in the vicinity of chemical plants, since attempts are being made to minimize the pressure on the environment by far-reaching absorption of nitrous gases NO_x. The absorption of the nitrous gases is commonly carried out by a double-stage procedure:

— stage 1, the acidic absorption, where the NO together with air is passed into a column containing a mixture of a NO_2^-/NO_3^- solution, and

— stage 2, the soda absorption, where the NO_2 coming from the end of stage 1 is passed into a column containing the soda solution. The discharge of stage 2 is simultaneously the inlet of stage 1.

In both stages many isotope exchange reactions take place. Two of the most important reactions are:

$$^{15}NO + H^{14}NO_3 \rightleftharpoons {}^{14}NO + H^{15}NO_3$$

$$^{15}NO_2 + H^{15}NO_3 \rightleftharpoons {}^{14}NO_2 + H^{15}NO_3,$$

which are well-known in isotope separation processes for the production of ^{15}N-enriched nitric acid (Spindel/Taylor process). The single-stage separation factor, α, is defined for the NO/NHO_3 system as:

$$\alpha = \frac{(^{15}N/^{14}N)_{liquid}}{(^{15}N/^{14}N)_{gas}} = \frac{(^{15}N/^{14}N)_{HNO_3}}{(^{15}N/^{14}N)_{NO}}$$

with an $\alpha = 1.055$ at room temperature and in a 10 molar nitric acid solution (Taylor and Spindel, 1958). For the other exchange reaction only a calculated α value exists, derived from spectroscopical data (Spindel, 1954). This α value equals about 1.030. Since there is a dependence not only upon the temperature but also upon the nitric acid concentration it is not an easy task to calculate the δ^{15}N value of the nitrogen oxides discharging from absorption columns nor in any other existing isotopic exchange reactions. But all exchange reactions show a general trend, namely that the gas phases (NO,

TABLE 9-5

Some further δ^{15}N values of NH_4^+ and NO_3^- in precipitations

$\delta^{15}N_{NH_4^+}$	$\delta^{15}N_{NO_3^-}$	Location
−9.7	+4.3	Tokyo, Japan, May 1966
−8.3	—	Tokyo, Japan, July 1966
−3.2	—	Tokyo, Japan, September 1966
−0.3; −0.1	−0.2; −0;2	Greppin, G.D.R.
+2.0; +2.0	−0.9; −0.9	Bitterfeld, G.D.R.
+6.2; +6.9	+3.0; +3.0	northern G.D.R.

NO_2) are depleted in ^{15}N content, i.e. the $\delta^{15}N$ values of the NO_x gases are becoming more and more negative depending on the effectiveness of the absorption process. Since the flue gases of nitrogen oxides can also be washed out by rain a further change in the isotope ratio will take place. Depending on the degree of absorption, values in the region of $-140^0/_{00}$ can be obtained, as for example in the area of Bitterfeld, G.D.R., where NO_x with $\delta^{15}N = -114^0/_{00}$ was determined (Maass and Weise, 1981; Weise and Maass, 1981).

Isotopic variations of nitrogen in fertilizers

In the case of farmland, use of nitrogenous fertilizers will have a great influence upon $\delta^{15}N$ values in the pedosphere, and comparisons between agricultural and non-agricultural areas are not infrequently undertaken (Broadbent et al., 1980). The $\delta^{15}N$ values of some fertilizers and chemicals are given in Table 9-6. Without doubt, it is quite difficult to derive a general trend from the $\delta^{15}N$ values given in this table because it would of course be necessary to consider the engineering and technological aspects of the production of fertilizers and chemicals. Depending on the "previous history" of branching in the chemical process stages of the chemistry of nitrogen, a large range of δ values may be observed in fertilizers; δ values ranging from $+2.06^0/_{00}$ to $-15.7^0/_{00}$ for KNO_3 can only be explained in terms of particular processes of production including isotope fractionating stages. There are only few references in literature where such careful analyses of the previous chemical history were made, e.g. by K. Clusius and his coworkers (Clusius and Piesbergen, 1960; Clusius and Poschet, 1962). Conclusive statements are possible, such as in the case of nitrate where very negative $\delta^{15}N$ values indicate absorption processes of nitrogenous gases. As mentioned above, the absorption of NO/NO_2 in special columns of the chemical industry yields nitric acid which can be neutralized by ammonia producing NH_4NO_3 or other nitrates. These nitrates show very variable $\delta^{15}N$ values depending upon the applied absorption process and its effectiveness, or depending upon the site of the column where the nitric acid is stripped. Furthermore the absorption processes can be recycled which causes additional complications in prediction of $\delta^{15}N$ values.

Attempts to give a global $\delta^{15}N$ value for the input of nitrogen through fertilizers, carry a relatively wide range of error because of the different origins involved. From Table 9-6, the following mean values are obtained (without chemicals):

$\delta^{15}N_{NH_4^+} = -0.91 \pm 1.88$ $(n = 39)$

$\delta^{15}N_{NO_3^-} = +2.75 \pm 1.76$ $(n = 28)$

$\delta^{15}N_{urea} = +0.18 \pm 1.27$ $(n = 8)$.

TABLE 9-6

$\delta^{15}N$ values of fertilizers, chemicals and some naturally occurring compounds

Compound	Mean value ± s.d.	Range	N^a	Origin/supplier	Reference
N_2	—	−2.5 to −3.5		superpure gas — pressure cylinder	Central Institute for Isotope and Radiation Research (1978)
N_2	$(-7.3 \pm 8.2)^b$	—		Linde — superpure (99.99%)	Fiedler and Proksch (1975)
N_2	−6.5			ultrapure gas cylinder —	Shearer et al. (1975)
N_2	2.57 ± 0.03	± 0.12 at single measurements		superpure gas cylinder	Zschiesche (1972)
NH_3	+1.4	—			Central Institute for Isotope and Radiation Research (1978)
NH_3	+3.7	—		fertilizer	Kohl et al. (1971)
NH_3	+3.6	—		Farm Bureau, fertilizers	Shearer et al. (1974a)
NH_3	−2.7	—		Matheson, fertilizer	Shearer et al. (1974a)
NH_3 liquid	−0.3	—		Ruhrstickstoff-Veba Chemie	Aly (1975)
NH_3 liquid	+0.9 ± 0.6	+1.3 to +0.5		fertilizer	Létolle and Mariotti (1974)
NH_3 liquid	−6.0			Edwards, fertilizer	Edwards (1973)
NH_3 liquid	+1.6			Farm Bureau, fertilizer	Shearer et al. (1974a)
NH_3 liquid	+0.2			fertilizer	Shearer et al. (1974a)
NH_4Cl	−6.1 ± 7.8	−15 to −2.2 to −1.0	3		Létolle and Mariotti (1974)
NH_4Cl	+11.5			salammoniac, Vesuvius 1868; 23% N	Parwel et al. (1957)
NH_4Cl	+11.0			Salammoniac, Etna 1886; 23% N	Parwel et al. (1957)
NH_4Cl	+13.0			Paracutin volcano (Mexico)	Hoering (1955)
NH_4Cl	+13.0			side crater of a volcano on Kamchatka	Wolinez et al. (1967)

Compound	δ value	Other	n	Source	Reference
NH_4Cl	-5.25 ± 5.25	$+2.0, -5.0, -8.0, -10.0$	4	sublimate on a stream of lava (Kamchatka)	Wolinez et al. (1967)
NH_4Cl	-1.4		1	Mallinckrodt Chemicals	Shearer et al. (1974a)
$Ca(NO_3)_2$	$+5.3$		1	natural salpetre Elnasr. Co. for fertilizers, Suez, Egypt	Létolle and Mariotti (1974)
$Ca(NO_3)_2$	$+1.0, +1.1$	with 15.5% N	2		Aly et al. (1981)
$Ca(NO_3)_2$	$+6.8$	6.2 to 7.3	2	Fisher Chemicals	Shearer et al. (1974a)
$(NH_4)_2SO_4$	11.12 ± 0.25		41	VEB Laborchemie Apolda Charge 020971 (1971)	Drechsler (1976)
$(NH_4)_2SO_4$	$+7.1 \pm 0.4$			VEB Laborchemie Apolda Charge 130780 (1980)	Central Institute for Isotope and Radiation Research (1980)
$(NH_4)_2SO_4$	$+0.26 \pm 1.06$		25	Mallinckrodt Chemicals	Shearer et al. (1974a)
$(NH_4)_2SO_4$	$+1.4 \pm 0.7$			Mallinckrodt Chemicals	Feigin et al. (1974)
$(NH_4)_2SO_4$	-2.71 ± 0.14			Baker and Adamson, reagent grade	Junk and Svec (1958)
$(NH_4)_2SO_4$	$(-20.0 \pm 8.2)^b$			Merck Chemicals, Darmstadt	Perschke et al. (1971)
$(NH_4)_2SO_4$	-2.4	-2.3 to -2.5	2	N-K fertilizer, Gebr. Fuilini Ltd.	Aly (1975)
$(NH_4)_2SO_4$	$+4.8$			fertilizer, formulation 21-0-0	Rennie et al. (1976)
$(NH_4)_2SO_4$	$+1.4 \pm 0.7$				Shearer et al. (1975)
$(NH_4)_2SO_4$	-27.3 ± 3.2	$-25.0, -26.0, -31.0$	3	fumaroles, southern Kamchatka	Wolinez et al. (1967)
$(NH_4)_2SO_4$	-1.1			Agrico, fertilizer	Shearer et al. (1974a)
$(NH_4)_2SO_4$	-1.9			Allied Chem. fertilizer	Shearer et al. (1974a)
$(NH_4)_2SO_4$	$+0.8$			Taylor's Farm, fertilizer	Shearer et al. (1974a)
$(NH_4)_2HPO_4$	$+0.5$				Central Institute for Isotope and Radiation Research (1978)
$(NH_4)_2HPO_4$	± 0			Merck Chemicals, Darmstadt	Shearer et al. (1974a)
$(NH_4)_2HPO_4$	$+1.0$			Agrico, fertilizer	Shearer et al. (1974a)

TABLE 9-6 (continued)

$\delta^{15}N$ values of fertilizers, chemicals and some naturally occurring compounds

Compound	Mean value ± s.d.	Range	N^a	Origin/supplier	Reference
$(NH_4)_2HPO_4$	−0.4			Allied Chem., fertilizer	Shearer et al. (1974a)
$(NH_4)_2HPO_4$	−1.6			Arco Chem., fertilizer	Shearer et al. (1974a)
$(NH_4)_2HPO_4$	+1.0			Cyanimid, fertilizer	Shearer et al. (1974a)
$(NH_4)_2HPO_4$	−1.0			Grace, fertilizer	Shearer et al. (1974a)
$(NH_4)H_2PO_4$	+1.3 ± 1.7			fertilizer, formulation 11-55-0	Rennie et al. (1976)
$(NH_4)_3PO_4$	+1.8				Central Institute for Isotope and Radiation Research (1978)
NH_4-poly-phosphate	−0.4			Allied Chem., fertilizer	Shearer et al. (1974a)
KNO_3	+0.94 ± 4.2	+4.0 to −6.3	5		Létolle and Mariotti (1974)
KNO_3	+2.06 ± 0.16				Aly (1975)
KNO_3	−15.7 ± 0.3			Merck Chemicals, Darmstadt VEB Berlin Chemie, DAB 7	Zschiesche (1972)
KNO_3	+1.13 ± 1.21		19	Mallinckrodt Chemicals, bottle 1	Shearer et al. (1974a)
KNO_3	−7.5 ± 0.72		12	Mallinckrodt Chemicals, bottle 2	Shearer et al. (1974a)
$NaNO_3$	+7.6 ± 4.8	+11.0 to +4.2	2	Mallinckrodt Chemicals	Létolle and Mariotti (1974)
$NaNO_3$	+15.3	+15.1 to +15.4	2	Caliche blanco, Tocopilla (Chile)	Shearer et al. (1974a)
$NaNO_3$	−2.3	with 6% N		Caliche blanco, Tocopilla (Chile)	Parwel et al. (1957)
$NaNO_3$	−5.0	with 4.5% N		Caliche blanco, Tocopilla (Chile)	Parwel et al. (1957)
$NaNO_3$	−1.2	with 1.2% N		Caliche blanco, Tocopilla (Chile)	Parwel et al. (1957)
$NaNO_3$	−1.4			Chilean nitrate	Shearer et al. (1974a)
$NaNO_3$	−5.4	with 1.1% N		Caliche colorado, Tocopilla (Chile)	Parwel et al. (1957)

Compound	$\delta^{15}N$	n	Description	Reference
NaNO$_3$	−22.75		soda nitre, 99.6%; BASF synthetically fixed, University of Illinois	Aly (1975)
NaNO$_3$	−4.7			Shearer et al. (1974a)
NH$_4$NO$_3$	NH$_4^+$: −0.6; NO$_3^-$: 3.6		fertilizer, formulation 34-0-0	Rennie et al. (1976)
NH$_4$NO$_3$	NH$_4^+$: 0.1; NO$_3^-$: 4.8		fertilizer, formulation 34-0-0	Karamanos and Rennie (1980a, b)
NH$_4$NO$_3$	NH$_4^+$: −2.7; NO$_3^-$: 2.0, 2.1		Fertilizer, nitrochalk 26% N, 75% NH$_4$NO$_3$; BASF	Aly (1975)
NH$_4$NO$_3$	NH$_4^+$: −1.5, −4.4			
NH$_4$NO$_3$	NO$_3^-$: 2.5, 1.3			
NH$_4$NO$_3$	NH$_4^+$: −4.9, −4.99; NO$_3^-$: −3.6, −4.1		−22.6 to −22.9	2		Létolle and Mariotti (1974)
NH$_4$NO$_3$	NH$_4^+$: −0.02, −0.20, −0.13	with 31% N	Elnasr Co. for fertilizers, Talkha, Egypt	Aly et al. (1981)
NH$_4$NO$_3$	NO$_3^-$: +0.8, +1.3; NH$_4^+$: −1.1, −1.2; NO$_3^-$: −0.68, −1.30	with 31% N	Elnasr Co. for fertilizers and chemicals, Helwan, Egypt	Aly et al. (1981)
NH$_4$NO$_3$	NH$_4^+$: +1.6 ± 0.2; NO$_3^-$: +5.7 ± 0.4	2	with 35% N	Kema, Aswan, Egypt	Aly et al. (1981)
NH$_4$NO$_3$	NH$_4^+$: −1.2; NO$_3^-$: +2.1	4	fertilizer, nitrochalk — 24% N; Ruhrstickstoff AG, 69% NH$_4$NO$_3$	Aly (1975)
NH$_4$NO$_3$	NH$_4^+$: −2.5; NO$_3^-$: +1.0	1	Ruhrstickstoff (RS) — Veba Chemie	Aly (1975)
NH$_4$NO$_3$	NH$_4^+$: −2.7; NO$_3^-$: +4.4		Agrico, fertilizer	Shearer et al. (1974a)
NH$_4$NO$_3$	NH$_4^+$: −0.7; NO$_3^-$: +1.6		Cyanimid, fertilizer	Shearer et al. (1974a)
NH$_4$NO$_3$	NH$_4^+$: −1.8; NO$_3^-$: +1.6		Grace, fertilizer	Shearer et al. (1974a)
NH$_4$NO$_3$	NH$_4^+$: −0.6; NO$_3^-$: +1.8		Kaiser, fertilizer	Shearer et al. (1974a)
NH$_4$NO$_3$	NH$_4^+$: −4.6; NO$_3^-$: +1.6		Hercules, fertilizer	Shearer et al. (1974a)
NH$_4$NO$_3$			MFA, fertilizer	Shearer et al. (1974a)

TABLE 9-6 (continued)

δ^{15}N values of fertilizers, chemicals and some naturally occurring compounds

Compound	Mean value ± s.d.	Range	N^a	Origin/supplier	Reference
$NH_4NO_3/$ $(NH_4)_2SO_4$	NH_4^+: −3.5 NO_3^-: +2.8			ammonium nitrate sulphate, fertilizer 26% N, Ruhrstickstoff AG, 39% NH_4NO_3, 60% $(NH_4)_2SO_4$	Aly (1975)
$NH_4NO_3/$ $(NH_4)_2SO_4$	NH_4^+: −1.0 NO_3^-: +2.2, 2.4			ammonium nitrate sulphate, fertilizer, 26% N, BASF, 40% NH_4NO_3, 56% $(NH_4)_2SO_4$	Aly (1975)
$NH_4NO_3/$ $(NH_4)_2SO_4$	NH_4^+: −2.4 NO_3^-: +2.4			N-K fertillizer, Ruhrstickstoff AG, 48% NH_4NO_3, 15% $(NH_4)_2SO_4$	Aly (1975)
$NH_4NO_3/$ $(NH_4H_2PO_4$	NH_4^+: −0.1 NO_3^-: +6.2			N-P-K fertilizer, Ruhrstickstoff AG, 38% NH_4NO_3, 9% $(NH_4)H_2PO_4$	Aly (1975)
$NH_4NO_3/$ $(NH_4)_2PO_4$	NH_4^+: −1.9, −2.3 NO_3^-: −0.4, −0.5		2	N-P fertilizer, Ruhrstickstoff AG, 51% NH_4NO_3, 12% $(NH_4)H_2PO_4$	Aly (1975) Aly (1975)
Uran	NH_4^+: −2.5 NO_3^-: +4.9 red. N: +3.3			Allied Chem., fertilizer	Shearer et al. (1974a)
Uran	NH_4^+: −1.5 NO_3^-: +2.1 red. N: −1.8			Edwards, fertilizer	Shearer et al. (1974a)
Uran	NH_4^+: −0.5 NO_3^-: +2.4 red. N: +2.5			Edwards, fertilizer	Shearer et al. (1974a)

Compound	Values		Source	Reference
Uran	NH_4^+: -0.2 NO_3^-: $+2.1$ red. N: $+2.7$		Grace, fertilizer	Shearer et al. (1974a)
Uran	NH_4^+: -0.6 NO_3^-: $+2.6$ red. N: $+3.0$		Hercules, fertilizer	Shearer et al. (1974a)
Uran	NH_4^+: $+1.8$ NO_3^-: $+2.6$ red. N: $+5.9$		Kaiser, fertilizer	Shearer et al. (1974a)
Urea	$+0.9$		Allied Chem., fertilizer	Shearer et al. (1974a)
Urea	$+1.9$		Cyanimid, fertilizer	Shearer et al. (1974a)
Urea	-1.7		Dupont, fertilizer	Shearer et al. (1974a)
Urea	$+4.4 \pm 1.6$	$+2.8$ to $+6.1$	natural	Létolle and Mariotti (1974)
Urea	0.0		synthetically	Létolle and Mariotti (1974)
Urea	$+0.8 \pm 0.2$		fertilizer	Kohl et al. (1973)
Urea	$+1.05 \pm 0.1$	3	VEB Stickstoffwerke Piesteritz (G.D.R.), fertilizer	Faust et al. (1980)
Urea	-2.7		Mallinckrodt Chemicals	Shearer et al. (1974a)
Urea	$-2.7, -2.5$		Abu Keir, Egypt	Aly et al. (1981)
Urea coated	$+0.1$		Grace, fertilizer	Shearer et al. (1974a)
Urea-formaldehyde	-1.6		Dupont	Shearer et al. (1974a)
Ca cyanimide	-0.1		University of Illinois	Shearer et al. (1974a)
Purine	$+23$			Létolle and Mariotti (1974)
Uric acid	$+12.5 \pm 0.1$		Serva Chemicals	Faust et al. (1980)
Hippuric acid	$+2.97 \pm 0.14$	1	Chimexport (U.S.S.R.)	Faust et al. (1980)
Creatinine	-2.77 ± 0.5		Merck Chemicals, Darmstadt	Faust et al. (1980)
Pyridine	$+1.2$			Central Institute for Isotopes and Radiation Research (1978)
α, α'-Dipyridyl	$+1.43 \pm 0.2$	20	Merck Chemicals Darmstadt	Drechsler (1976)
$(CH_3)_2NH$	$+11.6$			Central Institute for Isotopes and Radiation Research (1978)

TABLE 9-6 (continued)

$\delta^{15}N$ values of fertilizers, chemicals and some naturally occurring compounds

Compound	Mean value ± s.d.	Range	N^a	Origin/supplier	Reference
$(CH_3)_2N$	+28				Central Institute for Isotopes and Radiation Research (1978) Zschiesche (1972)
Peptone	+8,75 ± 0.1			Mixture of aminoacids nutrient medium, VEB Berlin Chemie, Charge 281067	
Serulate PK II	+10.61 ± 0.30			VEB Sächsisches Serumwerk, Dresden, after DAB 7 (D.L.)	Faust et al. (1980)
Casein hydrolysate	+4.6		1	Nutritional Biochemicals	Shearer et al. (1974a)
Fish oil	+15.4			Atlas fish meal	Shearer et al. (1974a)
Fish oil	+13.7			Atlas fish meal	Shearer et al. (1974a)
Blood meal	+8.4			Armour	Shearer et al. (1974a)
Blood meal	+9.5			Armour	Shearer et al. (1974a)
Manure	+12.7			Mulch Incorporated	Shearer et al. (1974a)
Liquid manure	NH_4^+: +14.1 NO_3^-: +38.4 red. N: +9.8			Scully	Shearer et al. (1974a)
Manure	NH_4^+: +17.3 NO_3^-: +21.0 red. N: +10.7			Scully (2 month comp.)	Shearer et al. (1974a)
Manure	NH_4^+: +17.3 NO_3^-: +11.1 red. N: +8.0			Scully (1 year comp.)	Shearer et al. (1974a)

[a] N = number of measurements.
[b] Not precision mass-spectrometrically measured.

If we cover only those fertilizer which represent the ammonium nitrate and their mixtures with ammonium sulphate then we obtain $\delta^{15}N_{NH_4^+} = -1.28 \pm 1.70$ ($n = 28$), with about the same standard deviation as the $\delta^{15}N_{NO_3^-}$ value.

Isotope effects in ion-exchange processes in the pedosphere

In a study performed in 1969 and dealing with the transport transformation of nitrogen in the pedosphere it was found that a shift of isotopic composition takes place as a result of an isotopic exchange. This effect could be as high as 4% (Reid et al., 1969). This is astonishing because it was as early as 1955 (Spedding et al., 1955) that nitrogen isotopes were successfully separated with ion exchangers and that, as with any separation process, attempts were made to maximize the separation factor. Using Dowex 50-X12 (Spedding et al., 1955), it was possible to determine α values of 1.0257 ± 0.0002 for ammonium ions, i.e., at most 2.6% of accumulation for each separation stage, with isotopic enrichment in the ion exchanger. Numerous studies of ion-exchange processes proceeding in the pedosphere have since been conducted. In this context, mention should be made of a paper by Delwiche and Steyn in 1970, with additional papers (see Table 9-7) on the subject having been published during recent years (Karamanos and Rennie, 1978).

The α value can also be defined as the quotient of the distribution coefficient, k_d:

$$\alpha = \frac{\overline{C_{^{15}NH_4R}}/\overline{C_{^{14}NH_4R}}}{C_{^{15}NH_4^+}/C_{^{14}NH_4^+}} = \left(\frac{\overline{C_{^{15}NH_4R}}}{C_{^{15}NH_4^+}}\right) : \left(\frac{\overline{C_{^{14}NH_4R}}}{C_{^{14}NH_4^+}}\right) = \frac{^{15}k_d}{^{14}k_d}$$

TABLE 9-7

α Values for ion-exchange processes

Ion exchanger	α Value[a]	Reference
Dowex 50-X12	NH_4^+: 1.0257 ± 0.0002	Spedding et al. (1955)
Dowex 50	NH_4^+: 1.00074 ± 0.00016	Delwiche and Steyn (1970)
Clay (kaolinite)	NH_4^+: 1.00078 ± 0.0003	Delwiche and Steyn (1970)
Clay (NH_4^+-saturated)	NH_4^+: 1.003	Karamanos and Rennie (1978)
Clay (K^+-saturated)	NH_4^+: 1.007	Karamanos and Rennie (1978)
Clay (Ca^{2+}-saturated)	NH_4^+: 1.001	Karamanos and Rennie (1978)
Dowex 1	NO_3^-: $1/\alpha = 1.0021 \pm 0.0005$	Delwiche and Steyn (1970)
Lateritic soil (oxisol, Australia)	NO_3^-: 'α' = 1.0047 (calculated) (between top and 6 m depth)	Black and Waring (1977, 1979)

[a] The α value is here defined so that ^{15}N enrichment will take place in the ion exchanger, with α thus being greater than 1. For α values lower than 1, accumulation will be in solution, as with NO_3^-.

since $k_d = \overline{C_{NH_4R}}/C_{NH_4^+}$. In this relations $C^{14}NH_4^+$ and $C^{15}NH_4^+$ are the concentrations of $^{14}NH_4^+$ and $^{15}NH_4^+$ in solution, respectively, and $\overline{C^{14}NH_4R}$ and $\overline{C^{15}NH_4R}$ are their concentrations in the solid (adsorbed) phase, where R is the negative charged rest of an polymer inorganic anion (e.g., silicates or clay particles). The concentrations are expressed in mol g^{-1}, mol mol^{-1} or meq 100 g^{-1} in order to introduce the cation exchange capacity (CEC) which is normally expressed in that way. This yields:

$$\overline{C_{14NH_4R}} + \overline{C_{15NH_4R}} = CEC$$

$$C_{14NH_4^+} + C_{15NH_4^+} = C_{NH_4^+}$$

and

$$\alpha = {}^{15}k_d \cdot \frac{C_{14NH_4^+}}{\overline{C_{14NH_4R}}} = {}^{15}k_d \cdot \frac{(C_{NH_4^+} - C_{15NH_4^+})}{(CEC - \overline{C_{15NH_4R}})}$$

or since $C_{15NH_4^+}$ and also $\overline{C_{15NH_4R}}$ are small as compared with the concentrations and the CEC respectively:

$$\alpha = {}^{15}k_d \cdot \frac{C_{NH_4^+}}{CEC}$$

Another transformation yields:

$$10^3 \ln \alpha = \delta^{15}N_{exch.} - \delta^{15}N_{sol.}$$

$$= 10^3 (\ln {}^{15}k_d + \ln C_{NH_4^+} - \ln CEC)$$

Since α can be nearly equal to 1.007, we have:

$$\delta^{15}N_{exch.} = \delta^{15}N_{sol.} + 7 + 10^3 (\ln {}^{15}k_d + \ln C_{NH_4^+} + \ln CEC)$$

This equation implies that the $\delta^{15}N$ values of NH_4^+ are dependent upon the concentration $C_{NH_4^+}$, the CEC, and the distribution coefficient, which may vary with the depth of the soil horizon. In failing of suitable values for $\delta^{15}N_{NH_4^+}$ and NH_4-N in solution an attempt shows for a forest soil the following correlation

$$\delta^{15}N\text{-Kjeldahl} = -4.44 \ln (\text{ppm } N_{total}) + 34.66$$

with $r = 0.9221$, applying measured values from Broadbent et al. (1980). Another soil (Rennie et al., 1976), namely chernozem, yielded:

$$\delta^{15}N = 0.122 \ln (\text{ppm N}) + 5.67$$

with $r = 0.0977$, i.e. no significant trend between $\delta^{15}N$ and the N content of the soil. However, in the same work a luvisol soil yielded:

$$\delta^{15}N = -1.18 \ln (\text{ppm N}) + 13.64$$

with $r = -0.8358$.

Isotope effects involved in biological processes

For the assimilation of nitrogen in species-specific nitrogen compounds, NH_4^+ can be taken up either directly or after prior reduction of higher-oxidized nitrogen compounds to NH_4^+, since, with few exceptions, nitrogen is present in organisms in its most strongly reduced form, namely, -3 valent ammonium. A review about the various forms of nitrogen uptake and metabolism of higher plants has been recently published by Haynes and Goh, 1978.

A special form of nitrogen assimilation is N_2 fixation, i.e., the ability of certain organisms (N_2 fixers) to take up dinitrogen, N_2, directly and convert it to species-specific nitrogen compounds. However, if dissimilation is understood to be the opposite of assimilation (incorporation) of nitrogen compounds, then this term can be used for the nitrogen metabolism, i.e., for the non-incorporation of nitrogen compounds in organisms. In this connection, it is necessary to distinguish between two cases:

— The nitrogen compound serves, in redox reactions, as a supplier of energy → nitrifying bacteria.

— The nitrogen compound serves as a electron acceptor in the oxidation of organic compounds → denitrifying bacteria.

These relations are schematically represented in Fig. 9-4.

Isotope effects involved in assimilation

Isotope effects of nitrogen resulting from assimilation are summarized in Table 9-8. For comparison, isotope effects determined during chemical reduction of nitrogen compounds have also been included.

Of the first four micro-organisms given in Table 9-8, $\overline{\beta} = 1.0052 \pm 0.0065$ is calculated as the mean which, in the model presented by Shearer et al. (1974b) has been taken as a value of isotope fractionation. Assimilation isotope effects of higher plants are still under discussion or as it has been recently expressed by Karamanos and Rennie (1980a, p. 59), "... while further experimentation will be necessary before a clear conclusion can be drawn concerning discrimination or lack of the N isotopes during plant uptake, it can be concluded from the evidence presented thus far that plant uptake does not significantly alter the isotope composition of the fertilizer or inorganic soil N remaining in the soil at harvest time. Rather, it does appear that the $\delta_a{}^{15}N$ values of plant material reflect changes in the isotopic composition of the soil N and in particular the fertilizer N pool."

It is at present not possible to give a representative value for the isotope effect of assimilation, since a number of factors (such as exposure to light, concentration of the nitrogen compound, phase of growth of micro-organisms, ventilation of cultures, agitation or quiescence of cultures, etc.) tend to have an influence upon the values determined. An example is the dependence of the isotope effect upon the rate of growth, μ, with:

$$\mu = \frac{dN}{dt} \cdot \frac{1}{N} \quad \text{or} \quad N_t = N_0 \cdot e^{\mu t}$$

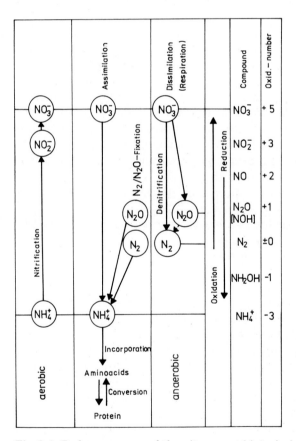

Fig. 9-4. Redox processes of the nitrogen at biological conversions (N-metabolism).

TABLE 9-8

Isotope fractionations of assimilation in micro-organisms

	Temperature (°C)	$\beta = {}^{14}k/{}^{15}k$	Reference
1. Assimilation of NH_4^+:			
Azotobacter vinelandii	30	1.0148	Delwiche and Steyn (1970)
Hansenula californica	36	1.0031	Delwiche and Steyn (1970)
Pichia terricola	36	1.0018	Delwiche and Steyn (1970)
Schwanniomyces alluvius	36	1.0011	Delwiche and Steyn (1970)
Chaetoceros sp.	23 ± 1	0.9903	Wada and Hattori (1978)
(~3000 lux)		0.9947	
2. Assimilation of NO_2^-:			
Phaeodactylum tricornutum	20 ± 1	1.0007	Wada and Hattori (1978)
(~1000 lux)			
Reduction of $NO_2^- \to NH_3$	110	1.029 ± 0.003	Brown and Drury (1967)
	25	1.035 ± 0.002	
3. Assimilation of NO_3^-:			
Chaetoceros sp.	23 ± 1	1.0009	Wada and Hattori (1978)
		1.0045	Wada and Hattori (1978)
Phaeodactylum tricornutum	23 ± 1	1.0076	Wada and Hattori (1978)
(~3000 lux)	23 ± 1	1.013	
	23 ± 1	1.014	
	23 ± 1	1.016	
Phytoplankton (44°N, 154°E)	2–8	1.005	Wada and Hattori (1978), Wada et al. (1975)
Reduction of $NO_3^- \to NH_3$	100	1.057 ± 0.004	Brown and Drury (1967)
	25	1.075 ± 0.004	

TABLE 9-9
Isotopic fractionations in assimilation of higher plants

Order/Family	Plant	e'	$\overline{e'}$
CLASS I: *Dicotyledoneae* ($\overline{e'} = -0.25 \pm 0.65$; $n = 38$)			
Subclass 1: *Archichlamydeae* ($\overline{e'} = -0.27 \pm 0.61$; $n = 20$)			
Urticales			
Urticaceae	*Urtica urens* L. (stinging nettle)	−0.1	
Polygonales			
Polygonaceae	*Rumex acetosa* L. (sorrel)	+0.3	
Centrospermae			
Caryophyllaceae	*Gypsophila* sp.	−0.7	
Papaverales			
Papaveraceae	*Eschscholtzia california* Scham.	−0.2	
Cruciferae (*Brassicaceae*)	*Raphanus sativus* L. (large radish)	−0.7	
	Brassica oleavaceae L. (cabbage)	0.0	−0.2 ± 0.4
	Brassica napus L. (rape)	+0.1	
Geraniales			
Tropaeolaceae	*Tropaeolum majus* L. (nasturtium)	−0.3	
Linaceae	*Linum usitatissimum* L. (linseed)	−2.2	−0.9 ± 1.1
Euphorbiaceae	*Ricinus communis* L. (castor-oil plant)	−0.2	
Myrtiflorae (*Myrtales*)			
Onagraceae (*Oenotheraceae*)	*Clarkia elegans* Dougl. (fuchsia)	+0.2	
Cucurbitales			
Cucurbitaceae	*Citrullus vulgaris* Schrader (watermelon)	−0.5	
	Cucumis melo L. (melon)	−0.1	
	Cucumis sativus L. (cucumber)	+0.1	+0.02 ± 0.34
	Cucumis sativus L. (gherkin)	+0.4	
	Cucumis pepo L. (pumpkin)	+0.2	
Umbelliflorae			
Umbelliferae (*Apiaceae*)	*Petroselinum sativum* Hoffm. (parsley)	−0.8	
	Apium graveolens L. (celery)	+0.3	−0.125 ± 0.5
	Cerefolium sativum Bess. (chervil)	−0.1	
	Daucus carota L. (carrot)	−0.1	

Subclass 2: *Sympetalae* ($\overline{\epsilon'} = -0.16 \pm 0.84; n = 12$)

Tubiflorae
    Polemoniaceae   *Cobaea scandeus* Cav.   −0.1
    Borraginaceae   *Myosotis alpestris* Sch. (forget-me-not)   −0.1
    Labiatae (Lamiaceae)   *Mentha piperata* L. (peppermint)   0.0
    Solanaceae   *Solanum lycopersicum* L. (tomato)   +0.5

Disacales
    Valerianaceae   *Valeriana* sp. (valerian)   +0.3

Campanutales
    Compositae (Asteraceae)   *Lactuca sativa* L. (garden lettuce)   −1.7
        Tagetes patula L. (French marigold)   −0.5
        Helianthus annuus L. (sunflower)   +0.3   −0.6 ± 1.1
        Dahlia sp. (dahlia)   +0.6
        Aster sp. (aster)   −1.7

CLASS II: *Monocotyledoneae* ($\overline{\epsilon'} = -0.38 \pm 0.32; n = 6$)

Lilliflorae
    Lilliaceae   *Allium porrum* L. (leek)   −0.4

Graminales (Glumiflorae)
    Gramineae (Poaceae)   *Avena sativa* L. (oats)   −0.7   } C_3
        Hordeum vulgare L. (barley)   −0.8   −0.75
        Zea mays L. (maize)   −0.1
        Pennisetum americanum K. Schum.   0.0   } C_4 −0.13 ± 0.15
        Sorghum vulgare Pers.   −0.3

were N is the number of cells per volume. For example, Wada and Hattori (1978) found that in the case of *Phaeodactylum tricornutum*, a marine diatom, the isotope effect may decrease from 1.015 to 1.00075 at the lowest rate of growth ($\sim 0.03\ d^{-1}$) and at a rate of growth, μ, of $1.0\ d^{-1}$, respectively. This has been determined for a 10^{-2} molar NO_3-N and at a temperature of $(22 \pm 1)°C$.

For a study of isotope effects of assimilation in higher plants it is necessary to use an experimental procedure that is different from that used for micro-organisms. Depending on the duration of the growing season, isotopically constant nutrient salts have to be added until an amount of vegetal material has grown which is sufficient for isotopic analysis. Extensive research in this field has been done by Mariotti et al. (1980) who examined a total of 42 different species of plants, of which 38 were non-legumes (Table 9-9) and four, leguminous plants. The isotopically constant NO_3^- solution supplied had a $\delta^{15}N_{NO_3^-} = 3.5\text{\textperthousand}$, and the data of the above authors were given in ϵ values which were defined as $\epsilon \simeq (\delta^{15}N_{plant} - \delta^{15}N_{nutrient}) \times 10^{-3}$. Introducing the ϵ value which is defined as:

$$\epsilon = \frac{1}{\beta} - 1$$

where $\beta = {}^{14}k/{}^{15}k$ is the kinetic isotope effect expressed by the corresponding rate constants ${}^{14}k$ and ${}^{15}k$, we will define the ϵ values in Table 9-9 by the relation:

$$\epsilon' = 10^3\ \epsilon$$

and this ϵ' value should be considered when data from Mariotti et al. (1980) are directly applied. Furthermore the relation:

$$\beta = \frac{{}^{14}k}{{}^{15}k} = \frac{1}{10^{-3}\ \epsilon' + 1} = \frac{1}{(\delta^{15}N_{plant} - \delta^{15}N_{nutrient}) \times 10^{-3} + 1}$$

is valid evaluating the desired β values for comparison with each other. From the data given in Table 9-9 a total β value can thus be calculated for all 38 plant species:

$$\overline{\epsilon'} = -0.25 \pm 0.65 = 10^3\ \epsilon = \left(\frac{1}{\beta} - 1\right) \times 10^3$$

or:

$$\overline{\beta} = 1.00025 \pm 0.00065$$

which is an even smaller isotope effect than that ($\overline{\beta} = 1.0052 \pm 0.006$) which was observed in the case of micro-organisms.

Comparing the isotope fractionation factors β for the uptake of NO_3^- into non-nodulating soybeans (*Glycine max* L. Merrill, variety Harosoy) with ryegrass (*Lolium perenne* L.) and marigold (*Tagetes erecta* L.), Kohl and Shearer (1980) have found almost the same β values:

Glycine max L. Merrille var. Harosoy (soybean):	1.0049 ± 0.0010
Lolium perenne L. (ryegrass):	1.0047 ± 0.0018
Tagetes erecta L. (marigold):	1.0045 ± 0.0011

Shearer et al. (1980) described a remarkable difference in the $\delta^{15}N$ values between the nodules and the whole plant of soybeans. Differences of up to 10—12 δ units can be reached. In another study recently published Kohl et al. (1982), have found that the ^{15}N enrichment of the nodules ($\Delta\delta^{15}N = \delta^{15}N_{nodules} - \delta^{15}N_{whole\ plant}$) was highly correlated ($r = 0.985$, $p < 0.001$) with N_2-fixing efficiency, calculated as the quantity of N_2 fixed per quantity of nitrogen in the nodules.

Lichens (crustate and foliose lichens) obtained from the Soviet Antarctic Station at Novolasarevskaya gave $\delta^{15}N$ values of from 2.0—2.7‰ (Central Institute for Isotope and Radiation Research, 1978) depending upon the particular location. Mosses showed similar values ranging from 2.6 to 3.2‰.

The classic work of Parwel et al. (1957) mentions data on spruce (*Picea abies*, near Ekolsund) where $\delta^{15}N = -3.8$‰ and birch (*Betula alba*, same location) with $\delta^{15}N = -2.2$‰ for the wood and $\delta^{15}N = -5.0$‰ for the bark.

The following conclusions can be drawn:

— A taxonomic assignment of the ϵ' values (i.e., isotope fractionation) is not possible, although there are minor differences.

— A possible different isotope fractionation with respect to nitrogen in C_3 and C_4 plants is not recognizable, although there are distinct differences in the degree of utilization of nitrogen (production of biomass relative to the same supply of nitrogen) (Brown, 1978). C_4 plants are characterized by a higher rate of utilization.

— Systematic studies of lower plants (lichens, mosses, and ferns) are necessary to form a more comprehensive picture of isotope effects of nitrogen assimilation, which could also give a new insight into the $\delta^{15}N$ values in coals (Parwel et al., 1957):

$\delta^{15}N = -0.9 \pm 1.2$‰ Carboniferous ($\sim 300 \times 10^6$ a); chiefly pteridophytes

$\delta^{15}N = 2.8 \pm 0.7$‰ Lower Cretaceous ($\sim 120 \times 10^6$ a); chiefly gymnosperms

Isotope effects involved in N_2 fixation

As early as 1960 Hoering and Ford published careful measurements of

TABLE 9-10
Isotope fractionation during N_2 fixation

Species	Temperature[a] (°C)	α[b]	Reference
Azotobacter agile	RT	1.0015 ± 0.0015	Hoering and Ford (1960)
Azotobacter chroococum	RT	1.00037 ± 0.001	Hoering and Ford (1960)
Azotobacter indicum	RT	0.9963 ± 0.003	Hoering and Ford (1960)
Azotobacter vinelandii	RT	1.0022 ± 0.0012	Hoering and Ford (1960)
Azotobacter vinelandii	—	1.0039 ± 0.0008	Delwiche and Steyn (1970)
No data	22—27	1.0003	Wada et al. (1975)
Anabaena cylindrica	—	0.997—0.999	Wada et al. (1975)
Leguminosae:			
Medicago sativa L., var. Mireille (alfalfa)		1.00092	Mariotti et al. (1980)
Trifolium pratense L., var. Alpilles (clover)		1.00088	Mariotti et al. (1980)
Trifolium pratense L. (clover)		0.99812 ± 0.00014	Kohl and Shearer (1980)
Lupinus luteus L., var. Sulfa		1.0007	Mariotti and Shearer (1980)
Lupinus luteus L., var. Sulfa		1.0009	Amarger et al. (1977)
Phaseolus vulgaris L., var. Contender (kidney bean)		1.00197 ± 0.00006	Mariotti et al. (1980)
Phaseolus vulgaris L., var. Hodgson (kidney bean)		1.00183 ± 0.00006	Mariotti et al. (1980)
Vicia faba L., var. Ascott (broad bean)		1.00063 ± 0.00015	Mariotti et al. (1980)
Pisum sativum L., var. Rondo (pea)		1.0010 ± 0.0001	Mariotti et al. (1980)
Glycine max L. Merrill, var. Harosoy (soybean)		0.99902 ± 0.00018	Kohl and Shearer (1980)
Glycine max L. Merrill, var. Hodgson (soybean)		1.0016	Mariotti et al. (1980)
Glycine max L. Merrill, var. Amsoy (soybean)		1.0012	Mariotti et al. (1980)

Glycine max L. Merrill, var. Chippewa (soybean)		1.0015	Mariotti et al. (1980)
Glycine max L. Merrill, var. Wells (soybean)		1.0013	Mariotti et al. (1980)
Soybean with Rhizobium japonicum		0.9986	Amarger et al. (1979)
Soybean (symbiot.)		1.0030	Bardin et al. (1977)
Reduction/rearrangement:			
azobenzene — reduction	RT	0.9994 ± 0.0026	Hoering and Ford (1960)
hydrazobenzene — rearrangement	O	1.0203 ± 0.0007	Shine et al. (1977)

[a] RT = room temperature.
[b] $\alpha = R_{\text{atmospheric nitrogen}}/R_{\text{fixed nitrogen}}$ according to Delwiche and Steyn (1970). In general the isotope fractionation is defined as difference between product and source, i.e. $\delta^{15}N$ content in plants and $\delta^{15}N$ content in the source N (e.g. according to Kohl and Shearer, 1980).

isotope effects during N_2 fixation and attempted to interpret the values obtained by comparison with the reduction of azobenzene (Hoering and Ford, 1960). Calculation of the total average of the 23 α values given in Table 9-10 yields:

$$\overline{\alpha_{N_2 \text{ fixation}}} = 1.00072 \pm 0.00177 \; (= R_{\text{air}}/R_{\text{fixed}})$$

This thermodynamic isotope effect between the air nitrogen (infinite supply) and the nitrogen fixed in micro-organisms or plants can be understood in the following way. Before the actual reduction of the dinitrogen molecule takes place the N_2 must be dissolved in the aqueous phase of the cells (in the cytoplasm). This first step is accompanied with an isotope effect α. The equation can be written as:

$$(N_2)_{\text{gas}} \overset{\alpha}{\rightleftharpoons} (N_2)_{\text{sol.}} + 6 \, \text{Fd}_{\text{red.}} (6 \, e^-) + n \, \text{ATP} + 8\text{H}^+ \xrightarrow[\text{Mg}^{2+}]{k}$$

$$\xrightarrow[\text{Mg}^{2+}]{k} 2\text{NH}_4^+ + 6 \, \text{Fd}_{\text{oxid.}} + n \, \text{ADP} + n \, \text{P}_i$$

where Fd = Ferredoxin, $\text{Fd}_{\text{red.}}$ = Ferredoxin in its reduced form, $\text{Fd}_{\text{oxid.}}$ = Ferredoxin in its oxidized form, ATP = adenosine triphosphate, ADP = adenosine diphosphate, P_i = inorganic phosphate anion; $n \geqslant 6$. But the α value for the solubility of N_2 is defined as:

$$\alpha_{\text{sol.}} = R_{\text{sol.}}/R_{\text{gas}}$$

and shows the expected direction namely the higher ^{15}N content in the liquid phase (normal isotope effect according to Waldmann's rule). The solubility isotope effect of dinitrogen obeys the temperature dependence equation:

$$\ln \alpha_{\text{sol.}} = -1.319 \times 10^{-3} + 0.586/T$$

in a temperature range of 273 K $< T <$ 373 K (Harting, 1981). For the temperatures concerned we have $\alpha_{20°C} = 1.00068$ and $\alpha_{25°C} = 1.00065$.

The next steps are the transformation of the dissolved N_2 into diimide as an intermediate product and into ammonia, ammonium ions and amino acids which are not accompanied with remarkable kinetic isotope effects of k (Rummel et al., 1976; 1981; Rummel, 1979). One can understand this observation if we accept a more or less synchronous mechanism of single steps:

$$\begin{pmatrix} N \\ \mathop{\mathrm{I\!I\!I}} \\ N \end{pmatrix}_{\text{air}} \xrightarrow{\alpha_{\text{sol.}}} \begin{pmatrix} N \\ \mathop{\mathrm{I\!I\!I}} \\ N \end{pmatrix}_{\text{sol.}} \xrightarrow{\beta_1} \begin{pmatrix} \text{NH} \\ \| \\ \text{NH} \end{pmatrix} \xrightarrow{\beta_2} \begin{matrix} \text{NH}_2 \\ | \\ \text{NH}_2 \end{matrix} \xrightarrow{\beta_3} \begin{matrix} \text{NH}_3 \\ \\ \text{NH}_3 \end{matrix} \xrightarrow{} \ldots \xrightarrow{\beta_4} \text{(proteins)}$$

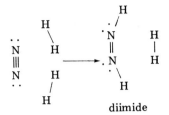
diimide

An estimation of the overall α value yields:

$$\alpha_{N_2 \text{ fixed}} = \alpha_{\text{sol.}}^{-1} \cdot \beta_1 \cdot \beta_2 \cdot \beta_3 \cdot \beta_4 \approx 1.001$$

with $\beta_1 = \beta_2 = 1.0$ (electron transfer reactions) and $\beta_5 = 1.002$ (bond breaking reaction).

This diimide is a well discussed intermediate (Winter and Burris, 1976; Nikonova et al., 1980) which can also suffer disproportionation reactions likewise accompanied with isotope effects. More precise assumptions about the transition state cannot be made since the detailed mechanism of the nitrogenase reaction is at present under way (Burns and Hardy, 1975; Mortenson and Thorneley, 1979).

In a recent paper, Delwiche et al. (1979) gave some results on the N_2 fixation by clover and compared it with that of grass. Converting the given atom % excess values* into $\delta^{15}N$ values one finds that the mean value of clover is $\overline{\delta^{15}N_{\text{clover}}} = 0.7\%_0$ and that of grass is $\overline{\delta^{15}N_{\text{grass}}} = 3.6\%_0$. They found that the $\overline{\delta^{15}N_{\text{clover}}}$ value is close to the a value of Kohl and Shearer (1980), i.e. $\alpha = 0.9993$. The other published data cannot easily be converted into $\delta^{15}N$ values because they are given as differences between soil N and 54 northern California plant species expressed also in atom % excess. Therefore, no isotope effects can be determined or have been considered within the framework of the commonly employed measurement technique.

Isotope effects of N_2 fixation and assimilation are essential to a determination of the proportions of nitrogen uptake through N_2 fixation (symbiotic N outputs), through the soil or a nitrogenous fertilizer unless a ^{15}N tracer technique is employed. Briefly, the principle is as follows:

$$\delta^{15}N_{\text{total}} = x\,(\delta^{15}N_{\text{air}} + \epsilon_1) + (1-x)\,(\delta^{15}N_F + \epsilon_2)$$

where $\delta^{15}N_{\text{total}} = \delta^{15}N$ value of the plant, stalk, fruit, etc.; $\delta^{15}N_F = \delta^{15}N$ value of the soil, fertilizer; $\epsilon_1 = (\alpha_1 - 1) \times 10^3$ of N_2 fixation; $\epsilon_2 = (\alpha_2 - 1) \times 10^3$

---

* x atom % excess is defined as 0.3663 atom % ^{15}N (= air) + x, mostly multiplied by 10^4.

of N assimilation; $\delta^{15}N_{air} = 0$ per definitionem; and x = proportion as a result of N_2 fixation.

By solving this relation for x, one obtains:

$$x = \frac{\delta^{15}N_{total} - \delta^{15}N_F - \epsilon_2}{\epsilon_1 - \epsilon_2 - \delta^{15}N_F}$$

This method was tested by Amarger et al. (1979; in comparison with the acetylene reduction method and the total nitrogen balance) and yielded promising results; this is illustrated by a numerical example from that work on the Amsoy variety of soybean:

Hothouse experiments yielded $\epsilon_1 = -1.2 \pm 0.2$ and field experiments, $\delta^{15}N_F = -4.6$. The plants grew up without prior inoculation with *Rhizobium japonicum*. For the assimilation, it was determined that $\epsilon_2 = 0$.

With no addition of fertilizer at the time of harvest

$$x = \frac{(3.1 \pm 0.8) - (4.6 \pm 0.4)}{(-1.2 \pm 0.2) - (4.6 \pm 0.4)} \times 100 = 25.8\%$$

With an addition of 80 kg of fertilizer per hectare at the time of harvest $x = 8.7\%$.

Slightly different values were obtained for other varieties of soybean, for the Chippewa variety $x = 28.12\%$ (without applying a fertilizer) and $x = 17.57\%$ (by applying the same amount of the fertilizer, 80 kg per hectare). Such figures can only be obtained using ^{15}N-labelled fertilizer. The general formula for evaluating the amount of a compound in a given mixture of compounds is the isotope dilution equation derived from the isotope balance equation:

$$xC_a + yC_b = (x + y)C_m$$

where x and y are amounts of a compound with the ^{15}N abundance of C_a and C_b, respectively, expressed in atom %. If C_a is the natural ^{15}N abundance (0.366 atom % ^{15}N) and C_b is the corresponding ^{15}N abundance of the labelled compound then we are able to determine the amount x without quantitative separation of the substance for analytical determination.

$$x\,(\%) = y\,\frac{C_b - C_m}{C_m - C_a} \times 100$$

or, by applying the excess values subtracting the natural abundance C_a we will obtain

$$y(C_b - C_a) = (x + y)(C_m - C_a)$$

$$yC'_b = (x + y)C'_m.$$

The quotient C'_m/C'_b is the relative abundance related to the applied ^{15}N labelling of the fertilizer. This quotient is named the nitrogen derived from the fertilizer (abbreviated Ndff) and is well-known in the literature concerning agriculture. Rearranging the last relation we find:

$$x \frac{C'_m}{C'_b} = y \left(1 - \frac{C'_m}{C'_b}\right)$$

or:

$$x = y \frac{(100 - \% \text{ Ndff})}{\% \text{ Ndff}}.$$

This x value is now called the A value and means the amount of plant-available nitrogen in the soil (kg ha^{-1}) and y is often called the B value and is the amount of fertilizer nitrogen applied (kg ha^{-1}). This equation

$$A = B \frac{1-y}{y} = B \frac{1-\text{Ndff}}{\text{Ndff}}$$

was derived by Fried and Dean in 1953 and since that time it has been applied many times (Hauck and Bremner, 1976; Filipović, 1980). Compared with ^{15}N-labelled tests performed under field conditions, where for beans and oats the proportion of N_2 fixation can be determined indirectly only (Fried and Broeshart, 1975), the determination by measuring the $\delta^{15}N$ values is a remarkably simple method (Kohl et al., 1979, 1980; Shearer et al., 1980).

Isotope effects during nitrification
Strictly speaking, nitrification involves two partial reactions, with the differences being indicated through different terms, namely:

nitrification $NH_4^+ \rightarrow NO_2^-$

for example, by species of *Nitrosomonas*; or:

nitrification $NO_2^- \rightarrow NO_3^-$

for example, by species of *Nitrobacter*. These two biological oxidation reactions proceed separately in the pedosphere.

The isotope effects are known for the reaction $NH_4^+ \to NO_2^-$ with *Nitrosomonas europaea* (Delwiche and Steyn, 1970; Freyer and Aly, 1975) and for the overall reaction $NH_4^+ \to NO_3^-$ in cultivated soils and para-brown earth (Aly, 1975), but not for the intermediate reaction $NO_2^- \to NO_3^-$, with the schematic representation being shown below:

$$NH_4^+ \xrightarrow{\beta_1} NO_2^- \xrightarrow{\beta_2} NO_3^-$$
$$\xrightarrow{\beta_3}$$

where β_3 must be equal to $\beta_1 \times \beta_2$.

The isotope effect of the intermediate stage, β_2, was calculated according to the assumptions made within the scope of the theory of kinetic isotope effects (Hübner, 1980). In the frame of the theory of kinetic isotope effects we will take the following route to calculate the β_2 (Bigeleisen and Wolfsberg, 1958; Collins and Bowman, 1970; Melander and Saunders, 1980). The basic equation for the partition function is:

$$\frac{\sigma_2}{\sigma_1} f = \frac{Q_2}{Q_1} = f(A) = \prod_i^N \frac{u_{2i}}{u_{1i}} \cdot \exp\left(\frac{u_{1i} - u_{2i}}{2}\right) \cdot \frac{1 - e^{-u_{1i}}}{1 - e^{-u_{2i}}}$$
$$= \text{VP} \times \text{ZPE} \times \text{EXC}$$

where VP = vibrational frequency product; ZPE = zero point energy; EXC = excitation factor; σ_2, σ_1 are the symmetry numbers; and u_{1i}, $u_{2i} = \omega_i hc/kT = 1.4388\, \omega_i/T$ with the wave number ω_i in cm^{-1} and T in K. The other constants are h = Planck's constant, k = Boltzmann's constant and c = light velocity. The product is over the 3 N-6 (3 N-5 for linear molecules) normal vibrational frequencies of the N-atomic molecule and β can be calculated by the equation:

$$\beta = \frac{k_1}{k_2} = \frac{f(A)}{f(A^\ddagger)} \cdot \frac{\nu_{1L}^\ddagger}{\nu_{2L}^\ddagger}$$

where the symbol "$\ddagger$" denotes the transition state of the reaction concerned. The ratio $\nu_{1L}^\ddagger/\nu_{2L}^\ddagger$, the temperature-independent factor, TIF, is treated in the manner of a Slater coordinate and we have:

$$\nu_{1L}^\ddagger / \nu_{2L}^\ddagger = (\mu_2/\mu_1)^{1/2}$$

with μ_1, μ_2 the reduced masses. For example of the transition state $[N \ldots O]^\ddagger$ we obtain:

$$\mu_1^\ddagger = \frac{1}{14} + \frac{1}{16} \quad \text{and} \quad \mu_2^\ddagger = \frac{1}{15} + \frac{1}{16}$$

$(\mu_2^{\ddagger}/\mu^{\ddagger})^{1/2} = 1.01827.$

For the transition state $[O_2N \ldots O]^{\ddagger}$ we obtain:

$(\mu_2^{\ddagger}/\mu_1^{\ddagger})^{1/2} = 1.002756.$

For calculating the β value there are different approximations possible:
— Case 1: $A^{\ddagger} \approx A$, i.e. the structure of the transition state $A^{\ddagger}$ is very similar to the structure of the educt A:

$$\beta = \frac{Q(\nu_{1L}^{\ddagger})}{Q(\nu_{2L}^{\ddagger})} \left(\frac{\mu_2^{\ddagger}}{\mu_1^{\ddagger}}\right)^{1/2} = 1.07447$$

with the accepted transition state $[N \ldots O]^{\ddagger}$ and

$\beta = 1.05810$

with the accepted transition state $[O_2N \ldots O]^{\ddagger}$.
— Case 2: High-temperature approximation $f(A)/f(A^{\ddagger}) \approx 1$:

$\beta = (\mu_2^{\ddagger}/\mu_1^{\ddagger})^{1/2} = 1.01827 \quad [N \ldots O]^{\ddagger}$

$\beta = (\mu_2^{\ddagger}/\mu_1^{\ddagger})^{1/2} = 1.002756 \quad [O_2N \ldots O]^{\ddagger}$

— Case 3: $f(A^{\ddagger}) \approx 1$:

$\beta = f(A) \cdot (\mu_2^{\ddagger}/\mu_1^{\ddagger})^{1/2} = 1.1045 \qquad [N \ldots O]^{\ddagger}$

$\beta = f(A) \cdot (\mu_2^{\ddagger}/\mu_1^{\ddagger})^{1/2} = 1.08768 \qquad [O_2N \ldots O]^{\ddagger}$

— Case 4: $A^{\ddagger} \approx P$, i.e. the structure of the transition state $A^{\ddagger}$ is very similar to the structure of the product P:

$$\beta = \frac{f(NO_2^-)}{f(NO_3^-)} \cdot \left(\frac{\mu_2^{\ddagger}}{\mu_1^{\ddagger}}\right)^{1/2} = 0.95981 \qquad [O_2N \ldots O]^{\ddagger}$$

As we can see, only in case 4 will we find inverse isotope effects $\beta < 1$. But all results are approximations and to find the exact value for β_2 we have to calculate the partition function for the transition state. We can see the coherence of this statement in the following way:

$[O_2N \ldots]^{\ddagger}$		$[O_2N \ldots O]^{\ddagger}$		$[O_2NO]^{\ddagger}$
1.018	>	β	>	0.9598
case 2		"real" structure		case 4

In Fig. 9-5 we can see the normal vibrations of the NO_2^- and NO_3^- anions and in the following compilation are the corresponding normal frequencies according to Monse et al. (1969); $\Delta\omega_i = \omega_{1i}(^{14}N) - \omega_{2i}(^{15}N)$

NO_2^-:

		ω_1	ω_2	ω_3
	cm^{-1}	1233.15	810.77	1324.75
	$\Delta\omega_i$	25.68	8.15	17.85

NO_3^-:

		ω_1	ω_2	$2 \times \omega_3$	$2 \times \omega_4$
	cm^{-1}	1049.46	830.98	717.05	1375.90
	$\Delta\omega_i$	0	21.66	2.15	31.83

For the structure of the transition state we have assumed that the γ-frequency (out-of-plane) ω_2 of the NO_3^- and one of the degenerate deformation vibration ω_3 are becoming imaginary frequencies. Under this condition we have calculated the partition functions and β_2 values (sin hyp calculations Table 9-11). For the zero point energies we obtained:

NO_2^- $\Delta E_0 = 147.665$ cal mol^{-1}
$[NO_3]^{\ddagger}$ $\Delta E_0 = 188.039$ cal mol^{-1}
NO_3^- $\Delta E_0 = 256.071$ cal mol^{-1}

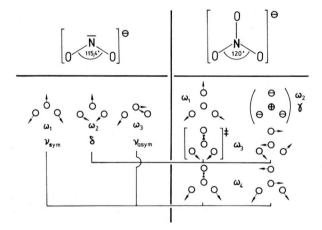

Fig. 9-5. Normal vibrations of NO_2^- and NO_3^- anions: ν = strecking vibrations; δ = bending vibrations; γ = out-of-plane vibration (ω_2 NO_3^-).

what leads to a difference between the educt NO_2^- and the transition state of 40.3737 cal mol^{-1}. In normalizing this difference to the calculated β_2 values at 25°C we obtain the equation for the temperature dependence of:

$$\beta = 1.06013 \exp\left(\frac{40.37375}{RT}\right)$$

and according to this equation there is an inversion point at 74.77°C which can be seen in Fig. 9-6. The activation energy of this reaction step is extremely sensitive to pH changes (Fig. 9-7); the relationship:

$$E_a \text{ (kJ mol}^{-1}) = 117.6 \times pH - 835.3$$

can be derived from the values reported by Wong-Chong and Loehr (1978).

Unfortunately, no data on temperature conditions were given when the β_1 values were determined, so that only certain assumptions can be made.

TABLE 9-11

Calculated partition functions and kinetic isotope effects for the reaction $NO_2^- \rightarrow NO_3^-$

Temperature (°C)	$\frac{\sigma_2}{\sigma_1} f(NO_2^-)$	$\frac{\sigma_2}{\sigma_1} f(NO_3^{-\ddagger})$	β
0	1.0967245	1.1320556	0.975664
10	1.0916231	1.1252503	0.9815646
20	1.0869133	1.1189635	0.9870794
25	1.0846925	1.1159975	0.9897028
30	1.0825549	1.1131413	0.9922422

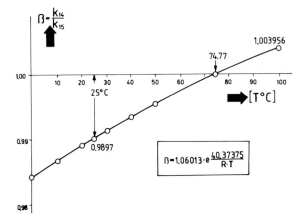

Fig. 9-6. Calculated temperature dependence of the kinetic isotope effect β of the $NO_2^- \rightarrow NO_3^-$ reaction with respect to the nitrogen isotopes.

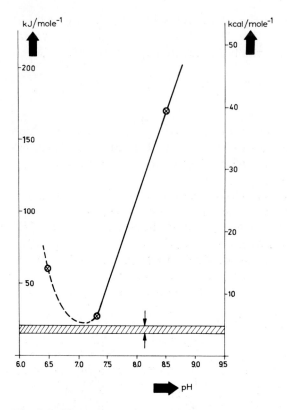

Fig. 9-7. Dependence of the activation energy on pH after Wong-Chong and Loehr (1978) The hatched interval corresponds to the activation energy of diffusion (about 3—4 kcal mol^{-1}).

If it is assumed that one β_1 value (1.026) was obtained at 30°C and that the other β_1 value (1.038) was obtained at 20°C (which follows from the temperature dependence of the isotope effects), then the following picture can be formed:

	β_1	β_2	β_3
20°C	1.031	(0.9871)	(1.0177)
25°C	1.0284	(0.9897)	1.0178
30°C	(1.026)	(0.9922)	1.018

The values in parentheses have been assumed, and the other ones may be calculated. From these the following temperature dependencies can be derived:

$\ln \beta_1 = + 43.2030/T - 0.11685$

$$\ln \beta_2 = -45.798/T + 0.14325$$

$$\ln \beta_3 = -2.5958/T + 0.0264$$

and, according to the conclusion that $\beta_3 = \beta_1 \times \beta_2$:

$$\ln \beta_3 = \ln \beta_1 + \ln \beta_2.$$

Given the value of 1.038 determined for β_1, then this should have been ascertained at a temperature of approximately 7°C, which is a possible, though experimentally not very probable test practice.

The isotope effects during nitrification are compiled in Table 9-12 which shows the calculated β values together with a few measured values.

Isotope effects during denitrification

The literature on isotope effects during denitrification, i.e., $NO_3^-/NO_2^- \rightarrow N_2/N_2O$ conversion is much more comprehensive than that on isotope effects of nitrification, although it contains quite a few inconsistencies, as is readily apparent from Table 9-13.

In the natural nitrogen cycle, a major role will be played by the obligate heterotrophic respiration of nitrate by *Pseudomonas denitrificans* rather than the chemo-autotrophic respiration by *Thiobacillus denitrificans*, the free energy production of which is based upon the oxidation of elemental sulfur. Because of the shortage of petrochemical products (and more particularly methanol), the last-mentioned reaction could of course rapidly assume considerable importance in sewage purification (third stage of purification) (Batchelor and Lawrence, 1978).

TABLE 9-12

Isotope effects during nitrification

Species or reaction, respectively		Temperature (°C)	β	Reference
Nitrosomonas europaea	β_1	—	1.026	Delwiche and Steyn (1970)
Nitrosomonas europaea		—	1.038 ± 0.003	Delwiche and Steyn (1970)
$NH_4^+ \rightarrow NO_2^-/NO_3^-$ (loess)	β_3	20	1.0177 ± 0.002	Freyer and Aly (1975), Aly (1975)
$NO_2^- \rightarrow NO_3^-$ (calculated)	β_2	0	0.9756	Hübner (1980)
		10	0.9815	
		20	0.9871	
		25	0.9897	
		30	0.9922	

TABLE 9-13
Isotope effects during denitrification

Species/reaction	Temperature (°C)	β	Reference
Pseudomonas denitrificans	—	1.0173 ± 0.0028	Delwiche and Steyn (1970)
Pseudomonas stutzeri (NCIB 9040)	—	1.019–1.02	Wellman et al. (1968)
Soil samples, Iowa:			
Nicollet ($n = 3$)	30	1.0123 ± 0.0015	Blackmer and Bremner (1977)
Webster ($n = 3$)	30	1.0133 ± 0.0032	Blackmer and Bremner (1977)
Harpster ($n = 3$)	30	1.0126 ± 0.0021	Blackmer and Bremner (1977)
$n = 9$	30	1.0128 ± 0.0021	Blackmer and Bremner (1977)
Catlin (silty clay):			
with bromegrass (permanent)	35	1.0065	Chien et al. (1977)
with corn/soybean (rotation)	35	1.0191	Chien et al. (1977)
Iowa	30	1.01424	Bremner (1977)
Denitrifier		1.03	Wada et al. (1975)
Denitrifier, Maizuru Bay, Japan		1.0144–1.0207	Wada et al. (1975)
Denitrifier, Lake Hamana, Japan		1.0019–1.0123	Wada et al. (1975)
Reactions:			
carbonyl elimination ($E_{CO}2$)	30	1.0196 ± 0.0007	Buncel and Bourns (1960)
reductions with Fe(II)			
NO_3^-	110	1.057 ± 0.004	Brown and Drury (1967)
	25	1.075 ± 0.004	Brown and Drury (1967)
NO_2^-	110	1.029 ± 0.003	Brown and Drury (1967)
	25	1.035 ± 0.002	Brown and Drury (1967)
NH_2OH	110	1.028 ± 0.001	Brown and Drury (1967)
	25	1.032 ± 0.002	Brown and Drury (1967)

Energetically, respiration of glucose with O_2 and with KNO_3 is nearly equal in value or effect:

$$5\ C_6H_{12}O_6 + 24\ KNO_3 \rightarrow 12\ N_2 + 24\ KHCO_3 + 6\ CO_2 + 18\ H_2O$$

$$\Delta G' = -645.5\ \text{kcal}$$

$$C_6H_{12}O_6 + 6\ O_2 \rightarrow 6\ CO_2 + 6\ H_2O$$

$$\Delta G' = -688.9\ \text{kcal}$$

This also explains that *Pseudomonas denitrificans* is counted among facultative anaerobic micro-organisms, with the switch to nitrate respiration not being made until an O_2 level less than 0.5 mg O_2 l^{-1} is reached. In the case of *Pseudomonas stutzeri*, the switch is also not made until an O_2 level of 0.25 mg O_2 l^{-1} is arrived at.

In general, denitrification takes place in deeper layers of soil where O_2 deficiency is caused by brisk soil respiration or where diffusion of atmospheric O_2 is prevented by flooding. This makes temperature conditions quite complex and difficult to predict, e.g., by solving the heat transport equation with due consideration of seasonal variations of temperature. Nevertheless, temperature effects have been described with particular reference to rates of denitrification in soils (Stanford et al., 1975); under conditions of flow, use can also be made of non-linear kinetic equations (Starr and Parlange, 1975).

A relation between the fractionation factor and the denitrification rate has been found by Mariotti et al., (1982). They studied the denitrification of NO_2^- anions in calcareous brown soil samples carried out at different temperatures (10°, 20°, and 30°C). The isotopic enrichment factor varied greatly, ranging from -33 to $-11^o/_{oo}$ expressed as $\epsilon_{NO_2^-/N_2O} = (1/\beta - 1) \times 10^3$, i.e. $\beta = 1.034-1.011$. On the one hand, an increase in the incubation temperature induced a decrease in the isotope fractionation, on the other hand there was a reciprocal relation between the denitrification rate and the isotope fractionation. Mariotti et al. calculated the fit of the experimentally determined reaction rate to that expected from assumptions of the first order ($Q = Q_0 \cdot e^{-k_1 t}$, where Q = substrate concentration). There is an excellent correlation of:

$$\epsilon_{NO_2^-/N_2O} = 5.78 + 5.14\ \ln k_1$$

with $n = 9, r = 0.949$ ($p < 0.01$).

Other pedosphere processes resulting in isotope variations

Besides biological processes proceeding in the pedosphere, complex interactions naturally result in additional processes taking place in the soil-bearing layer of the earth's surface, which can lead to isotopic variations of nitrogen. These include above all NH_3 evaporation that is associated with a major isotopic effect:

$$^{15}NH_3 + {}^{14}NH_4^+ \rightleftharpoons {}^{14}NH_3 + {}^{15}NH_4^+$$

$\alpha = 1.031-1.034$ at $25°C$ (Urey, 1947; Scalan, 1958; Wlotzka, 1972). Table 9-14 shows the α values which, according to Scalan (1958), can be given for this isotopic exchange reaction.

The last column in this table gives the α values calculated by the equation ($r^2 = 0.9937$):

$$\ln \alpha = 491.3156/T^2 + 8.7753/T - 0.0049$$

The difficulty of including such effects lies in their assessment which is significantly affected by many factors such as the pH value of the soil, the Ca^{2+} content, and the form of NH_4^+ compounds. NH_4F, $(NH_4)_2SO_4$, $(NH_4)_2HPO_4$, NH_4Cl, and NH_4I tend to lose very variable amounts of NH_3, which is of considerable importance when using nitrogenous fertilizers and has been frequently investigated (Fenn and Kissel, 1973, 1974, 1975a,b; Is-

TABLE 9-14

Equilibrium constants (α values) calculated for the exchange equilibrium $^{15}NH_3 + {}^{14}NH_4^+ \rightleftharpoons {}^{14}NH_3 + {}^{15}NH_4^+$ according to Scalan (1958)

Temperature (K)	α Value (Scalan, 1958)	α Value (correlated)
273.15	1.0341	1.0343
278.15		1.0335
283.15		1.0327
288.15		1.0319
293.15		1.0312
298.15	1.0311	1.0305
303.15	1.0292	1.0298
400	1.0216	1.0203
500	1.0136	1.0147
600	1.0104	1.0111
800	1.0077	1.0068
1000	1.0042	1.0043

lam and Parsons, 1979). The mineralization of urea and urea derivatives is also usually marked by differences and therefore can lead to corresponding variable losses of NH_3 (Smith and Chalk, 1980). Quite naturally, particularly high concentrations of NH_3 in the air can be measured in the vicinity of barns, i.e., buildings for stabling farm livestock (Stewart, 1970); individual measuring points ($\delta^{15}N_{NH_3}$) have been reported by Freyer (1978a).

Quite difficult to assess is also the share of chemo-denitrification in the nitrogen cycle in the pedosphere, i.e., the non-biological reaction of NO_2^- (from nitrification) with NH_4^+ (from mineralization) to form N_2.

Finally, mention should be made of a process that may best be termed NH_3 evapotranspiration, similar to the model of H_2O evapotranspiration of plants. This NH_3 loss was recently observed in aging corn leaves (Farquhar et al., 1970) and can be as high as 7 g N ha^{-1} d^{-1}, which may be of considerable consequence as compared with NH_3 evaporation from soil in Great Britain of 9 g N ha^{-1} d^{-1} (Healey et al., 1970).

Besides the evaporation from soil an isotopic exchange between ammonia gas and ammonium ions is also possible. The fractionation of nitrogen isotopes appears to commence, for example, soon after animals wastes are deposited on the ground. This has been discussed by Gormley and Spalding, 1979. The $\delta^{15}N$ values of pig urine was + 2.9 ± 0.9 and the values of pig manure + 4.0. The same pattern could be found with cattle urine + 1.7 ± 0.5 and cattle manure + 4.8 ± 1.4.

INTERPRETATION OF ISOTOPIC VARIATIONS OF NO_3-N IN THE PEDOSHPERE, GROUNDWATER, AND SURFACE WATERS

Generally, authors either refer to the difficulties of determining variations of ^{15}N in small quatities of nitrate and give no detailed information (Broadbent et al., 1980) or avoid considering isotope effects of the numerous transformation processes proceeding in the pedosphere: ". . . A quantitative correction for these isotope fractionations is not yet possible . . ." (Kohl et al., 1971) or ". . . It is difficult to evaluate and make quantitative correcting for the effect of biological, chemical, and physical isotope fractionation processes in soils on the ^{15}N concentrations of soil isolates . . ." (Hauck et al., 1972). It is in connection with nitrification that even the following formulation has been used: ". . . It is difficult to imagine a practical situation where isotope effects during conversion of NH_4^+ to NO_3^- could result in a lower ^{15}N concentration in the nitrate than in the ammonium form . . ." (Edwards, 1975). Actually, the partial differential equations:

$$\frac{\partial C}{\partial t} = D \cdot \frac{\partial^2 C}{\partial z^2} - W \frac{\partial C}{\partial z} + P - R$$

$$\frac{\partial C'}{\partial t} = D' \cdot \frac{\partial^2 C'}{\partial z^2} - W \frac{\partial C'}{\partial z} + P' - R'$$

should be solved for the nuclides ^{14}N and ^{15}N. In these equations, C and C' are the corresponding ^{14}N and ^{15}N concentrations of the nitrate, D and D' are the diffusion coefficients, W is the advection or convection, P and P' are the rates of formation of $^{15}NO_3^-$ and $^{14}NO_3^-$, and R and R' are the corresponding rates of conversion with formation of products. The z direction points here positively from the top of the soil to deeper horizons. Assuming that certain boundary and initial conditions are satisfied, the system of equations could then be solved, provided of course all other constants are known. This system of equations is far too difficult to be solved even under real conditions, so that approximations are virtually indispensable. Initially, a distinction should be made between $\delta^{15}N_{NO_3^-}$ values in the aerobic and anaerobic pedosphere. Thereafter, $\delta^{15}N_{NO_3^-}$ values in groundwaters and in surface waters (receiving streams) will be considered.

$\delta^{15}N_{NO_3^-}$ values in the pedosphere

Fig 9-8 shows $\delta^{15}N_{NO_3^-}$ values reported by Rennie and coworkers (Rennie et al., 1974; 1976; Karamanos and Rennie, 1980a, b) for soils from Saskatchewan, Canada. They are brown and black earths. An attempt was made to interpret the decrease of ^{15}N in the upper soil horizon as follows (Hübner, 1980). In the aerobic soil horizon, assimilation of NO_3^- by plant roots is taking place in addition to a conversion of NH_4^+ to NO_2^- to NO_3^- brought about above all by nitrifying bacteria, because NH_4^+ is thermodynamically unstable in the presence of O_2 and is converted to NO_3^- by the catalyst "enzyme of nitrifying bacteria" but no free enzyme system has yet been isolated which oxidizes ammonia (Painter, 1970). The reason is thought to be due to close association of the enzyme of nitrifying bacteria with the cell wall. The first step has been postulated as the formation of hydroxylamine:

$$2NH_4^+ + O_2 \rightarrow 2NH_2OH + 2H^+$$

which is an endergonic reaction and needs energy-rich phosphate for activation. The actual gain of free energy for the chemo-autotrophic nitrifying bacteria comes from the oxidation of hydroxylamine to nitrite and further on to nitrate. On the other hand, however, only a very small proportion can be considered really free NH_4^+ ion, the vast majority being bound on account of the high ion-exchange capacity. Consequently, the isotope effect proper will occur not during nitrification but during release of NH_4^+ from the soil acting as an ion exchanger:

$$\overline{NH_4^+} \rightleftharpoons NH_4^+ \xrightarrow{k_1} NO_2^- \xrightarrow{k_2} NO_3^-$$

From the theory of separation processes follows that:

$\alpha^n = R_n/R_0$

where n is the number of repeated adjustments of the isotope balance and R_n and R_0 are the $^{15}N/^{14}N$ ratios at the end (R_n) and at the beginning (R_0), respectively. Furthermore, it can be shown that:

$n = l/\text{HETP}$

where l is the length of the separation column and HETP is the height equivalent to a theoretical plate. By using these equations and employing 0.99924 for α (Table 9-7), we obtain:

$\ln R_n = n \ln \alpha + \ln R_0$ and $n = \text{length}/\text{HETP}$

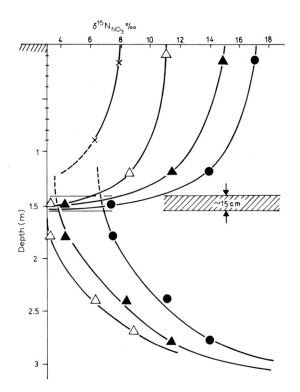

Fig. 9-8. Measured $\delta^{15}N_{NO_3^-}$ values in dependence on the depth: ● = Rennie et al. (1974); ▲ = Rennie and Paul (1975); △ = Rennie et al. (1976); × = Karamanos and Rennie (1980b).

(length equals the z-direction in cm). For the $\delta^{15}N_{NO_3^-}$ values given by Rennie et al., (1974) and Rennie and Paul (1975), HETP = 12.5 cm and 12.6 cm, respectively, while for those given by Rennie et al. (1976) and Karamanos and Rennie (1980b) HETP = 15.5 cm and 28.71 cm (two measuring points only), respectively. In all cases the soils were chernozemic dark brown. Below certain soil horizons an adequate supply of O_2 (and also CO_2) will no longer be available, so that a chemo-autotrophic process can no longer take place. However, a heterotrophic respiration utilizing NO_3^- oxygen (denitrification) is still possible and the decreasing NO_3^- will become enriched in ^{15}N.

A correlation of the values given in Fig. 9-8 (applying the data from Rennie et al., 1974, 1976) leads to $r = -0.9952$ and -0.9959 with β values of denitrification of $\beta = 1.00466$ and 1.00511, respectively, i.e., smaller isotope effects than those which are measured in batch systems. It has been proposed that these smaller isotope effects should under natural, open conditions be referred to as a diluted or a mixing isotope effect (Hübner, 1980); besides, it is possible for an ion-exchange effect to be involved in variations of NO_3^- concentrations. This latter effect may, however, be considered to be of secondary importance because of the extremely low anion capacities of soils.

$\delta^{15}N_{NO_3^-}$ in the groundwater

$\delta^{15}N$ values reported by Aly (1975) for groundwaters (15—30 m deep) are plotted in Fig. 9-9 as a function of the concentration of NO_3^-. The oxygen concentration mg O_2 l^{-1} decreases with the increasing depth. Supposing that the decrease of the NO_3^- concentration is caused by denitrification processes in the deeper horizons then two concurrent reactions take place between the two isotopes associated by an isotope effect. The corresponding reaction rates can be generally written as follows:

$$^{14}NO_3^- + B + C + \ldots \xrightarrow{^{14}k} {}^{14}P + Q + \ldots$$

$$^{15}NO_3^- + B + C + \ldots \xrightarrow{^{15}k} {}^{15}P + Q + \ldots$$

where ^{14}P, ^{15}P are the reaction products containing the nuclides ^{14}N and ^{15}N, respectively. The corresponding differential equations are:

$$-\frac{d[^{14}NO_3^-]}{dt} = {}^{14}k\,[^{14}NO_3^-]\,[B]^b\,[C]^c \ldots$$

$$-\frac{d[^{15}NO_3^-]}{dt} = {}^{15}k\,[^{15}NO_3^-]\,[B]^b\,[C]^c \ldots$$

and the simultaneous solution of these differential equations yields:

$$\frac{d[^{14}NO_3^-]}{d[^{15}NO_3^-]} = \frac{^{14}k}{^{15}k} \frac{[^{14}NO_3^-]}{[^{15}NO_3^-]} = \beta \frac{[^{14}NO_3^-]}{[^{15}NO_3^-]}$$

or

$$\frac{d[^{15}NO_3^-]}{[^{15}NO_3^-]} = \frac{1}{\beta} \frac{d[^{14}NO_3^-]}{[^{14}NO_3^-]}$$

By introducing the abbreviation $R_f = (^{15}N/^{14}N)_{NO_3^-}^{f}$ after the fraction f of the substrate NO_3^- has reacted and $R_0 = (^{15}N/^{14}N)_{NO_3^-}^{0}$, the initial isotope ratio, we find

$$\ln \frac{R_f}{R_0} = \left(\frac{1}{\beta} - 1\right)\left(\frac{1 + R_0}{1 + R_f}\right) \ln(1-f).$$

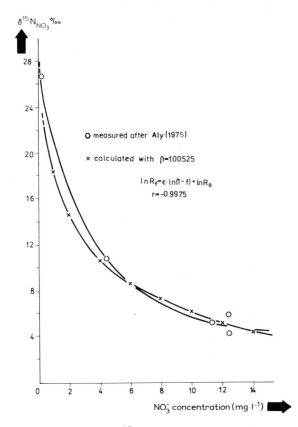

Fig. 9-9. Measured $\delta^{15}N_{NO_3^-}$ values in groundwaters (after Aly, 1975).

The quotient can be neglected since $(1 + R_0)/(1 + R_f) \approx 1$, because R_0 and R_f are much smaller than unity ($R_0 = 0.003663$ in air nitrogen). Thus we obtain:

$\ln R_f = \epsilon \cdot \ln(1-f) + \ln R_0$ with $(1/\beta - 1) = \epsilon$.

If the starting substrate has a isotope ratio R_0 and the substrate has reacted to a fraction f then the remaining substrate will have the isotope ratio R_f with:

$R_0 = R_{standard} \cdot (1 + 10^{-3} \delta A_0)$

$R_f = R_{standard} \cdot (1 + 10^{-3} \delta A_f)$.

If the starting substrate has an isotope ratio R_0 and the substrate has reacted troduce the abbreviation $\lambda \equiv \ln(R/R_{standard})$ since then $\lambda = \ln(1 + 10^{-3}\delta) \sim \delta$ as can be seen for instance from Gat and Gonfiantini (1981) we can directly convert the measured δ values into R values. Since the δ values refer to specially defined standards the R_0 and R_f ratios can be divided by the $R_{standard}$ values but the $R_{standard}$ are cancelled out by applying the above derived equation for the correlation of R_f values with the remaining fraction $(1-f)$ of the substrate corresponding with the degree of conversion.

Using this formula for the data by Aly (1975) the following values can be obtained:

$\beta = 1.00525 \qquad r = -0.9975$

$t = \sqrt{\dfrac{(n-2)r^2}{1-r^2}} = 26.63$ with $n = 5$

$t = 3.18\ (95\%)$ and $t = 5.84\ (99\%)$,

with extremely high significance of the β and t values. In this correlation the degree of conversion is related to a concentration of 12.5 mg NO_3-N l^{-1} because this was the highest value measured in the groundwater (Fig. 9-9). We can consider this highest concentration of NO_3^- as a reference point, i.e. the redox condition is reversed and as the oxygen content is not sufficient to act as an electron acceptor nitrate is assuming this function. In this way the highest concentration acts as an initial point to ascertain the degree and the extent of the denitrification which lowers concentrations of NO_3^- provided there are anaerobic conditions. Further data to fix this reference concentration are needed in special case studies, e.g. the pϵ values of the horizon considered to make sure that the turning-point from the oxidation to the reduction zone is crossed.

Surprisingly, the β value is relatively small, and this may again be considered to be due to a dilution effect. The migration of the NO_3^- anion from the upper horizons to the lower ones is accompanied by diffusion and dispersion resulting in a dilution of the concentration. The decrease in regard of the denitrification of the NO_3^- concentrations due to denitrification is therefore superimposed with this diffusion effects. Even if more data were available about the migration velocity of the pore water and about the apparent diffusion coefficient some other effects which are still playing and are connected with the matrix of the soil. It is well-known that the soil nitrate is easily leached but most temperate soils possess an over all net negative charge on their colloids which repels the nitrate ion. That means that the nitrate may move downward through the soil faster than the drainage water (Haynes and Goh, 1978).

We can only try to eliminate the dilution/diffusion effect supposing the following relation:

$$R_f/R_0 = (C_{NO_3-N})_f^\epsilon / (C_{NO_3-N})_0^\epsilon$$

if, as usual, the degree of conversion is expressed as mg NO_3-N l^{-1}. However,

TABLE 9-15

Calculated and measured $\delta^{15}N$ values for groundwater samples according to values reported by Aly (1975)

C_{NO_3-N}	$\delta^{15}N$ (⁰/₀₀) measured	$\delta^{15}N$ (⁰/₀₀) calculated	$1-f$	'l'
0.01	—	42.99	—	1.08767
0.1	—	30.53	—	1.05855
0.2	26.7	26.81	0.016	—
1	—	18.21	—	1.03022
2	—	14.54	—	1.02183
4	—	10.87	—	1.01352
4.5	10.7	10.25	0.36	—
6	—	8.73	—	1.00869
8	—	7.22	—	1.00527
10	—	6.05	—	1.00263
11.4	5.0	5.36	0.912	—
12	—	5.09	—	1.00048
12.3	5.8	4.96	0.984	—
12.4	4.1	4.92	0.992	—
12.5	—	$4.877 = \delta A_0$	1.0000	1.00000
20	—	2.41	—	—
31.74	—	0	—	—
50	—	−2.37	—	—
100	—	−5.97	—	—

if NO$_3$-N is diffusing out of the "liter" or water serving as a solvent is diffusing into the "liter", then there will be changed the concentration but not the isotopic composition. In Table 9-15 use is made of a fixed value of β of 1.017 (Delwiche and Steyn, 1970) for denitrification, but a change in reference volume (liter) is permitted, with the exact concentration of 12.5 mg NO$_3$-N per liter holding in this case.

If changes in reference volume as great as 5% are permitted, then it is possible to make direct use of a β value obtained from laboratory tests performed under batch conditions. This statement is of major importance to a definition of the degree of denitrification and has already been discussed in connection with the ozone shield of the earth (Hübner, 1981).

We can also consider the above mentioned two partial differential equations for the ^{14}N and ^{15}N nuclides as a diagenetic equation (so designated in the literature of sedimentology, e.g. by Berner, 1971), which can be solved by separation into two functions namely into a dispersion function depending only upon the spatial coordinates x, y, z and into a time-dependent function expressed in the following way:

$$C_{\text{reaction, dispersion}}(x, y, z, t) = C_{\text{dispersion}}(x, y, z) \cdot C(t)$$

or in only one dimension:

$$C_{\text{reaction, dispersion}}(z, t) = C_{\text{dispersion}}(z) \cdot C(t)$$

But this approach carries the implicit assumption that these two processes are independent of each other. Now assuming that the denitrification reaction obeys a first (or pseudo-first) -order reaction law then we can write:

$$C(t) = \exp(-kt) = \exp(-kz/W)$$

$$C_{\text{reaction, dispersion}}(z, t) = C_{\text{dispersion}}(z) \cdot \exp(-kz/W)$$

where k may be the first-order reaction constant and W may be the migration velocity directed downward in z direction. Since too many assumptions have to be made about data needed for an exact solution of this partial differential equation we can form the following equation by analogy with this product of two independent functions:

$$\begin{pmatrix} \text{mixing or} \\ \text{diluted} \\ \text{isotope effect} \end{pmatrix} = \begin{pmatrix} \text{dispersion} \\ \text{or} \\ \text{dilution} \end{pmatrix} \times \begin{pmatrix} \text{isotope effect} \\ \text{measured under} \\ \text{batch conditions} \end{pmatrix}.$$

Through the presentation of the reference volume as the diluted "liter" we gain an insight into the low β values obtained from the correlation of the $\delta^{15}N_{NO_3^-}$ data and the nitrate concentration. With this relation the dispersion or dilution term is determined experimentally under natural conditions.

This increase of the $\delta^{15}N_{NO_3^-}$ values with the decrease of the nitrate concentration is not unique. In a study carried out during 1976—1977 Gormly and Spalding (1979) observed the same significant negative correlation between $\delta^{15}N_{NO_3^-}$ and the NO_3^- concentration and between $\delta^{15}N_{NO_3^-}$ and the depth to water table. The study area covered 2119 km² in the Central Plate Region of Nebraska especially the Counties Buffalo, Hall and Merrick. Altogether 256 groundwater samples were collected and the authors concluded that the net positive isotope effect within the unsaturated layer appears to be denitrification. The correlation was carried out in a linear relation between $\delta^{15}N_{NO_3^-}$ and the nitrate concentration. Unfortunately, the measured data are not explicitly compiled in figures. There is one extreme $\delta^{15}N_{NO_3^-}$ value of + 44.0‰ found at an active feedlot soil at a depth of 23—30 cm and with a concentration of 3 mg NO_3-N l^{-1}.

In a further study of the $\delta^{15}N_{NO_3^-}$ values in groundwater nitrate on Long Island, New York, 39 samples were collected by Kreitler et al. (1978). Long Island has two different aquifers: one is the unconfined, upper glacial aquifer and the other, the Magothy aquifer, underlies the former one. Water in the Magothy aquifer is older than the water from the upper glacial aquifer (up to hundreds of years). The investigation revealed a trend of increasing $\delta^{15}N_{NO_3^-}$ values in the upper glacial aquifer from Suffolk County ($\delta^{15}N_{NO_3^-}$: 5.21 ± 1.82‰, n = 13 samples) through Nassau County ($\delta^{15}N_{NO_3^-}$: 8.32 ± 2.41‰, n = 10 samples) to Queens County ($\delta^{15}N_{NO_3^-}$: 12.1‰ and 21.3‰ only 2 samples). A correlation between the $\delta^{15}N_{NO_3^-}$ and the corresponding NO_3^- concentrations is not existing besides from the direction of the decrease in NO_3^- concentration from Suffolk County (C_{NO_3-N} = 52.1 ± 19.6) through Nassau County (C_{NO_3-N} = 30.7 ± 17.7) to Queens County (30 mg NO_3-N l^{-1} and 20 mg NO_3-N l^{-1} were found in the 2 samples). But it must be mentioned that the variations in the NO_3^- concentrations are remarkably high. Kreitler et al. (1978) are discussing three hypotheses why the $\delta^{15}N_{NO_3^-}$ range of groundwater nitrate is heavier than the fertilizer range: (1) more soil nitrogen is mineralized into nitrate than is applied as fertilizer; (2) the applied nitrate fertilizer in Suffolk County may be isotopically heavier than average fertilizer; and (3) nitrate from fertilizer is being altered isotopically by denitrification, ammonia volatilization, or nutrient assimilation.

Since the land use in vicinity of the investigated wells at Suffolk County area is mainly agricultural and since the $\delta^{15}N_{NO_3^-}$ of this area is lighter than any other ground waters samples on Long Island, the authors concluded that this is an indication for a predominantly agricultural rather than an animal-waste source of the nitrate nitrogen. Since the two samples from Queens County fall into the range of animal waste ($\delta^{15}N$: + 10 to + 20‰) the authors have argued that this may be the nitrate nitrogen source. Such conclusions require that the original $\delta^{15}N_{NO_3^-}$ of the different inputs are not changed or that the "individuality" of the natural tracer is maintained.

A very similar conclusion has been recently made by Aly et al. (1982) investigating natural variations of nitrogen isotope ratios of nitrate in ground and surface waters in the Wadi El-Natrun in Egypt. Since anaerobic conditions in the Wadi El-Natrun area are not met the denitrification process for changing the isotope ratio can be excluded. The basic nature of the soil in the investigated area (pH range 8.1—9.4) may favour the process of volatilization accompanied with an isotope exchange reaction between ammonia and the ammonium ion.

Very high NO_3-N concentrations measured in groundwater are described in two investigations. Super et al. (1981) found a strong correlation between nitrate intake and methaemoglobin levels of infants in South West Africa/Namibia. An extreme case was one infant with the methaemoglobin level of 35%. The well-water which this infant had been drinking contained 56 mg NO_3-N l^{-1} and his calculated daily intake of nitrate was 7.6 mg kg^{-1} body weight. The other study about nitrate concentrations in groundwater was carried out by Robertson (1979) who collected and analyzed over 800 well samples in Sussex County, Delaware. The nitrate concentration ranged from zero to 225 mg NO_3 l^{-1} corresponding 49.8 mg NO_3-N l^{-1}.

$\delta^{15}N_{NO_3^-}$ in surface waters

The most well-known study on $\delta^{15}N$ values in surface waters was published in 1971 (Kohl et al., 1971) and was subject to severe criticism (Hauck et al., 1972). If a $\delta^{15}N_{NO_3^-}$ value of + 7.5 is observed, then two factors may have contributed to it. One of them is derived from mineralized soil nitrogen, with an assumed mean value of 13‰. The other comes from fertilizer nitrogen, mainly nitrified NH_3, and has a mean value of 3‰. Using the isotope balance equation, one obtains:

$$7.5 = x \cdot 3 + (1-x) \cdot 13$$

or

$$x = \frac{13-7.5}{10} \times 100 = 55\% \text{ fertilizer nitrogen}$$
$$= 45\% \text{ soil nitrogen}$$

This is a simple two-component mixing system in which any isotope fractionation was excluded. Fig 9-10 shows the measuring points with a linear correlation to:

$$\delta^{15}N_{NO_3^-} = -0.4455(C_{NO_3\text{-}N}) + 12.168$$

with $r = -0.776$. It may be obvious to apply the power function derived above correlating the data measured in surface waters. The numerical expression is:

$$\ln R_f = -0.004758 \ln (C_{NO_3\text{-}N}) + 0.018576$$

with the same meaning of

$$R_f = 1 + 10^{-3} \delta A_f$$

$$R_0 = 1 + 10^{-3} \delta A_0, \quad \delta A_0 = 18.75\%_{00}$$

$$\epsilon = 1/\beta - 1 = -0.004758; \quad \beta = 1.00478$$

and, finally, the correlation coefficient $r = -0.8543$:

$$t = \sqrt{\frac{(n-2)r^2}{1-r^2}} = 5.45 \text{ with } t_{\nu=11} = 3.106 \quad (99\%)$$

The f values used in the correlation are again the converted fractions of nitrate, i.e.:

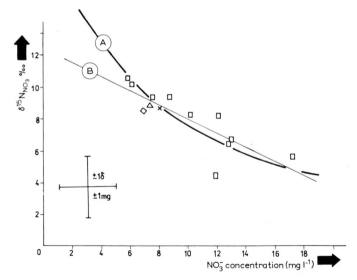

Fig. 9-10. Measured $\delta^{15}N_{NO_3^-}$ values in surface waters (after Kohl et al., 1971): □, △, + = measurements from A: calculated with $\ln R_f = \epsilon \cdot \ln C_{NO_3\text{-}N} + \ln R_0$; B: calculated with $\delta^{15}N_{NO_3^-} = a \cdot C_{NO_3\text{-}N} + b$.

$$f = \frac{(NO_3)_0 - (NO_3)_f}{(NO_3)_0} \quad \text{or} \quad 1 - f = \frac{(NO_3)_f}{(NO_3)_0}$$

i.e.:

$$\ln R_f = \epsilon \cdot \ln[(NO_3)_f/(NO_3)_0] + \ln R_0$$

$$= \epsilon \cdot \ln(NO_3)_f - \epsilon \cdot \ln(NO_3)_0 + \ln R_0.$$

Abbreviating for $-\epsilon \cdot \ln(NO_3)_0 + \ln R_0 = \ln R_0'$ we obtain:

$$\ln R_f = \epsilon \cdot \ln(NO_3)_f + \ln R_0'.$$

As might be expected f must be related to an initial value $(NO_3)_0$; the intercept $\ln R_0$ corresponds to $\ln(NO_3)_f = 0$ or $(NO_3)_f = 1$ mg NO_3-N l^{-1}, an arbitrary reference quantity which cannot necessarily lead to the degree of conversion. From the introduced abbreviation for $\ln R_0' = -\epsilon \cdot \ln(NO_3)_0 + \ln R_0$ we can show in the normal δ presentation that:

$$\ln(1 + 10^{-3} \delta A_0') = -\epsilon \cdot \ln(NO_3)_0 + \ln(1 + 10^{-3} \delta A_0)$$

or:

$$(NO_3)_0 = \left(\frac{1 + 10^{-3} \delta A_0}{1 + 10^{-3} \delta A_0'} \right)^{1/\epsilon},$$

i.e., an accurate determination is made either of $(NO_3)_0$ or of the associated δA_0 value, the numerical expressions being:

δA_0 = 5.0‰, with $(NO_3)_0$ = 17.512
or:
$(NO_3)_0$ = 17.2, with δA_0 = 5.051‰.

When these values are known, the degree of conversion can be determined.

Posing greater difficulties is the interpretation of the β value determined by correlation, which can be an assimilatory isotope effect (see Table 9-8) or a mixed isotope effect of denitrification. This can be decided only by additional data, namely, dissolved O_2, pϵ value, autotrophy, photosynthesis, etc.

As has recently been shown by using a shallow and impounded rivulet as an example (Gorsler, 1980), the nitrogen balance in flowing river water and in impounded river water is influenced by the velocity of the drain alone.

The NO_3-N content follows the relation:

$$\text{mg } NO_3\text{-N } l^{-1} = 12.331 \cdot [m^3 \text{ s}^{-1}]^{0.614},$$

while the $NH_4^+/N_{org.}$ content follows the hyperbolic relation:

$$\text{mg TKN } l^{-1} = \frac{1.265}{[m^3 \text{ s}^{-1}]} + 2.674$$

where TKN = total Kjeldahl N, i.e., NH_4^+ and organic nitrogen.

Data reported by Létolle (1980) were interpreted in the same manner. They were collected during studies performed in the Yerres River and Mélarchez Basin (Mariotti and Létolle, 1977) in a catchment area to the east and southeast of Paris, France. For the Yerres River, the following relationships between NO_3^--contents and $\delta^{15}N_{NO_3^-}$ are obtained and shown in Fig. 9-11:

Curve A:

$\ln R_f = -0.00607 \ln (C_{NO_3\text{-}N}) + 0.0236$
with $\beta = 1.0061$, $\delta\ A_0' = 23.89\%_{00}$ $\qquad r^2 = 0.7079$

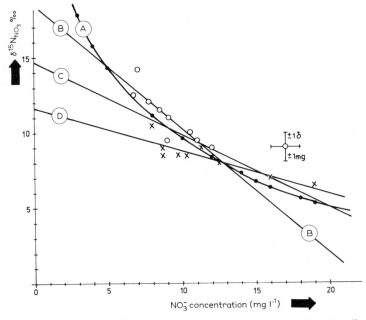

Fig. 9-11. Measured $\delta^{15}N_{NO_3^-}$ values in surface water (after Létolle, 1980). A: calculated with $\ln R_f = \epsilon \cdot \ln C_{NO_3\text{-}N} + \ln R_0$; B, C, D: calculated with $\delta^{15}N_{NO_3^-} = a \cdot C_{NO_3\text{-}N} + b$. B = Yerres River, upper part. C = Yerres River, upper and lower part. D = Yerres River, lower part.

Curve B (upper course of the river):

$$\delta^{15}N_{NO_3^-} = -0.832\,(C_{NO_3\text{-}N}) + 18.55 \qquad r^2 = 0.7772$$

Curve D (lower course of the river):

$$\delta^{15}N_{NO_3^-} = -0.278\,(C_{NO_3\text{-}N}) + 11.61 \qquad r^2 = 0.7847$$

Curve C (upper and lower course of the river combined):

$$\delta^{15}N_{NO_3^-} = -0.5003\,(C_{NO_3\text{-}N}) + 14.867 \qquad r^2 = 0.6295.$$

For a critical estimation of the coefficient of determination, r^2, obtained for single segments of the river (curves B and C), it is necessary to compare the values for curves A and C, since they take account of the entire river course. Actually, the single segments show a higher r^2 (B: $r^2 = 0.7772$ and C: $r^2 = 0.7847$) than the r^2 of the curve A (0.7079) but the last should be compared with the linear correlation of curve C. Curve A covers the found measuring points in a better way. On the other hand, it should be considered that so far it has not been possible for the regression coefficients (B: -0.832; C: -0.278; D: -0.5003) to be interpreted in a linear correlation in comparison with the exponential statement where the regression coefficient can be understood as the result of isotope effects ($\beta = 1.0061$) taking place.

OUTLOOK

It is not an easy task to analyse and obtain a proper understanding of the numerous processes proceeding in the pedosphere. The question as to the three R's (route, rate, reservoir) can be answered only in rough outlines for nitrogen and its nuclides. Of the large number of questions that have not as yet been addressed, mention should be made of the following ones:

— The non-stationary case that involves a time record of the isotopic variations of nitrogen (be it with respect to the different nitrogen compounds in natural soils, be it with respect to plants). In 1974, a paper was published by Feigin et al. supporting the idea on the dependence of the $\delta^{15}N$ values upon time and depth.

— The entire field of sediments in rivers and inland lakes which cannot be handled without giving detailed consideration to the aquatic ecosystems and its food chains.

— Finally, at the end of the food chains are the animals (graminivorous and carnivorous) and man as well. Therefore, it is not surprising that a few papers have been published dealing with the $\delta^{15}N$ variations in animals. It is obvious that predictive results can only be obtained if experiments are performed under constant conditions with respect to the $\delta^{15}N$ values of the diet. A further complication stems from the fact that different tissues have different δ^{15}N values, e.g. relatively the δ^{15}N values of tissues of mice brains are more positive than those of kidneys, hairs or livers which was recently shown by De Niro and Epstein (1981). On the other hand the microflora in rumens will be responsible for the δ^{15}N values in tissues of ruminants which was observed by Steele and Daniel (1978).

Two other, but older papers by Gaebler et al. (1963) should also be mentioned in this connection. As a rough estimate one can show that our planet earth is presently populated with about 4.5×10^9 people which equals a biomass of about 6.44×10^7 t (Lieth and Whittaker, 1975) and supposing a C/N ratio of about 7 this yields about 4.6×10^6 t N. Comparing this value with figures in Table 9-2 one realizes that this nitrogen pool does not play an important role in the biogeochemical cycles of chemical elements and their isotopes. Therefore these publications tend towards another direction and aim to contribute to the understanding of living conditions and food sources of mankind during the past.

REFERENCES

Abrahamsen, G., 1980. Acid precipitation, plant nutrients and forest growth in ecological impact of acid precipitation. In: *Proceedings, SNSF Project, Oslo*. pp. 58—63.

Aly, A.I.M., 1975. *N-15 Untersuchungen zur anthropogenen Störung des natürlichen Stickstoffzyklus*. Dissertation, TH Aachen, 89 pp.

Aly, A.I.M., Mohamed, M.A. and Hallaba, E., 1981. Mass-spectrometric determination of the N-15 content of different Egyptian fertilizers. *J. Radioanal. Chem.*, 67: 55—60.

Aly, A.I.M., Mohamed, M.A. and Hallaba, E., 1982. Natural variations of ^{15}N content of NO_3 in ground and surface waters and total nitrogen of soil in the Wadi El-Natrun area in Egypt, in: H.-L. Schmidt, H. Förstel and K. Heinzinger (Editors), *Stable Isotopes. Proceedings of the 4th International Conference, Jülich, March 23—26, 1981. Analytical Chemistry Symposia Series, 11*. Elsevier, Amsterdam, pp. 475—481.

Amarger, N., Mariotti, A. and Mariotti, F., 1977. Essai d'estimation du taux d'azote fixe symbiotiquement chez le Lupin par le traçage isotopique naturel N-15. *C.R. Acad. Sci. Paris, Sér. D*, 284: 2179—2182.

Amarger, N., Mariotti, A., Mariotti, F., Durr, J.C., Bourguignon, C. and Lagacherie, B., 1979. Estimate of symbiotically fixed nitrogen in field grown soybeans using variations in N-15 natural abundance. *Plant Soil*, 52: 269—280.

Ayers, G.P. and Gras, J.L., 1980. Ammonia gas concentrations over the Southern Ocean. *Nature*, 284: 539—540.

Bardin, R., Domenache, A.M. and Chalamet, A., 1977. Rapports isotopiques naturels de l'azote II Application à la mesure de la fixation symbiotique de l'azote in situ. *Rev. Ecol. Biol. Sol.* 14: 395—402.

Batchelor, B. and Lawrence, A.W., 1978. A kinetic model for autotrophic denitrification using elemental sulfur. *Water Res.*, 12: 1075—1084.

Berner, R.A., 1971. *Principles of Chemical Sedimentology*. McGraw-Hill, New York, N.Y., pp. 86—113.

Bigeleisen, J. and Wolfsberg, M., 1958. Theoretical and experimental aspects of isotope effects in chemical kinetics. In: I. Prigogine (Editor), *Advances in Chemical Physics, 1*. Interscience, New York, N.Y., pp. 15—76.

Black, A.S. and Waring, S.A., 1977. The natural abundance of N-15 in the soil-water system of a small catchment area. *Austr. J. Soil Res.*, 15: 51—57.

Black, A.S. and Waring, S.A., 1979. Effect of nitrate leaching in oxisol columns of N-15 abundance and nitrate breakthrough curves. *Commun. Soil Sci. Plant Anal.*, 10: 521—529.

Blackmer, A.M. and Bremner, J.M., 1977. N-isotope discrimination in denitrification of nitrate in soils. *Soil Biol. Biochem.*, 9: 73—77.

Böttger, A., Ehhalt, D.H. and Gravenhorst, G., 1978. Atmosphärische Kreisläufe von Stickoxiden und Ammoniak. Kernforschungsanlage Jülich GmbH, Institut für Chemie, Jül-1558, 158 pp.

Bremner, J.M., 1977. Role of organic matter in volatilization of sulphur and nitrogen from soils. In: *Soil Organic Matter Studies, 2*. IAEA, Vienna, pp. 229—240.

Broadbent, F.E., Rauschkolb, R.S., Lewis, A.Kr. and Chang, G.Y., 1980. Spatial variability of N-15 and total N in some virgin and cultivated soils. *Soil Sci. Soc. Am. J.*, 44: 524—527.

Brown, L.L. and Drury, J.S., 1967. N-isotope effects in the reduction of NO_3^-, NO_2^-, and NH_2OH to NH_3, I. In sodium hydroxide solution with Fe(II). *J. Chem. Phys.*, 46: 2833—2837.

Brown, R.H., 1978. A difference in N use efficiency in C_3 and C_4 plants and its implications in adaptation and evolution. *Crop Sci.*, 18: 93—97.

Buncel, E. and Bourns, A.N., 1960. The mechanism of the carbonyl elimination reaction of benzyl nitrate. *Can. J. Chem.*, 38: 2457—2466.

Burns, R.C. and Hardy, R.W.F., 1975. Nitrogen fixation in bacteria and higher plants. In: A. Kleinzeller, G.F. Springer and H.G. Wittmann (Editors), *Molecular Biology Biochemistry and Biophysics, 21*. Springer-Verlag, Berlin, 189 pp.

Central Institute for Isotope and Radiation Research, 1978. *Laboratory Report 1978*. Academy of Sciences of the G.D.R., Leipzig.

Central Institute for Isotope and Radiation Research, 1980. *Laboratory Report 1980*. Academy of Sciences of the G.D.R., Leipzig.

Chang, T.Y., Norbeck, J.M. and Weinstock, B., 1979. An estimate of the NO_x removal rate in an urban atmosphere. *Environ. Sci. Technol.* 13: 1534—1537.

Cheng, H.H., Bremner, J.M. and Edwards, A.P., 1964. Variations of N-15 abundance in soils. *Science*, 146: 1574—1575.

Chien, S.H., Shearer, G. and Kohl, D.H., 1977. The N-isotope effects associated with nitrate and nitrite loss from waterlogged soils. *Soil Sci. Soc. Am. J.*, 41: 63—69.

Clusius, K. and Piesbergen, U., 1960. Reaktionen mit ^{15}N, XXXIV. Trenneffekte bei den Isotopen ^{14}N und ^{15}N während der technischen Fabrikation von Salpetersäure. *Helv. Chim. Acta*, 43: 1562—1569.

Clusius, K. and Poschet, H.J., 1962. Reaktionen mit ^{15}N, XXXVII. Trenneffekte bei den Isotopen ^{14}N und ^{15}N während der Herstellung wasserfreier Salpetersäure nach dem direkten Verfahren. *Helv. Chim. Acta*, 45: 1—4.

Collins, J. and Bowman, N.S. (Editors), 1970. *Isotope Effects in Chemical Reactions*. Van Nostrand Reinhold, New York, N.Y., 435 pp.

Dean-Raymond, D. and Alexander, M., 1976. Plant uptake and leaching of dimethylnitrosamine. *Nature*, 262: 394; *Naturwiss. Rundsch.*, 30: 224.

Delwiche, C.C. and Steyn, P.L., 1970. Nitrogen isotope fractionation in soils and microbial reactions. *Environ. Sci. Technol.*, 4: 929—935.

Delwiche, C.C., Zinke, P.J., Johnson, C.M. and Virginia, R.A., 1979. Nitrogen isotope distribution as a presumptive indicator of nitrogen fixation. *Bot. Gaz. (Suppl.)*, 140: 65—69.

De Niro, M.J. and Epstein, S., 1981. Influence of diet on the distribution of nitrogen isotopes in animals. *Geochim. Cosmochim. Acta*, 45: 341—351.

Denison, W.C., 1973. Life in tall trees. *Sci. Am.*, 228 (6): 75—80.
Domenach, A.M., Chalamet, A. and Pachiaudi, C., 1977. Rapports isotopiques naturels de l'azote, I. Premiers résultats: sols de Dombes. *Rev. Ecol. Biol. Sol*, 14: 279—287.
Drechsler, M., 1976. *Entwicklung einer probenchemische Methode zur Präzisionsisotopenanalyse am N biogener Sedimente; Aussagen der N-Isotopenvariationen zur Genese des N_2 in Erdgasen*. Dissertation, AdW der DDR, Leipzig, 87 pp.
Edwards, A.P., 1973. Isotope tracer technique for identification of source of nitrate pollution. *J. Environ. Qual.*, 2: 382.
Edwards, A.P., 1975. The interpretation of $^{15}N/^{14}N$ ratios in tracer studies. In: *Isotope Ratios as Pollutant Source and Behaviour Indicators*. IAEA, Vienna, pp. 455—468.
Faust, H., Bornhak, H., Hirschberg, K. and Birkenfeld, H., 1980. Methode zur Bestimmung natürlicher N-15 Häufigkeiten an geringen Mengen medizin.-biol. Materials. *ZfI-Mitt.*, 29, pp. 217—225.
Farquhar, G.D., Wetselaar, R. and Firth, P.M., 1979. Ammonia volatilization from senescing leaves of maize. *Science*, 203: 1257—1258.
Feigin, A., Shearer, G., Kohl, D.H. and Commoner, B., 1974. The amount and N-15 content of nitrate in soil profiles from two Central Illinois fields in a corn-soybean rotation. *Soil Sci. Soc. Am. Proc.*, 38: 465—471.
Fenn, L.B. and Kissel, D.E., 1973. Ammonia volatilization from surface applications of ammonium compounds on calcareous soils, I. General theory. *Soil Sci. Soc. Am. Proc.*, 37: 855—859.
Fenn, L.B. and Kissel, D.E., 1974. Ammonia volatilization from surface applications of ammonium compounds on calcareous soils, II. Effects of temperature and rate of ammonium nitrogen application. *Soil Sci. Soc. Am. Proc.*, 38: 606—610.
Fenn, L.B. and Kissel, D.E., 1975a. Ammonia volatilization from surface applications of ammonium compounds on calcareous soils, III. Effects of mixing low and high loss ammonium compounds. *Soil Sci. Soc. Am. Proc.*, 39: 366—368.
Fenn, L.B. and Kissel, D.E., 1975b. Ammonia volatilization from surface applications of ammonium compounds on calcareous soils, IV. Effect of $CaCO_3$ content. *Soil Sci. Soc. Am. Proc.*, 39: 631—633.
Fiedler, R. and Proksch, G., 1975. The determination of N-15 by emission and mass spectrometry in biochemical analysis: a review. *Anal. Chim. Acta*, 78: 1—62.
Filipović, R., 1980. Fertilizer-nitrogen residues: useful conservation and pollutant potential under maize. In: *Soil Nitrogen as Fertilizer or Pollutant*. IAEA, Vienna, pp. 47—60.
Freyer, H.D., 1978a. Seasonal trends of NH_4^+ and NO_3^- nitrogen isotope composition in rain collected at Jülich, Germany. *Tellus*, 30: 83—92.
Freyer, H.D., 1978b. Preliminary ^{15}N studies of atmospheric nitrogenous trace gases. *Pageoph*, 116: 393—404.
Freyer, H.D. and Aly, A.J.M., 1975. N-15 studies on identifying fertilizer excess in evironmental systems. In: *Isotope Ratios as Pollutant Source and Behaviour Indicators*. IAEA, Vienna, pp. 21—33.
Fried, M. and Broeshart, H., 1975. An independent measurement of the amount of nitrogen fixed by a legume crop. *Plant Soil*, 43: 707—711.
Fried, M. and Dean, L.A., 1953. A concept concerning the measurement of available soil nutrient. *Soil Sci.* 73: 263—271.
Gaebler, O.H., Choitz, H.C., Vitti, T.G. and Vukmirovich, R., 1963. Significance of ^{15}N excess in nitrogenous compounds of biological origin. *Can. J. Biochem. Physiol.*, 41: 1089—1097.
Gaebler, O.H., Vitti, T.G. and Vukmirovich, R., 1966. Isotope effects in metabolism of ^{14}N and ^{15}N from unlabelled dietary proteins. *Can. J. Biochem.*, 44: 1249—1257.
Gat, J.R. and Gonfiantini, R. (Editors), 1981. *Stable Isotope Hydrology, Deuterium and Oxygen-18 in the Water Cycle*. IAEA Tech. Rep. Ser., 210: 21.

Gormly, J.R. and Spalding, R.F., 1979. Sources and concentrations of NO_3-N in ground water of the Central Plate region, Nebraska. *Ground Water*, 17: 291—301.

Gorsler, M., 1980. Ein graphisches Verfahren zur Abschätzung des minimalen relativen Sauerstoffgehaltes in Stauhaltungen und Flußstauseen. *Wasser/Boden*, 32: 559—562.

Harting, P., 1981. *Theoretische und experimentelle Untersuchungen über thermodynamische Isotopieeffekte bei Lösungsprozessen vom Typ Gas-Wasser.* Dissertation B, Akademie der Wissenschaften der DDR, Berlin, 98 pp.

Hauck, R.D. and Bremner, J.M., 1976. Use of tracers for soil and fertilizer nitrogen research. *Adv. Agron.*, 28: 219—266.

Hauck, R.D., Bartholomew, W.V., Bremner, J.M., Broadbent, F.E., Cheng, H.H., Edwards, A.P., Keeney, D.R., Legg, J.O., Olsen, S.R. and Porter, L.K., 1972. Use of variations in natural nitrogen isotope abundance for environmental studies: a questionable approach. *Science*, 177: 453—454.

Haynes, R.J. and Goh, K.M., 1978. Ammonium and nitrate nutrition of plants. *Biol. Rev. Cambridge Philos. Soc.*, 53(4): 465—510.

Healy, T.V., McKay, H.A.C., Pilbeam, A. and Scargill, D., 1970. Ammonia and ammonium sulfate in the troposphere over the United Kingdom. *J. Geophys. Res.*, 75: 2317—2321.

Heinrichs, H. and Mayer, R., 1977. Distribution and cycling of major and tracer elements in two central European forest ecosystems. *J. Environ. Qual.*, 6: 402—407.

Hoering, T.C., 1955. Variations in N-15 abundance in naturally occurring substances. *Science*, 122: 1233—1234.

Hoering, T.C. and Ford, T.H., 1960. The isotope effect in the fixation of nitrogen by Azotobacter. *J. Am. Chem. Soc.*, 82: 376—378.

Hübner, H., 1980. Über einige Isotopieeffekte des Stickstoffs in der Bodenzone. *ZfI-Mitt.*, 30: pp. 141—162.

Hübner, H., 1981. Wird der Ozonschild der Erde durch den zunehmenden Stickstoff-Düngemittelverbrauch zerstört? Ein Beitrag der Isotopenforschung. *Isotopenpraxis*, 17(4): 140—150.

Islam, M.S. and Parsons, J.W., 1979. Mineralization of urea and urea derivatives in anaerobic soils. *Plant Soil*, 51: 319—330.

Junk, G. and Svec, H.J., 1958. The absolute abundance of the N isotopes in the atmosphere and compressed gas from various sources. *Geochim. Cosmochim. Acta*, 14: 234—243.

Karamanos, R.E. and Rennie, D.A., 1978. N isotope fractionation during NH_4^+-exchange reactions with soil clay. *Can. J. Soil Sci.*, 58: 53—60.

Karamanos, R.E. and Rennie, D.A., 1980a. Variations in natural N-15 abundance as an aid in tracing fertilizer nitrogen transformations. *Soil Sci. Soc. Am. J.*, 44: 57—62.

Karamanos, R.E. and Rennie, D.A., 1980b. Changes in natural N-15 abundance associated with pedogenic processes in soil, I. Changes associated with saline seeps; II. Changes on different slope positions. *Can. J. Soil Sci.*, 60: 337—344; 365—372.

Kohl, D.H., Shearer, G.B. and Commoner, B., 1971. Fertilizer nitrogen: contribution to nitrate in surface water in a corn belt watershed. *Science*, 174: 1331—1334.

Kohl, D.H., Shearer, G.B. and Commoner, B., 1973. Variation of N-15 in corn and soil following application of fertilizer N. *Soil Sci. Soc. Am. Proc.*, 37: 888—892.

Kohl, D.H., Shearer, G. and Harper, J.E., 1979. The natural abundance of N-15 in nodulating and non-nodulating isolines of soybeans. In: E.R. Klein and P.D. Klein (Editors), *Stable Isotopes, Proceedings 3rd International Conference.* Academic Press, New York, N.Y., pp. 317—325.

Kohl, D.H. and Shearer, G., 1980. Isotopic fractionation associated with symbiotic N_2 fixation and uptake of NO_3^- by plants. *Plant Physiol.*, 66: 51—56.

Kohl, D.H., Bryan, B.A., Feldman, L., Brown, P.H. and Shearer, G., 1982. Isotopic fractionation in soybean nodules. In: H.-L. Schmidt, H. Förstel and K. Heinzinger (Editors), *Stable Isotope, Proceedings of the 4th International Conference, Jülich, March 23—26, 1981. Analytical Chemistry Symposia Series, 11*. Elsevier, Amsterdam, pp. 451—457.

Kreitler, C.W., Ragone, S.E. and Katz, B.G., 1978. $^{15}N/^{14}N$ ratios of ground-water nitrate, Long Island, N.Y. *Ground Water*, 6: 404—409.

Laue, K.-H., Sander, Th.A. and Wagener, D., 1980. Das DHG-Verfahren — ein NO_x-Reduzierungsverfahren für Abgase. *Umwelt*, 5/80: 517—522.

Létolle, R., 1980. N-15 in the natural environment. In: P. Fritz and J.Ch. Fontes (Editors), *Handbook of Environmental Isotope Geochemistry, 1. The Terrestrial Environment, A*. Elsevier, pp. 407—433.

Létolle, R. and Mariotti, A., 1974. Utilisation des variations naturelles d'abondance de l'azote-15 comme traceur en hydrogeologie. In: *Isotope Techniques in Groundwater Hydrology, 2*. IAEA, Vienna, pp. 209—220.

Lieth, H. and Whittaker, R.H. (Editors), 1975. *Primary Productivity of the Biosphere*. Springer-Verlag, Berlin, 339 pp.

Maass, I. and Weise, G., 1981. Untersuchungen zur regionalen Ausbreitung von Schadstoffen mit Hilfe technogener Isotopenvariationen. *Isotopenpraxis*, 17: 156—159.

Machta, L. and Hughes, E., 1970. Atmospheric oxygen in 1967 to 1970. *Science*, 168: 1582—1584.

Mariotti, A. and Létolle, R., 1977. Application de l'étude isotopique de l'azote en hydrologie et en hydrogéologie — Analyse des résultants obtenus sur un exemple précis: le bassin de Mélarchez (Seine-et-Marne, France). *J. Hydrol.*, 33: 157—172.

Mariotti, A., Mariotti, F., Amarger, N., Pizelle, G., Ngambi, J.-M., Champigny, M.-L. and Moyse, A., 1980. Fractionnements isotopiques de l'azote lors de processus d'absorption de nitrates et de fixation de l'azote atmosphérique par les plantes. *Physiol. Veg.*, 18: 163—181.

Mariotti, A., Germon, J.C., Leclerc, A., Catroux, G. and Létolle, R., 1982. Experimental determination of kinetic isotope fractionation of nitrogen isotopes during denitrification. In: H.-L. Schmidt, H. Förstel and K. Heinzinger (Editors), *Stable Isotopes, Proceedings of the 4th International Conference, Jülich, March 23—26, 1981. Analytical Chemistry Symposia Series, 11*. Elsevier, Amsterdam, pp. 459—464.

McConnell, J.C., 1973. Atmospheric ammonia. *J. Geophys. Res.*, 78: 7812—7821.

Melander, L. and Saunders, W.H., Jr., 1980. *Reaction Rates of Isotopic Molecules*. J. Wiley and Sons, New York, N.Y., 331 pp.

Monse, E.U., Spindel, W. and Stern, M., 1969. Analysis of isotope effect calculations illustrated with exchange equilibria among oxynitrogen compound. In: Isotope Effects in Chemical Processes. *Am. Chem. Soc., Adv. Chem. Ser.*, 89: 148—184.

Moore, H., 1977. The isotopic composition of NH_3, NO_2 and NO_3^- in the atmosphere, *Atmos Environ.*, 11: 1239—1243.

Mortenson, L.E. and Thorneley, R.N.F., 1979. Structure and function of nitrogenase. *Ann. Rev. Biochem.*, 48: 387—418.

Nikonova, L.A., Rummel, S., Shilov, A.E. and Wahren, M., 1980. Is diimide an intermediate in dinitrogen reduction by vanadium (II) complexes in protic media? *Nouveau J. Chim.*, 4: 427—430.

Orel, A.E. and Seinfeld, J.H., 1977. Nitrate formation in atmospheric aerosols. *Environ. Sci. Technol.*, 11: 1000—1007.

Painter, H.A., 1970. A review of literature on inorganic nitrogen metabolism in microorganisms. *Water Res.*, 4: 393—450.

Parwel, A., Ryhage, R. and Wickman, F.E., 1957. Natural variations in the relative abundance of the nitrogen isotopes. *Geochim. Cosmochim. Acta*, 11: 165—170.

Perschke, H., Keroe, E.A., Proksch, G. and Muehl, A., 1971. Improvements in the determination of N-15 in the low concentration range by emission spectroscopy. *Anal. Chim. Acta*, 53: 459—463.
Reid, A.S., Webster, G.R. and Krouse, H.R., 1969. Nitrogen movement and transformation in soils. *Plant Soil*, 31: 224—237.
Remy, H., 1973. *Lehrbuch der Anorganischen Chemie, II.* Geest & Portig, Leipzig, 12-13th ed., p. 846, table 108.
Rennie, D.A. and Paul, E.P., 1975. Nitrogen isotope ratios in surface and sub-surface soil horizons. In: *Isotope Ratio as Pollutant Source and Behaviour Indicators.* IAEA, Vienna, pp. 441—453.
Rennie, D.A., Paul, E.A. and Johns, L.E., 1974. Isotope traceraided research on the nitrogen cycle in selected Saskatchewan soils. In: *Effects of Agricultural Production on Nitrates in Food and Water with Particular Reference to Isotope Studies.* IAEA, Vienna, pp. 77—90.
Rennie, D.A., Paul, E.A. and Johns, L.E., 1976. Natural N-15 abundance of soil and plant samples. *Can. J. Soil Sci.*, 56: 43—50.
Riga, A., Van Praag, H.J. and Brigode, N., 1971. Rapport isotopique naturel de l'azote dans quelques sols forestiers et agricoles de belgique soumis à divers traitements culturaux. *Geoderma*, 6: 213—222.
Robertson, F.N., 1979. Evaluation of nitrate in the ground water in the Delaware coastal plain. *Ground Water*, 17: 328—337.
Rummel, S., 1979. *Bestimmung von Stickstoffisotopie-effekten zum Nachweis kurzlebiger Zwischenstufen im Prozeß der komplexkatalytischen Reduktion von molekularem Stickstoff.* Dissertation B, Akademie der Wissenschaften der DDR, Leipzig, 126 pp.
Rummel, S., Bayerl, B., Johansen, H., Lorenz, B. and Wahren, M., 1976. ^{15}N-Isotopieeffekte in modellierten Teilschnitten der komplexkatalytischen Stickstoffixierung. *Z. Chem.*, 16: 193—195.
Rummel, S., Bayerl, B., Heine, I. and Wahren, M., 1981. Isotopenfraktionierungseffekte in Disproportionierungsreaktionen von Diimin- und Hydrazinderivaten. *Isotopenpraxis*, 17(4): 176—178.
Scalan, R.S., 1958. *The isotopic composition, concentration, and chemical state of the nitrogen in igneous rocks.* Thesis, University of Arkansas, Fayetteville, Ark., 79 pp.
Shearer, G.B. and Legg, J.O., 1975. Variations in the natural abundance of N-15 of wheat plants in relation to fertilizer nitrogen applications. *Soil Sci. Soc. Am. Proc.*, 39: 896—901.
Shearer, G.B., Kohl, D.H. and Commoner, B., 1974a. The precision of determinations of the natural abundance of nitrogen-15 in soils, fertilizer, and shelf chemicals. *Soil Sci.*, 118: 308—316.
Shearer, G., Duffy, J., Kohl, D.H. and Commoner, B., 1974b. A steady-state model of isotopic fractionation accompanying N transformation in soil. *Soil Sci. Soc. Am. Proc.*, 38: 315—322.
Shearer, G., Kohl, D.H. and Chien, Sen-Hsiung, 1978. The N-15 abundance in a wide variety of soils. *Soil Sci. Soc. Am. J.*, 42: 899—902.
Shearer, G., Kohl, D.H. and Harper, J.E., 1980. Distribution of ^{15}N among plant parts of nodulating and non-nodulating isolines of soybeans. *Plant Physiol.*, 66: 57—60.
Shine, H.J., Hendersen, G.N., Cu, A. and Schmid, P., 1977. Benzidine rearrangement, 14. The N-kinetic isotope effect in the acid-catalyzed rearrangement of hydrazobenzene, *J. Am. Chem. Soc.*, 99: 3719—3723.
Smith, C.J. and Chalk, P.M., 1980. Comparison of the efficiency of urea, aqueous NH_3 and $(NH_4)_2SO_4$ as N-fertilizers. *Plant Soil*, 55: 333—337.
Söderlund, R. and Svensson, B.H., 1976. The global nitrogen cycle, SCOPE Report 7. *Ecol. Bull. (Stockholm)*, 22: 23—73.

Spedding, F.H., Powell, J.E. and Svec, H.J., 1955. A laboratory method for separating N isotopes by ion exchange. *J. Am. Chem. Soc.*, 77: 6125—6132.
Spindel, W., 1954. The calculation of equilibrium constants for several exchange reactions of N-15 between oxy compounds of nitrogen. *J. Chem. Phys.*, 22: 1271—1272.
Stanford, G., Dziencia, St. and Vander Pol, R.A., 1975. Effect of temperature on denitrification rate in soils. *Soil Sci. Am. Proc.*, 39: 867—870.
Starr, J.L. and Parlange, J.-Y., 1975. Nonlinear denitrification kinetics with continuous flow in soil columns. *Soil Sci. Am. Proc.*, 39: 875—880.
Steele, K.W. and Daniel, R.M., 1978. Fractionation of nitrogen isotopes by animals: a further complication to the use of variations in the natural abundance of ^{15}N for tracer studies. *J. Agric. Sci.*, 90: 7—9.
Stewart, B.A., 1970. Volatilization and nitrification of N from urine under simulated cattle feedlot conditions. *Environ. Sci. Technol.*, 4: 579—582.
Sulser, H. and Sager, F., 1976. Identification of uncommon amino acids in the lentil seed (*Lens culinaris* Med.) *Experimentia*, 32: 422.
Super, M., Heese, H. De V., MacKenzie, D., Dempster, W.S., Plessis, Du J. and Ferreira, J.J., 1981. An epidemological study of well-water nitrates in a group of South West African/Namibian infants. *Water Res.*, 15: 1265—1270.
Taylor, T.I. and Spindel, W., 1958. Preparation of highly enriched N-15 by chemical exchange of NO with HNO_3. In: J. Kistemaker, J. Bigeleisen and A.O.C. Nier (Editors), *Proceedings of the Symposium on Isotope Separation*. North-Holland, Amsterdam, pp. 158—177.
Urey, H.C., 1947. The thermodynamic properties of isotopic substances. *J. Chem. Soc.*, pp. 562—581.
Wada, E. and Hattori, A., 1978. N-isotope effect in assimilation of inorganic nitrogenous compounds by marine diatoms. *Geomicrobiol. J.*, 1: 85—101.
Wada, E., Kadonaga, T. and Matsuo, S., 1975. N-15 abundance in nitrogen of naturally occuring substances and global assessment of denitrification from isotopic viewpoint. *Geochem. J.*, 9: 139—148.
Weise, G. and Maass, I., 1980. Untersuchungen zur regionalen Ausbreitung von Schadstoffen mit Hilfe technogener Isotopenvariationen. *ZfI-Mitt.*, 30: 267—276.
Wellman, R.P., Cook, F.D. and Krouse, H.R., 1968. N-15 microbiological alteration of abundance. *Science*, 161: 269—270.
Winter, C.H. and Burris, R.H., 1976. Nitrogenase. *Ann. Rev. Biochem.*, 45: 409—426.
Winteringham, F.P., 1980. Nitrogen balance and related studies: a global review. In: *Soil Nitrogen as Fertilizer or Pollutant*. IAEA, Vienna, pp. 307—344.
Wlotzka, F., 1972. Nitrogen, 7-B. Isotopes in nature. In: K.H. Wedepohl (Editor), *Handbook of Geochemistry, II/3*. Springer-Verlag, Berlin, pp. 7-B-1 to 7-B-9.
Wolinez, W.F., Sadoroshny, J.K. and Florenski, K.P., 1967. Über die isotope Zusammensetzung des Stickstoffs in der Erdkruste. *Geochimija*, 5: 587—592 (in Russian).
Wong-Chong, G.M. and Loehr, R.C., 1978. Kinetics of microbial nitrification: nitrite-nitrogen oxidation. *Water Res.*, 12: 605—609.
Zschiesche, M., 1972. *Entwicklung probenchemischen Verfahren zur Bestimmung der isotopen Zusammensetzung des N aus Gasen, Gesteinen und N-Lieferanten für bakteriologische Modelluntersuchungen.* Dissertation, Karl-Marx-Universität, Leipzig, 76 pp.

Chapter 10

CHLORINE-36 IN THE TERRESTRIAL ENVIRONMENT

HAROLD W. BENTLEY, FRED M. PHILLIPS and STANLEY N. DAVIS

INTRODUCTION

Chlorine possesses only one unstable isotope with a half-life greater than one hour: ^{36}Cl. The half-life of ^{36}Cl, $(3.01 \pm 0.04) \times 10^5$ years, renders it suitable for a wide range of geologic applications (half-life averaged from data of Bartholomew et al. (1955) and Goldstein (1966), as revised by Endt and Van der Leun (1973)). The geochemistry of chlorine is also advantageous for many types of geological and hydrological studies. Chlorine has a higher electron affinity than any element except fluorine, and thus in virtually all natural circumstances exists as the chloride anion. Chloride is, therefore, an excellent natural tracer, passing through various systems with only minimal chemical interaction. Chloride is also one of the most hydrophilic substances, rendering it especially suitable for hydrologic tracing.

Analysis of ^{36}Cl

A major problem in the application of ^{36}Cl to geological studies has always been the combination of a low natural abundance with a long half-life, resulting in a very low specific activity. ^{36}Cl activity in natural samples ranges from 10^{-5} to 10 disintegrations per minute (dpm) per gram chloride (equivalent to 1.36×10^{-16} to 1.36×10^{-10} atoms ^{36}Cl/Cl), with most samples near the lower end of the range. The activity of sea salt and bedded salts is even lower than 10^{-5} dpm per gram chloride. The history of ^{36}Cl application is, therefore, closely related to analytical developments.

^{36}Cl was initially measured in screen-wall counters similar to those used for ^{14}C in the 1950's (Davis and Schaeffer, 1955). This technique had a sensitivity of about one ^{36}Cl in 10^{12} chlorine, an abundance considerably above that of most natural samples. This sensitivity was substantially improved in the 1960's with the application of liquid scintillation to about 5×10^{-13} ^{36}Cl/Cl (Bagge and Willkom, 1966; Ronzani and Tamers, 1966; Tamers et al., 1969). Some additional increase in sensitivity has been achieved through the improvements to the liquid scintillation analysis described by Roman and Airey (1981), bringing many near-surface rocks and modern waters within range of the technique, but not deep rocks and old groundwater. The most recent advance has been through the application of tandem accelerator mass spectrometry (TAMS) (Bentley, 1978; Elmore et al., 1979).* This mass spectrometric approach is a radical departure from the direct counting procedures previously used and combines the advantages of great sensitivity (down to 5×10^{-16} ^{36}Cl/Cl) with small sample size (10 mg of chloride). With this instrument, all natural samples (with the exception of marine salts) are now within the range of analysis. The introduction of ^{36}Cl analysis by tandem accelerator mass spectrometry has already expanded the scope of ^{36}Cl application, discussed below.

NATURAL PRODUCTION OF ^{36}Cl

Most ^{36}Cl is produced by three reactions: spallation of heavier nuclei, principally Ar, K and Ca, by energetic cosmic rays, slow neutron activation of ^{36}Ar, and neutron activation of ^{35}Cl. Cosmic radiation is attenuated by the atmosphere and lithosphere; therefore, the first two reactions occur primarily in the atmosphere and in the upper layers of the earth's crust. The third is important throughout the lithosphere.

Meteoric ^{36}Cl

About 30% of the near-surface global ^{36}Cl is of atmospheric origin, termed "meteoric ^{36}Cl". Virtually all of this ^{36}Cl is derived from cosmic ray spallation of ^{40}Ar by the reaction ^{40}Ar$(x,x'\alpha)$ ^{36}Cl and neutron activation of ^{36}Ar through the reaction ^{36}Ar(n,p) ^{36}Cl. The neutrons are cosmic ray secondaries. Neutron activation of ^{35}Cl and spallation of K and Ca are very minor due to scarcity of the parent isotopes in the atmosphere. Lal and Peters (1967) calculated that spallation of ^{40}Ar produces a global ^{36}Cl fallout of 11 atoms m^{-2} s^{-1}, while Oeschger et al. (1969) calculated a fallout varying between 17 and 26 atoms m^{-2} s^{-1} depending on the sunspot cycle. In addition to this, neutron activation of ^{36}Ar results in a fallout of about 5 atoms m^{-2} s^{-1} (Onufriev, 1968). The variation of fallout with latitude has been calculated from the data of Lal and Peters (1967), and is illustrated in Fig. 10-1. The heaviest fallout is in the mid-latitudes of both hemispheres.

^{36}Cl is produced by both spallation and neutron activation throughout the

---

* All TAMS ^{36}Cl measurements to date have been performed on the M.P. Van der Graaff accelerator at the Nuclear Structure Research Laboratory at the University of Rochester, under the direction of David Elmore and Harry Gove. However, many groups worldwide are now constructing TAMS apparatus, and we expect that the application of ^{36}Cl to problems of isotope geology will substantially increase in the near future.

atmosphere. About 40% of the total atmospheric production is in the troposphere and 60% in the stratosphere (Oeschger et al., 1969). Stratospherically produced ^{36}Cl enters the troposphere during periods of mixing. The troposphere also contains stable chloride derived from sea spray. The mixed aerosol of ^{36}Cl and stable chloride is quickly washed out of the troposphere, or becomes associated with aerosols which fall out. The mean residence time is about one week (Turekian et al., 1977). The meteoric stable chloride fallout decreases exponentially from coastal areas toward continental interiors, with some modification due to prevailing winds and orographic effects (Eriksson, 1960). The ^{36}Cl fallout is superimposed on this pattern, with only a latitudinal variation. The long-term average ratio of ^{36}Cl to total chloride in the atmospheric fallout may thus be predicted at any particular location. Bentley and Davis (1982) have calculated the variation in this ratio over the United States. Fig. 10-2 was produced by dividing the ^{36}Cl fallout from Fig. 10-1 by the chloride deposition from Eriksson (1960). The chloride in precipitation shown by Eriksson (1960) was multiplied by 1.3 to account for dry fallout, based on discussion in the text.

The ^{36}Cl/Cl ratio of modern, but pre-bomb, groundwater is compared to

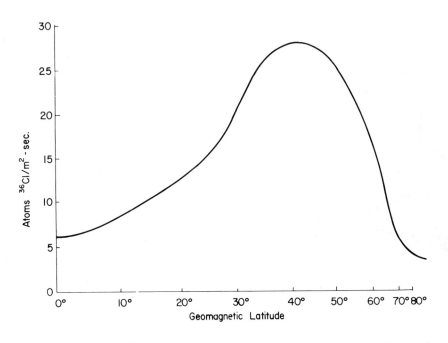

Fig. 10-1. Meteoric ^{36}Cl fallout with latitude. Latitude dependence from Lal and Peters (1967) and total fallout from Lal and Peters (1967) and Onufriev (1968).

the predicted ratio in Table 10-1. The pre-bomb status of the water was verified by ^{14}C or ^{3}H dating. Chloride delivery to the Madrid Basin was extrapolated using data from Northern Europe reported by Eriksson (1960). The mean ^{36}Cl/Cl for each location is in good agreement with the predicted ratio. These data support the smaller ^{40}Ar spallation production estimate of Lal and Peters (1967), used to calculate the predicted ^{36}Cl/Cl, over the larger production estimate of Oeschger et al. (1969).

Knowledge of the ^{36}Cl/Cl ratio is valuable for hydrologic tracing using ^{36}Cl, inasmuch as the ratio will remain constant despite evaporation or transpiration. The ^{36}Cl concentration will increase under these circumstances. Nearly all of the meteoric ^{36}Cl will enter the hydrologic system and remain there, due to its hydrophilic behavior. Probably most of this ^{36}Cl is transported to the ocean relatively rapidly via groundwater and/or rivers. A significant portion of the ^{36}Cl in groundwater may decay in the subsurface. Some is also trapped in closed basins. The determination of the exact figures for these various fluxes would seem to be an ideal method of tracing the fate of meteoric solutes in the hydrologic cycle. Unfortunately, ^{36}Cl from atmo-

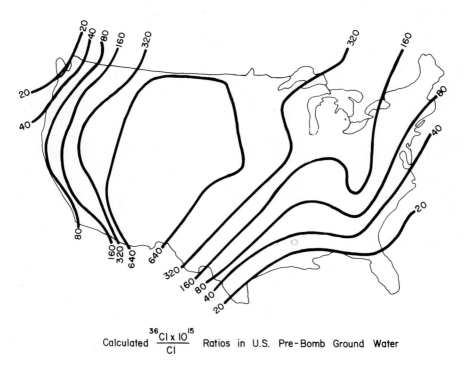

Calculated $\frac{^{36}Cl \times 10^{15}}{Cl}$ Ratios in U.S. Pre-Bomb Ground Water

Fig. 10-2. Calculated ^{36}Cl/Cl ($\times 10^{15}$) ratios in precipitation and dry fallout over the United States.

spheric nuclear weapons testing has greatly complicated the task of such studies.

Epigene production of ^{36}Cl

Spallation of ^{40}Ar and neutron activation of ^{36}Ar are relatively minor sources of ^{36}Cl at the surface of the lithosphere and oceans due to the low abun-

TABLE 10-1

A comparison of observed and predicted $^{36}Cl/Cl$ ratios in modern (but pre-bomb) groundwater (from Bentley and Davis, 1982)

	$^{36}Cl/Cl$ ($\times 10^{15}$)	$^{36}Cl\ l^{-1}$ ($\times 10^{-7}$)	Average $^{36}Cl/Cl$ ($\times 10^{15}$)	Predicted $^{36}Cl/Cl$ ($\times 10^{15}$)
Southeast Arizona				
St. David	400 ± 40	6		—
Tucson Well	320 ± 50	6		—
Monkey Springs	478 ± 19			—
Tucson Well B-8	365 ± 18	11.7	364.5 ± 78	400
C-13	379 ± 22	9.6		—
A-31	245 ± 16	6.6		—
Madrid Basin				
535-7-b	231 ± 21	—		—
535-7-9	295 ± 12	—	254 ± 36	250
535-5-c	235 ± 7	—		—
Southeast Texas				
Al 68-51-803	32 ± 3		32 ± 3	30
South Alberta				
M.R. 42	538	14.4		—
M.R. 41	327	3.7	453 ± 111	500
M.R. 26	494	50.3		—
Central New Mexico				
Socorro				
SS-6	652 ± 32	—		—
SS-7	780 ± 39	—		—
SS-8	723 ± 36	—		—
SS-9	718 ± 36	—	707 ± 37	>640
SS-10	620 ± 37	—		—
SS-11	722 ± 43	—		—
SS-12	737 ± 37	—		—
Southern Nevada				
Nevada Test Site	634 ± 60	—	634 ± 60	640

dance of argon in these environments. The major processes are spallation of abundant K and Ca, and neutron activation of ^{35}Cl. ^{35}Cl is not particularly abundant in most rocks, but its large neutron cross-section nevertheless ensures measurable ^{36}Cl production. The two largest sources of neutrons at the earth's surface are evaporation neutrons from cosmic ray spallation and neutrons associated with various decay processes of elements in the uranium and thorium decay series. Lal and Peters (1967) calculated a sea-level, surface capture rate for thermal neutrons from cosmic ray interactions of 3×10^{-2} per kilogram rock per second. This is about two orders of magnitude above the neutron production rate due to U and Th series decay.

Buildup of ^{36}Cl from cosmic ray processes begins as soon as a rock is exposed at the surface of the earth. Eventually the ^{36}Cl/Cl ratio will arrive at secular equilibrium with the production rate and the ^{36}Cl half-life. The equilibrium ^{36}Cl/Cl ratio may be calculated, given the altitude and the chemical composition of the rock. Secular equilibrium ^{36}Cl/Cl ratios for most rocks at sea level are on the order of 10^{-11}. Observed ratios in rocks are generally at least an order of magnitude less than this, an indication of weathering rates. A detailed discussion of these processes may be found in the next section.

Hypogene ^{36}Cl

Most cosmic rays are attenuated in the atmosphere and the first few meters of the earth's crust or ocean. Spallation is not common below this depth. The major source of neutrons below about 30 m (Zito et al., 1982) is decay of U and Th series elements. The U and Th decay series produce both direct fission neutrons and secondary neutrons from α,n reactions on light isotopes. The secondary neutrons are more numerous in most rocks. Feige et al. (1968) presented methods for the calculation of the subsurface neutron flux.

^{36}Ar is not common in the subsurface, and therefore neutron activation of ^{35}Cl is the only significant production mechanism. Although most rocks contain only small concentrations of chloride, the large activation cross section of ^{35}Cl allows the production of measurable ^{36}Cl under even the modest neutron fluxes in the deep subsurface.

Global inventories and fluxes

Lal and Peters (1967) estimated a global ^{36}Cl inventory of 15 tons, with 70% in the oceans. However, they did not take into account the hypogene production. The total subsurface inventory is difficult to calculate because most chlorine is in the mantle, and the mantle concentration is problematic. Estimates of the chlorine concentration in the mantle range from about 1000 ppm (Mason, 1958) to about 100 ppm (Rösler and Lange, 1972). The

U and Th concentrations necessary for calculating the thermal neutron density are also uncertain, but assuming U is 0.1 ppm and Th 0.2 ppm (Rogers and Adams, 1978), the equilibrium ^{36}Cl/Cl ratio would be about 5×10^{-16}. The total subsurface ^{36}Cl inventory would thus lie between 300 tons (assuming a mantle chlorine of 100 ppm) and 3000 tons (assuming 1000 ppm). In contrast, the oceanic inventory is only about 10 tons. Measurements have not demonstrated that most surface rocks have achieved the high ^{36}Cl/Cl ratios calculated for cosmic ray equilibrium, thus the estimate of 5 tons by Lal and Peters (1967) should be lowered. Assuming an average ^{36}Cl/Cl ratio of 10^{-13} in the upper two meters of the land area of the earth, the total inventory is about 10 kg.

The total flux of ^{36}Cl from the atmosphere to the continents is about 4 g a^{-1} (assuming an average fallout of 15 atoms m^{-2} s^{-1}). Dividing this by the global meteoric chloride flux to the ocean of 1.06×10^{14} g a^{-1} (Meybeck, 1979) gives an average runoff ^{36}Cl/Cl ratio of 40×10^{-15}. The remaining chloride flux to the ocean, 1.15×10^{14} g a^{-1}, is derived from weathering. Assuming that half of this is derived from bedded evaporites with a ^{36}Cl/Cl ratio below 10^{-16} and the other half of near-surface rocks with a ratio of 10^{-13}, the weathering-derived ^{36}Cl flux to the ocean would be about 6 g a^{-1}.

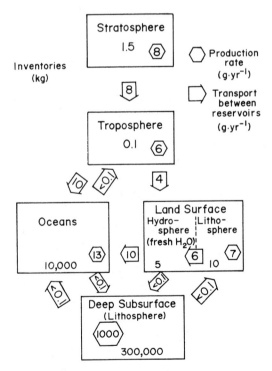

Fig. 10-3. Estimated global inventories and fluxes of ^{36}Cl.

Fluxes between the deep subsurface and the land and ocean are quite minor in comparison with those discussed above. The global inventories and fluxes are summarized in Fig. 10-3. The numbers shown should be regarded as preliminary. They are mainly intended to illustrate the relative magnitudes of the processes involved.

STUDIES OF ^{36}Cl IN THE LITHOSPHERE

Epigene ^{36}Cl

Davis and Schaeffer (1955) first suggested that the buildup of ^{36}Cl in minerals could be a useful means of determining the period of surface exposure experienced by rocks and studying rates of erosion and weathering. They reasoned that because of the low subsurface neutron flux the ^{36}Cl produced there should be much less than the amounts which would build up after exposure to cosmic rays. However, they did not take into account K and Ca spallation production, and thus their calculations should be revised.

^{36}Cl from both spallation and neutron activation accumulates at a rate proportional to the exposure time. The same ^{36}Cl also decays at a rate proportional to its decay constant. Thus the buildup with time is given by the differential equation:

$$dN_{36} = (\psi + \phi_n f) \, dt - \lambda_{36} N_{36} \, dt \qquad (1)$$

where N_{36} is the number of ^{36}Cl atoms, ψ the rate of spallation production of ^{36}Cl, ϕ_n the neutron flux, f the fraction of the neutrons absorbed by ^{35}Cl, t time, and λ_{36} the ^{36}Cl decay constant (2.3×10^{-6} a^{-1}). The solution of this equation, integrated between $t = t$, and $t = 0$, is:

$$N_{36} = \frac{\psi + \phi_n f}{\lambda_{36}} (1 - e^{-\lambda_{36} t}) \qquad (2)$$

The method of calculating ψ is treated in Yokoyama et al. (1973) and representative values are given in Yokoyama et al. (1977). The fraction of neutrons absorbed by ^{35}Cl may be calculated from the formula:

$$f = N_{35} \sigma_{35} / \sum_i N_i \sigma_i \qquad (3)$$

where N is the concentration of an isotope in atoms kg^{-1}, σ_{35} is the neutron activation cross-section of ^{35}Cl for ^{36}Cl, and σ_i is the neutron absorption cross section of an isotope (in barns). The sea-level neutron flux is about 3×10^{-2} neutrons kg^{-1} s^{-1} (Lal and Peters, 1967). This flux must be corrected for differences in atmospheric attenuation dependent upon elevation using the tables of Montgomery and Tobey (1949) or Yokoyama et al. (1977). If the sample is not actually on the surface an additional correction for the depth, given by Lal and Peters (1967), must be applied. These correction factors may be designated E, for elevation, and D, for depth. The most convenient way of representing the amount of ^{36}Cl is as the $^{36}Cl/Cl$ ratio. In this case the number of ^{36}Cl atoms must be divided by the chlorine concentration of the rock, in atoms kg^{-1}. In addition, the original ^{36}Cl content of the rock must be accounted for. The final formula is thus:

$$\frac{^{36}Cl}{Cl} = \frac{ED(\psi + \phi_n f)}{\lambda_{36} N_{Cl}} (1 - e^{-\lambda_{36} t}) + \left(\frac{^{36}Cl}{Cl}\right)_0 e^{-\lambda_{36} t} \qquad (4)$$

where N_{Cl} is the number of atoms of chloride.

Davis and Schaeffer (1955) performed the first rock ^{36}Cl measurement on a "pre-Wisconsin" exposure phonolite from Bull Cliff near Cripple Creek, Colorado; obtaining a $^{36}Cl/Cl$ of 1.6×10^{-12}. They calculated an exposure time of 24,000 years. Consideration of spallation would not significantly decrease this age since phonolite is a chlorine-rich rock. An attempt to measure ^{36}Cl in a nepheline-sodalite syenite from New Hampshire was not successful due to lack of sufficient analytical sensitivity.

Later attempts to measure the ^{36}Cl activity of solid salt samples from Dugway Salt Flat, Utah Salt Flat, and the Forty Mile Desert (all in Utah) by Bonner et al. (1961) were also unsuccessful. This indicated a $^{36}Cl/Cl$ ratio below 10^{-13}. Saline waters from Mono Lake in California, Pyramid Lake in Nevada, and Great Salt Lake in Utah were also found to have $^{36}Cl/Cl$ ratios below 10^{-13}. However, Bagge and Willkom (1966) were able to measure ^{36}Cl in a salt lake from Turkey, and calculated an age of $0.9-1.4 \times 10^6$ years. They did not consider the meteoric ^{36}Cl input, which would probably lower the calculated age. Onufriev and Soifer (1968) measured an average $^{36}Cl/Cl$ ratio of 3×10^{-14} in surface sediments, in conformity with the results of other investigations.

A novel and innovative approach to ^{36}Cl measurement was taken by Sinclair and Manuel (1974) who measured the buildup of ^{36}Ar from ^{36}Cl decay in a sodalite from Dungannon, Ontario. They divided the excess ^{36}Ar from ^{36}Cl by the K-Ar age of the sodalite to obtain a production rate of 560 ± 210 ^{36}Ar (g Cl)$^{-1}$ a^{-1}. This is equivalent to a secular equilibrium $^{36}Cl/Cl$ ratio of 1.5×10^{-14}, well below the direct detection sensitivity of the time. From the ^{36}Ar (or ^{36}Cl) production and the chlorine content of the rock a typical hypogene neutron flux of 2×10^{-4} neutrons kg^{-1} s^{-1} may be calculated. The rock was 3.9×10^8 years old, and presumably spent most of this time at a greater depth than the near-surface exposure in which it was collected, accounting for the low average flux.

All of the studies discussed above were hampered by the necessity of selecting high-chlorine rocks, of considerable age, in order to be able to detect the ^{36}Cl. Their interpretations were also clouded by ignorance of the meteoric ^{36}Cl input and of spallation production of ^{36}Cl. The advent of TAMS, which can measure ^{36}Cl in most rocks at all expected concentrations, and increased understanding of ^{36}Cl production should encourage greatly expanded application of cosmic ray ^{36}Cl buildup for dating minerals. This application should be particularly valuable for rocks which are relatively suddenly emplaced at the earth's surface, such as lava flows, caliches in soils, and landslide scarps. As an example, Fig. 10-4 shows ^{36}Cl buildup in a typical basalt and caliche. The caliche has much higher initial ^{36}Cl/Cl due to incorporation of meteoric ^{36}Cl during accretion. ^{36}Cl buildup appears to be a possible method of dating both these rock types in the range 10^3 to 5×10^5 years.

In most circumstances rocks are not suddenly emplaced at the earth's surface. They are usually slowly exposed by erosion. If the erosion rate is reasonably constant, cosmic-ray-produced ^{36}Cl buildup may be used to calculate the erosion rate. The basic differential equation is the same as equation (1). However, the rate of ^{36}Cl production varies with depth of burial. The variation of the cosmic ray flux may be closely approximated by an exponential function:

$$\psi + \phi_n f = (\psi_0 + \phi_{n_0} f) e^{-\alpha z} \tag{5}$$

where ψ_0 and ϕ_{n_0} are the production rates at the surface and α is the cosmic

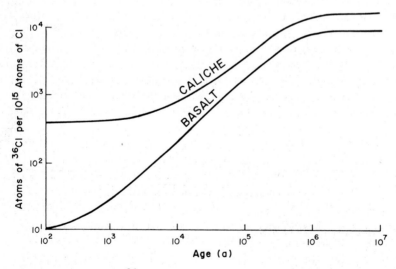

Fig. 10-4. Calculated ^{36}Cl buildup with time in typical basalt and caliche suddenly emplaced at the earth's surface at an altitude of 1500 m. Higher initial ^{36}Cl/Cl ratio in the caliche is due to presumed incorporation of meteoric ^{36}Cl at the time of concretion.

ray attenuation constant. The time of exposure may be related to the depth because the erosion rate is constant:

$$dt = -\beta dz \tag{6}$$

where β is the reciprocal of the erosion rate. The differential equation then becomes:

$$dN = -\beta(\psi_0 + \phi_{n_0} f)e^{-\alpha z} dz + \lambda_{36}\beta N_{36} dz \tag{7}$$

The solution of this equation is:

$$N_{36} = C e^{\lambda_{36}\beta z} + \frac{(\psi_0 + \phi_{n_0} f)\beta}{\lambda_{36}\beta + \alpha} e^{-\alpha z} \tag{8}$$

where C is the integration constant. The integration constant may be assigned a numerical value by using the boundary condition that at depth (below 15 m) the cosmic ray ^{36}Cl production is zero. For nearly all realistic combinations of erosion rates and ^{36}Cl production rates the first term in equation (8) is negligible above a depth of about 15 m. Therefore the equation, expressed in terms of the ^{36}Cl/Cl ratio, becomes:

$$\frac{^{36}Cl}{Cl} = \frac{(\psi_0 + \phi_{n_0} f)\beta}{(\lambda_{36}\beta + \alpha) N_{Cl}} e^{-\alpha z} \tag{9}$$

At the surface the exponential term goes to one. The total ^{36}Cl at the surface will therefore be:

$$R_s = \frac{(\psi_0 + \phi_{n_0} f)\beta}{(\lambda_{36}\beta + \alpha) N_{Cl}} + R_{se} \tag{10}$$

where R_s is the ^{36}Cl/Cl at the surface and R_{se} is the secular equilibrium ^{36}Cl/Cl due to neutrons from U and Th decay. This equation may be solved for the erosion rate:

$$\text{erosion rate} = \frac{1}{\beta} = \frac{\psi_0 + \phi_{n_0} f}{(R_s - R_{se}) N_{Cl}\alpha} - \frac{\lambda}{\alpha} \tag{11}$$

If the ^{36}Cl content of the latite phonolite from Cripple Creek, Colorado, reported by Davis and Schaeffer (1955) is assumed to be from steady erosion, the erosion rate, calculated by equation (11), is 4.25 cm ka^{-1}, very close to the mean global erosion rate of 5 cm ka^{-1} (Statham, 1977).

Hypogene production

Production of ^{36}Cl in the subsurface may be calculated using the methods presented above for surface production. Assuming production by spallation to be zero, the buildup with time is given by:

$$N_{36} = \frac{\phi_n f}{\lambda_{36}} (1 - e^{-\lambda_{36} t}) \tag{12}$$

If our interest is in the ^{36}Cl content of the groundwater, we must take into account porosity, chloride concentration, and the neutron absorption of water. In such a case, f in equation (12) is replaced by f^w, defined as

$$f^w = N^w_{35} \sigma_{35} / \sum_i N_i \sigma_i \tag{13}$$

where N^w_{35}, the number of ^{35}Cl atoms in the water per kilogram of saturated porous media, is given by:

$$N^w_{35} = 0.7577 \frac{n_e}{\rho_\beta} C_{Cl} A \tag{14}$$

where n_e is porosity, ρ_β is rock bulk density, C_{Cl} is chloride concentration in the water (mol l^{-1}), A is Avogadro's number, and 0.7577 is the isotopic abundance of ^{35}Cl.

We can expand the denominator in equation (13):

$$\sum_i N_i \sigma_i = \sum_j N_j \sigma_j + N^w_{35} \sigma_{35} + N^w_{H_2O} \sigma_{H_2O} \tag{15}$$

where j refers to those elements in the rock. Dividing each side of equation (12) by the concentration of chloride atoms in the water and including the initial ^{36}Cl/Cl ratio yields:

$$R = \frac{0.7577 \phi_n \sigma_{35} (1 - e^{-\lambda_{36} t})}{\lambda_{36} (\sum_j N_j \sigma_j + N^w_{35} \sigma_{35} + N^w_{H_2O} \sigma_{H_2O})} + \left(\frac{^{36}Cl}{Cl}\right)_0 e^{-\lambda_{36} t} \tag{16}$$

The secular equilibrium ratio, as $e^{-\lambda_{36} t} \to 0$, is:

$$R_{se} = \frac{0.7577 \phi_n \sigma_{35}}{\lambda_{36} (\sum_j N_j \sigma_j + N^w_{35} \sigma_{35} + N^w_{H_2O} \sigma_{H_2O})} \tag{17}$$

A major difficulty is determination of ϕ_n, the subsurface neutron flux. This may be calculated using the procedures of Pine and Morrison (1953), Morrison and Pine (1955), and Feige et al. (1968). Neutron fluxes calculated by the method of Feige et al. (1968) and the resultant secular equilibrium ^{36}Cl/Cl ratios, for low-chlorinity waters, calculated using equation (17) are presented in Table 10-2 (from Bentley, 1978).

The possibility of subsurface ^{36}Cl production was first mentioned by Jonte (1956). He attempted to measure ^{36}Cl activity in water from Arkansas hot springs. The activity he obtained was far above expected values, and he attributed it to analytical difficulties. It may have reflected bomb-^{36}Cl contamination.

Kuroda et al. (1957) and Kuroda and Kenna (1960) measured ^{36}Cl in uranium ores. The object of their investigation was similar to that of Sinclair and Manuel (1974): the determination of the subsurface neutron flux. The neutron flux was needed to calculate the ratio of induced to spontaneous fission in uranium ores. Kuroda and Kenna (1960) reported a ^{36}Cl/Cl ratio of 9×10^{-11} for a pitchblende from the Belgian Congo (Zaire) and a ratio of 5×10^{-11} for a pitchblende from the Great Bear Lake in Canada.

A number of samples have been obtained from formations where the groundwaters can reasonably be expected to have resided for millions of years. Analyses of these samples are summarized in Table 10-3. These pro-

TABLE 10-2

Calculated neutron production and secular equilibrium ^{36}Cl/Cl ratios for typical granite, sandstone, limestone, and shale, and groundwaters in these rocks[a] (from Bentley and Davis, 1982)

Rock type	U (ppm)	Th (ppm)	Neutron production (neutrons kg^{-1} a^{-1})		
			natural fission	uranium-induced	thorium-induced
Granite	2.8	11	1306	6776	7755
Sandstone	1.0	3.9	467	935	1482
Shale	4.5	13	2100	4110	5579
Limestone	2.2	0.2	1027	1408	57
Rock type	Total production (neutrons kg^{-1} a^{-1})		Steady state ^{36}Cl ratio ($\times 10^{15}$) at $t = \infty$		
Granite	15837		30.1		
Sandstone	2884		4.68		
Shale	11789		12.5		
Limestone	2482		10.9		

[a] Chloride in rock and water assumed to be less than 1000 ppm.

TABLE 10-3

Hypogene ^{36}Cl in very old groundwaters

Lithology	Chloride content (mg l^{-1})	Number of samples	Mean ^{36}Cl/Cl ($\times 10^{15}$)	Range ($\times 10^{15}$)
Sandstone	brine	5	3.6 ± 0.5	3— 4
Sandstone	< 2000	11	6.4 ± 1.2	4— 8
Limestone	brine	4	11.5 ± 2.5	8—15

vide some measure of the validity of the predicted results presented in Table 10-2.

DATING SALINE SEDIMENTS WITH ^{36}Cl

Saline lake deposits are of major economic and scientific significance. They are among the most important sources of lithium, boron, bromine, and other elements. During high-water epochs the lakes which filled their basins attracted game, and early man, and therefore their ancient shorelines contain numerous archeological sites. The character of the sediments beneath the lakes is frequently a sensitive indicator of climatic conditions. Studies of sediment cores from saline lakes in various parts of the world have shown their value as paleoclimatic records.

One of the most serious impediments to the study of these saline lakes is the difficulty of dating the sediments. If sufficient carbon is present they may be dated by ^{14}C, but only back to about 5×10^4 years. ^{230}Th dating has a somewhat longer range (several hundred thousand years), but frequently, large uncertainties are associated with the initial Th/U ratio. Paleomagnetic dating does not have an age limit, but numerous samples are required in order to identify the reversal patterns, and the method provides no age control for samples younger than 7×10^5 years.

^{36}Cl was proposed as a means of dating these deposits by Bagge and Willkom (1966). Chloride is hydrophilic and most chloride in a closed drainage basin will relatively quickly be transported to the terminal lake or playa. The half-life is suitable for dating late Pleistocene deposits.

Phillips et al. (1983) have measured the ^{36}Cl/Cl ratio in five halite samples from Searles Lake, California, U.S.A. Searles Lake is now a nearly dry playa, but during pluvial periods in the Pleistocene, Searles Lake became part of a drainage system which extended from Mono Lake in the Sierra Nevada to Death Valley. During dry periods the connections with upstream and downstream basins were severed. These episodes produced alternating layers of evaporites and muds in the lake sediments, which form a detailed record of climatic fluctuations.

TABLE 10-4

^{36}Cl in Searles Lake core KM-3 (from Phillips et al., 1983)

Depth (m)	^{36}Cl/Cl (× 10^{15})	Age[a] (ka)	Uncorrected ^{36}Cl age (ka)
14.2	55 ± 5%	10	—
32.5	55 ± 6%	28	10 ± 27
190.7	8.9 ± 12%	800	802 ± 50
206.5	7.7 ± 13%	900	865 ± 53
304.3	6.6 ± 23%	1300	922 ± 113

[a] "Actual" ages are based on ^{14}C, ^{230}Th, and magnetostratigraphy.

^{36}Cl from meteoric fallout, weathering of rocks containing epigene ^{36}Cl, and hypogene ^{36}Cl carried by groundwater discharge all entered the lake along with ordinary chloride. ^{36}Cl could also be produced within the lake sediments by cosmic rays if they remained near the land surface for a sufficient time after deposition. After hundreds of thousands of years, measurable ^{36}Cl will be produced by hypogene processes, if sufficient uranium or thorium are present. Successful sediment dating by ^{36}Cl depends on two conditions: constancy of the input ^{36}Cl/Cl ratio, and negligible cosmic ray production after deposition.

Measured ^{36}Cl/Cl ratios of the halite samples from various depths are listed in Table 10-4. The "actual" ages were determined by ^{14}C, ^{230}Th, and magnetostratigraphy. ^{36}Cl ages were calculated using the simple radiometric decay equation, and 55×10^{-15} as the initial ratio. Phillips et al. (1983) calculated that only half of this initial ^{36}Cl is of meteoric origin.

The ^{36}Cl ages are in agreement with the independently determined ages, with the exception of the deepest sample (Fig. 10-5). This discrepancy was ascribed most probably to contamination. The concordancy of the other dates indicates that the ^{36}Cl/Cl input ratio was probably relatively constant over the last 10^6 years. Thus ^{36}Cl appears to show considerable potential for the dating of continental saline sediments.

DATING OLD GROUNDWATER WITH ^{36}Cl

Chloride is one of the most desirable natural groundwater tracers. Highly soluble, it normally does not exist in a solid phase until brine concentrations are reached, nor does it participate in redox reactions. Due to its negative charge and small radius it is among the least sorbed ions on solid surfaces. Circulating water readily leaches chloride from soils and aquifers. Chloride therefore usually lacks sources as well as sinks in shallow groundwater systems. All of these geochemical properties suggest application of natural

^{36}Cl for dating groundwater. Due to its long half-life, ^{36}Cl is most suitable for dating old (> 50,000 years) groundwater. These advantages have been recognized for the past 20 years (Davis and DeWiest, 1966), but could not be utilized until the advent of TAMS analysis.

In order to calculate the groundwater age, the initial ^{36}Cl must be known. The ^{36}Cl concentration may be expected to vary widely as meteoric ^{36}Cl in precipitation is concentrated through evaporation and transpiration. The ^{36}Cl/Cl ratio in precipitation should remain constant despite reduction in the volume of the precipitation. Fig. 10-2 may therefore be used as a general guide to expected initial ^{36}Cl/Cl ratios in groundwaters in the United States.

^{36}Cl decay may be described by a typical exponential equation. For old groundwater the buildup of ^{36}Cl due to the subsurface neutron flux must also be accounted for. The secular equilibrium ^{36}Cl/Cl ratio may be calculated using the methods outlined under "*Hypogene production*". The change in the ^{36}Cl/Cl ratio with time is then described by:

$$R = R_0 e^{-\lambda_{36} t} + R_{se} (1 - e^{-\lambda_{36} t}) \tag{18}$$

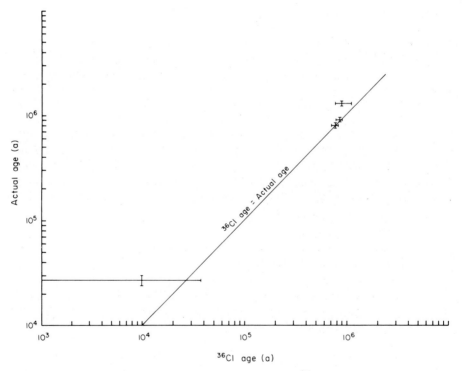

Fig. 10-5. A comparison of "actual" with uncorrected ^{36}Cl ages for halite samples from Searles Lake core KM-3.

where R is the measured ^{36}Cl ratio and R_0 the initial ratio. This equation may be solved for the water "age":

$$t = -\frac{1}{\lambda_{36}} \ln \frac{R - R_{se}}{R_0 - R_{se}} \tag{19}$$

The normal range of values of R_0 and R_{se} limits the groundwater age determinable by equation (19) to a maximum of approximately 2.5 Ma.

Although chloride typically has no sources or sinks in shallow groundwater systems (arid regions excepted), this is not necessarily true of deep groundwater. Chloride concentrations are frequently observed to increase substantially with depth, and along flow paths (Freeze and Cherry, 1979). This increase may be attributed to several possible sources. Chloride may be released during mineral dissolution or by diffusion from minerals. In some areas there are halite deposits in the deep subsurface. Even if solid halite is not present, high chloride water from either connate sources or previous salt dissolution may remain in less permeable zones. Thus either cross-aquifer leakage or diffusion of salt from aquitards may contribute chloride to relatively permeable aquifer systems. Finally, chloride initially present in shallow groundwater may be concentrated at depth by ion filtration (Graf, 1982).

These different possible sources of increased chloride concentration suggest different strategies for correction of ^{36}Cl groundwater "ages". If the increased concentration is due to solution of bedded salt, which may be expected to have a ^{36}Cl/Cl ratio below 10^{-16}, the ^{36}Cl concentration, rather than the ^{36}Cl/Cl ratio, may be used to calculate the age:

$$t = -\frac{1}{\lambda_{36}} \ln \frac{C^{36}}{C_0^{36}} = -\frac{1}{\lambda_{36}} \ln \frac{R\,C}{R_0\,C_0} \tag{20}$$

where C^{36} is the measured ^{36}Cl concentration (in atoms ^{36}Cl l^{-1}) C_0^{36} the initial concentration, C is the measured chloride concentration (in atoms Cl l^{-1}), and C_0 the original Cl concentration. Addition of "dead" chloride will not alter the ^{36}Cl concentration.

If the increased chloride concentration is from mixing with a high-chloride groundwater or from mineral dissolution, the expected ^{36}Cl concentration of the chloride source may be calculated (assuming secular equilibrium). The observed ^{36}Cl concentration may then be corrected by using the increase in chloride concentration in a mixing equation which also accounts for ^{36}Cl buildup from the subsurface neutron flux:

$$C^{36} = RC = R_0 C_0 e^{-\lambda_{36} t} + R_{se} C_0 (1 - e^{-\lambda_{36} t}) + R_{se}(C - C_0) \tag{21}$$

R_{se} is the secular equilibrium ^{36}Cl/Cl. R_{se} can be determined by measurement of the ^{36}Cl/Cl ratio of the source, either directly or after extraction from the mineral, by chemical analysis or the mineral U, Th, and high cross-section elements, or by direct measurement of the subsurface neutron flux (Zito et al., 1982). From equation (21) the water age may be obtained. The solution is given by:

$$t = -\frac{1}{\lambda_{36}} \ln \frac{C\,(R-R_{se})}{C_0\,(R_0-R_{se})} \tag{22}$$

The dating age limit is again about 2.5 Ma. Finally, if the increase in concentration is a result of ion filtration, equation (19) may be used without correction. Ion filtration will increase the chloride concentration without affecting the ^{36}Cl/Cl ratio, inasmuch as little fractionation of chloride may be expected during ion filtration.

Groundwater ^{36}Cl field studies

The Great Artesian Basin, Australia. The Great Artesian Basin covers about 1.7×10^6 km^2 in northeastern Australia, about one-fifth of the continent. It is one of the largest artesian aquifer systems in the world. The physical hydrology of the basin has been described by Habermehl (1980) and the isotope hydrology by Airey et al. (1979). The basin contains a multiple aquifer system in Triassic, Jurassic, Cretaceous, and Tertiary sandstones. The ^{36}Cl distribution in the Jurassic-Cretaceous "J" aquifer has been investigated by Bentley et al. (1983).

The aquifers are recharged around the margin of the structural basin and discharge in the south-central portion, near Lake Eyre. The study was conducted along two flowlines extending from a recharge area on the western flanks of the Great Dividing Range, northeast of Charleville, west and southwest to Innamincka, in South Australia. The study area and the distribution of the ^{36}Cl/Cl ratio in the "J" aquifer are illustrated in Fig. 10-6. The ^{36}Cl/Cl ratio decreases smoothly from the recharge area in the direction of flow and can be contoured consistently.

The chloride concentrations along these flowlines in the "J" aquifer are relatively constant. There is no consistent increase in chloride which would indicate subsurface sources. Groundwater ages were therefore calculated using equation (19). The ^{36}Cl/Cl ratios in the recharge area show little variation, ranging from about 95×10^{-15} to 125×10^{-15}. Fig. 10-6 therefore contains isochrons which show ages assuming an initial ^{36}Cl/Cl ratio of 110×10^{-15} and a secular equilibrium ratio of 2×10^{-15} (calculated from an estimated sandstone content of 1 ppm U and 3 ppm Th). These isochrons are relatively evenly spaced and indicate groundwater velocities which are reasonable given the known hydraulic parameters of the system.

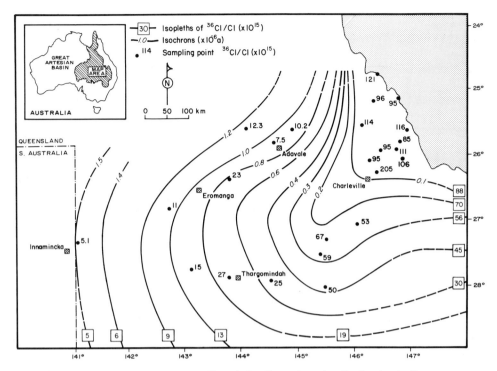

Fig. 10-6. Chlorine-36 in the "J" aquifer of the Great Artesian Basin, Australia.

Isochrons calculated by the Australian Bureau of Mineral Resources digital simulation model of the Great Artesian Basin model, GABHYD, are illustrated in fig. 1 of Airey et al. (1979). With the exception of the area near Adavale, these isochrons are in substantial agreement with the ^{36}Cl isochrons. This concordance between the ^{36}Cl ages and independently calculated, physically based ages in this relatively simple, well understood aquifer system is a strong endorsement of the fundamental assumptions of ^{36}Cl groundwater dating.

Milk River aquifer, Alberta. The Milk River Sandstone crops out in southern Alberta and northern Montana, near the United States border. The formation, a moderately cemented marine sandstone of Cretaceous age, is confined above by the shale of the Pakowki Formation and below by the Alberta Group, also shale. The aquifer recharges south of the Milk River in the Sweetgrass Hills, where the formation crops out. The formation dips gently to the north and about 100 km north of the border undergoes a facies change to a shaley sand. At this point it is no longer an aquifer.

Present-day head in the Milk River Sandstone exceeds that in the overlying and underlying units, and discharge from the Milk River aquifer is apparently entirely by leakage outward to both these units. Under these conditions chloride is unlikely to have entered the Milk River aquifer from other formations, and its water is expected to be very slow moving. Therefore, the Milk River aquifer was selected as a test of the ^{36}Cl dating method. Another motivation for a ^{36}Cl investigation of this aquifer was the possibility of field verification of ion filtration. Ion filtration in a system with this unusual head distribution would be expected to concentrate ^{36}Cl but leave the ^{36}Cl/Cl ratio unaffected. Thus, under the conditions of ion filtration, the ratio would decay with residence time in the aquifer, but the ^{36}Cl concentration would increase.

The chloride concentration distribution in the Milk River aquifer is shown

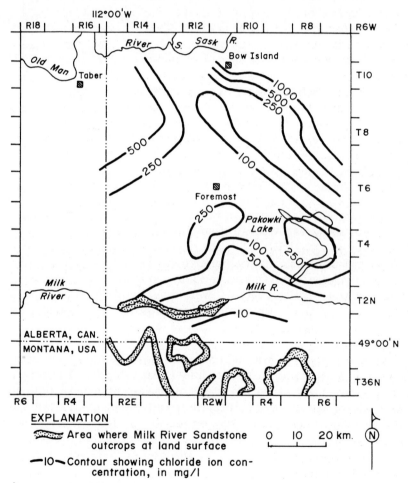

Fig. 10-7. Chloride concentration (mg l^{-1}) in the Milk River Sandstone aquifer, southern Alberta, Canada.

in plan view in Fig. 10-7. The outcrop of the Milk River Sandstone is shown in the lower center of the figure. The outcrop area intercepted by the Milk River is a recent feature resulting from post-glacial down-cutting of the overlying Pakowki shale during the Wisconsin Glacial Stage (Meyboom, 1960). Distribution of ions or isotopes in the groundwater has not yet been affected at our sampling points by the relatively recent influx of water from the Milk River in this area. This has been shown by Swanick (1982), on the basis of ^{14}C dating and numerical modeling of the groundwater system. The Milk River aquifer exhibits strong changes in chloride concentration, which increases by about two orders of magnitude between the recharge area and the distal portion of the aquifer (Fig. 10-7). Schwartz and Muehlenbachs (1979) and Schwartz et al. (1981) explain the chloride distribution, as well as the stable isotopes of hydrogen and oxygen and other chemical constituents, on the basis of a macroscopic dispersion process. They envision a mixing of "connate solutes", with infiltrating meteoric water displacing the connate water. A major difficulty with this hypothesis is that, given the measured hydraulic conductivities of the aquifer, all connate water should have long since been flushed out. Schwartz et al. (1981) explain this discrepancy by noting that flow through the aquifer system may have been interrupted by glaciers, which have covered the aquifer for about 75% of the Pleistocene. Schwartz and Muehlenbach (1979) also considered an ion filtration hypothesis to explain the solute increase. Although they did not favor this explanation, it remains a logical alternative in view of the aquifer discharge by leakage through confining shale strata. In fact, the hypotheses may be complementary, inasmuch as the ion filtration would help explain retention of connate solutes within the aquifer.

The distribution of the $^{36}Cl/Cl$ ratio is illustrated in Fig. 10-8. Notable features in Fig. 10-8 include a sharp decrease of the $^{36}Cl/Cl$ ratio away from the recharge zone and a large area of $^{36}Cl/Cl$ less than 10×10^{-15} at the distal end of the aquifer. These latter values apparently reflect secular equilibrium. Isochrons drawn in Fig. 10-8 were determined using equation (19), the observed $^{36}Cl/Cl$ ratios, an input $^{36}Cl/Cl$ ratio estimated from Fig. 10-2 to be 500×10^{-15}, and a secular equilibrium ratio of 6×10^{-15}. The isochrons indicate that the ^{36}Cl "age" rapidly increases to that limited by secular equilibrium, about 2.5 Ma. The large number of secular equilibrium $^{36}Cl/Cl$ ratios is clear evidence that the aquifer chloride in the distal portion of the aquifer has been in residence in the system for more than 2.5 Ma. However, by means of a numerical model and various sensitivity analyses, Swanick (1982) calculated a maximum age of 500,000 years for Milk River groundwater. The discrepancy between the calculated and experimental maximum ages might be due to dilution of the initial $^{36}Cl/Cl$ by low-^{36}Cl ancient chloride, or, as pointed out by Schwartz et al. (1981), might be explained by effects of Pleistocene glaciation on recharge and groundwater flow.

The isochrons show younger waters along lines north and northwest of the center of the outcrop area. Swanick (1982) showed that the age of the waters in this system depends almost entirely on leakage through the confining shales. One implication of the data, then, is higher leakage in the area about 10 km southwest of Bow Island and in Township 6, Range 15. These areas coincide, respectively, with folds in the Milk River Sandstone along the Bow Island Arch and a northward-trending syncline (Meyboom, 1960). Increased fracturing of the confining shales in the vicinity of these folds might well be associated with these structural features. The Bow Island Arch is

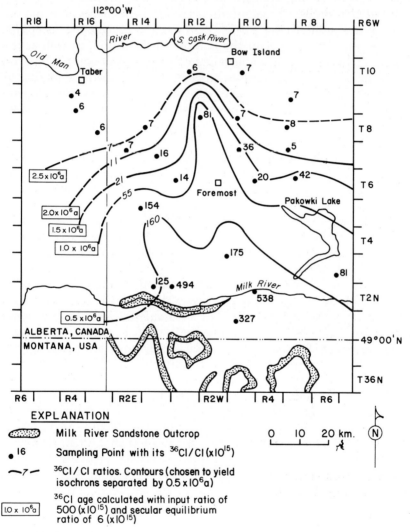

Fig. 10-8. ^{36}Cl/Cl ($\times 10^{15}$) ratio in the Milk River Sandstone aquifer and ^{36}Cl isochrons calculated using equation (19).

accompanied by a thinning of overlying Pakowki shale (Swanick, 1982) which should also enhance upward leakage. The isochrons thus appear to conform to the known hydrodynamic controls on the system.

^{36}Cl is also helpful in determining whether the solutes in the distal portion of the aquifer originated through ion filtration or connate water retention. The ^{36}Cl concentration (atoms ^{36}Cl l^{-1}), mapped in Fig. 10-9 does not show the dramatic decrease of the ^{36}Cl/Cl ratio shown in Fig. 10-8. The contours in Fig. 10-9 show *increasing* ^{36}Cl concentration down flow lines in three zones south of the area in which the chloride has attained secular equilibrium: northward from the 49°00' parallel to the Milk River, north of the outcrop cut by the Milk River, and an area northwest of Pakowki Lake.

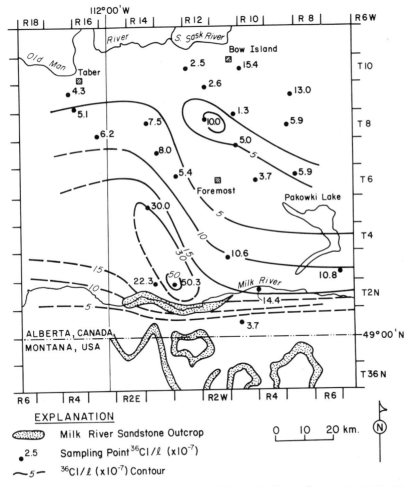

Fig. 10-9. ^{36}Cl concentration (atoms ^{36}Cl l^{-1} ($\times$ 10^{-7})) in the Milk River Sandstone aquifer.

These increases in the ^{36}Cl concentration may only be explained by ion filtration and are persuasive arguments against a mixing with connate salts which would at best leave the ^{36}Cl concentration unaffected. The increasing concentration of ^{36}Cl is demonstrated even more clearly in Fig. 10-8 where the sampling points near the Milk River (ratios 494 and 538 $\times$ 10^{-15}), show chloride concentrations more than 10 times those at the recharge zone but ratios close to the predicted input of 500 $\times$ 10^{-15}. A further comparison of Fig. 10-8 and 10-9 indicates that the increases in ^{36}Cl concentration are associated with the higher groundwater velocity, and thus less rapidly aging waters in the system, while Figs. 10-7 and 10-8 show that the low ^{36}Cl/Cl ratios are associated with the high-chloride areas. Although the presence of residual ancient chlorides in the high-chloride zones cannot be completely discounted, these waters are more likely so slow moving that ^{36}Cl decay exceeds its concentration by ion filtration. On balance, then, chloride concentration by ion filtration seems to be more consistent with the ^{36}Cl data than does the connate salt hypothesis. The ^{36}Cl ages shown in Fig. 10-8 are therefore probably a reasonable representation of historical groundwater flow in the Milk River aquifer.

The Fox Hills—Basal Hell Creek aquifer

The Fox Hills—Basal Hell Creek aquifer in southwestern North Dakota presents an example of how ^{36}Cl may be used to delineate groundwater cir-

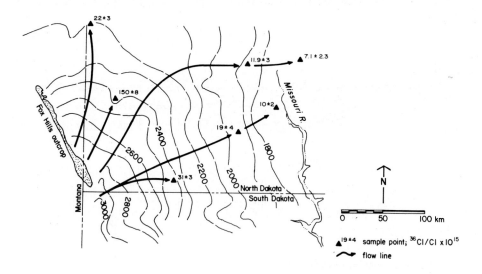

Fig. 10-10. Potentiometric surface of the Fox Hills—Basal Hell Creek aquifer, North Dakota. ^{36}Cl sampling points are indicated by triangles.

culation (Bentley and Davis, 1982). The hydrogeology and geochemistry of the aquifer have been described by Thorstenson et al. (1979). The aquifer is composed of the late Cretaceous Fox Hills marine deltaic sandstone, grading upward into the fluvial and deltaic Hell Creek Formation. The Fox Hills Formation conformably overlies the thick Pierre Shale, which in turn overlies the Dakota Sandstone, a saline aquifer in this area. The potentiometric surface of the aquifer and ^{36}Cl sampling points are shown in Fig. 10-10. The potentiometric data indicate that leakage, if present, is downward near the recharge area and upward from the Dakota near the Missouri River, illustrated in cross-section in Fig. 10-11.

The chloride concentration is observed to increase substantially from an initial value of about 30 mg l^{-1} in the "transition zone" (Thorstenson et al., 1979). This zone separates areas of widely differing chemical composition at the recharge and discharge ends of the aquifer. Thorstenson et al. (1979) presented two basic hypotheses to account for this water quality transition: various chemical reactions within a single water body moving down the aquifer, or replacement of the Fox Hills—Basal Hell Creek water by water moving upward from the Dakota Sandstone through the Pierre Shale. They stated that the second hypothesis is supported by a major helium concentration increase in the transition zone, without a concomitant temperature increase which would indicate in-situ production. However, they favored the first hypothesis because the Fox Hills—Basal Hell Creek aquifer contains water with a chloride concentration 20 times less than the Dakota Sandstone and thus chloride mass balance arguments appear to preclude an origin from the Dakota.

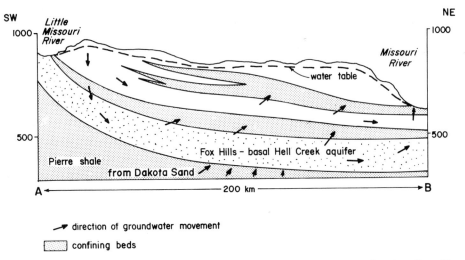

Fig. 10-11. Cross-section of the Fox Hills—Basal Hell Creek aquifer, showing flow directions.

Residual meteoric ^{36}Cl in the aquifer was determined by subtracting the product of the chloride concentration and the estimated Dakota Sandstone secular equilibrium ^{36}Cl/Cl ratio (5×10^{-15}) from the observed ^{36}Cl concentration. The result, plotted against age calculated from hydrodynamic parameters (Fig. 10-12), shows that meteoric ^{36}Cl linearly decreases to zero at the transition zone.

The ^{36}Cl/Cl data clearly support the idea that the meteoric chloride is forced out of the aquifer and replaced by chloride in secular equilibrium with the Dakota Sandstone. However, the Fox Hills—Basal Hell Creek water has chloride concentrations 20 times less than that in the Dakota. Use of chloride mass balance arguments may not be valid due to ion filtration holding back the chloride during the water's transit through the Pierre Shale. If so, the shale would have to possess a filtration efficiency of about 95%. The line of reasoning suggested by the ^{36}Cl data leads to the possibility that the water resource available from the Fox Hills—Basal Hell Creek aquifer may include that from the underlying Dakota Sandstone.

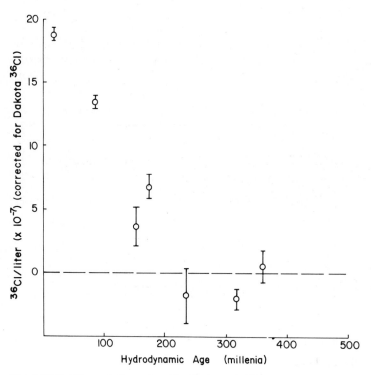

Fig. 10-12. The variation of residual meteoric ^{36}Cl in the Fox Hills—Basal Hell Creek with hydrodynamic age.

The Carrizo aquifer

^{36}Cl data for the Carrizo aquifer in southern Texas have been presented and discussed by Bentley and Davis (1982). This aquifer illustrates the difficulties of interpretation when the assumption of constant meteoric chloride input is not upheld. The outcrop of the Carrizo Sand in the study area is 5—12 km wide (Fig. 10-13). The outcrop is about 200 km northwest of the coastline, toward which the formation dips at about 25 m km^{-1}. The Carrizo is composed of well-sorted sand which unconformably overlies the shale, sand, and clay of the Eocene Willcox Group. Flow in the Carrizo Sand is controlled by leakage through the overlying Reklaw and Queen City Formations, composed of shale and sand.

The chloride concentrations down gradient in the Carrizo (Fig. 10-14) show an unusual pattern of decrease, followed by increase, a slight decrease, and finally, a rapid increase. These isochlors were based on chloride data from only those wells which exhibited little chemical change over an extended sampling period. The groundwater in the Carrizo aquifer has been dated both by ^{14}C (Pearson and White, 1967) and by hydrodynamic modeling

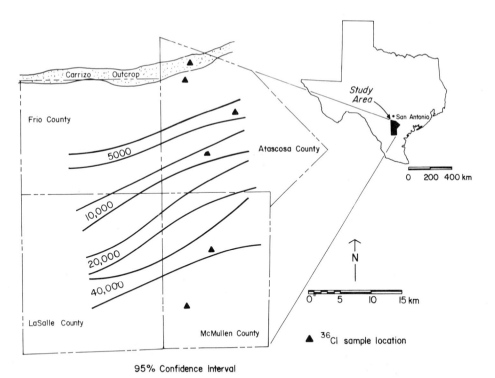

Fig. 10-13. The Carrizo Sand aquifer, southern Texas, showing hydrodynamic groundwater ages calculated by Brinkman (1981) within 95% confidence intervals.

(Brinkman, 1981), and these methods are in reasonable agreement. In Fig. 10-15 the ^{36}Cl concentration has been plotted against the hydrodynamic age. This pattern of chloride concentration increase may be compared with the chloride concentration vs. hydrodynamic age in Fig. 10-16a. Both curves show an over-all increase in concentration with age, more evident in the ^{36}Cl concentration curve. The increase in the ^{36}Cl content suggests chloride concentration by ion filtration. On the basis of the hydrodynamic age, the ^{36}Cl may be presumed to have experienced little decay or subsurface production. Therefore the increase in the ^{36}Cl concentration may be used to estimate the extent of ion filtration. In Fig. 10-16b the calculated increase in

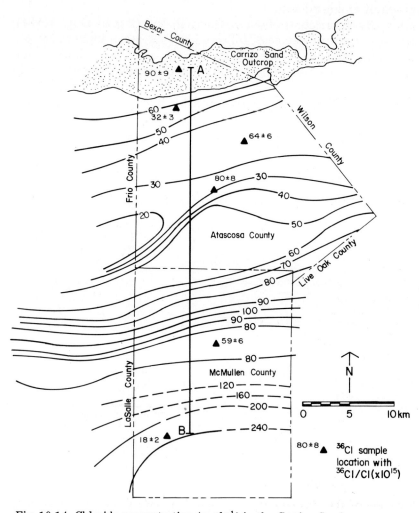

Fig. 10-14. Chloride concentration (mg l^{-1}) in the Carrizo Sand.

chloride concentration has been subtracted from the observed concentration. A residual chloride concentration variation still remains. This variation is probably related to some climatic control on the relative chloride and water inputs to the aquifer. In Fig. 10-16c the variation of sea level is plotted on the same temporal scale as the hydrodynamic water age. There appears to be a close correspondence between the curves in Fig. 10-16b and c. This correlation may be explained by temporal variation in the distance of the coast from the Carrizo outcrop. This distance was almost doubled by the maximum Wisconsinan sea level decline. Meteoric chloride fallout drops off exponentially with distance from the coast (Eriksson, 1960), and thus variations in the distance to the coast may be expected to have had a major influence on the aquifer chloride input.

^{36}Cl is not really appropriate for groundwater dating in the case of an

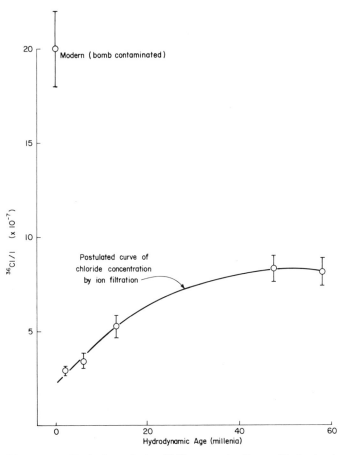

Fig. 10-15. Variation of the ^{36}Cl concentration with hydrodynamic age in the Carrizo Sand.

aquifer with water as young as the Carrizo Sand. However, under the assumption of relatively constant ^{36}Cl input, (as demonstrated in the Searles Lake and Australian studies cited above), it has proved useful in separating variations in concentration caused by subsurface processes from those caused by input variations. ^{36}Cl should be viewed not only as a groundwater dating tool, but also as a tracer for subsurface mixing and concentration.

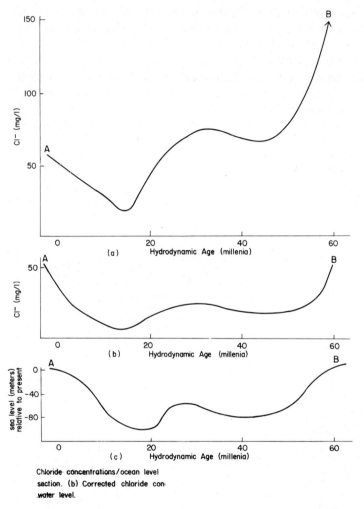

Fig. 10-16. (a) Variation of chloride concentration (mg l^{-1}) with calculated hydrodynamic age in the Carrizo Sand. (b) Corrected chloride concentration (mg l^{-1}) with hydrodynamic age. The chloride concentration was corrected by subtracting the chloride increase due to ion filtration (determined from Figs. 10-16a and 10-15). (c) Sea-level variation with time.

^{36}Cl measurements in chloride brines

Pockets of brine are occasionally encountered during mining operations in salt domes. Their origin is crucial to the determination of domal stability and hydrologic isolation if such a structure is to be considered an acceptable location for a nuclear waste repository. On the basis of brine geochemistry and ^2H and ^{18}O composition, Kumar and Martinez (1978) and Knauth et al. (1980) concluded that the fluids in Louisiana salt domes were most likely entrapped during the salt's diapiric rise as it pushed through the overlying formations. Knauth et al. further suggested that the isotopic data indicate a residence time in the salt of 10—13 Ma.

Fabryka-Martin et al. (1983) reported ^{36}Cl analyses of brines and minerals collected in the vicinity of three salt domes in southern Louisiana (Fig. 10-17). The samples fall into four categories: (1) "trapped formation waters", based on ^2H and ^{18}O measurements by Knauth; (2) meteoric leaks derived from the overlying fresh-water aquifer; (3) oil-field brines from producing wells adjacent to the domes; and (4) mineral samples collected in the domes. Table 10-5 describes the location and ^{36}Cl content of each sample.

Meteoric water for this area is calculated to have a $^{36}Cl/Cl$ ratio of 20 × 10^{-15} and is assumed to represent the initial concentration for the fresh-water aquifer overlying the domes. Because the chloride content of the leaks

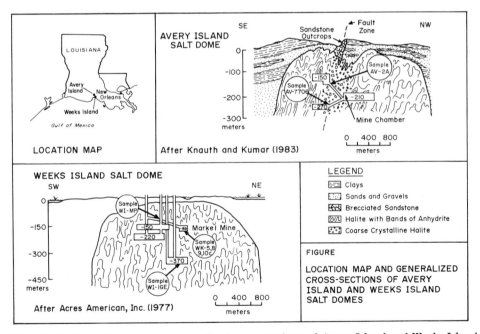

Fig. 10-17. Location map and generalized cross-sections of Avery Island and Weeks Island salt domes.

TABLE 10-5

^{36}Cl results for Louisiana Salt Dome and oil-field brine samples (from Fabryka-Martin et al., 1983)

Sample No.	Lab. No.	Sample description	Approximate depth (m.a.s.)	Chloride concentration (g l^{-1})	$^{36}Cl/Cl$ ($\times 10^{15}$)	^{36}Cl content (10^{-8} atoms l^{-1})
"Trapped brines" in salt domes						
1	AV-7706	active leak, Avery Is.	−270	228	3 ± 60%	116
2	WI-1GE	active leak, Weeks Is.	−370	201	1 ± 50%	34
3	WK-5	active leak, Markel mine, Weeks Is.[a]	−150	201	<0.5	<15
4	WK-8	active leak, Markel mine, Weeks Is.[b]	−150	204	<0.5	<15
5	WK-9	isolated pool near WK-8, Weeks Is.[a]	−150	201	<0.5	<15
6	WK-10c	pool near WK-5, Weeks Is.[b]	−150	201[c]	<0.5	<15
Meteoric leaks in salt domes						
7	WI-MP	Markel pool, Weeks Is.	−120	190	<0.5	<15
8	AV-2A	active ceiling leak, Avery Is.	−150	190	<0.5	<15
Oil-field brines adjacent to salt domes						
9	Gulf PC-3	Miocene sand, Weeks Is. oil field	−3540	90	3 ± 70%	46
10	Gulf PC-8	Miocene sand, Weeks Is. oil field	−3460	95	3 ± 30%	48
11	WC-W122	Miocene sand, White Castle oil field	−2105	83	4 ± 30%	56
12	WC-W141	Miocene sand, White Castle oil field	−950	71	3 ± 25%	36
13	WC-W155	Miocene sand, White Castle oil field	−2690	79	4 ± 21%	53
14	WC-W245	Miocene sand, White Castle oil field	−2020	81	4 ± 25%	55

Mineral samples from salt domes

15	H-1b	halite, Markel mine, Weeks Is.	−150	—	<0.5
16	LA-8b	soluble part of "sandstone" inclusion, Weeks Is.	−370	—	0.5
17	LA-2b	soluble part of "sandstone", from vicinity of AV-7706, Avery Is.	−270	—	<0.5

[a] Provided by M. Kumar, Institute for Environmental Studies, Louisiana State University, Baton Rouge, La.
[b] Provided by L.P. Knauth, Department of Geology, Arizona State University, Tempe, Ariz.
[c] Estimated.

that derive from this shallow aquifer increases by three orders of magnitude as the fresh water dissolves Jurassic, ^{36}Cl-free halite (such as sample 15) in its path, the original ^{36}Cl tag of the source is overwhelmed, giving rise to zero values (samples 7 and 8).

The initial ratio of ^{36}Cl/Cl for the trapped brines, which would be the secular equilibrium value, may be inferred from that of the oil-field brines for the area (samples 9—14). These average 4×10^{-15}. Mass balance arguments indicate that the trapped brines in the salt domes, if derived from adjacent formations, have been diluted with "dead" chloride by a factor of 2.5. The initial result of such dilution would be a decrease in the observed ratio to $1-2 \times 10^{-15}$, after which the ratio would continue to decrease through decay. A ratio of less than 0.5×10^{-15}, the practical detection limit, would be reached after a residence time of 300,000 years or more.

Most of the trapped brines have values below the detection limit, suggesting a residence time greater than 300,000 years (samples 3—6). This result is consistent with Knauth et al. (1980). However, two samples (1 and 2) have been little changed from the original ^{36}Cl content, suggesting a relatively short residence time. These two leaks have therefore been isolated from the source formation less than 300,000 years.

ATMOSPHERIC NUCLEAR WEAPONS TESTING ^{36}Cl AS A HYDROLOGIC TRACER

Radioisotopes began to be widely used as groundwater tracers after Libby proposed naturally produced tritium as a tracer in 1953. The constant input and short (12.43 a) half-life appeared to make tritium ideal for radiometric dating of young water. However, within a short time the significance of tritium added to the hydrosphere by nuclear weapons testing was recognized (Bergmann and Libby, 1957). This "pulse" of bomb-tritium has subsequently proved invaluable in identifying water introduced into hydrologic systems during the years 1955—1970.

The use of bomb-tritium as an indication of rapid groundwater circulation is now so common that a survey of applications would be a major undertaking. Notable recent examples are provided by Hufen et al. (1974), Poland and Stewart (1975), Rabinowitz et al. (1977), and Hobba et al. (1979). A somewhat less common, but equally significant, application is the determination of recharge rates from the penetration depth of bomb-tritium in the vadose zone. The technique was pioneered by Schmalz and Polzer (1969) and Smith et al. (1970). The most important result has been the demonstration of appreciable recharge in even very arid regions (Dinçer et al., 1974; Verhagen et al., 1979).

The presence of the tritium bomb pulse in the hydrologic system is, from the standpoint of the hydrologist, unfortunately a temporary phenomenon. In the Southern Hemisphere, where the tritium fallout was much less than north of the equator, the bomb pulse has already decayed to within 15 tritium units (1 TU = 1 ^{3}H in 10^{18} ^{1}H) of natural background, or less. Even in the areas of greatest fallout in the Northern Hemisphere, bomb-tritium will be difficult to detect in 30—40 years. A bomb-produced isotope with a longer half-life and chemical behavior similar to the conservative tracer properties of tritium would be highly desirable.

Bentley et al. (1982) and Phillips et al. (1983) pointed out that bomb-^{36}Cl appears to meet the requirements for such an environmental tracer, and even to have certain advantages over tritium. ^{36}Cl produced by a limited number of nuclear weapons tests, fell out from the atmosphere relatively rapidly at rates far above the natural background, and is among the most conservative of chemical species in groundwater. Its half-life of 301,000 years precludes disappearance due to radioactive decay for the forseeable future.

Although minor amounts of ^{36}Cl may be produced in nuclear explosions directly by fission, or by neutron activation of ^{36}Ar, the only major production mechanism is neutron activation of ^{35}Cl. Thus, only explosions which have occurred near large amounts of chloride and whose radioactive clouds penetrated the stratosphere have contributed to the global fallout of ^{36}Cl. The only reported explosions meeting these criteria are certain weapons tests carried out by the United States in the South Pacific between late 1952 and mid-1958. These were able to activate chloride in seawater because they were detonated on barges or small islands. (French tests in the South Pacific may have injected small amounts of ^{36}Cl since 1966).

^{35}Cl absorbs a large proportion of the neutrons released into seawater because of its very large neutron cross-section (44 barns). The major competitor for neutrons is hydrogen. Dyrsson and Nyman (1955) calculated that 30.40% of the neutrons stopped by seawater produce ^{36}Cl.

Prediction of ^{36}Cl fallout

Phillips et al. (1983) modeled ^{36}Cl fallout from nuclear tests by using estimates of ^{36}Cl injection into the stratosphere for individual explosions as input for an atmospheric box model. The dates, location, and explosive yield of tests were taken from Zander and Araskog (1973) and Carter and Moghissi (1977). The proportion of the neutrons which were absorbed by seawater and then injected into the stratosphere was obtained by calibrating the fallout model using the ^{36}Cl concentrations in rainfall reported by Schaeffer et al. (1960).

Injection amounts and elevations derived from these data were entered into a modification of the atmospheric box model of Krey and Krajewski (1970). Global fallout of ^{36}Cl from nuclear weapons tests, calculated by this model is illustrated in Fig. 10-18. The scale on the right-hand side of the graph is fallout in atoms ^{36}Cl m^{-2} s^{-1} between 30°N and 50°N latitude. The conversion from global fallout was made according to Peterson (1970). For comparison, natural fallout is about 30 atoms m^{-2} s^{-1}.

In order to test these predictions ^{36}Cl concentrations in a Greenland ice core, a well-understood shallow aquifer, and the vadose zone in an arid region were measured.

Bomb-^{36}Cl in a Dye 3 ice core

A 100-m ice core was drilled at Dye 3, Greenland (65°11'N, 43°5"W) in 1980. The top 23 m of the core was divided in the field into 29 approximately yearly samples. These were subsequently determined by ^{18}O/^{16}O measurement to span the period 1950—1978. Chloride was extracted from the individual samples and analyzed for ^{36}Cl. The chloride extraction technique and ^{36}Cl concentration calculations are presented in Elmore et al. (1982).

The fallout of ^{36}Cl as a function of time at Dye 3 compared with the predicted fallout is shown in Fig. 10-19. The predicted fallout at 65°N was as-

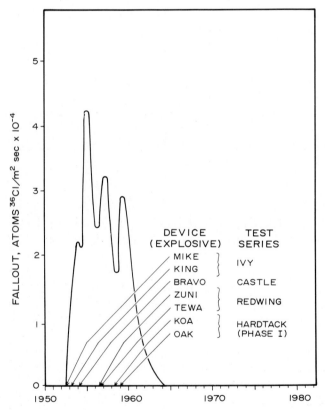

Fig. 10-18. Fallout of ^{36}Cl from atmospheric nuclear weapons testing (in atoms ^{36}Cl m^{-2} s^{-1}) between 30°N and 50°N latitude, as predicted by the HASL atmospheric box model.

sumed to be 20% of the fallout between 30°N and 50°N. July 1958 ^{185}W fallout from the Hardtack explosions (major ^{36}Cl-producing tests in May and June 1958) at 65°N was about 4% of that at 30°N to 50°N (Lockhart et al., 1959), but this proportion is probably representative of only the early stages of fallout. Natural ^{36}Cl fallout at 65°N is about 30% of that between 30°N and 50°N (Bentley and Davis, 1981). The atoms ^{36}Cl kg^{-1} data in Elmore et al. (1982) was converted to fallout rate (atoms ^{36}Cl m^{-2} s^{-1}) assuming a constant annual precipitation of 500 kg m^{-2} a^{-1} (Reeh et al., 1978). Natural background ^{36}Cl fallout could not be determined from the ice core.

The predicted and measured ^{36}Cl fallout show excellent agreement, except for the slope of the fallout curve. This difference may be due to additional contributions by unreported tests, a difference between the mean atmospheric lifetimes of ^{36}Cl and ^{90}Sr and ^{105}Cd, or a small degree of mixing between years by melting or blowing snow.

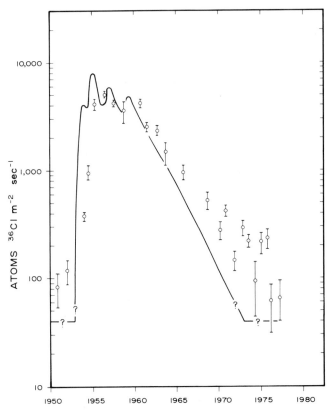

Fig. 10-19. Comparison of predicted ^{36}Cl fallout at Dye-3, Greenland (assuming fallout at 65°N is 20% of that between 30°N and 50°N) with measured ^{36}Cl fallout (assuming a constant annual precipitation of 500 kg m^{-2} a^{-1}).

Bomb-^{36}Cl in shallow groundwater

Bentley et al. (1982) sampled a "plume" of recent recharge beneath an abandoned landfill at the Borden Canadian Forces Base in Ontario, Canada. The hydrogeology of the site has been thoroughly studied and described by McFarlane et al. (1982) and Sudicky et al. (1982). Flow beneath the landfill is largely vertical, and thus a suite of samples at different depths was collected from nested piezometers T5-A and T5-B, near the center of the landfill.

These data are diagrammed in Fig. 10-20 with tritium data from Egboka (1980) included for comparison. The sample depth cannot be definitely correlated with recharge data, but the sequence of the ^{36}Cl peak followed by the tritium peak agrees with the timing of the fallout peak.

The Borden ^{36}Cl peak shows (from deep to shallow) a steep rise, followed by a more gentle tailing, similar to the ^{36}Cl fallout peak in Fig. 10-19. The maximum fallout rate, calculated from the maximum ^{36}Cl concentration using a yearly precipitation of 83 cm (McFarlane et al., 1982) and evapotranspiration of 50% (Egboka, 1980), is 2.4×10^4 atoms ^{36}Cl m^{-2} s^{-1}. This is somewhat less than the predicted maximum of 4.2×10^4 atoms ^{36}Cl m^{-2}

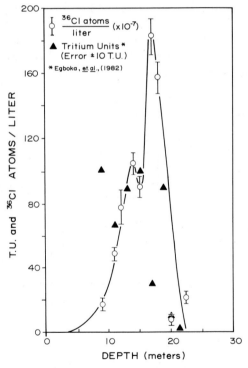

Fig. 10-20. ^{36}Cl concentration with depth in nested piezometers T-5A and T-5B at the Borden CFB landfill, Ontario. Tritium concentrations are included for comparison.

s^{-1}, perhaps due to hydrodynamic dispersion or the limited number of sampling points. The total ^{36}Cl in the profile, 5.75×10^{12} atoms ^{36}Cl m^{-2}, is only slightly smaller than the predicted total, 7.1×10^{12} atoms ^{36}Cl m^{-2}.

The bomb-^{36}Cl pulse in the vadose zone

The bomb-^{36}Cl fallout has provided a tracer for downward movement of soil moisture through the vadose zone. In humid regions with shallow water tables this ^{36}Cl pulse has probably entered the groundwater. In arid regions, however, it should still be in transit through the vadose zone, and thus provide a measure of recharge rates.

In order to test this hypothesis soil samples were collected from a vertical auger hole drilled 5 m deep in a relatively homogenous sandy loam near Socorro, New Mexico (Trotman et al., 1983). The soil samples were leached with distilled, deionized water, and chloride precipitated from the leachate as AgCl. The ^{36}Cl/Cl ratio of the chloride was measured by TAMS.

The variation of the ^{36}Cl/Cl ratio with depth is illustrated in Fig. 10-21. The ^{36}Cl/Cl peak was encountered at about 0.75 m and the maximum penetration of the bomb pulse was about 1.75 m. The average soil-water velocity over the last 30 years was thus about 4 cm a^{-1}. The average volumetric water content was approximately 10%. From this a mean annual infiltration rate (to a depth of 125 cm) of 4 mm a^{-1} may be calculated, about 2% of the annual precipitation. The actual recharge flux is probably less than this, due to extraction to soil moisture by roots below 125 cm.

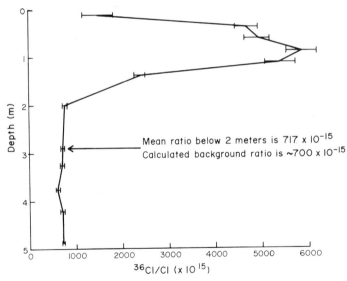

Fig. 10-21. The ^{36}Cl/Cl ratio of soil-water solutions in the vadose zone from a vertical borehole near Socorro, New Mexico.

Identification of salt source in saline shallow groundwaters

One possible use of the bomb-^{36}Cl pulse is the identification of the source of salt in shallow saline groundwaters. The technique has been applied by Bentley et al. (1983) in a preliminary study in northeast Brazil.

The 8×10^6 km^2 area between latitudes 1° and 17°S and longitudes 34°30' and 49°W is characterized by semi-arid climatic conditions, forming the "Drought Polygon". Water supply to the area's approximately 22 million inhabitants is primarily shallow groundwater from weathered bedrock or minor amounts of alluvial fill. The bedrock aquifer groundwater is common-

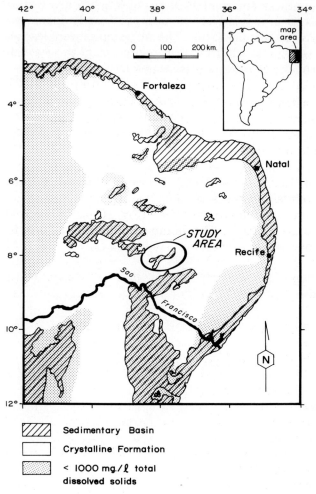

Fig. 10-22. Groundwater salinity and shallow aquifer lithology in the "Drought Polygon" of northeast Brasil.

ly saline, often unpotable. Fig. 10-22 maps the areas of salinity and shallow-aquifer lithology; samples were taken from shallow wells in crystalline rock in the outlined study area.

Schoff (1972) suggested that the origin of the salinity is residual seawater from an upper Cretaceous marine invasion of the land. Others he quoted considered rock weathering to be the principal salt source. Conversely, on the basis of isotope techniques and chloride balance, Salati et al. (1974), Stolf et al. (1979), and Salati et al. (1980) concluded that the salts are probably meteoric in origin, and resulted from evapotranspiration of rainwater. The controversy is more than academic; the limited reservoir of either a residual marine salt or mineralization due to weathering could possibly be removed by various pumping schemes. A meteoric, constant input source could not (Salati et al., 1980).

In general, the shallow crystalline-rock groundwaters contain significant amounts of bomb-origin tritium, most with mean residence times less than 100 years (Salati et al., 1974). These would be expected to contain bomb-origin ^{36}Cl as well.. If the salts were residual marine or weathering products, the ^{36}Cl ratio would be a mixture of the bomb-pulse chloride with a chloride which would have reached a secular equilibrium ratio of about 30×10^{-15} (see Table 10-2). If the groundwater salinity were of meteoric origin, the observed ^{36}Cl ratio would be a time-averaged mixture of meteoric input ratio plus the bomb-pulse ratio. The natural meteoric ratio is easily calculated by dividing yearly ^{36}Cl fallout by 1.3 times rainfall chloride. In the study area yearly rainfall is 55 cm, rainfall chloride is about 1.9 mg l^{-1} (Salati et al., 1974), and, at 8°S, ^{36}Cl fallout is 7.5 atoms m^{-2} s^{-1} (Fig. 10-1). The natural meteoric ratio is therefore 10×10^{-15}. The bomb-pulse input at 10°N to 10°S is estimated from that which occurred at 30—50°N (7.5×10^{12} ^{36}Cl atoms) by multiplying by 25% (Peterson, 1970). The average yearly input ratio over the 12-year period from 1952 to 1964, 8800×10^{-15}, is obtained by dividing the bomb-pulse input by both 12 and the yearly rainfall chloride multiplied by 1.3.

Under the assumption of complete mixing of the shallow groundwaters, one can develop plots of expected ^{36}Cl/Cl ratios, which depend on percentage of meteoric chloride in the groundwater chloride and residence time of the meteoric (and bomb) chloride in the system. In Fig. 10-23a, the case of a residual marine or weathering salt, and Fig. 10-23b, the case for no nonmeteoric salt, plots of expected ratios are compared with experimental data from wells in the study area. Fig. 10-23a shows the two lowest observed ratios falling below the minimum ratio possible for a residual salt. On the other hand, the highest observed ratio indicates an admixture of between 15% and 100% meteoric chloride. In Fig. 10-23b all ratios fall in the possible range. The minimum observed value is composed of meteoric water, about 5% of which contains the bomb pulse. The maximum observed ratio is composed of between 15% and 100% of bomb-pulse water. Thus the data

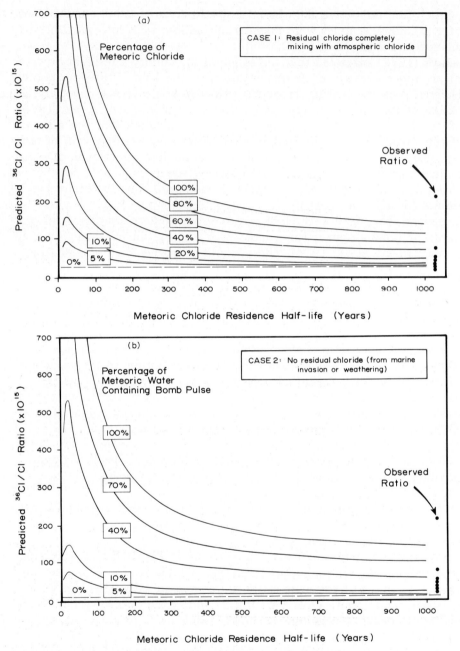

Fig. 10-23. (a) Expected $^{36}Cl/Cl$ ratios ($\times 10^{15}$) in the Drought Polygon as a function of meteoric chloride residence time and the percentage of meteoric chloride (as opposed to residual marine or weathering-derived chloride). (b) Expected $^{36}Cl/Cl$ ratios ($\times 10^{15}$) as a function of the percentage of recently recharged water containing bomb-^{36}Cl (as opposed to pre-bomb recharge free of marine or weathering-derived chloride).

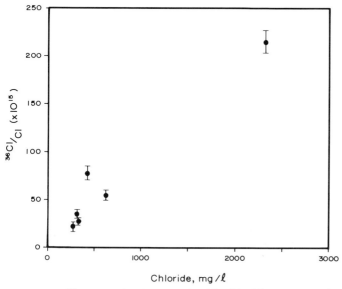

Fig. 10-24. ^{36}Cl/Cl ratios compared with chloride concentration for groundwaters from the Drought Polygon of northeast Brazil.

are more consistent with Fig. 10-23b, the meteoric salt hypothesis. Even under the residual salt hypothesis, one sample contains a minimum of 15% meteoric salt. The case for meteoric salt is strengthened by a comparison of ^{36}Cl/Cl ratio with chloride concentration, plotted in Fig. 10-24. The data are obviously consistent with a mixing of low-chloride/low-^{36}Cl ratio (pre-bomb) water with high-chloride/high-^{36}Cl ratio (bomb-pulse) water. If the main source of chloride were a residuum, the expected slope of such a plot would be negative. One interesting implication of the plot (the data are too few to draw definite conclusions) is that the pre-bomb water, which has a ^{36}Cl/Cl ratio of about 10, has less than 100 mg Cl l^{-1}.

A hypothesis consistent with these ^{36}Cl data is that at least two sources of groundwater exist in the study area, both meteoric in origin. The first, a shallow, short-residence-time groundwater which is highly concentrated by evapotranspiration processes before recharge; the second, a deeper, longer-residence-time water with a much lower salinity. If this hypothesis is correct, (a more extensive study is required for proper evaluation), a possible solution to the water supply problem in the Drought Polygon would be to tap the deeper, regional source, avoiding the higher-salinity shallow waters by suitable well construction.

Other ^{36}Cl fallout data

Elmore et al. (1979) and Bentley and Davis (1982), have measured ^{36}Cl

TABLE 10-6

^{36}Cl in surface water and groundwater dated before, during, or after the atmospheric nuclear weapons testing period, and in mixtures of water from different periods (from Bentley et al., 1982)

Location		Date collected	Precipitation evapotranspiration (mm)	Measured concentration (10^{-7} atoms ^{36}Cl l^{-1})	Calculated fallout (atoms ^{36}Cl m^{-2} s^{-1})
Pre-1953 water (before nuclear testing)					
Arizona[a]					
Tucson City well	B8	January 1979	300/98%	11.7	23
	C13	January 1979		9.6	18
	A31	January 1979		6.6	12
Monkey Springs		July 1979		8.1	15
Benson well		July 1979		7.4	14
St. David well		July 1979		6.5	12
			Mean:	8.4	16
Madrid Basin wells			400/98%		
535-7a				7.1	18
535-7b				8.3	22
535-5a				5.6	15
			Mean:	7.0	18
Texas well			660/98%		
AL-68-59-504		December 1979		2.9	12
Atmospheric nuclear testing period (1953–1964)					
New York Long Island[b]			1150/0%		
rainfall		August 1957		77	28000
		September 1957		183	67500
		April 1959		24	8900
		May 1959		1.02	370
		February 1960		26.3	9600
			Mean:	62	23000
shallow well		August 1957		64	23500

Post-1964 water

Arizona[a]			
San Pedro River	July 1979	300/98%	9	17
Arizona (Tucson)		300/0%		
rainfall	February 1982		0.26	25
rainfall	March 1982		0.31	29
California (La Jolla)[c]				
rainfall	March 1979	250/0%	0.2	16

Mixed samples

Arizona				
Campbell Ave well	January 1979	300/98%	158	310
Texas well				
AL-61-58-803	December 1979	660/98%	20	85
New York				
Lake Ronkonkoma[b]	September 1960	1150/50%	26.8	4600
Lake Ontario[a]	May 1978	750/50%	9.5	1100
	June 1978	750/50%	5.4	640
Vermont[b]				
stream	September 1957	1150/50%	26.8	4600
Nevada[d]				
Truckee River	August 1958	380/98%	167.5	420

Sources: [a] Elmore et al. (1979); [b] Schaeffer et al. (1960); [c] Finkel et al. (1980); [d] Bonner et al. (1962).

concentration in both surface and groundwater. In addition, earlier investigators have measured ^{36}Cl in rain and surface water (Schaeffer et al., 1960; Bonner et al., 1961). The available data are presented in Table 10-6. The data may be divided into four categories: water dating from prior to nuclear detonations, water from the testing period, water from after atmospheric testing ceased, and mixtures of water from different periods.

The "pre-bomb" well water from Arizona and Texas and the well water from the Madrid Basin are all from wells close enough to the recharge zones of the aquifers that measurable ^{36}Cl decay should not have taken place. Average ^{36}Cl fallout was computed by multiplying the ^{36}Cl concentrations by the effective yearly precipitation. The effective precipitation in the arid or semi-arid areas (Arizona, Texas, Madrid Basin, and Nevada) was assumed to be 2% of the actual precipitation (Bentley et al., 1982). For the humid areas (Lake Ontario, Vermont, and New York) it was assumed to be 50%. The pre-bomb ^{36}Cl fallouts calculated in this way ranged from 12 to 18 atoms ^{36}Cl $m^{-2} s^{-1}$.

The only precipitation samples assuredly from the atmospheric nuclear testing period are five rainfall samples collected by Schaeffer et al. (1960) from 1957 to 1960 on Long Island, New York. Fallout fluxes calculated from these samples (composites of several weeks of rainfall) ranged from 370 to 67,500 ^{36}Cl $m^{-2} s^{-1}$, with an average of 23,000 ^{36}Cl $m^{-2} s^{-1}$. This was the average fallout used to calibrate the model presented above. The ^{36}Cl concentration of the shallow groundwater sample from Long Island is close to the precipitation mean, indicating that it was probably composed of recently recharged precipitation with little evaporation. The samples from Lake Ronkonkoma and the Vermont stream may be representative of contemporary precipitation, but they may also contain groundwater or the sources may have a mixing time large enough that the measured concentration is a composite of several years precipitation.

The only samples certain of being representative of post-atmospheric-testing precipitation are from the San Pedro River (Arizona), and Tucson (Arizona) and La Jolla (California) rainfall. The San Pedro is an ephemeral river and is not fed by any lakes, therefore mixing should not be a problem. However, there may be some groundwater inflow. This runoff sample and the rainfall data indicate that ^{36}Cl fallout has returned to natural background rates.

The remaining samples were from sources in which precipitation from several years was mixed. The "post-bomb" groundwater samples from Arizona and Texas were collected in the recharge areas of the aquifers where the bomb pulse has mixed with pre- and post-bomb recharge. The sample from Lake Ontario shows the effect of dilution of the ^{36}Cl bomb pulse in the very large volume of the Great Lakes. Similarly, the Truckee River sample reflects dilution in Lake Tahoe, another large body of water.

A comparison of bomb-^{36}Cl with bomb-tritium in various hydrological applications

Determination of groundwater velocity. The present (in the Southern Hemisphere) and future (Northern Hemisphere) difficulties for isotope hydrologists caused by the decay of bomb-tritium have been discussed above. The long half-life of ^{36}Cl obviously renders it immune to this problem. Present-day bomb-tritium levels in the Southern Hemisphere are low due to the Arctic location of the major tritium-producing tests. ^{36}Cl was produced by equitorial tests, therefore its distribution should be approximately hemispherically symmetrical, as evidenced by ^{185}W fallout from the Hardtack explosions (Lockhart et al., 1959).

Tritium has been demonstrated to be retarded relative to chloride in short-duration experiments. This has been interpreted as temporary retention of tritium in an immobile water phase around clay particles (Kaufman and Orlob, 1956; Biggar and Nielsen, 1962; Krupp et al., 1972). Longer-term experiments have also shown that tritium may actually exchange with clay hydroxyls or become incorporated in the clay lattice, resulting in a more or less permanent fixation (Stewart, 1967; Rabinowitz et al., 1971). On all time scales ^{36}Cl appears to be at least as satisfactory as tritium for tracer applications.

Field-scale determination of dispersivity. Dispersivity is a notoriously difficult parameter to determine the field. Wood (1981) has suggested the approach of measuring the widening of a peak or front of solute naturally introduced into an aquifer. However, the input function must be known and moreover, an abrupt square or step function shape for the input function is desirable.

The bomb-^{36}Cl fallout meets this requirement far better than tritium fallout. Bomb-^{36}Cl has returned to natural background levels, while tritium has been incorporated in the atmospheric gas reservoir and persists at concentrations several times natural background (Fig. 10-25). Furthermore, soil organisms may maintain tritium as $^3H^1H$ in the unsaturated zone at levels well above atmospheric concentrations (Ehhalt, 1973). Finally, decrease of tritium concentrations due to dispersion is difficult to distinguish from decrease of concentration due to radioactive decay. ^{36}Cl should be much more useful than tritium for the determination of hydrodynamic dispersion.

Determination of recharge rates in the vadose zone. Several parameters must be known if the penetration depth of an anthropogenic tracer is used to estimate water transport through the vadose zone. These include the input function and the behavior after entry into the soil. The tritium input is a function of both the global tritium fallout and the local precipitation (Rabinowitz et al., 1977). The tritium entering the deep vadose zone is furthermore a function of the degree of evapotranspiration in the upper soil, which in turn depends on the season in which the precipitation occurred and the distribution of tritium between the seasons. Allison and Hughes (1974) discussed these considerations in detail.

Unlike tritium, a significant fraction of bomb-^{36}Cl fell out in dry form, and was thus much less sensitive to local precipitation variations. Once fallout-^{36}Cl has entered the soil it is conservative; only the position in the soil profile changes with evapotranspiration, and not the amount. The result of the much simpler behavior of bomb-^{36}Cl should be greater ease in interpretation.

The conservative properties of ^{36}Cl in soils bestow the further advantage of ease of sample extraction. Tritium samples are normally extracted by vacuum distillation. If electrolytic concentration is required, or if the soils are dry, inordinately large volumes of soil may be necessary. Great care must be taken to avoid loss of soil moisture or contamination by tritiated water vapor. In contrast, sufficient chloride for ^{36}Cl analysis may be extracted from a few kilograms of soil by simple leaching with distilled water, and neither loss of ^{36}Cl nor contamination is a serious problem.

Finally, the transport properties of ^{36}Cl in the vadose zone should, in most cases, be superior to those of tritium. Tritium is subject to vapor, as well as liquid, phase transport, which tends to broaden the bomb peak to a

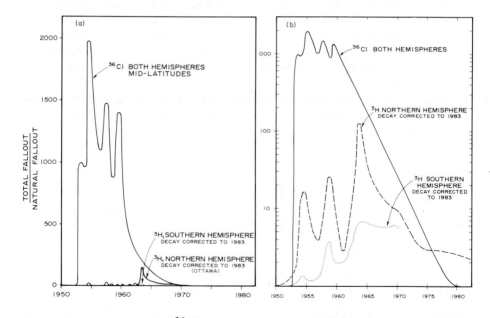

Fig. 10-25. A comparison of ^{36}Cl fallout with tritium fallout (decay-corrected to 1983) in the Northern and Southern Hemispheres, on arithmetic (a) and logarithmic (b) scales.

greater extent than ordinary dispersion. Although inclusion of vapor transport may be desirable in extremely dry soils where it constitutes a major portion of the total water flux, in most cases it will result in loss of information, or even loss of the signal if the tritium concentrations are already low. Chloride should move only with the liquid phase, eliminating the complications associated with vapor transport.

Analytical considerations. In order to be able to fill the role of bomb tritium, ^{36}Cl should be comparable in terms of the difficulty and cost of sampling, preparation and analysis. Sampling is equally simple for both species, except for vadose zone sampling, which is simpler for ^{36}Cl. Due to the high bomb fallout of ^{36}Cl, carrier techniques allow small water samples even when chloride concentrations are low. Preparation of ^{36}Cl involves several wet chemical purification steps. ^{36}Cl preparation is somewhat more complex than the preparation of tritium samples for direct liquid scintillation counting, but is much less difficult and time consuming than electrolytic enrichment of tritium. Actual counting time for ^{36}Cl on the accelerator mass spectrometer is normally less than an hour, much less than the usual liquid scintillation or gas proportional counting times for tritium.

Bomb-^{36}Cl will probably not be competitive with bomb-tritium as a tracer for shallow groundwater in the Northern Hemisphere as long as bomb-tritium concentrations are high enough to measure without electrolytic concentration. However, bomb-^{36}Cl should compare favorably with bomb tritium at the present time for ordinary tracing in the Southern Hemisphere.

Bomb-^{36}Cl demonstrates several important advantages over bomb-tritium for specialized applications such as determination of aquifer dispersivity or tracing in the vadose zone. In these cases ^{36}Cl is probably preferable to tritium in both hemispheres. At the present time the bomb-^{36}Cl pulse can usefully supplement the bomb-tritium pulse. As the tritium pulse decays to background levels, ^{36}Cl may eventually become more important than tritium as a hydrologic tracer.

REFERENCES

Acres America, Inc., 1977. *U.S. Federal Energy Administration, National Strategic Oil Storage Program.* Weeks Island Mine Geotechnical Study, Contract FEA-1251-75 to Gulf Interstate Engineering Company, Houston, Texas, 2 volumes.

Airey, P.L., Calf, G.E., Campbell, B.L., Hartley, P.E., Roman, D. and Habermehl, M.A., 1979. Aspects of the isotope hydrology of the Great Artesian Basin, Australia. In: *Isotope Hydrology 1978*, 2. IAEA, Vienna, pp. 205—219.

Airey, P.L., Davidson, M.R. and Roman, D., 1981. Solute transport and dispersion in very old ground water: potential applications of chlorine-36. In: S.N. Davis (Editor), *Workshop on Isotope Hydrology Applied to the Evaluation of Deeply Buried Nuclear Waste Repositories.* University of Arizona, Tucson, Ariz., pp. 68—86.

Allison, G.B. and Hughes, M.W., 1974. Environmental tritium in the unsaturated zone: Estimation of recharge to an unconfined aquifer. In: *Isotope Techniques in Groundwater Hydrology, 1*. IAEA, Vienna, pp. 57—69.

Bagge, E. and Willkom, H., 1966. Geologic age determination with ^{36}Cl. *Atomkernenergie*, 11: 176—184 (in German).

Bartholomew, R.M., Boyd, A.W., Brown, F., Hawkings, R.C., Lounsbury, M. and Morritt, W.F., 1955. The half-life of ^{36}Cl. *Can. J. Phys.*, 33: 43—48.

Bentley, H.W., 1978. Some comments on the use of chlorine-36 for dating very old ground water. In: S.N. Davis (Editor), *Workshop on Dating Old Ground Water*. Divisions, by Department of Hydrology and Water Resources, University of Arizona, Tucson, Ariz., 138 pp.

Bentley, H.W. and Davis, S.N., 1982. Application of AMS to hydrology. In: M. Kutchera (Editor), *2nd Annual Symposium on Acceleration Mass Spectrometry*. Argonne National Laboratories.

Bentley, H.W., Phillips, F.M., Davis, S.N., Gifford, S., Elmore, D., Tubbs, L.E. and Gove, H.E., 1982. Thermonuclear ^{36}Cl pulse in natural water. *Nature*, 300: 737—740.

Bentley, H.W., Phillips, F.M., Davis, S.N., Airey, P.L., Calf, G.E., Habermehl, M.A., Torgenson, T., Elmore, D. and Gove, H.G., 1983. Chlorine-36 dating of very old ground water: the Great Artesian Basin, Australia (in preparation).

Bentley, H.W., Salati, E., Menezes, M.O.A., Elmore, D. and Gove, H., 1983. Use of thermonuclear ^{36}Cl fallout to identify source of salinity in shallow groundwaters (in preparation).

Bergmann, F. and Libby, W.F., 1957. Continental water balance, groundwater inventory and storage times, surface ocean mixing rates, and worldwide water circulation patterns from cosmic ray and bomb tritium. *Geochim. Cosmochim. Acta*, 12: 277—296.

Biggar, J.W. and Nielsen, D.R., 1962. Miscible displacement, II. Behavior of tracers. *Soil Sci. Soc. Am., Proc.*, 26: 125—128.

Bonner, F.T., Roth, E., Schaeffer, O.A. and Thompson, S.O., 1961. Chlorine-36 and deuterium study of Great Basin lake waters. *Geochim. Cosmochim. Acta*, 25: 261—266.

Brinkman, J.E., 1981. *Water age dating of the Carrizo Sand*. M.S. Thesis, University of Arizona, Tucson, Ariz.

Carter, M.W. and Moghissi, A.A., 1977. Three decades of nuclear testing. *Health Phys.*, 33: 55—71.

Davis, R., Jr. and Schaeffer, O.A., 1955. Chlorine-36 in nature. *Ann. N.Y. Acad. Sci.* 62: 105—122.

Davis, S.N. and DeWiest, R.J.M., 1966. *Hydrogeology*. John Wiley and Sons, New York, N.Y., 463 pp.

Dinçer, T., Al-Mugrin, A. and Zimmerman, U., 1974. Study of the infiltration and recharge through the sand dunes in arid zones, with special reference to the stable isotopes and thermonuclear tritium. *J. Hydrol.*, 23: 79—109.

Dyrsson, D. and Nyman, P., 1955. Slow-neutron-induced radioactivity of seawater. *Acta Radiol.*, 43: 421—427.

Egboka, B.C.E., 1980. *Bomb tritium as an indication of dispersion and recharge in shallow sand aquifers*. Ph.D. Dissertation, University of Waterloo, Waterloo, Ont., 190 pp.

Ehhalt, D., 1973. On the uptake of tritium by soil water and ground-water. *Water Resour. Res.*, 9: 1073—1074.

Elmore, D., Fulton, B.R., Clover, M.R., Marsden, J.R., Gove, H.E., Naylor, H., Purser, K.H., Kilius, L.R., Beukins, R.P. and Litherland, A.E., 1979. Analysis of ^{36}Cl in environmental water samples using an electrostatic accelerator. *Nature*, 277: 22—25.

Elmore, D., Tubbs, L.E., Newman, D., Ma, X.Z., Finkel, R., Nishiizumi, K., Beer, J., Oeschger, H. and Andree, M., 1982. The ^{36}Cl bomb pulse measured in a shallow ice cove from Dye 3, Greenland. *Nature*, 300: 735—737.

Endt, P.M. and Van der Leun, C., 1973. Energy levels of $A = 21-44$ nuclei, V. *Nucl. Phys. A* 214: 1—625.

Eriksson, E., 1960. The yearly circulation of chloride and sulfur in nature, II. Meteorological, geochemical, and pedological implications, *Tellus*, 12: 63—109.

Fabryka-Martin, J., Bentley, H., Davis, H.S., Elmore, D. and Gove, H.E., 1983. ^{36}Cl measurements in Gulf Coast salt domes (in preparation).

Feige, Y., Oltman, B.G. and Kastner, J., 1968. Production rates of neutrons in soils due to natural radioactivity. *J. Geophys. Res.*, 73: 3135—3142.

Finkel, R.C., Nishiizumi, K., Elmore, D., Ferraro, R.D. and Gove, H.E., 1980. ^{36}Cl in polar ice, rain, and seawater. *Geophys. Res. Lett.*, 7: 983—986.

Freeze, R.A. and Cherry, J.A., 1979. *Groundwater*. Prentice-Hall, N.J.

Goldstein, G., 1966. Partial half-life for β-decay of ^{36}Cl. *J. Inorg. Nucl. Chem.*, 28: 937—939.

Graf, D.L., 1982. Chemical osmosis, reverse chemical osmosis, and the origin of subsurface brines. *Geochim. Cosmochim. Acta*, 46: 1431—1448.

Habermehl, M.A., 1980. The Great Artesian Basin, Australia. *J. Aust. Geol. Geophys.*, 5: 9—38.

Hobba, W.A., Jr., Fisher, D.W., Pearson, F.J., Jr. and Chemerys, J.C., 1979. Hydrology and geochemistry of thermal springs of the Appalachians. *U.S. Geol. Surv. Prof. Pap.*, 1044-E: 36 pp.

Hufen, T.H., Buddenmeier, R.W. and Lau, L.S., 1974. Isotopic and chemical characteristics of high-level groundwaters on Oahu, Hawaii. *Water Resour. Res.*, 10: 366—370.

Jonte, J.H., 1956. *Studies of radioelement fractionation in hydrothermal transport processes and of the contribution of some nuclear reactions to hydrothermal activity*. Ph.D. Dissertation, University of Arkansas, Fayetteville, Ark., 72 pp.

Kaufman, J.L. and Orlob, G.T., 1956. Measuring groundwater movement with radioactive and chemical tracers. *J. Am. Water Works Assoc.*, 48: 559—572.

Knauth, L.P. and Kumar, M.B., 1983. Isotopic character and origin of brine leaks in the Avery Island salt mine, South Louisiana, U.S.A. *J. Hydrol.* (in press).

Knauth, L.P., Kumar, M.B. and Martinex, J.D., 1980. Isotope geochemistry of water in Gulf Coast salt domes. *J. Geophys. Res.*, 85: 4863—4871.

Krey, P.W. and Krajewski, B., 1970. Comparison of atmospheric transport model calculations with observation of radioactive debris. *J. Geophys. Res.*, 75: 2901—2908.

Krupp, H.K., Biggar, S.W. and Nielsen, D.R., 1972. Relative flow rates of salt and water in soil. *Soil Sci. Soc. Am., Proc.*, 36: 412—417.

Kumar, M.B. and Martinex, J.D., 1978. Sources of subsurface leaks in Belle Isle and Weeks Island salt mines. In: *Annual Report, 1*. Institute for Environmental Studies, Louisiana State University, Baton Rouge, La., pp. 311—324.

Kuroda, P.K., Edwards, R.R., Robinson, B.L., Jonte, J.H. and Goolsby, C., 1957. Chlorine-36 in pitchblende. *Geochim. Cosmochim. Acta*, 11: 194—196.

Kuroda, P.K. and Kenna, B.T., 1960. The ratio of induced fission vs. spontaneous fission in pitchblende, and natural occurrence of radiochlorine. *J. Inorg. Nucl. Chem.*, 16: 1—7.

Lal, D. and Peters, B., 1967. Cosmic ray produced radioactivity on the earth. In: K. Sitte (Editor), *Handbuch der Physik, 46/2* (S. Flugge, General Editor), Springer-Verlag, Berlin, pp. 551—612.

Libby, W.F., 1953. The potential usefulness of natural tritium. *Proc. Natl. Acad. Sci. U.S.A.*, 39: 245—247.

Lockhart, L.B., Jr., Baus, R.A., Patterson, R.L., Jr. and Saunders, A.W., Jr., 1959. Contamination for the air by radioactivity from the 1958 nuclear tests in the Pacific. *Science*, 130: 161—162.

Mason, B., 1958. *Principles of Geochemistry*. John Wiley and Sons, New York, N.Y., pp. 51 and 202.

McFarlane, D.S., Cherry, J.A., Gilham, R.W., Sudicky, E.A., 1983. Migration of contaminants in groundwater at a landfill: a case study, 1. Groundwater flow and plume delineation. *J. Hydrol.*, 63: 1—30.

Meybeck, M., 1979. Concentrations des eaux fluviales en éléments majeurs et apports en solution aux océans. *Rev. Géol. Dyn. Géogr. Phys.*, 21: 215—246.

Meyboom, P., 1960. Geology and groundwater resources of the Milk River sandstone in southern Alberta. *Alta. Resour. Counc. Mem.*, 2.

Montgomery, C.G. and Tobey, A.R., 1949. Neutron production at mountain altitudes. *Phys. Rev.*, 76: 1478—1481.

Morrison, P. and Pine, J., 1955. Radiogenic origin of helium isotopes in rocks. *Ann. N.Y, Acad. Sci.*, 62: 69—92.

Oeschger, H., Houtermans, J., Loosli, H. and Wahlen, M., 1969. The constancy of radiation from isotope studies in meteorites and on the Earth. In: I.U. Olssun (Editor), *Radiocarbon Variations and Absolute Chronology, 12th Nobel Symposium*. Wiley-Interscience, New York, N.Y., pp. 471—496.

Onufriev, V.G., 1968. Formation of chlorine-36 in nature. *Yad. Geofiz., Geokhim. Izot. Metody Geol.*, 1968: 364—369. (In Russian).

Onufriev, V.G. and Soifer, V.N., 1968. Use of some cosmogenic isotopes for geochronological studies. *Tr. Vses. Naucho-Issled. Inst. Yad. Geofiz. Geokhim.*, 4: 331—340 (in Russian).

Pearson, F.J., Jr. and White, D.E., 1967. Carbon-14 ages and flow rates of water in the Carrizo Sand, Atacosa County, Texas. *Water Resour. Res.*, 3: 251—261.

Peterson, K.R., 1970. An empirical model for estimating worldwide deposition from atmospheric nuclear detonations. *Health Phys.*, 18: 357—378.

Phillips, F.M., Bentley, H.W., Davis, S.N., Gifford, S., Elmore, D., Tubbs, L. and Gove, H.E., 1982. Chlorine-36 from atmospheric nuclear weapons testing: a potential successor to bomb tritium as a hydrologic tracer (in preparation.)

Phillips, F.M., Smith, G.I., Bentley, H.W., Elmore, D. and Gove, H.E., 1983. ^{36}Cl dating of saline sediments: preliminary results from Searles Lake, California (in preparation).

Pine, J. and Morrison, P., 1952. The neutron source strength of granitic rocks. *Phys. Rev.*, 86: 606.

Poland, J.F. and Stewart, G.T., 1975. New tritium data on movement of groundwater in western Fresno County, California. *Water Resour. Res.*, 11: 716—724.

Rabinowitz, D.D., Holmes, C.R. and Gross, G.W., 1971. Forced exchange of tritiated water with clays. In: A.A. Moghissi and M.W. Carter (Editors), *Proc. Conf., Las Vegas, Nevada, 1971.* pp. 471—485.

Rabinowitz, D.D., Gross, G.W. and Holmes, C.R., 1977. Environmental tritium as a hydrometeorologic tool in the Roswell Basin, New Mexico, I. Tritium input function and precipitation-recharge relation; II. Tritium patterns in groundwater; III. Hydrologic parameters. *J. Hydrol.*, 32: 3—46.

Reeh, N., Clausen, H.B., Dansgaard, W., Gundestrup, N., Hammer, C.U. and Johnsen, S.J., 1978. Secular trends of accumulation rates at three Greenland stations. *J. Glaciol.*, 20: 27—30.

Rogers, J.J.W. and Adams, J.A.S., 1978a. Thorium. In: K.H. Wedepohl (Editor), *Handbook of Geochemistry, II/5*, Springer-Verlag, Berlin, pp. 90-E-2 to 90-E-4 (Chapter 90).

Rogers, J.J.W. and Adams, J.A.S., 1978b. Uranium. In: K.H. Wedepohl (Editor), *Handbook of Geochemistry, II/5*. Springer-Verlag, Berlin, pp. 92-E-3 to 92-E-5 (Chapter 92).

Roman, D. and Airey, P.L., 1981. The application of environmental chlorine-36 to hydrology, I. Liquid scintillation counting. *Int. J. Appl. Radiat. Isot.*, 32: 287—290.

Ronzani, C. and Tamers, M.A., 1966. Low-level chlorine-36 detection with liquid scintillation techniques. *Radiochem. Acta*, 6: 206—210.
Rösler, H.J. and Lange, H., 1972. *Geochemical Tables*. Elsevier, Amsterdam, 468 pp.
Salati, E., Leal, J.M. and Campos, M.M., 1974. Environmental isotopes used in a hydrogeological study of northeastern Brazil. In: *Isotope Techniques in Groundwater Hydrology, 1*. IAEA, Vienna, pp. 259—283.
Salati, E., Matsui, E., Leal, J.M. and Fritz, P., 1980. Utilization of natural isotopes in the study of salination of the waters in the Pajeu River Valley, northeast Brazil. In: *Isotope Hydrology*. IAEA, Vienna, pp. 133—150.
Schaeffer, O.A., Thompson, S.O. and Lark, N.L., 1960. Chlorine-36 radioactivity in rain. *J. Geophys. Res.*, 65: 4013—4016.
Schmalz, B.L. and Polzer, W.L., 1969. Tritiated water distribution in unsaturated soils. *Soil Sci.*, 108: 43—47.
Schoff, S.L., 1972. Origin of mineralized water in Precambrian rocks of the upper Paraiba Basin, Paraiba, Brazil. *U.S. Geol. Surv. Water Supply Pap.*, 1663-H: 92 pp.
Schwartz, F.W. and Muehlenbachs, K., 1979. Isotope and ion geochemistry of groundwaters in the Milk River Aquifer, Alberta. *Water Resour. Res.*, 15: 259—268.
Schwartz, F.W., Muehlenbachs, K. and Chorley, D.W., 1981. Flow systems control of the chemical evolution of groundwater. *J. Hydrol.*, 54: 225—243.
Sinclair, D.E. and Manuel, O.K., 1974. Argon-36 from neutron capture on chlorine in nature. *Z. Naturforsch.*, A, 29: 488—492.
Smith, D.B., Wearn, P.L., Richards, H.J. and Rowe, P.C., 1970. Water movement in the unsaturated zone of high and low permeability strata by measuring natural tritium. In: *Isotope Hydrology*. IAEA, Vienna, pp. 73—87.
Statham, I., 1977. *Earth Surface Sediment Transport*. Oxford University Press, Oxford.
Stewart, G.L., 1967. Fractionation of tritium and deuterium in soil water. In: G.E. Stout (Editor), *Isotope Techniques in the Hydrologic Cycle*. American Geophysical Union, Washington, D.C., pp. 159—167.
Stolf, R., Leal, J.M., Fritz, P. and Salati, E., 1979. Water budget of a dam in the semi-arid region of the northeast of Brazil based on oxygen-18 and chlorine contents. In: *Isotopes in Lake Studies*. IAEA, Vienna, 57 pp.
Sudicky, E.A., Cherry, J.A. and Frind, E.O., 1982. Hydrogeological studies of a sandy aquifer at an abandoned landfill, 4. A natural gradient tracer test. *J. Hydrol.* (in press).
Swanick, G., 1982. *The hydrochemistry and age of groundwater in the Milk River aquifer*. Thesis, University of Arizona, Tucson, Ariz.
Tamers, M.A., Ronzani, C. and Scharpenseel, H.W., 1969. Naturally occurring chlorine-36. *Atompraxis*, 15: 433—437.
Thorstenson, D.C., Fisher, D.W. and Croft, M.G., 1979. The geochemistry of the Fox Hills-Basal Hell Creek aquifer in southwestern North Dakota and northwestern South Dakota. *Water Resour. Res.*, 15: 1479—1498.
Trotman, K.N., Phillips, F.M., Bentley, H.W., Davis, S.N., Elmore, D. and Gove, H.E., 1983. ^{36}Cl, an environmental tracer for soil water (in preparation).
Turekian, K., Nozaki, Y. and Benninger, L.K., 1977. Geochemistry of atmospheric radon and radon products. *Annu. Rev. Earth Planet. Sci.*, 5: 227—255.
Verhagen, B.T., Smith, P.E., McGeorge, I. and Dziembowski, Z., 1979. Tritium profiles in Kalahari sands as a measure of rain-water recharge. In: *Isotope Hydrology 1978, 2*. IAEA, Vienna, pp. 733—751.
Wood, W.W., 1981. A geochemical method of determining dispersivity in regional groundwater systems. *J. Hydrol.*, 54: 209—224.
Yokoyama, Y., Reyss, J.-L. and Guichard, F., 1977. Production of radionuclides by cosmic rays at mountain altitudes. *Earth Planet. Sci. Lett.*, 36: 44—50.

Yokoyama, Y., Sato, J., Reyss, J-L. and Guichard, F., 1973. Variations of solar cosmic-ray flux deduced from ^{22}Na—^{26}Al data in lunar samples. *Proc. 4th Lunar Sci. Conf., Geochim. Cosmochim. Acta, Suppl.*, 4.

Zander, I. and Araskog, R., 1973. Nuclear explosions 1945—1972 — basic data. *FOA-4, Rep.*, A-4505-A.

Zito, R.R., Davis, S.N., Bentley, H.W. and Kuhn, M.W., 1982. Water dating and nuclear radionuclide production by subsurface neutrons. *Geol. Soc. Am. Abstr. Prog.*, 14(7): 653.

Chapter 11

RADIOACTIVE NOBLE GASES IN THE TERRESTRIAL ENVIRONMENT

T. FLORKOWSKI and K. RÓŻANSKI

INTRODUCTION

An increased interest in the radioactive noble gases in the environment has been observed for nearly two decades. The majority of noble gas isotopes are produced in the atmosphere by cosmic rays. Several of them are of anthropogenic origin, i.e. they are produced as a consequence of nuclear fission. The half-lives of these radioisotopes ranges from 5.27 days for ^{133}Xe to 2.1×10^5 years for ^{81}Kr allowing for using them as tracers to study various geophysical processes such as mixing of the atmosphere and the oceans, age determination of groundwater and ice layers as well as studying the constancy of the cosmic ray flux. Noble gases are particularly useful as

TABLE 11-1

Radioactive noble gases existing in the environment

Isotope	Type of decay, energy	Half-life	Average specific activity in the troposphere[a]	Origin
^{81}Kr	K-capture, 13.5 keV	2.1×10^5 years	0.07 dpm l^{-1} Kr (1.17×10^{-3} Bq l^{-1} Kr)	natural
^{85}Kr	β^-, 675 keV	10.76 years	3.3×10^4 dpm l^{-1} Kr (550 Bq l^{-1} Kr)	mainly anthropogenic
^{37}Ar	K-capture, 2.8 keV	35.1 days	2.5×10^{-3} dpm l^{-1} Ar (4.17×10^{-5} Bq l^{-1} Ar)	natural and anthropogenic
^{39}Ar	β^-, 565 keV	269 years	0.112 dpm l^{-1} Ar (1.87×10^{-3} Bq l^{-1} Ar)	mainly natural
^{133}Xe	β^-, 347 keV	5.27 days	—	anthropogenic

[a] Many authors use dpm mmol^{-1} Kr or dpm l^{-1} Kr (dpm: disintegration per minute) as the unit of specific activity. These units will be, therefore, used in the text parallel to the SI unit of radioactivity (Bq).

tracers in the earth's atmosphere or hydrosphere because of the absence of chemical reactions with the reservoir. The release rate, radioactive decay, parameters of dissolution and dispersion process are the only factors governing their concentration in the environment. Moreover, their abundance can be measured in relation to their stable isotope rather than to the carrier from which they are extracted (i.e. air, water, rock, ice). Therefore, if isotope fractionation during extraction is avoided, it is not necessary to know the extraction efficiency which simplifies the sampling procedure. Their concentration is also much less sensitive to changes in meteorological or geochemical conditions as in the case of other environmental tracers (e.g. ^{32}Si, ^{7}Be, ^{14}C). The concentration levels of these radioisotopes in the environment are, however, very low. Detection of radioactivity requires sometimes enormous sample volumes (up to 10^6 litres H_2O in the case of ^{81}Kr) and sophisticated low-level counting techniques. Therefore, the applicability of the majority of the radioactive noble gases present in the environment is limited to special studies. Measurement of their concentration can only be done in a few laboratories in the world.

Table 11-1 summarises the essential properties of the isotopes discussed in this chapter.* ^{222}Rn, the heaviest noble gas, is omitted here because it is discussed elsewhere. Less important for environmental applications or not permanently present in the environment, the radioactive isotopes of ^{41}Ar and ^{42}Ar as well as ^{131m}Xe are also excluded.

^{85}Kr

The natural level of ^{85}Kr in the atmosphere is negligible as compared with its total world inventory. Processes responsible for natural ^{85}Kr production are the (n,γ) reactions of cosmic ray neutrons with the stable ^{84}Kr in the atmosphere as well as spontaneous fission of uranium and thorium in the earth's crust (neutron-induced fission of those elements contributes several orders of magnitude less). Estimations made by Diethorn and Stockho (1972) and Suzuki and Inoue (1972) provide the value of 14 Ci (5.18×10^{11} Bq) as a steady-state natural inventory of ^{85}Kr in the atmosphere before the nuclear age. This value is about six orders of magnitude lower than the actual activity (cf. Table 11-3).

For the first time ^{85}Kr activity in the atmosphere increased considerably as a result of nuclear tests in the atmosphere. In the early sixties specific activity of ^{85}Kr in the atmosphere reached the value of about 300 dpm mmol^{-1}

---

* Conversion factors:

1 dpm mmol^{-1} Kr = 0.0201 pCi cm^{-3} Kr = 0.0229 pCi m^{-3} air = 7.44×10^{-4} Bq cm^{-3} Kr.
1 Bq mmol^{-1} Kr = 60 dpm mmol^{-1} Kr = 4.46×10^{-2} Bq cm^{-3} Kr = 1.206 pCi cm^{-3} Kr.

Kr (5 Bq mmol^{-1} Kr). In the next years nuclear facilities (reactors and fuel-reprocessing plants) became the predominant source of this isotope in the environment.

Man-made ^{85}Kr originates in nuclear fuel as a product of different fission reactions. In Table 11-2 the ^{85}Kr production yields for various fuel materials and neutron energies are summarised. The production yield is defined as the ratio of fission events leading to formation of ^{85}Kr atoms to the total number of fissions in unit time, expressed in percent.

Data presented in Table 11-2 support the estimation of ^{85}Kr production in nuclear weapon tests and in nuclear reactors. According to Hilbert (1974) and Różanski (1979a) the production yield of ^{85}Kr in nuclear explosions amounts to 25 kCi Mt^{-1} TNT (9.25 × 10^{14} Bq Mt^{-1}) for ^{235}U and 10 kCi Mt^{-1} TNT (3.7 × 10^{14} Bq Mt^{-1}) for ^{239}Pu. The ^{85}Kr production in nuclear reactors depends on the type of reactor and the kind of fuel used. It varies from 166 kCi GW$_{(e)}^{-1}$ y^{-1} for ^{239}Pu and boiling-water reactors (BWR) to 973 kCi GW$_{(e)}^{-1}$ y^{-1} for ^{233}U and high-temperature gas-cooled reactors (HTGR) (6.14 × 10^{15} Bq GW$_{(e)}^{-1}$ y^{-1} and 3.6 × 10^{16} Bq GW$_{(e)}^{-1}$ y^{-1}, respectively) (Hilbert, 1974; Schröder and Roether, 1974). During normal operation of nuclear power reactors only about 1‰ of the total ^{85}Kr production is released into the atmosphere. Main emission of ^{85}Kr takes place in nuclear fuel reprocessing plants.

The earliest measurements of atmospheric ^{85}Kr activity have been carried out in the late fifties (de Vries, 1956; Delibrias and Jehanno, 1959; Griesser and Sittkus, 1961; Kigoshi, 1962). In the sixties several laboratories in Europe started with systematic measurements of ^{85}Kr in the atmosphere. At present, such regular measurements are continued in Freiburg (Federal

TABLE 11-2

Production yields of ^{85}Kr in various fission reactions

Fuel material	Type of neutrons	Production yield (%)	Reference
^{235}U	thermal	0.293	Katcoff (1969)
^{235}U	thermal	0.273	Katcoff and Rubinson (1965)
^{235}U	thermal	0.290	Chitwood (1973)
^{235}U	fission	0.32	Chitwood (1973)
^{235}U	fast	0.34	NCRP (1975)
^{233}U	thermal	0.58	Katcoff (1969)
^{238}U	fission	0.17	NCRP (1975)
^{239}Pu	thermal	0.099	Katcoff and Rubinson (1965)
^{239}Pu	thermal	0.14	Chitwood (1973)
^{239}Pu	fission	0.14	Chitwood (1973)
^{239}Pu	fast	0.17	NCRP (1975)
^{239}Pu	fast	0.076	Kirk (1972)

Republic of Germany), in Debrecen (Hungary), in the U.S.A. (network of the U.S. Environmental Protection Agency stations) and in the European part of the U.S.S.R.

The technique most often used for measurement of ^{85}Kr activity in the atmosphere are liquid scintillation counting and low-level gas counting. Prior to the radioactivity measurement, the krypton gas is extracted from the atmosphere by fractional distillation of liquid air, or by selective absorption of krypton on porous materials (e.g. active charcoal or molecular sieve). Two excellent reviews on the analytical methods and techniques for ^{85}Kr monitoring in the atmosphere have been published by Budnitz (1973) and Jaquish and Moghissi (1973). More or less comprehensive data on ^{85}Kr activity in the atmosphere obtained by various laboratories are also available in the literature (e.g. Schröder and Münnich, 1971; Farges et al., 1974; Csongor, 1977; Heller et al., 1977; Patti and Bourgeon, 1980). Moreover, several literary reviews on various aspects of ^{85}Kr presence in the earth's atmosphere have been published in the last ten years (Kirk, 1972; Hilbert, 1974; NCRP, 1975; Sittkus and Stockburger, 1976; Różanski, 1979b).

Fig. 11-1 summarises measurement data of ^{85}Kr activity in the surface atmosphere published until 1978 (de Vries, 1956; Delibrias and Jehanno, 1959; Griesser and Sittkus, 1961; Kigoshi, 1962; Ehhalt et al., 1963, 1964; Pannetier, 1968; Shuping et al., 1969; Schröder and Münnich, 1971; Shearin et al., 1971; Jaquish and Johns, 1972; Csongor, 1973, 1977; Wardaszko,

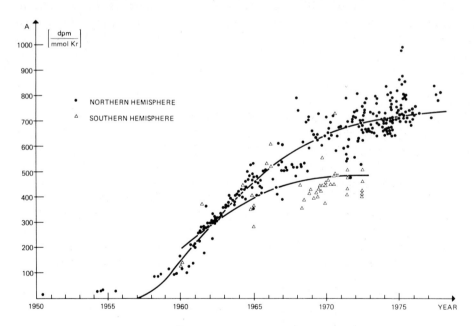

Fig. 11-1. ^{85}Kr activity in the surface air 1950—1977.

1973; Andrews and Wrubble, 1973; Farges et al., 1974; Kishida, 1974; U.S. EPA, 1974a, 1974b; Stockburger and Sittkus, 1975; Telegadas and Ferber, 1975; Gudkov et al., 1976; Sittkus and Stockburger, 1976; Różanski and Ostrowski, 1977; Ferber et al., 1977; Tertyŝnik et al., 1977; Stockburger et al., 1977; Wardaszko and Nidecka, 1978). The points in Fig. 11-1 are arithmetic montly means of the published data. In most cases krypton has been continuously sampled from the atmosphere.

The rather sharp increasing trend observed in the early sixties is followed by a significantly slower build-up of the ^{85}Kr activity in the last years. The scatter of the measurement data can be partly explained by the lack of standardization of applied analytical methods and partly by the influence of local ^{85}Kr sources. After 1965, the fuel reprocessing plants established in Europe have probably introduced considerable scatter (see, e.g., Schröder and Münnich, 1971; Sittkus and Stockburger, 1976; Heller et al., 1977). High ^{85}Kr concentrations recorded in the near-ground atmosphere coincide with peaks in the tritium concentration in precipitation (Weiss et al., 1979) indicating the same origin of tritium and ^{85}Kr discharged in gaseous form from nuclear installations. The influence of local surface sources of ^{85}Kr is seen in Fig. 11-2 where the ^{85}Kr concentration is shown of weekly composite samples collected in the period 1977—1981 at two stations in the Federal Republic of Germany (Weiss et al., 1982) differing in altitude by about 1000 m. Numerous peaks of ^{85}Kr concentration are more pronounced in the low station as compared with the high station. An interesting feature of the data presented in Fig. 11-2 is the apparent seasonal variability of the ^{85}Kr concentration with an amplitude of 40—80 dpm mmol^{-1} Kr (0.67—1.34 Bq mmol^{-1} Kr). The reason for this seasonal variability is not clear, several explanations are possible. Most likely it is connected with the seasonal variations in the vertical mixing of the atmosphere (Weiss et al., 1982).

As seen from Fig. 11-1 the discharge of ^{85}Kr into the atmosphere is slowed down considerably during the last decade. The emission rate of ^{85}Kr has dropped from about 80 dpm mmol^{-1} Kr y^{-1} (1.33 Bq mmol^{-1} Kr y^{-1}) in the late sixties to about 50 dpm mmol^{-1} Kr y^{-1} (0.83 Bq mmol^{-1} Kr y^{-1}) in 1977 with a tendency to remain constant (Heller et al., 1977; Różanski, 1979a). ^{85}Kr release from reprocessing plants is likely to be substantially reduced in future as krypton tapping systems are being introduced during the fuel reprocessing process. However, the most recent data (Fig. 11-2) indicate a small increase of about 25 dpm mmol^{-1} Kr y^{-1} (0.42 Bq mmol^{-1} Kr y^{-1}) which corresponds to an emission rate of about 75 dpm mmol^{-1} Kr y^{-1} (1.25 Bq mmol^{-1} Kr y^{-1}).

The spatial distribution of ^{85}Kr in the earth's atmosphere is primarily controlled by three factors: the rate of radioactive decay, the source configuration (area distribution and emission rates), and the global scale dispersion processes. Since removal processes other than radioactive decay are negligible for ^{85}Kr, this isotope is a potential atmospheric tracer and can be applied in verification of the existing global atmospheric dispersion models. Published ^{85}Kr concentration measurements indicate that the distribution of this isotope in the atmosphere is far from homogeneous. Practically all important sources of ^{85}Kr are situated at present in the north temperate latitudes. One can expect, therefore, the meridional inhomogeneity in ^{85}Kr concentration. There are two published meridional profiles of ^{85}Kr activity in the surface atmosphere. In 1964, during the cruise of the French ship "Thala Dan" from Le Havre to Adelie Coast (Antarctica) a set of air samples was collected for ^{85}Kr analysis (Pannetier, 1968). Few years later, a new north-south ^{85}Kr profile over the Atlantic Ocean has been measured (Gudkov, 1976). The most recent north-south profiles of ^{85}Kr were obtained by Weiss et al. (1982) in 1980—81 on board R/V "Meteor" during the cruise over the Atlantic Ocean. Fig. 11-3 summarises all data. Pannetier's results show a slight decrease of ^{85}Kr concentration in the south direction up to 40°S latitude followed by a sharp increase at 60°S latitude. This is probably

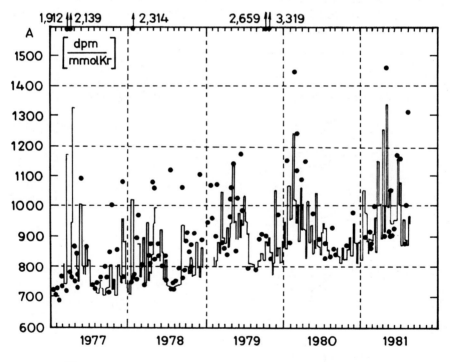

Fig. 11-2. ^{85}Kr concentration of weekly composite samples collected in 1977—1981 at two stations in the Federal Republic of Germany: Schaninsland, altitude 1240 m; Freiburg, altitude 240 m. Geographical position: 47.9°N, 7.8°E (modified from Weiss et al., 1982).

due to impact of stratospheric air with high ^{85}Kr concentration from nuclear tests in the early sixties. Stratospheric injections are facilitated by a low average tropopause level in these regions, as well as by the tropopause discontinuity occurring over 60°S. The profile of Gudkov in 1972 (Gudkov, 1976) shows a more regular north-south trend, although small jumps can be seen between 5°N and 10°N. The meridional gradient of ^{85}Kr concentration in the Northern Hemisphere seems to be larger than in 1964 and a constant distribution within the Southern Hemisphere is observed. The recent profiles of Weiss (Weiss et al., 1982) reveal similar patterns. At the respective geographical positions of the Intertropical Convergence (ITC), e.g. 5°N and 10°N, a concentration jump is observed amounting to more than 10 times the analytical error. The global mean surface air concentration of ^{85}Kr calculated from Pannetier's data reaches the value of 88% of the concentration measured in the Northern Hemisphere (30—50°N). In 1972 this value was about 76% (after Gudkov's profile). The Northern Hemispheric residence time of ^{85}Kr evaluated on the basis of its meridional distribution measured in 1964 and a two-box model of the atmosphere equals about 1.5 years (Pannetier, 1970). Estimates by Weiss et al. (1982) give the values 1.0 and 1.7 years for profile 1 and 2 respectively (Fig. 11-3). The authors claim

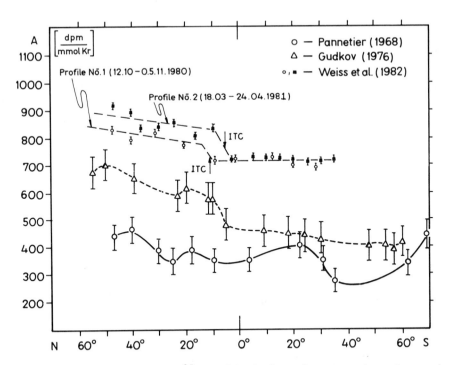

Fig. 11-3. Meridional profiles of ^{85}Kr activity in the surface atmosphere. Arrows show the position of the intertropical convergence zone during sampling.

that the difference is significant and assuming the model used indicates the seasonal variation of the interhemispheric exchange.

The meridional concentration gradient of ^{85}Kr concentration allows for estimation of mixing intensity of the atmosphere in the north-south direction over the Northern Hemisphere. The value of the meridional eddy diffusion coefficient reported by Weiss et al. (1982) for the region between 40°N and 10°N equals 1.6×10^{10} cm^2 s^{-1} which is in reasonable agreement with estimates based on ^{37}Ar distribution.

The vertical distribution of ^{85}Kr in the atmosphere represents integrated effects of diffusion, vertical mixing and decay processes as well as variable source configuration over the time period of the last thirty years. The vertical profile obtained by Pannetier (1968) in 1964 over France (Fig. 11-4) exhibits a generally constant ^{85}Kr concentration up to 24 km altitude with relatively high stratospheric values as compared with later measurements by

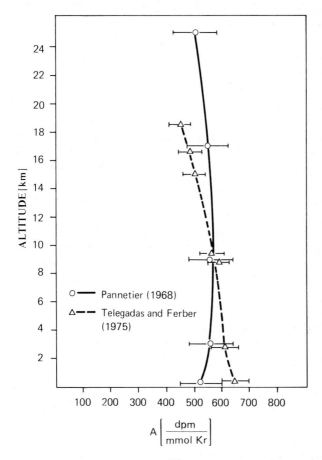

Fig. 11-4. Vertical profiles of ^{85}Kr activity in the atmosphere.

Telegadas and Ferber (1975). This effect again can be attributed to large stratospheric discharge of ^{85}Kr during nuclear bomb tests in the early sixties. The average profile of ^{85}Kr presented by Telegadas and Ferber (1975) is based on numerous measurements over the region 35—45°N. In this case a decrease of ^{85}Kr concentration with altitude of 13.5 dpm mmol^{-1} Kr per 1 km can be observed.

Telegadas and Ferber (1975) report also the results of ^{85}Kr concentration measurements in the tropopause and lower stratosphere (13—20 km altitude) over the region 80°N to 60°S. The stratospheric distribution of ^{85}Kr concentrations shows a maximum in the lower equatorial region of each hemisphere and a minimum concentration in the polar region. This distribution tends to confirm the mean stratospheric circulation as deduced from other radioactive tracer data (List and Telegadas, 1969).

The results of ^{85}Kr measurements in the atmosphere provide the basis for the global appraisal of the amount of fissioned nuclear fuel. These data provide also a unique opportunity for an indirect estimate of the amount of plutonium produced until now for military purposes. The estimate is based on the total atmospheric inventory of ^{85}Kr and the inventory accounted for by bomb tests and nuclear power production. The difference between these two time-dependent functions is thought to be caused by the ^{85}Kr liberated during reprocessing of the nuclear fuel from military reactors producing ^{239}Pu. The efficiency of plutonium production (about 100 fissions of ^{235}U are necessary to produce 60 nuclei of ^{239}Pu) and the efficiency of ^{85}Kr production (cf. Table 12-2) together yield the conversion factor of 0.63 MCi (2.33×10^{16} Bq) of ^{85}Kr per 1 tonne of ^{239}Pu produced. This factor is further used to convert the atmospheric excess of ^{85}Kr in a given year to the corresponding plutonium production. The calculations are complicated to some extent by a certain delay between production and release of ^{85}Kr into the atmosphere. This time lag is about 3 years for civil nuclear fuel (Schröder and Roether, 1974) and about 1.5 years for military fuel (Sittkus and Stockburger, 1976). According to Schröder and Roether (1974) about 130 tonnes of ^{239}Pu were produced up to 1970. More recent estimations (Sittkus and Stockburger, 1976) give the figure of about 160 tonnes up to the end of 1974. The estimated curve of the yearly production rate of ^{239}Pu is shown in Fig. 11-5. As seen from Fig. 11-5 this production in 1975 was still kept on the level of 7—8 t ^{239}Pu per year, with a slight decreasing tendency. Table 11-3 shows the contribution of various sources to the world atmospheric inventory of ^{85}Kr as estimated for the end of 1977 (Różanski 1979a).

The regular time trend observed in the atmospheric ^{85}Kr activity enables prognosis of expected ^{85}Kr activities in the future. Several of such forecasts have been published together with the expected impacts of ^{85}Kr to the human environment (Kirk, 1972; Bernhardt et al., 1973; Hilbert, 1974; NCRP, 1975; Csongor, 1977). All mentioned forecasts are based on the expected growth of nuclear energy production. Comparison of the projected and measured ^{85}Kr concentrations in the atmosphere (Fig. 11-6) show that all expected ^{85}Kr levels seem to be over-estimated. The expected cumulative build-up of ^{85}Kr in the atmosphere will contribute to the future doses. The calculated beta doses to the skin from the projected atmospheric ^{85}Kr concentration levels in the year 2000 are of the order of several millirem per year. This dose is far below the actual MPC (maximal permissible concentration) value. Therefore, the health hazard for the total population is rather negligible. Radiation protection problems caused by ^{85}Kr could only be important in the vicinity of the big reprocessing plants.

TABLE 11-3

World ^{85}Kr inventory as estimated for the end of 1977

Source	Estimated ^{85}Kr activity in the atmosphere	
Natural production	1.4×10^{-5} MCi	(5.18×10^{11} Bq)
Nuclear explosions	2.0 MCi	(7.4×10^{16} Bq)
Nuclear power	28.3 MCi	(10.4×10^{17} Bq)
Plutonium production for military purposes	30.3 MCi	(11.2×10^{17} Bq)

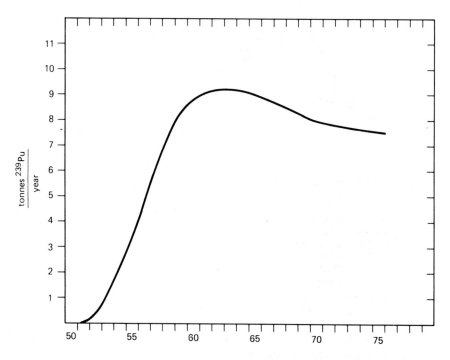

Fig. 11-5. Estimated yearly production rate of ^{239}Pu (for military purposes).

The high ^{85}Kr concentrations expected in the future can be considered as a potential climatic factor. Boeck (1976) argues that the expected atmospheric ^{85}Kr concentrations in the year 2000 may result in an increase of about 10% of the electrical conductivity of the entire atmosphere. Such considerable change of this important atmospheric parameter could visibly affect the climatic processes on the earth.

The atmosphere can be assumed as the main ^{85}Kr reservoir in the earth's ecosphere, where the ^{85}Kr level is primarily controlled by radioactive decay and emission rates from reprocessing plants. However, a certain amount of ^{85}Kr enters the hydrosphere with precipitation and directly in the gas exchange processes across the atmosphere-hydrosphere boundary layer. This transport reaches yearly about 0.1% of the total ^{85}Kr atmospheric inventory

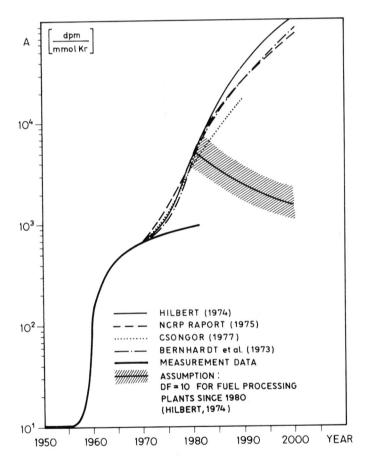

Fig. 11-6. Forecasts of ^{85}Kr activity in the atmosphere until the year 2000. DF = decontamination factor in nuclear reprocessing cycle equal to the ratio of ^{85}Kr produced and released to the environment.

(Różanski, 1979a). The removal process by direct diffusion of ^{85}Kr into the ground is less effective (about 0.0596% per year) (NCRP, 1975). As a radioactive noble gas with fairly well known atmospheric concentration history, ^{85}Kr is also a promising short-lived tracer in the hydrosphere. It has been formerly proposed as a new dating tool for shallow groundwaters (Lal et al., 1970; Oeschger et al., 1974). The first successful application of ^{85}Kr has been found, however, in oceanography in the study of mixing patterns in the ocean. Two depth profiles of ^{85}Kr concentration in the North Atlantic obtained by Schröder (1975) allowed to derive a vertical eddy diffusion coefficient in the main termocline of that region (300—1000 m depth — 5 $cm^2\ s^{-1}$). Recently new attempts have been made using ^{85}Kr in isotope hydrology (Różanski and Florkowski, 1979; Różanski, 1979a; Etzweiler, 1980; Salvamoser, 1981).

Despite the relatively high radioactivity of atmospheric krypton (at present about 920 dpm $mmol^{-1}$ Kr (0.684 Bq cm^{-3} Kr) in the Northern Hemisphere), its low atmospheric mixing ratio (1.11 ppmv) and low solubility in water (~8% by volume at 10°C) yield the activity of about 3.1 dpm ^{85}Kr in 1 m^3 water (23.0 × 10^{-4} Bq m^{-3} H_2O). Thus, a large sample volume and a rather sophisticated analytical procedure is necessary to measure this activity with sufficient accuracy. Prior to radioactivity measurement, the krypton gas is extracted from 0.1 to 0.3 m^3 of water and further separated from other gases by selective absorption on active charcoal and chemical purification with hot barium or calcium. The ^{85}Kr radioactivity is further measured in a miniature proportional counter (usually several cubic centimetres in volume) placed in the anticoincidence and heavy shields. Special caution is needed during the sampling of water due to the fact that atmospheric air usually represents significantly higher ^{85}Kr activity than it is observed in groundwaters. Typical error of analysis reaches at present several percent of the actual atmospheric ^{85}Kr activity (Schröder, 1975; Różanski, 1979a; Etzweiler, 1980; Salvamoser, 1981).

The first measurements of ^{85}Kr activity in groundwaters of various origin already confirm a potential utility of this isotope as a dating tool for shallow groundwater systems (Różanski and Florkowski, 1979; Etzweiler, 1980). Moreover, an attempt to develop a theoretical basis for interpretation of ^{85}Kr data together with other radioactive tracers used in groundwater studies has been undertaken (Maloszewski and Zuber, 1982; Grabczak et al., 1982). It should be noted, however, that the gaseous nature of this tracer can cause serious problems in its application to phreatic aquifers, where direct diffusion of atmospheric ^{85}Kr in the gaseous phase of the unsaturated zone to the water table can result in under-estimation of the mean transit time of water in such aquifers. Further developments in counting technique and interpretation of ^{85}Kr data are necessary before this method can be used routinely in isotope hydrology.

For further application of ^{85}Kr data in studies of large-scale meridional, interhemispheric and vertical mixing of the atmosphere, continuous monitoring of ^{85}Kr concentration in air is of high importance. Sampling stations situated in clean areas, i.e. far from nuclear installations, and also repeated meridional and vertical ^{85}Kr concentration profiles should yield base data for such studies.

^{81}Kr

^{81}Kr is the second radioactive isotope of krypton permanently present in the environment. As its half-life (2.5×10^5 years) is long compared to the atmospheric mixing time, it should be homogeneously distributed throughout the atmosphere and its concentration should be in steady state if its production has been constant during several half-lives. Eventual changes in the cosmic ray intensity during this time would yield the difference between the present activity and the present production rate of ^{81}Kr.

^{81}Kr is produced by the cosmic ray nucleons from stable isotopes of krypton (^{82}Kr, ^{83}Kr, ^{84}Kr, ^{86}Kr) by spallation reactions and also from stable ^{80}Kr by ^{80}Kr (n,γ) ^{81}Kr reaction by thermal neutrons in the atmosphere. A theoretical calculation made by Loosli et al. (1970) gives the total average production rate of 0.035 dpm l^{-1} Kr or 0.13 dpm l^{-1} Kr (5.83×10^{-4} Bq l^{-1} Kr or 2.17×10^{-3} Bq l^{-1} Kr, respectively) depending on the assumed cross-section of ^{80}Kr (n,γ) reaction. More recent calculation based on new cross-section value leads to production rate of 0.04 dpm l^{-1} Kr (6.67×10^{-4} Bq l^{-1} Kr) (Barabanov and Pomansky, 1977).

Experimental determination of ^{81}Kr activity in the present atmosphere is impossible because of contamination by anthropogenic ^{85}Kr. Several determinations were made on pre-bomb krypton samples. Loosli et al. (1970) report the value of 0.10 ± 0.01 dpm l^{-1} Kr ($1.67 \pm 0.16 \times 10^{-3}$ Bq l^{-1} Kr). Barabanov and Pomansky (1977) measured in samples from the year 1944 an ^{81}Kr activity of 0.046 ± 0.01 dpm l^{-1} Kr which is in good agreement with the estimated production rate value. New, more precise measurements done on the same krypton sample (Kuzminov and Pomansky, 1980) yield, however, a significantly higher value of ^{81}Kr activity equal to 0.069 ± 0.003 dpm l^{-1} Kr ($1.15 \pm 0.05 \times 10^{-3}$ Bq l^{-1} Kr), which is also higher than the best presently available estimate of the production rate of this isotope in the atmosphere. Obviously, much more experimental data is needed to draw final conclusions on the constancy of the cosmic ray flux during the last several hundred thousand years on the basis of ^{81}Kr data.

Another promising potential application of the ^{81}Kr isotope is its use in dating of groundwaters having ages in the range of 50,000—800,000 years (Oeschger, 1978). Advantages in using this isotope are similar to the case of ^{85}Kr, namely its atmospheric activity is known and the lack of complications due to chemical interactions in the aquifer or climatic variations. Measurement of this isotope is, however, complicated by extremely low concentration levels of ^{81}Kr in modern waters. This concentration is about 1000 atoms l^{-1} of water. Low-level counting spectroscopy requires in the case of ^{81}Kr a sample volume of about 10^6 litres (Oeschger, 1978). Adopting the new accelerator counting technique would diminish this volume by a factor of ten to twenty (Mast, 1978). The ^{81}Kr dating method would eventually be used as a comparative technique to the ^{36}Cl method (half-life = 3.08×10^5 years) (Elmore et al., 1979).

^{39}Ar AND ^{37}Ar

^{39}Ar and ^{37}Ar are the most extensively studied radioactive isotopes of the noble gases. Their activity levels observed in the environment are summarised in Table 12-4.

^{37}Ar has been largely used in the investigations of atmospheric mixing. ^{39}Ar, like the ^{81}Kr isotope, is a good indicator of constancy in the cosmic ray flux as well as potential tracer for dating groundwaters. The ^{37}Ar/^{39}Ar activity ratio in meteorites provides information about the spatial gradient of the cosmic ray flux.

The ^{37}Ar and ^{39}Ar isotopes are produced in many nuclear reactions (Table 11-5). However, only a few reactions are of importance for the natural production of these isotopes in the earth's atmosphere and lithosphere. The

TABLE 11-4

^{37}Ar and ^{39}Ar activity levels in the environment

	^{37}Ar	^{39}Ar
Stratosphere	0.029—0.037 dpm l^{-1} Ar (Machta, 1973)	no data available
Troposphere	0.001—1 dpm l^{-1} Ar (Loosli et al., 1973)	0.112 ± 0.012 dpm l^{-1} Ar (Loosli and Oeschger, 1979)
Lithosphere:		
surface	no data available[a]	0.03—0.07 dpm kg^{-1} rock (granite and biotit-gneis rocks; Hebert and Fröhlich, 1979)
deep layers	no data available[a]	no data available
Meteorites	4—29 dpm kg^{-1} (Heusser and Schaeffer, 1977a)	16—28 dpm kg^{-1} (Heusser and Schaeffer, 1977a)

[a] Spannagel and Fireman (1972) measured the production rate of ^{37}Ar by the reaction ^{39}K $(\bar{\mu},\nu_\mu,2n)$ ^{37}Ar. At a depth of 2 m water equivalent the equilibrium activity of ^{37}Ar was 0.065 dpm kg^{-1} ^{39}K. At depths of 40 and 830 m water equivalent this value reaches 2.9×10^{-3} dpm kg^{-1} and 2.4×10^{-4} dpm kg^{-1}, respectively (Kirsten and Hampel, 1977).

production rate of ^{39}Ar in the atmosphere has been discussed in detail by Lal and Peters (1967) and Loosli and Oeschger (1968). Only reactions on ^{40}Ar are of importance and the estimated average global production rate is about 0.10 dpm l^{-1} Ar (1.67 × 10^{-3} Bq l^{-1} Ar). Also, for the ^{37}Ar isotope the spallation reaction on ^{40}Ar in the atmosphere is the basic source and the average global tropospheric production rate of 2.5 × 10^{-3} dpm l^{-1} Ar (1.17 × 10^{-4} Bq l^{-1} Ar) has been derived from theoretical calculations. This value, being valid for periods of medium solar activity, is uncertain by at least a factor of 2 (Loosli et al., 1970). During high solar activity, dumping the galactic component of the cosmic ray flux, the tropospheric production rate of ^{37}Ar is lower by about 20% (Oeschger et al., 1970). The corresponding stratospheric value is equal to 6 × 10^{-2} dpm l^{-1} Ar (1 × 10^{-3} Bq l^{-1} Ar).

Tropospheric activity for ^{37}Ar has been measured since 1967. At present, numerous data for both hemispheres and a few determinations in the stratosphere are available (Loosli et al., 1970, 1973; Currie and Lindstrom, 1973; Rutherford et al., 1976; Oeschger and Loosli, 1977). In Fig. 11-7 ^{37}Ar activities measured at ground level in Bern, Switzerland, by Oeschger and Loosli are shown. The observed sudden increase of ^{37}Ar activity by two orders of magnitude is attributed to release from underground nuclear tests with venting. The estimations show that sufficient amounts of ^{37}Ar can be produced by the reaction ^{40}Ca(n,α)^{37}Ar to account for the observed activities (Oeschger and Loosli, 1977). The lowest ^{37}Ar activities observed in the atmosphere are attributed to natural production by cosmic radiation.

The distribution of ^{37}Ar in the atmospheric reflects the balance between cosmic ray production, radioactive decay and transport processes. If the first two processes are known, the global scale atmospheric transport can be determined. The estimated production rate and measured tropospheric activities of ^{37}Ar have been used as the input parameters in the global models of atmospheric mixing (Machta et al., 1970; Machta, 1973; Loosli et al., 1973). Machta (1973), on the basis of the two-dimensional eddy diffusion

TABLE 11-5

Production reactions of ^{37}Ar and ^{39}Ar isotopes (Loosli and Oeschger, 1968; Loosli et al., 1970; Schell, 1970; Kirsten and Hampel, 1977)

^{37}Ar	^{39}Ar
^{37}Cl (p,n) ^{37}Ar	^{39}K (n,p) ^{39}Ar
^{37}Cl (d,2n) ^{37}Ar	^{39}K ($\bar{\mu},\nu_\mu$) ^{39}Ar
^{39}K (d,α) ^{37}Ar	^{38}Ar (n,γ) ^{39}Ar
^{39}K ($\bar{\mu},\nu_\mu$, 2n) ^{37}Ar	^{40}Ar (n,2n) ^{39}Ar
^{40}Ca (n,α) ^{37}Ar	
^{36}Ar (n,γ) ^{37}Ar	
^{40}Ar (n,α) ^{37}Ar	

model and ^{37}Ar data, derived a value for the vertical eddy diffusion coefficient for the entire troposphere as 5×10^4 cm^2 s^{-1}. The multibox eddy diffusion model developed by Loosli et al. (1973) provides meridional eddy diffusion coefficients of the order of 1×10^{10} to 3×10^{10} cm^2 s^{-1} for the latitude range between 20° and 70°N. Both values agree well with previous determinations based on other tracers.

The first determinations of ^{39}Ar activity in the atmosphere have been made by Loosli and Oeschger (1968). The average activity of ^{39}Ar obtained in these measurements amounts to 0.10 dpm l^{-1} Ar (1.67 × 10^{-3} Bq l^{-1}). Further measurements done by these authors (Loosli and Oeschger, 1979) allowed them to obtain the more precise value of 0.112 ± 0.012 dpm l^{-1} Ar (1.87 × 10^{-3} Bq l^{-1}). Man-made contribution to this value due to nuclear weapon tests is presumably lower than 0.005 dpm l^{-1} Ar (8.35 × 10^{-5} Bq l^{-1}) (Loosli and Oeschger, 1979). A good agreement between the estimated production rate and measured activity of ^{39}Ar seems to indicate a constancy of the cosmic ray flux in the past (the time scale considered here is shorter than in the case of ^{81}Kr because of the 269-year half-life of ^{39}Ar).

The first application of ^{39}Ar was the dating of glaciological samples. The residual activity of ^{39}Ar in ice samples, when compared to modern atmospheric activity of this isotope, enables one to determine the time elapsed

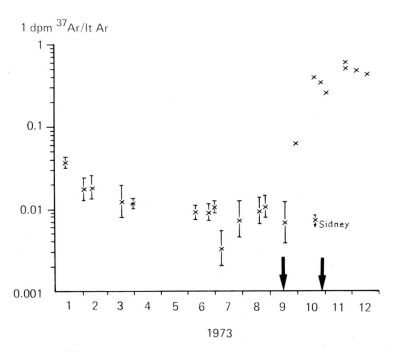

Fig. 11-7. ^{37}Ar activities in the surface air measured in Bern, Switzerland. Large, underground nuclear explosions are marked by arrows (Oeschger and Loosli, 1977).

since the argon-containing air was occluded in the ice. The low specific activity of ^{39}Ar requires, however, a large sample volume (a few tons of ice must be processed). An interesting in situ extraction technique has been developed by the group of scientists from the University of Bern, Switzerland (Oeschger et al., 1972, 1977; Oeschger and Loosli, 1977) enabling them to extract the Ar, CO_2 and dissolved particulate matter from large quantities of ice. Ice is melted in situ at a desirable depth of the bore hole and the extracted gases are pumped up to the surface. After CO_2 absorption on a molecular sieve (for ^{14}C dating), the air sample is stored for further ^{39}Ar analysis. Part of the melting water is pumped to the surface for precipitation of ^{32}Si for ^{32}Si dating and collection of dissolved particulate matter. The first determinations of ^{39}Ar activity in ice samples showed good agreement of the ^{39}Ar ages with the ages obtained from ^{32}Si measurement and with $^{18}O/^{16}O$ ages derived by counting of the annual layers of ice marked by seasonal variations of the $^{18}O/^{16}O$ isotope ratio (see Fig. 11-8).

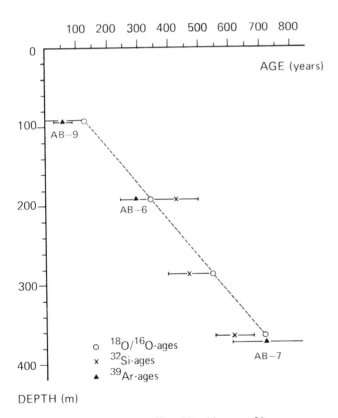

Fig. 11-8. Comparison of ^{39}Ar-$^{18}O/^{16}O$ and ^{32}Si ages for ice samples from the bore hole in Dye-3 station, Greenland (Oeschger and Loosli, 1977). For ^{32}Si a half-life of 650 years was used.

The technique developed for ^{39}Ar dating of ice has been further adopted to groundwater samples (Oeschger et al., 1974). Dating of groundwater with the ^{39}Ar isotope has similar advantages as the ^{85}Kr method; regular input function and roughly constant atmospheric concentration of ^{39}Ar within the dating range of this method, i.e. in the last thousand years, and lack of chemical interaction with the matrix of the aquifer. Furthermore, due to the half-life of ^{39}Ar (269 years) this method could eventually fill in the gap between the dating ranges offered by the ^{3}H and ^{14}C dating methods. The first determinations of ^{39}Ar activity in groundwater showed, however, a large discrepancy between ^{39}Ar ages and uncorrected ^{14}C ages (Oeschger et al., 1974). In some groundwater springs the ^{39}Ar activities significantly higher than those actually observed in the atmosphere have been measured. Further measurements showed that ^{39}Ar ages of the groundwaters studied are systematically younger compared with ^{14}C data (Loosli and Oeschger, 1979). One possible explanation of this effect could be the underground production of ^{39}Ar by the ^{39}K(n,p)^{39}Ar reactions with neutrons originating mainly from (α,n) processes, as well as the ^{39}K($\bar{\mu}$,v_μ)^{39}A reaction. A rough estimate showed that this underground production could lead to ^{39}Ar activities of the order of 0.4 dpm l^{-1} Ar (6.68 × 10^{-3} Bq l^{-1}), i.e. four times higher then the present atmospheric level (Loosli and Oeschger, 1979). Another reason for the discrepancy between ^{39}Ar and ^{14}C data could lie in the method of evaluating the ^{14}C ages. The simple correction factor adopted in that case by Loosli and Oeschger (1979) does not account for all processes which can modify the ^{14}C activity ratios within the aquifer (for details see, e.g., Fontes and Garnier, 1979; Mook, 1980; Neretnieks, 1980). Combined measurements of ^{39}Ar, ^{32}Si and ^{14}C activities in various aquifers also extensively studied by other methods should provide enough data to clarify this problem.

The radioactivity of ^{39}Ar produced on the surface of the earth's lithosphere by cosmic ray neutrons and muons (^{39}K(n,p)^{39}Ar, ^{39}K($\bar{\mu}$,v_μ)^{39}Ar) can be used for the determination of the erosion rate (Fröhlich and Lübbert, 1973). The ^{39}Ar activity level on the surface rocks depends on the ^{39}K content, on the elevation of the given area above the sea level and on its geographical position. Hebert and Fröhlich (1979) measured the ^{39}Ar activity in several different rock samples. They found values between 0.03 and 0.07 dpm kg^{-1} of rock. Calculated erosion rates are in fair agreement with geological estimations (Hebert and Fröhlich, 1979). The technique developed for the determination of ^{39}Ar activity in rocks could help to explain a present difficulty in the interpretation of ^{39}Ar measurements in groundwater.

Another interesting field of application of ^{37}Ar and ^{39}Ar isotopes are their measurements in meteorites. The ^{37}Ar activity in the Fe-Ni phase of meteorite reflects, due to its short half-life, the cosmic ray flux intensity near the earth shortly before the meteorite enters the atmosphere; whereas ^{39}Ar activity represents the average intensity of the cosmic radiation in the whole orbit of the meteorite. The ^{37}Ar/^{39}Ar activity ratio in the meteorites with known orbit can be then used for the determination of the spatial gradient of the cosmic radiation in the solar system (for details see, e.g., Fireman and Goebel, 1970; Huesser and Schaeffer, 1977a,b; Lavrukhina, 1977).

All the above-mentioned applications of the ^{37}Ar and ^{39}Ar isotopes require suitable sample preparation procedures, as well as sophisticated low-level counting techniques. In the case of ^{39}Ar dating of groundwater about 15 m^3 of water must be degassed in a closed system to get a sufficient amount of argon for analysis. The extracted gases are further separated by combined techniques of chemical purification, fractional distillation and gas chromatography (Oeschger et al., 1974). The ^{39}Ar activity of the pure argon sample is then measured by low-level proportional counting. A typical arrangement of counters and shield for low-level measurements of ^{37}Ar, ^{39}Ar and ^{81}Kr is shown in Fig. 11-9. Moreover, in the case of ^{37}Ar the rise time discrimination technique is often used parallel to anticoincidence and heavy shieldings in order to gain better background reduction (Rutherford et al., 1976; Heusser and Schaeffer, 1977; Oeschger and Loosli, 1977). Solid samples (rocks, meteorites) require special melting procedures followed by a gas purification step. An example of such apparatus is shown in Fig. 11-10. Since the amount of argon extracted from solid samples is usually very small (especially for meteorites, where the size of sample is limited) special miniature proportional counters must be used for radioactivity measure-

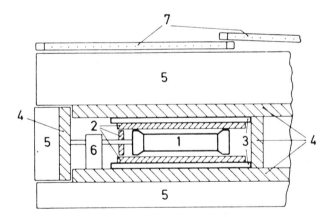

Fig. 11-9. Arrangement of counters and shield for low-level measurements of ^{37}Ar, ^{39}Ar and ^{85}Kr isotopes (Oeschger et al., 1970). *1* = main counter; *2* = 2 cm of radioactively very pure old lead reducing especially the background component induced in the shield (*4* and *5*) by muons and nucleons; *3* = anti-coincidence ring counter reducing the background component of charged particles; *4* = 5 cm old lead; *5* = 10—20 cm lead; *6* = preamplifier and filter for high voltage; *7* = flat anti-coincidence counters.

TABLE 11-6

Counting procedures used in detection of environmental activities of ^{37}Ar and ^{39}Ar

	Origin of the sample	
	atmosphere, hydrosphere	rocks, meteorites
1. Volume of argon gas used in analysis	2–5 l	~1 cm^3 (with carrier gas added)
2. Type of counter (volume)	proportional (0.02–1 l)	proportional (few cm^3)
3. Methods of the background reduction	(a) energy discrimination (b) anticoincidence (c) heavy shieldings (d) rise time discrimination	(a) energy discrimination (b) anticoincidence (c) heavy shieldings (d) rise time discrimination
4. Background levels	0.06–1 cpm (^{39}Ar)	
5. Time of analysis	10^4 minutes	few days
6. Typical error of analysis (1σ)	10–15%	10–15%

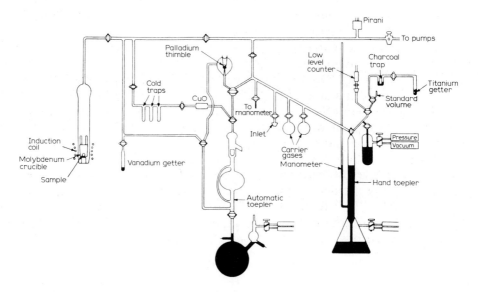

Fig. 11-10. Apparatus for melting samples and processing the released gas for measurement of ^{37}Ar and ^{39}Ar isotopes (Spannagel and Fireman, 1972).

ment. Background reduction techniques are the same as for larger counters and argon sample volumes.

Table 11-6 summarises essential parameters of the counting procedures being actually used for determination of environmental activities of ^{37}Ar and ^{39}Ar.

^{133}Xe

^{133}Xe is a product of nuclear fission. Detectable amounts of ^{133}Xe have been produced during nuclear bomb tests. Because of its relatively short half-life (5.27 days) a uniform distribution in the troposphere cannot be reached. First measurements were carried out by Ehhalt et al. (1963) and Schölch et al. (1966) over the area of Federal Republic of Germany and a good time correlation between high ^{133}Xe concentrations and time of nuclear tests has been observed. Like in the case of ^{85}Kr isotope, the ^{133}Xe activity detected in the atmosphere can be used to estimate the amount of fissioned material introduced by nuclear explosions (Ehhalt et al., 1963).

Nowadays the only important sources for ^{133}Xe are nuclear reactors. ^{133}Xe production yield in the fission reactions is about 30-fold higher than the relevant value for ^{85}Kr (Chitwood, 1973). Due to its short half-life ^{133}Xe decays completely in the nuclear fuel before the reprocessing process starts (mean storage time 3 years). Important atmospheric emission of ^{133}Xe occurs, therefore, only from operating nuclear reactors. Simultaneous measurements of ^{85}Kr and ^{133}Xe activities in the atmosphere (Stockburger et al., 1977) did not reveal any significant correlation between these isotopes. It confirms the different origin of both isotopes: reprocessing plants in the case of ^{85}Kr and nuclear reactors in the case of ^{133}Xe.

A short half-life of ^{133}Xe isotope makes it suitable to study regional dispersion processes in the atmosphere. Kunz and Paperiello (1976) made such studies in the New York State area in the U.S.A. The average measured ^{133}Xe activity in the surface air was 5.55 dpm m^{-3} air (9.27 × 10^{-2} Bq m^{-3}). Model calculations based on the regional dispersion model proposed by Machta et al. (1974) provide the value of 7.33 dpm m^{-3} air (0.122 Bq m^{-3}). The average value derived by Stockburger et al. (1977) from their regular measurements in Freiburg (Federal Republic of Germany) amounts only to 0.42 dpm m^{-3} air (0.7 × 10^{-2} Bq m^{-3}). It indicates a significantly higher discharge of gaseous effluents from nuclear facilities on the area studied in the U.S.A. than in Western Europe.

REFERENCES

Andrews, V.E. and Wrubble, D.T., 1973. Noble gas surveillance network, April 1972 through March 1973. In: R.E. Stanley and A.A. Moghissi (Editors), Noble Gases. pp. 281—289.

Barabanov, I.R. and Pomansky, A.A., 1977. Atmospheric abundance of ^{81}Kr and cosmic ray intensity. In: P. Povinec and S. Usaĉev (Editors), Low-Radioactivity Measurements and Applications, Int. Conf., High Tatras, 6—10 October, 1975. pp. 405—408.

Bernardt, D.E., Moghissi, A.A. and Cochran, J.A., 1973. Atmospheric concentrations of fission product noble gases. In: R.E. Stanley and A.A. Moghissi (Editors), Noble Gases. pp. 4—19.

Boeck, W.L., 1976. Meteorological consequences of atmospheric Kr-85. Science, 193: 195—198.

Budnitz, J.R., 1973. A review of instrumentation for environmental monitoring. In: R.E. Stanley and A.A. Moghissi (Editors), Noble Gases. pp. 192—198.

Chitwood, R.B., 1973. Production of noble gases by nuclear fission. In: R.E. Stanley and A.A. Moghissi (Editors), Noble Gases. pp. 69—80.

Csongor, E., 1973. Measurements of atmospheric ^{85}Kr activity at Debrecen (Hungary), II. Acta Phys. Acad. Sci. Hung., 34: 249—253.

Csongor. E., 1977. ^{85}Kr activity in the environment. In: P. Povinec and S. Usaĉev (Editors), Low-Radioactivity Measurements and Applications, Int. Conf., High Tatras, 6—10 October 1975. pp. 471—474.

Currie, L.A. and Lindstrom, R.M., 1973. The NBS measurement system for natural Argon-37. In: R.E. Stanley and A.A. Moghissi (Editors), Noble Gases. pp. 40—57.

Delibrias, G. and Jehanno, C., 1959. Activité de l'atmosphère due au Krypton-85. Bull. Inf. Sci. Technol., 30: 13.

de Vries, H., 1956. Purification of CO_2 for use in a proportional counter for C-14 age measurement. Appl. Sci. Res., B5: 387.

Diethorn, W.S. and Stockho, L., 1972. The dose to man from atmospheric Kr-85. Health Phys., 23: 653—667.

Ehhalt, D., Münnich, K.O., Roether, W., Schölch, J. and Stich, W., 1963. Artificially produced radioactive noble gases in the atmosphere. J. Geophys. Res., 68: 3817—3821.

Ehhalt, D., Münnich, K.O., Roether, W., Schölch, J. and Stich, W., 1964. Krypton-85 in the atmosphere. In: Proceedings, 3rd International Conference on Peaceful Uses of Atomic Energy, Geneva, 31 August—9 September. pp. 45—48.

Elmore, D., Fulton, B.R., Clover, M.R., Marsden, J.R., Gove, H.E., Naylor, H., Purser, K.H., Kilius, L.R., Beukens, R.P. and Litherland, A.E., 1979. Analysis of ^{36}Cl in environmental water samples using an electrostatic accelerator. Nature, 277: 22—25.

Etzweiler, A., 1980. Kr-85 als radioaktive Tracer in der Hydrosphäre. Thesis, Physikalisches Institut, Universität Bern.

Farges, L., Patti, F., Gros, R. and Bourgeon, P., 1974. Activité du Krypton-85 dans l'air des hémisphères nord et sud. J. Radioanal. Chem., 22: 147—155.

Ferber, G.J., Telegadas, K., Heefter, J.L. and Smith, M.E., 1977. Air concentrations of Krypton-85 in the midwest United States during January—May 1974. Atmos. Environ., 11: 379—385.

Fireman, E.L. and Goebel, R., 1970. Argon 37 and argon 39 in recently fallen meteorites and cosmic-ray variations. J. Geophys. Res., 75: 2115.

Fontes, J.Ch. and Garnier, J.M., 1979. Determination of the initial ^{14}C activity of the total dissolved carbon: A review of the existing models and a new approach. Water Resour. Res., 15: 399.

Fröhlich, K. and Lübbert, J., 1973. Ueber die Möglichkeit der Messung von Erosionsgeschwindigkeiten an Oberflächengesteinen mit dem natürlichen Radionukliden ^{42}Ca und ^{39}Ar. Z. Angew. Geol., 19: 550.

Grabczak, J., Zuber, A., Maloszewski, P., Rożanski, K., Weiss, W. and Sliwka, I., 1982. New mathematical models for the interpretation of environmental tracers in groundwaters and the combined use of tritium, C-14, Kr-85, He-3 and Freon-11 for groundwater studies. *Beitr. Geol. Schweiz, Hydrol.*, 28: 395—406.

Griesser, O. and Sittkus, A., 1961. Bestimmung des ^{85}Kr — Gehaltes der Luft. *Z. Naturforsch.*, 16a: 620—621.

Gudkov, A.M., Ovanov, V.I., Kard, I.L., Kolobashkin, V.M., Leipunskij, O.I., Nekrasov, V.I., Norichkov, V.P., Serbulov, A. and Ushakova, N., 1976. Latitudinal distribution of ^{85}Kr in the near-ground atmosphere (Results of the expeditions on the research ship "Academician Kurchatov" in the Atlantic Ocean in 1970—1973). *Akad. Nauk, Litovskoj SSR, Prikl. Yadern. Fizi.*, 5: 5—14.

Hebert, D. and Fröhlich, K., 1979. Messung von durch kosmische Strahlung in terrestrischen Gesteinen gebildeten Argon-39, 2. *Tagung Nucleaire Analysenverfahren, 19—23 März 1979, Dresden.*

Heller, D., Roedel, W. and Münnich, K.O., 1977. Decreasing release of ^{85}Kr into the atmosphere. *Naturwissenschaften*, 64: 383.

Heusser, G. and Schaeffer, O.A., 1977a. Spatial cosmic ray gradient by meteorites. In: P. Povinec and S. Usačev (Editors), *Low-Radioactivity Measurements and Applications, Int. Conf., High Tatras, 6—10 October 1975*, pp. 418—422.

Heusser, G. and Schaeffer, O.A., 1977b. ^{37}Ar and ^{39}Ar in meteorites and the spatial cosmic ray gradient. *Earth Planet. Sci. Lett.*, 33: 420—427.

Hilbert, F., 1974. Erzeugung und Freisetzung von radioaktiven Krypton- und Xenonisotopen durch Kernreaktoren und Wiederaufbereitungsanlagen und die voraussichtliche radiologische Belastung bis zum Jahr-2000. Kernforschungszentrum Karlsruhe, KFK 2035.

Jaquish, R.E. and Johns, F.B., 1972. Concentrations of krypton-85 in air. In: *Proceedings, Symposium on Natural Radioactivity of the Environment, Houston, Texas.*

Jaquish, R.E. and Moghissi, A.A., 1973. Survey of analytical methods for environmental monitoring of krypton-85. In: R.E. Stanley and A.A. Moghissi (Editors), *Noble Gases.* pp. 169—174.

Katcoff, S., 1969. Fission-product yields from neutron-induced fission. *Nucleonics*, 18: 201—204.

Katcoff, S. and Rubinson, W., 1965. Yields of ^{85}Kr in thermal neutron fission of U-235 and Pu-239. *J. Inorg. Nucl. Chem.*, 27: 1447—1459.

Kigoshi, K., 1962. Krypton-85 in the atmosphere. *Bull. Chem. Soc. Jpn.*, 35: 1014—1016.

Kirk, W.P., 1972. *Krypton-85; A Review of the Literature and Analysis of Radiation Hazards.* Environmental Protection Agency, Office of Research and Monitoring.

Kirsten, T. and Hampel, W., 1977. Weak radioactivities induced by cosmic ray muons in terrestrial minerals. In: P. Povinec and S. Usačev (Editors), *Low-Radioactivity Measurements and Applications, Int. Conf., High Tatras, 6—10 October 1975.* pp. 427—435.

Kishida, M., 1974. Measurement of Kr-85 concentration in atmosphere. *Symp. Radioactive Noble Gases, Osaka.*

Kunz, C.O. and Paperiello, C.J., 1976. Xenon-133: Ambient activity from nuclear power stations. *Science* 192: 1235—1237.

Kuzminov, V.V. and Pomansky, A.A., 1980. New Measurements of the ^{81}Kr atmospheric abundance. In: *Proceedings, 10th International Radiocarbon Conference, Bern-Heidelberg. Radiocarbon*, 22: 311—317.

Lal, D. and Peters, B., 1967. Cosmic ray produced radioactivity on the earth. In: S. Flugge (Editor), *Handbuch der Physik, XL VI/2*, pp. 551—612.

Lal, D., Nijampurkar, V.N. and Rama, S., 1970. Silicon-32 hydrology. In: *Isotope Hydrology 1970.* IAEA, Vienna, pp. 847—869.

Lavrukhina, A.K., 1977. Meteorites are space probes. In: P. Povinec and S. Usačev (Editors). *Low-Radioactivity Measurements and Applications, Int. Conf., High Tatras, 6—10 October 1975.* pp. 409—415.

List, R.J. and Telegadas, K., 1969. Using radioactive tracers to develop a model of the circulation of the stratosphere. *J. Atmos. Sci.*, 26: 1128—1136.

Loosli, H.H. and Oeschger, H., 1968. Detection of ^{39}Ar in atmospheric argon. *Earth Planet. Sci. Lett.*, 5: 191—198.

Loosli, H.H. and Oeschger, H., 1979. Argon-39, carbon-14 and krypton-85 measurements in groundwater samples. In: *Isotope Hydrology.* IAEA, Vienna, pp. 931—947.

Loosli, H.H., Oeschger, H. and Wiest, W., 1970. Argon-37, argon-39, and krypton-81 in the atmosphere and tracer studies based on these isotopes. *J. Geophys. Res.*, 75: 2895—2900.

Loosli, H.H., Oeschger, H., Studer, R., Wahlen, M. and Wiest, W., 1973. Atmospheric concentrations and mixing of argon-37. In: R.E. Stanley and A.A. Moghissi (Editors), *Noble Gases.* U.S. Government Printing Office, Washington, D.C., pp. 24—39.

Machta, L., 1973. Ar-37 as a measure of vertical mixing. In: R.E. Stanley and A.A. Moghissi (Editors), *Noble Gases.* pp. 58—68.

Machta, L., List, R.J., Smith, M.E. and Oeschger, H., 1970. Use of natural radioactivities to estimate large-scale precipitation scavenging. In: *Precipitation Scavenging 1970. Proc. Symp., Richland 2—4, June 1970. AEC Symp. Ser.*, 22: 465—474.

Machta, L., Ferber, G.J. and Heffer, J.L., 1974. Regional and global scale dispersion of ^{85}Kr for population-dose calculations. In: *Physical Behaviour of Radioactive Contaminants in the Atmosphere.* IAEA, Vienna, pp. 411—425.

Maloszewski, P. and Zuber, A., 1982. Determining the turnover time of groundwater systems with the aid of environmental tracers, I. Models and their applicability. *J. Hydrol.*, 57: 207—231.

Mast, T.S., 1978. Measuring low level concentrations of ^{14}C, ^{36}Cl and ^{81}Kr with the Berkeley 88-inch cyclotron. *Workshop on Dating Old Ground Water, University of Arizona, March 16—18, Rep.*, Y/OWI/SUB-78/55412.

Mook, W.G., 1980. Carbon-14 in hydrogeological studies. In: P. Fritz and J.Ch. Fontes (Editors), *Handbook of Environmental Isotope Geochemistry, 1. The Terrestrial Environment, A.* Elsevier, Amsterdam, pp. 49—71.

Neretnieks, J., 1980. Diffusion in the rock matrix: an important factor in radionuclide retardation? *J. Geophys. Res.*, 85 (B8): 4379—4397.

NCRP, 1975. Krypton-85 in the atmosphere — accumulation, biological significance and control technology. *NCRP Rep.*, 44.

Oeschger, H., 1978. *Workshop on Dating Old Ground Water, University of Arizona, March 16—18, Rep.*, Y/OWI/SUB-78/55412.

Oeschger, H. and Loosli, H.H., 1977. New developments in sampling and low level counting of natural radioactivity. In: P. Povinec and S. Usačev (Editors), *Low-Radioactivity Measurements and Applications. Int. Conf., High Tatras, 6—10 October 1975.* pp. 13—22.

Oeschger, H., Houtermans, J., Loosli, H.H. and Wahlen, M., 1970. The constancy of cosmic radiation from isotope studies in meteorites and on the Earth. In: I. Olsson (Editor), *Radiocarbon Variations and Absolute Chronology, XII Nobel Symposium.* Almquist and Wiksell, Stockholm, pp. 471—498.

Oeschger, H., Stauffer, B., Frommer, H., Möll, M., Langway, C.C., Hansen, B.L. and Clausen, H., 1972. ^{14}C and other isotope studies on natural ice. In: *Proceedings, 8th International Radiocarbon Conference, Lower Hutt, 18—25 October, 1972.* pp. D70—D92.

Oeschger, H., Gugelman, A., Loosli, H., Schotterer, U., Siegenthaler, U. and Wiest, W., 1974. ^{39}Ar dating of groundwater. In: *Isotope Techniques in Groundwater Hydrology.* IAEA, Vienna, pp. 179—190.

Oeschger, H., Stauffer, B., Bucher, B. and Loosli, H.H., 1977. Extraction of gases and dissolved and particulate matter from ice in deep boreholes. In: *Isotopes and Impurities in Snow and Ice, Proc. Symp. Grenoble, August—September 1975. IAHS Publ.*, 118: 307—311.
Pannetier, R., 1968. Distribution, transfer atmosphérique et bilan du Krypton-85. *CEA Rapp.*, CEA-R-3591.
Pannetier, R., 1970. Original use of the radioactive tracer gas Krypton-85 to study the meridian atmospheric flow. *J. Geophys. Res.*, 75: 2985—2989.
Patti, F. and Bourgeon, P., 1980. Concentrations atmospheriques de ^{85}Kr dans les hemispheres Nord et Sud. *J. Radioanal. Chem.*, 56: 221—228.
Rózanski, K., 1979a. Application of krypton-85 in groundwater dating. *Rep.*, INT 137/1.
Rózanski, K., 1979b. Krypton-85 in the atmosphere 1950—1977. a data review. *Env. Int.*, 2: 139—143.
Rózanski, K. and Ostrowski, A., 1977. Measurements of atmospheric krypton radioactivity. *Nukleonika*, 22: 343—347.
Rózanski, K. and Florkowski, T., 1979. ^{85}Kr dating of groundwater. In: *Isotope Hydrology 1978.* IAEA, Vienna, pp. 949—961.
Rutherford, W.M., Evans, J. and Currie, L.A., 1976. Isotopic enrichment and pulse shape discrimination for measurement of atmospheric Ar-37. *Anal. Chem.*, 48: 607—612.
Salvamoser, J., 1981. Vergleich der Altersbestimmung von Grundwasser mit Hilfe des Tritium- und Krypton-85-Gehalts. *Naturwissenschaften*, 68: 328—329.
Schell, W.R., 1970. Investigation and comparison of radiogenic argon, tritium and carbon-14 in atmospheric reservoirs. In: I. Olsson (Editor), *Radiocarbon Variation and Absolute Chronology, XII Noble Symposium.* Almquist and Wiksell, Stockholm, pp. 447—459.
Schölch, J., Stich, W. and Münnich, K.O., 1966. Measurement of radioactive xenon in the atmosphere. *Tellus*, 18: 298—300.
Schröder, J., 1975. Krypton-85 in the Ocean. *Z. Naturforsch.*, 30a: 962—967.
Schröder, J. and Münnich, K.O., 1971. Kr-85 in the troposphere. *Nature*, 233: 614.
Schröder, J. and Roether, W., 1974. The releases of krypton-85 and tritium to the environment and krypton-85 to tritium ratios as source indicators. In: *Isotope Ratios as Pollutant Source and Behaviour Indicators.* IAEA Vienna, pp. 231—252.
Shearin, R.L., Porter, S.L. and Cummings, C.R., 1971. Study of the feasibility of measuring ^{85}Kr through a national surveillance system. In: *Rapid Methods for Measuring Radioactivity in the Environment.* IAEA, Vienna, pp. 649—653.
Shuping, R.E., Phillips, C.R. and Moghissi, A.A., 1969. Low-level counting of environmental Krypton-85 by liquid scintillation. *Anal. Chem.*, 41: 2082—2083.
Sittkus, A. and Stockburger, H., 1976. Krypton-85 als Indikator des Kernbrennstoffverbrauches. *Naturwissenschaften*, 63: 266—272.
Spannagel, G. and Fireman, E.L., 1972. Stopping rate of negative cosmic-ray muons near sea level. *J. Geophys. Res.*, 77: 5351—5359.
Stockburger, H. and Sittkus, A., 1975. Messung der Krypton-85 Aktivität der atmosphärischen Luft. *Z. Naturforsch.*, 30a: 959—961.
Stockburger, H., Sarorius, H. and Sittkus, A., 1977. Messung der Krypton-85 und Xenon-133-Aktivität der atmosphärischen Luft. *Z. Naturforsch.*, 32a: 1249—1253.
Suzuki, T. and Inoue, K., 1972. Radiokrypton in atmosphere. *J. Nucl. Sci. Technol.*, 9: 55.
Telegadas, K. and Ferber, G.J., 1975. Atmospheric concentrations and inventory of Krypton-85 in 1973. *Science*, 190: 882—883.
Tertyŝnik, E.G., Siverin, A.A. and Barbanov, W.G., 1977. Koncentracja Kr-85 w atmosfere nad teritoriej CCCR 1971—1975. *At. Energy*, 42: 5.
U.S. EPA, Eastern Environmental Radiation Facility, 1974a. Krypton-85 in air, July 1970 to December 1970. *Radiat. Data Rep.*, 15: 133.

U.S. EPA, Office of Radiation Programs, 1974b. ERAMS krypton-85 in air component, January—Juny 1973. *Radiat. Data Rep.*, 15: 721—722.

Wardaszko, T., 1973. Contamination of the atmosphere with krypton-85 in Poland. In: R.E. Stanley and A.A. Moghissi (Editors), *Noble Gases.* pp. 20—23.

Wardaszko, T. and Nidecka, J.. 1978. Krypton-85 in atmospheric air: Measurement method and results for 1975—1977. *Nukleonika*, 23: 833—850.

Weiss, W.. Bullacher, J. and Roether, W., 1979. Evidence of pulsed discharges of tritium from nuclear energy installations in Central European precipitations. IAEA Conf. *Behaviour of Tritium in the Environment.* IAEA, Vienna, pp. 17—30.

Weiss, W., Sittkus, A., Stockburger, H., Sartorius, H. and Münnich. K.O., 1983. Large-scale atmosphere mixing derived from meridional profiles of Kr-85. *J. Geophys. Res.*, 88: 8574—8578.

Reprinted from: P. Fritz and J.Ch. Fontes (Editors),
Handbook of Environmental Isotope Geochemistry, Volume 2.
The Terrestrial Environment, B.

© 1986 Elsevier Science Publishers B.V.

Chapter 12

ISOTOPES AND FOOD

G. HILLAIRE-MARCEL

INTRODUCTION

The isotopic composition of plants depends on their environment and physiology with the result that the derived food-products may be isotopically labelled. For example, the ^{18}O and ^{2}H contents of fruit juices depend on their geographical origin. Similarly, different photosynthetic pathways result in different ^{13}C contents in natural sugars. This allows the checking of the authenticity of natural products by means of chemical isotope techniques.

The pioneers in this research were Smith and Epstein (1971), Bricout et al. (1973) and Nissenbaum et al. (1974). Since the publication of these works, numerous applications of the isotope techniques have been developed in food science. Most of them are derived from the same principles although recent developments in the knowledge of oxygen isotopes in organic matter have yielded new possible applications (Hillaire-Marcel et al., 1976; Epstein et al., 1977).

In the present chapter, we will successively consider each chemical element. A short summary on the distribution of its isotopes in nature will enable us to examine various applications in food control which have been proposed or which may be considered.

CARBON AND PLANTS

General remarks

The carbon used in the biosynthetic processes of continental plants is taken from the atmospheric CO_2 reservoir. In contrast, aquatic plants use H_2CO_3 and HCO_3^-. One should note that the atmospheric carbon reservoir is extremely homogeneous: its isotopic composition $\delta^{13}C$ varies very little from the $-7^0/_{00}$ average found by Keeling (1961). In practice, it can be considered constant. Any terrestrial plant thus uses an isotopically determined carbon source. To a first-order approximation, the isotopic compositions of the biosynthesized products will depend solely on the photosynthetic mechanism used and to a much lesser extent on environmental aspects. Indeed, an isotopic segregation takes place during the photosynthetic processes; an isotopic fractionation is produced between the atmospheric CO_2 and the plant: the plant is depleted in the heavier isotope (^{13}C). This segregation differs, however, according to the photosynthetic pathway followed.

The isotopic composition $^{13}C/^{12}C$: its relation to the photosynthetic cycle

Wickman (1952) and Craig (1954a) first showed that the carbon isotopic composition of plants fell into two sets of values (Fig. 12-1), one centered around -12 to $-14^0/_{00}$, the other around -26 to $-28^0/_{00}$. Later on, studies by Bender (1968, 1971), Smith and Epstein (1971), Lerman and Raynal (1972) and Troughton (1972) permitted linking those different compositions to different processes of photosynthesis. The most common of these processes used the Calvin cycle (Calvin and Bassham, 1962), which directly fixes the CO_2 to a molecule of 3-phosphoglyceric acid. The other photosynthetic process uses the Hatch and Slack cycle (cf. Hatch et al., 1967); there HCO_3^- is fixed by carboxylation of phosphenol pyruvate and formation of oxaloacetic acid, which, after reaction with either a C_2 or a C_5 molecule, produces respectively pyruvic acid or phosphoglyceric acid.

The first cycle is often called the C_3 cycle since the first synthesized sugars are 3 carbon sugars. In the second cycle, the C_4 cycle, the first sugars formed

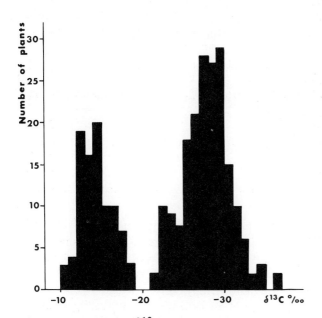

Fig. 12-1. Variation in $\delta^{13}C$ values among 250 plant species (from Troughton, 1972). The bimodal distribution reflects the C_3 and C_4 photosynthetic pathways.

are 4 carbon sugars. The latter cycle can be considered a "shorter" cycle, since the first synthesis products are closer related to the end products. The C_4 cycle is known to exist in corn (Bender, 1968), while the C_3 cycle is present in wheat. This explains, amongst other factors, the differences in yield per hectare in these two cereals, which is much higher for corn.

Coming back to the fractionation of carbon isotopes, one can thus understand why the shorter C_4 cycle results in a smaller isotopic fractionation and explains the average $\delta^{13}C$ value of -12 to $-14‰$ for these plants. The "longer" C_3 cycle causes a greater isotopic segregation, with $\delta^{13}C$ values ranging approximately from -25 to $-28‰$.

Another photosynthetic cycle is also known to exist. It occurs only infrequently in nature and has not been studied extensively. It is found especially in the Crassulaceae family. Consequently, it is called the "Crassulaceae acid metabolism" (CAM) cycle. In general, this cycle is active in plants which stock water beyond their immediate needs (cactuses, epithytes, orchids). The magnitude of the ^{13}C depletion associated with it is in between that of the C_4 and C_3 cycles (cf. Bender, 1971; Lerman, 1972).

Therefore, all continental plants will fall into one of the three $^{13}C/^{12}C$ categories described above according to the type of photosynthesis followed; their carbon will be drawn from the CO_2 atmospheric reservoir, a relatively isotopically stable source because of tropospheric mixing and equilibrium with the oceanic reservoir (cf. Deines, 1980).

It is desirable, however, to examine all the possible factors capable of slightly modifying these isotopic compositions, before using the property described above to distinguish the biosynthesis products of each plant category.

Other sources of fractionation in plants
The type of inorganic carbon caboxylized. The differences in the isotopic depletion magnitude between C_4 and C_3 plants is evidently due to the different structure of the enzymes involved (RUDP-carboxylase for the C_3 cycle, PEP-carboxylase for the C_4 cycle). However, in the C_4 cycle, the carbon is fixed in the HCO^- ionic form (Cooper and Wood, 1971; Troughton, 1972) while in the C_3 cycle, it is directly fixed in the CO_2 form. This fact, of course, changes the isotopic composition of the real source of inorganic carbon.

The dissolution of the CO_2 and its transformation into the HCO^- ionic form is indeed accompanied by an isotopic fractionation (cf. Vogel et al., 1970):

$$^{13}CO_2 + H^{12}CO_3^- \stackrel{\alpha}{\rightleftharpoons} {}^{12}CO_2 + H^{13}CO_3^-$$

where $10^3 \log \alpha = (9.552 \times 10\ T^{-1}) - 24.10$. Thus $\alpha \simeq 1.0075$ (25°C) and $\epsilon \simeq (\alpha - 1) \times 10^3 \simeq 7.5‰$; $d\epsilon/dt \simeq -0.1‰\ °C^{-1}$.

Thus, admitting isotopic equilibrium between atmospheric CO_2 and plant-bicarbonate, the composition of the inorganic carbon used by a C_4 plant is in the order of $0‰$. This cycle would therefore be associated with a total fractionation of -12 to $-14‰$.

In the C_3 cycle, the dissolved CO_2 is slightly depleted in ^{13}C with respect to the CO_2 gas (Vogel et al., 1970):

$$^{12}CO_2 \text{ (gas)} + {}^{13}CO_2 \text{ (diss.)} \overset{\alpha}{\rightleftharpoons} {}^{13}CO_2 \text{ (gas)} + {}^{12}CO_2 \text{ (diss.)}$$

where $\alpha \simeq 1.001$ and $\epsilon \simeq (\alpha - 1) \times 10^3 \simeq +1‰$; $d\epsilon/dt \simeq -0.004‰ \,°C^{-1}$.

Assuming again isotopic equilibrium, the total fractionation for the C_3 cycle is thus on the order of $-19‰$. Therefore, the real fractionation difference between the two cycles is only $\simeq 5-7‰$.

Furthermore, kinetic factors can also step in during the diffusion of the CO_2 through the cellular membranes and its liquid phase transfer to the chloroplasts (Troughton, 1972). Furthermore, PEP- and RUDP-carboxylase are located respectively on the outside and on the inside of the chloroplasts. In the latter case, an additional diffusion is involved.

Existence of isoenzymes. Hatch et al. (1967) have demonstrated the existence of isoenzymes of PEP-carboxylase and other enzymes, in *Atriplex* for example. Thus, slight differences in the chemical behavior of the carbon isotopes can be expected, depending on the types of enzymes involved. This factor, however, only moderately influences the $^{13}C/^{12}C$ fluctuations in plants.

The influence of plant respiration. The CO_2 expired by the plant during the night has an isotopic composition close to that of the plant tissues. This CO_2 will be mixed into the surrounding air and under certain conditions will slightly modify the isotopic composition $\delta^{13}C$ of the CO_2 used in the photosynthesis. During the day, the CO_2 formed by oxydation and respiration can be directly used again by the plant for the chlorophyll function. This process will modify the isotopic composition of the total CO_2 source of the organic carbon.

In general, these factors have roughly the same effect from one plant to the next. Two interesting incidental factors should be noted, however: (1) differences in the respiration rates of plants, i.e. in the production of CO_2 depleted in ^{13}C, can introduce small variations in the isotopic composition of the CO_2 used for photosynthesis; and (2) the length and intensity of the light period can also be a factor: an increase in the CO_2 assimilation rate will tend to enrich the surrounding air in ^{13}C.

Isotopic composition variations in between the organs, the tissues and the various constituents of the plant. Several authors have studied ^{13}C variations in plant organs. Troughton (1972) observed a 2‰ enrichment between the leaves and the tubers of the potato. Evans and Dunstow (1970) found an enrichment of approximately 5‰ in wheat grains with respect to the leaves.

Actually, the variations in the isotopic compositions of the tissues can be shown to correspond to the isotopic differences present in between the various organic compounds of the plant (Abelson and Hoering, 1961).

An example, taken from Whelan et al. (1970) and which concerns sorgho and cotton (Fig. 12-2) may illustrate this aspect. Lipids are always depleted in ^{13}C with respect to other compounds, especially sugars. The numerous analyses done on samples of vegetable and even animal origin show that this observation can be generalized (Hillaire-Marcel et al., unpublished). On a smaller scale, each cellular chemical compound is characterized by its own fractionation factor (Fig. 12-3; see also Deines, 1980 and Galimow, 1979).

The external factors
Variations of the CO_2 concentration. Park and Epstein (1960) have shown that a strong increase in the atmospheric CO_2 concentration (1.5% instead of 0.035%) could be correlated, in the tomato, with an isotopic depletion of approximately 2‰. The natural variations of the CO_2 concentration are not as great of course. They can intervene especially near urban and industrial regions. This factor, nevertheless has only a very small influence.

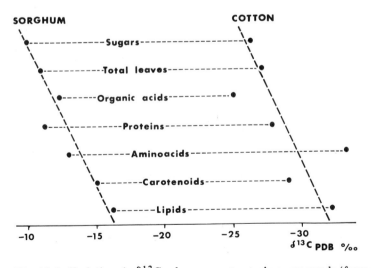

Fig. 12-2. Variations in $\delta^{13}C$ values among organic compounds (from Whelan et al., 1970). Lipids are usually the most depleted in heavy isotope.

Variations in the isotopic composition of atmospheric CO_2. As a rule, the CO_2 atmospheric reservoir is well homogenized isotopically (Keeling, 1961; Craig and Keeling, 1963).

In reality, however, slight fluctuations can be observed more or less in correlation to the CO_2 concentration in prairie or forest regions (Keeling, 1961). These diurnal fluctuations are evidently related to the metabolism

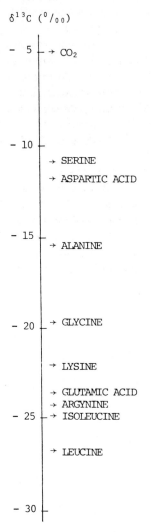

Fig. 12-3. Example of distinct carbon isotope fractionation between some amino-acids in *Chlorella pyrenoides* (from Abelson and Hoering, 1961).

(respiration—chlorophyll synthesis) of plants. During the day, the local air reservoir is enriched in $^{13}CO_2$, due to the preferential assimilation of $^{12}CO_2$. During the night, the CO_2 previously assimilated is released ($\delta^{13}C \simeq -25‰$) and the local air reservoir is thus depleted in $^{13}CO_2$ (Fig. 12-4).

One can therefore predict annual variations in the CO_2 concentrations and in the $^{13}C/^{12}C$ ratios, especially in deciduous forest regions. Lowdon and Dyck (1974) have observed differences of up to 6‰ in the isotopic composition $\delta^{13}C$ of maple leaves between spring and autumn. In springtime, during the growth period of plants, chlorophyll synthesis is very active especially since the days are long. The $^{13}CO_2/^{12}CO_2$ ratio of the local air then tends to increase. This effect lessens during the summer and the homogenization of the atmospheric reservoir brings the $^{13}CO_2/^{12}CO_2$ ratio back to normal. The relative ^{13}C enrichment, at the start of the growth season, in the maple leaves analyzed by Lowdon and Dyck (1974), corresponds to this imbalance, related to the metabolism of the maple.

The effect of temperature variations. The data concerning the influence of temperature on carbon isotopic compositions are generally contradictory (cf. Degens et al., 1968; Berry et al., 1972; Troughton, 1972). The isotopic variations observed and related to temperature do not exceed 2—3‰ and, in general, do not show any specific tendency. The effect of temperature on

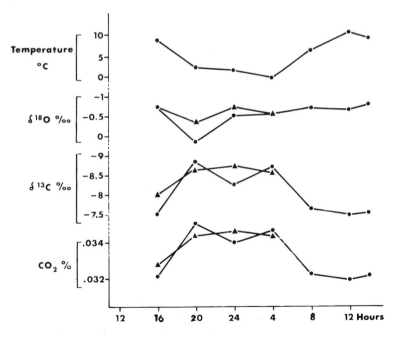

Fig. 12-4. Diurnal fluctuations, above a forest of CO_2 and corresponding changes in its isotopic composition in relation to the rate of photosynthesis (from Keeling, 1961).

the activity of a plant and therefore on its CO_2 assimilation rate, can possibly lessen the thermo-dependent variations of the plant's isotopic composition, especially those related to the equilibrium reactions preceding the fixation of the CO_2:

$$CO_{2(gas)} \rightleftharpoons CO_{2(diss.)} \rightleftharpoons HCO_3^-.$$

Remarks on aquatic plants. Although not directly related to the subject, it might be useful to note a peculiarity of autotroph aquatic plants which used HCO_3^- ions as a carbon source. These plants exhibit different $\delta^{13}C$ values, depending on their environment, marine or continental (fluvial and lacustrine). In well homogenized oceanic basins, the $\delta^{13}C$ value of the HCO_3^- ions is quite constant, approximately $0^0/_{00}$. Continental waters, on the other hand, are enriched in ^{12}C (percolation through the humus and consequent influence of organic carbon with a very low $\delta^{13}C$). Thus, since the carbon source is depleted in ^{13}C, lower $\delta^{13}C$ values can be expected in the plants of these continental waters. Furthermore, organic activity (especially of bacterial origin) can considerably diminish the ^{13}C supply in poorly mixed and poorly oxygenized basins. Thus, considerable differences in isotopic compositions can exist between one lake and the next.

Some applications

It is easy to understand, considering the natural variations of the ^{13}C content of organic matter, why the first applications of carbon geochemistry in the control of the authenticity of alimentary products were done on molecular substances (especially on sugars) rather than on mixtures.

The analytical differentiation between cane and beet sugars

The European laws on sugar implied the possibility to distinguish between beet and cane saccharose. Even though characteristic substances exist, for example in beets (betaine, betanines; cf. Cenci and Cremonini, 1973), the purification of commercial sugars is such that it is practically impossible to detect these substances. This is why Bricout and Fontes (1974) have suggested ^{13}C contents as a means of identification. Beets *(Beta vulgaris)* can indeed be classified as a C_3 plant whereas sugar cane *(Saccharum officinarum)* is a C_4 plant.

Therefore, the respective ^{13}C contents enable the distinction of the biosynthesized sugars: the sugar cane $\delta^{13}C$ value being on the order of $-11.5^0/_{00}$; that of beets on the order of $-25.5^0/_{00}$ (Table 12-1).

The following example will enable us to go into more detail and especially (a) to specify the natural variations of the isotopic composition of a sugar of determined origin, and (b) to fix the detection thresholds of mixtures.

TABLE 12-1

Isotopic composition $^{13}C/^{12}C$ ($^0/_{00}$ vs. PDB) of sugars (from Carro et al., 1980).

Sample	Origin	$\delta^{13}C$	Sample	Origin	$\delta^{13}C$
Maple (Acer saccharum) syrup[a]			V-4	Vermont University	−24.28
Q-1	Québec Min. Agric.	−23.78	V-5	Vermont University	−24.40
Q-2	Québec Min. Agric.	−23.30	NV3.9	Vermont University	−23.35
Q-3	Québec Min. Agric.	−23.15	NVB.14	Vermont University	−24.10
Q-4	Québec Min. Agric.	−23.53	VA	Vermont University	−24.30
Q-5	Québec Min. Agric.	−23.84	NV3.16	Vermont University	−24.33
Q-6	St-Maurice County	−24.81			
Q-7	Dorchester County	−22.77	*Maple (Acer saccharum) syrup, lyophilized*		
Q-8	Wolfe County	−22.88	B180B	Vermont University	−24.06
Q-9	Laviolette County	−22.94	B184A	Vermont University	−22.61
Q-10	St-Hyacinthe County	−24.54	NV3.16	Vermont University	−23.06
Q-11	Levis County	−22.78	B180B'	Vermont University	−24.57
Q-12	Missisquoi County	−22.65	NV3.9	Vermont University	−23.18
Q-13	Bagot County	−22.62			
Q-14	Megantic County	−22.43	*Maple (Acer saccharum) sap, lyophilized*		
Q-15	L'Islet County	−23.38	NV3.9.77	Vermont University	−23.69
Q-16	Beauce County	−23.23	NV3.16.77	Vermont University	−24.53
Q-18	Nicolet County	−23.63			
Q-19	Montmorency County	−23.75	*Maple (Acer saccharum) sugar*		
Q-20	Shefford County	−23.41	SE.1	Commercial	−22.71
Q-21	Montcalm County	−23.33			
Q-22	Stanstead County	−23.02	*Cane (Saccharum officinarum) Sugar*		
Q-23	Brome County	−22.83	AUS.1	Australia	−11.67
Q-24	Vaudreuil-Soulange County	−23.92	CUB.1	Cuba	−11.54
			NAT.1	Natal	−11.20
Q-25	Rouville County	−23.93	NAT.2	Natal	−10.53
Q-26	Champlain County	−23.37	—	Antilles	−10.50[b]
Q-27	Sherbrooke County	−22.37	—	Réunion	−11.40[b]
Q-29	Portneuf County	−24.08	—	Madagascar	−12.20[b]
Q-30	L'Assomption County	−22.63			
Q-31	Deux-Montagnes County	−22.52	*Beet (Beta vulgaris) sugar*		
			Qué.40	Québec 1974	−26.63
			Qué.41	Québec 1975	−25.70
Q-33	Argenteuil County	−22.60	Qué.42	Québec 1976	−25.52
Q-34	Drummond County	−22.92	—	France	−25.20[b]
Q-35	Montcalm County	−23.82	—	Italy	−25.10[b]
Q-36	Richmond County	−23.36	—	Germany	−26.00[b]
Q-37	Frontenac County	−23.78			
Q-38	Athabaska County	−22.89	*Corn (Zea mays) sugar*		
Q-39	Compton County	−23.77	Qué.43	Québec	−10.39
Q-44	Huntington County	−22.76	—	Wisconsin	−12.20[c]
Q-45	Bellechasse County	−23.27	—	Arizona	−12.40[c]
V-1	Vermont University	−23.63	—	Oklahoma	−12.40[c]
V-2	Vermont University	−23.79			
V-3	Vermont University	−23.78			

[a] $\delta^{13}C$ values for samples Q-6 through NV3.16 are the means for 2—5 samples.
[b] Bricout and Fontes (1974).
[c] Bender (1968).

The identification of maple sugars and syrups

The usual techniques of the Association of Official Analytical Chemists (AOAC: 1975 sections 31-147, 31-188) used to determine the grading of maple syrup (color classification, ash analysis, conductivity, lead determination, etc.) were often poorly effective in the detection of adulterations. One of these adulterations, hardly detectable, is the addition of commercial sugars, mainly corn syrup or cane sugar. Such a situation can be explained by the fact that the sugars in maple products are mainly sucrose with a variable percentage of hexose (usually less than 11%) (Morselli, 1975).

Hillaire-Marcel et al. (1976) proposed a technique for the identification of maple products (sugars and syrups) based on their ^{13}C content. Insofar as this specific study has given details on several aspects of the application of carbon isotopic geochemistry, in the control of the adulteration of natural products, greater attention will be given in this paper to this specific work.

The isotopic composition $\delta^{13}C$ of the maple sugars. The maple (*Acer saccharum*) is a plant with a C_3 photosynthetic cycle. Therefore, its isotopic composition $\delta^{13}C$ will show a marked depletion in ^{13}C. All the factors described in the previous section can cause slight variations in this isotopic composition. However, a quite restricted $\delta^{13}C$ interval can be predicted; first because the production region is restricted to Quebec, Ontario and northeastern U.S.A., and thus corresponds to specific climatic conditions; secondly because the collected sugars have been synthesized during the entire preceding growth season and thus reflect an average isotopic composition integrating all diurnal and seasonal variations discussed above. An additional factor should be considered: does the technique of concentration of the maple syrup by means of an evaporator modify its isotopic composition?

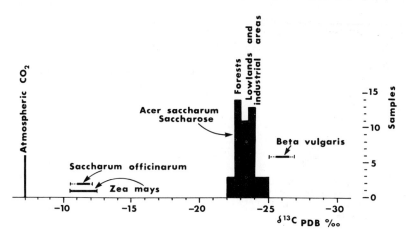

Fig. 12-5. Isotopic range of maple-sugars vs beet, cane and corn sugars (from Carro et al., 1980).

The sugars extracted by traditional means and those obtained by lyophilization are characterized by similar $\delta^{13}C$ values (Table 12-1).

Maple sugars have been experimentally found to have $\delta^{13}C$ values between -22.4 and $-25.5‰$ (Table 12-1 and Fig. 12-5). Thus, the range of isotopic compositions is indeed very restricted. The mean $\delta^{13}C$ value is $-23.33‰$ and the associated standard deviation is $0.6‰$. Some minor tendencies emerge from the $\delta^{13}C$ values found. Forested regions seem to correlate with the highest ^{13}C contents observed. The slight imbalance in the concentration and in the $^{13}C/^{12}C$ ratio of the atmospheric CO_2 (associated with plant metabolism; cf. above) can be held responsible. A mean $\delta^{13}C$ value of $-22.8‰$ is observed in syrups from "forested" regions whereas those from "open" or industrial regions have a mean $\delta^{13}C$ of $-23.5‰$ (Fig. 12-5). In industrial regions, one should probably consider the ^{13}C depleted CO_2 ejected by factories, as a cause for the slight isotopic depletion of the sugars.

Carro et al. (1980) specify that no coherent correlation seems to exist between the mean temperatures of the production regions and the observed ^{13}C variations. The thermo-dependent fluctuations of the $^{13}C/^{12}C$ ratio are probably concealed by other factors. Hillaire-Marcel et al. (1976) observe, for example, that a ^{13}C depletion, generally less than $1‰$, appears in the extracted sugars towards the end of the collecting season. This relative depletion can be caused by the progressive temperature increase or by a gradual isotopic depletion in the sugar reserves.

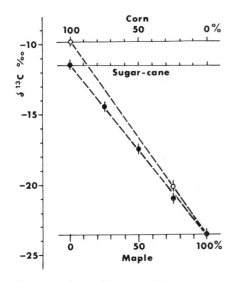

Fig. 12-6. Isotopic composition of mixtures of maple and cane and maple and corn sugars (from Carro et al., 1980).

Differentiation between maple, corn and cane sugars. As seen before p. 514), the sugar cane, with its short C_4 photosynthetic cycle, is characterized by a higher ^{13}C content, which varies between -10.5 and $-12.2‰$.

Corn also has a C_4 photosynthetic cycle (cf. Bender, 1968). This type of carboxylation is reflected by an isotopic composition $\delta^{13}C$ comprised between -9.5 and $-12.5‰$ (cf. Hillaire-Marcel et al., 1976). This greater isotopic range can be explained by the spread of the corn production regions over several different climatic zones. In contrast, the sugar cane, as well as the maple, are associated with much more specific climates. The detection of cane-maple or corn-maple mixtures is thus possible on the basis of ^{13}C contents (Fig. 12-6).

Carro et al. (1980) have thought it useful to calibrate standard curves enabling the quantification of mixtures. Given the natural variations of the $\delta^{13}C$ value of a particular sugar, the analytical error ($0.2‰$ in routine conditions; $0.1‰$ in optimum conditions) can be considered negligible. Since maple sugars have compositions varying between -22.4 and $-24.5‰$, the detection of cane sugar (-10.5 to $-12‰$) can be made at varying percentage levels (Fig. 12-7). A mixture with a $\delta^{13}C$ value of $-21‰$ can point to an addition of 10—28% of cane sugar, since the exact isotopic compositions of the original maple and cane sugars are not known a priori. Statistically, such a composition would correspond to an addition of about 20%.

Even though the reasoning based on a population defined by a standard deviation is valid from a scientific standpoint (in this case the mean detec-

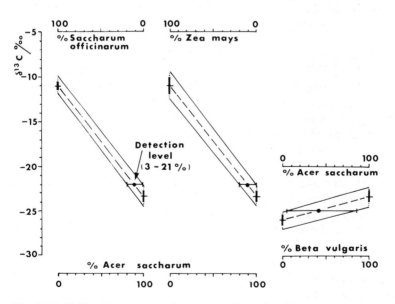

Fig. 12-7. Calibration curves of mixtures of maple with corn and cane sugars (from Carro et al., 1980).

tion threshold of this type of adulteration is of the order of 5%), its legal validity probably remains in question. Therefore, due to the natural distribution of $^{13}C/^{12}C$ ratios in maples, one will be able to suspect an adulteration only with isotopic compositions higher than $-22.4‰$. A limiting value of $-22‰$ seems reasonable, given the analytical error and an additional error margin for more safety. In this case, the detection threshold will vary between 3 and 21%, depending on the compositions of the original products, and will have a mean of 12%.

Commonly speaking, this means that on the average, one will be able to detect an addition of 12% of cane sugar to the maple syrup (or sugar). However, any possible defrauder runs the risk of being discovered if he relies on this percentage. Indeed, if the maple and cane sugars have isotopic compositions near the upper limits (respectively $-22.4‰$ and $-10‰$) a 3% adulteration becomes detectable. Any sophisticated fraud would therefore imply repeated and systematic isotopic analysis of the original products (cane sugar and maple syrup) by the defrauder. In ideal conditions, the adulteration could reach 21%, but on the average, it could not exceed 12%.

This technique has therefore been adopted in the control of the adulteration of maple syrup by cane sugar or corn syrup, since it can potentially detect, in optimum conditions, a 3% adulteration. This value is in fact a fraud-deterring threshold. The same limits on the detection ability are encountered in most applications of isotopic geochemistry in the control of natural products. The maple is an ideal case since its restricted geographic location limits the natural spread of the $\delta^{13}C$ values. The detection threshold is rarely less than 20% for most other products.

The identification of honeys

Doner and White (1977) in the United States and Zeigler et al. (1977, 1979) in Europe have suggested the identification of honeys by means of their ^{13}C contents. In these two regions, the bees gather honey from plants belonging for the most part to the long C_3 photosynthetic cycle. Thus, the honeys produced have ^{13}C contents on the order of $-25‰$ (Doner and White, 1977). Only a few particular honeys from tropical regions exhibit different $\delta^{13}C$ values (Doner and White, 1977). Some possible adulterations are the addition of cane sugar (to the honey) or its feeding to the bees, and the addition of high-fructose corn syrup (HFCS), now produced in large quantities. Previously, these adulterations were difficult to detect. HFCS was especially difficult to detect, because of the similarity of its major components and minor oligo-saccharides with those of honey. Let us recall that both cane sugar and corn syrup are characterized by high ^{13}C contents. Therefore, their admixture with honey (Fig. 12-8) is easily recognizable from $\delta^{13}C$ isotopic compositions.

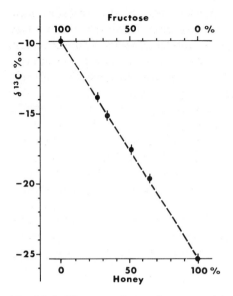

Fig. 12-8. Mixtures of corn fructose with natural honey and corresponding isotopic composition (data from Doner and White, 1977 and Hillaire-Marcel et al., 1976).

Differentiation of natural and synthetic vanillas

Bricout et al. (1974a) and Hoffman and Salb (1979) have made a study of the isotopic composition $\delta^{13}C$ of natural vanilla (*Vanilla planifolia*) and synthetic vanilla extracted from lignin. The latter, produced from trees with a C_3 cycle, is characterized by a mean $\delta^{13}C$ of $-27‰$. *Vanilla planifolia* on the other hand is an Orchid and, as such, uses the special CAM photosynthetic cycle. The mean isotopic composition of *Vanilla* is $-20.5‰$. The differentiation between the two products is therefore very easy.

The adulteration of fruit juices

The addition of foreign sugars or the admixture of synthetic fruit juices to certain fruit juices are not easily detectable. It is possible, however, to recognize the addition of certain sugars (corn or sugar cane), due to their high ^{13}C contents (cf. pp. 514, 518). Thus, studies have been effected on the characterization of lemon juices (Schmid et al., 1978), or are in preparation for orange juices (P. Fritz, personal communication) and apple juices (Doner and Hillaire-Marcel, unpublished). We will see later that it is possible to detect the addition of beet sugar to certain juices (orange, lemon, grapefruit, etc.) by means of oxygen and hydrogen isotope composition.

Isotopic identification of certain alcohols

Even though fermentation processes induce occasionally a depletion in ^{13}C (Hillaire-Marcel et al., unpublished data), the ^{13}C contents of alcohols reflect for the most part the photosynthetic cycle of the original plant. Rum and cognac can be easily distinguished one from the other, even though they might be depleted by a few permil with respect to cane sugar (for rum) and grapes (for cognac). The first alcohol (rum) is richer in ^{13}C and reflects the short C_4 cycle of the sugar cane, whereas the second, depleted in ^{13}C, reflects the long C_3 cycle of grapes.

Other applications can be considered (see also Rauschenbach et al., 1979). Whiskies, for example, according to local laws, are produced from a determined percentage of the following grains: corn, rye and barley. The corn, recognized by its greater ^{13}C content, can be quantified by means of the isotopic composition of the mixture whilst taking into account the slight ^{13}C depletion due to fermentation. The example illustrating this applica-

TABLE 12-2

Carbon isotope composition of various natural alcohols

			$\delta^{13}C$ (‰ vs. PDB)
Rums			
#1			−12.5
#2			−13.8
#3			−13.3
#4			−13.6
#5	(1977)		−13.4
#6	(1979)		−14.2
Vodkas			
#1	from corn		−11.4
#2	from corn (1977)		−13.4
#3	from corn		−12.8
#4	from potatoes		−26.0
#5	from potatoes		−25.6
Cognac			
#1			−31.8
#2			−29.3
#3			−30.0
Whiskies			
#1	(Scotch	70% corn)	−17.0
#2	(Rye	95% corn)	−12.8
#3	(Rye	95% corn)	−13.2
#4	(Bourbon	80% corn)	−14.7
#5	(Bourbon	80% corn)	−13.7
Industrial alcohol wines			
#1			−27.6
#1	White Bordeaux		−26.5
#2	Red Bordeaux		−25.9

tion (Table 12-2) shows that the three samples of Rye Bourbon and Scotch confirm this theory. It is also possible to certify the authenticity of a Polish vodka, produced from potatoes (C_3 photosynthetic cycle) instead of other sources, especially corn (Table 12-2).

Finally, any admixture of industrial ethanol produced from ethylene (derived from petroleum or from coal distillation) to rums or to corn alcohols should be relatively easy to detect considering the characteristically low ^{13}C content of industrial ethanol (Table 12-2). Petroleums have indeed mean ^{13}C contents on the order of $-29.4‰$ (Craig, 1953; Silverman and Epstein, 1958) whereas most types of coal have $\delta^{13}C$ values associated with C_3 plants (-23 to $-28‰$; Craig, 1953; Silverman and Epstein, 1958). However, the first distillation products, amongst which the ethylene of our interest (for the synthesis of ethanol), are depleted in ^{13}C by 5–10‰ with respect to unrefined oil for example (cf. Silverman, 1964).

Ethylene and its derivatives will thus show a characteristic deficit in ^{13}C. However, one should note that ethylene can also be obtained from the cracking of the heavier fractions of the oil, which have been relatively enriched in ^{13}C during the distillation. Ethanol thus prepared from hydrocarbons can be found on the market. Their ^{13}C content would be close

TABLE 12-3

Stable carbon isotope ratios in animals vs. their diet

	$\delta^{13}C$ (‰ vs. PDB)
Eggs from hens fed with mixtures of corn and wheat	
1 wheat	−23.7
2 corn	−11.0
3 egg (100% wheat)	−25.9
4 egg (100% wheat)	−25.3
5 egg (75% wheat)	−16.8
6 egg (75% wheat)	−16.7
7 egg (80% wheat)	−17.7
8 egg (80% wheat)	−17.8
Milk solids	
1 Diet of the cow	−24.6
2 Milk solids	−25.9
Tallow	
1	−27.5
2	−27.5
3	−28.3
Bacon	
# 1	−18.7
# 2	−22.8

to that of the ethanol produced by fermentation of the sugars of C_3 plants. Nevertheless, we will see later on that ^{14}C and ^3H will be the conclusive isotopes.

Differentiation between vinegars and industrial acetic acid

Whatever its origin may be (malt, cider or wine), vinegar is characterized by an isotopic composition close to that of its corresponding alcohol and, thus, close to that of the C_3-cycle plants.

Industrial acetic acid is either synthesized from petroleum derivatives or produced by the distillation of wood. In the first case, the acid is recognized by a ^{13}C depletion with respect to vinegar. Schmid et al. (1978) have therefore suggested this method to detect the adulteration of vinegars. Here again, we will see that ^3H and ^{14}C are more practical isotopes to use.

Isotopic differentiation of animal proteins and fats

Heterotrophic organisms have isotopic compositions essentially corresponding to that of their food regime (Minson and Ludlow, 1975; De Niro and Epstein, 1976, 1978), except for a slight enrichment in ^{13}C.

De Niro and Epstein (1976, 1978) attribute this enrichment to a relative depletion of the rejected CO_2. This rule, however, seems to have many exceptions, for example in bacteria (Hillaire-Marcel et al., unpublished). Nevertheless, the $\delta^{13}C$ value of proteins is very close to that of the food regime and can enable the characterization of proteins and other organic constituents. We will give a few examples.

A corn-fed chicken will be quite recognizable from an industrially raised chicken nourished with feed, since feed has a lower ^{13}C content than corn (Table 12-3). Similarly, the pig, an omnivorous animal, is characterized by a higher ^{13}C content than the ox, the latter animal being fed for the most part on fodder from C_3-cycle plants, is generally very depleted in ^{13}C (Table 12-3). The (already observed) ^{13}C depletion of lipids with respect to other organic constituents is again seen here. The isotopic composition of the food regime and of the animal enables, for example, the identification of the types of plants (C_3 or C_4) present in the pastures (cf. Minson and Ludlow, 1975).

In the same manner, it is sometimes possible to reconstruct the food regimes of past civilizations from the ^{13}C contents of organic remains. De Niro and Epstein (1978) give the example of the Indians in Peru whose bones became enriched in ^{13}C between 5000 and 2000 B.P., due to an increased consumption of corn (a C_4 plant). Hillaire-Marcel and Plumet (unpublished) have also established the hunting possibilities of paleoeskimos in Hudson Strait, by means of the ^{13}C contents of the carbonized fats. The caribou, more depleted in ^{13}C than the marine mammals (walrus, seal, whale) is easily recognizable (isotopically) from the latter. These have isotopic compositions corresponding to the marine food chain, whose $\delta^{13}C$ is close to $-22‰$. It is thus possible to recognize the fats of these mammals from those of oxen or other mammals whose food regime has an isotopic composition close to that of C_3 plants.

Isotopic differentiation of corn oil

Due to the short C_4 cycle of corn, corn oil can be easily recognized from the greater majority of vegetable oils (Table 12-4). Its ^{13}C content can thus be considered as an effective control of its purity. We will see later on that oxygen isotope compositions permit the characterization of several other types of oils.

Concluding remarks

In our study, carbon ^{13}C is found to be a relatively precise tracer for a wide range of products. Its concentrations are related to the type of photosynthetic cycle in the plant at the start of the food chain. They enable the rather accurate characterization of certain constituents, sugars for example. The same property is kept with respect to the isotopic identification of the elementary derivatives of these constituents (e.g. alcohols). Even though generally not as efficient, the isotopic characterization of complex mixtures, tissues, organs, etc., can sometimes solve certain problems. Finally,

TABLE 12-4

Stable carbon isotope composition of various oils. The corn oil may be easily identified by its $\delta^{13}C$ value

Type of oil	Geographical origin	$\delta^{13}C$ ($^0/_{00}$ vs. PDB)
Soya	Iowa	−29.6
Soya	unknown	−30.1
Soya	unknown	−29.5
Soya	unknown	−29.7
Cotton seeds	Arkansas	−27.8
Peanut	North Carolina	−28.3
Peanut	unknown	−27.9
Cambra	Saskatchewan	−29.0
Coconut	Sri Lanka	−27.4
Coconut	Phillipines	−27.2
Palmtree	Indonesia	−29.8
Palmtree	Malaysia	−30.4
Olive	Italy	−29.5
Olive	unknown	−28.8
Olive	unknown	−28.7
Olive	unknown	−29.1
Paraffin	industrial	−28.3
Paraffin	industrial	−28.6
Corn	unknown	−15.3
Corn	unknown	−14.5

^{13}C may be a good means to distinguish natural vs. synthetic products (cf. Bommer et al., 1976).

^{18}O AND ^{2}H IN PLANTS

Oxygen and hydrogen, and carbon of course, represent the most abundant elements in organic matter. It could therefore be of interest to use ^{18}O/^{16}O and ^{2}H/^{1}H ratios as well as ^{13}C concentrations, to characterize certain food substances. We will see later that there is a certain similarity of behaviour between the oxygen and hydrogen isotopes, in part related to their common cycle in natural waters. We will, however, put the emphasis on oxygen, which seems to possess a more regulated behaviour, and consequently is more useful in the characterization of organic matter.

Even though carbon isotope variations in plants attracted the attention of researchers very early and are by now almost completely explained (as can be seen in the preceding section), oxygen isotope variations still remain a problem. Two reasons can explain this time lag. First of all, the oxygen pathway is far from being as linear as the carbon one (CO_2 photosynthesis), since oxygen has a more complex metabolism and a greater diversity of sources (H_2O, atmospheric CO_2). Furthermore, the technical difficulties associated with the extraction of the oxygen of organic matter have for a long time stalled the research.

Biochemical studies were nevertheless done quite soon with tracing substances enriched in ^{18}O (cf. Cohn, 1964). An extraction technique of organic oxygen was available indeed since the work of Rittenberg and Ponticorvo (1956). Hardcastle and Friedman (1974) have proposed another extraction mehod which equally does not introduce any isotopic fractionation. Both techniques have enabled several laboratories to pursue research in this field.

Even though some hypotheses, on the control of the ^{18}O/^{16}O composition in plants by the feeding waters, were suggested by Kamen (1946), it is only since the work of Hardcastle and Friedman (1974) that these hypotheses have been partly justified.

Fehri et al. (1975), Fehri and Létolle (1976, 1977), Bricout (1977) and Epstein et al. (1977) have tried to determine more precisely the role of the various variables involved. Gray and Thompson (1976) and Libby et al. (1976) have more or less demonstrated a relationship between the ^{18}O/^{16}O fractionation and the temperature. According to the former authors, the principal variables involved seem to be, on the one hand, the isotopic composition of the feeding water and, on the other hand, the evapotranspiration and the photosynthetic cycle. Since temperature does indeed modulate in many ways the preceding variables, the two approaches are not basically contradictory.

The $^{18}O/^{16}O$ isotopic fractionation: the elements of the problem

As early as 1941, Ruben et al. had shown that the oxygen released by chlorophyll synthesis was that of the feeding water. The oxygen of organic matter therefore probably comes from atmospheric CO_2, as Gray and Thompson (1976) thought. Epstein et al. (1977), however, consider a shared contribution ($\frac{2}{3} : \frac{1}{3}$ of CO_2 and water).

As a simplification, Fig. 12-9 summarizes the pathway of oxygen during photosynthetic cycle.

One can well see that even if HCO_3^- ions take the place of the dissolved CO_2, from an isotopic viewpoint, exchange reactions will occur between the CO_2 (or HCO_3^- ions) and the cellular water, independently of the fractionation caused by carboxylation.

These exchange reactions are of course temperature dependent. To complete this scheme, variables of a second order must be taken into consideration: do isotopic exchanges occur between the CO_2-H_2O system and dissolved molecular oxygen? How does respiration modify the isotopic budget of the cell?

As said previously, the oxygen problem is not as simple theoretically as that of carbon. The various factors one must consider are presented below without any determined order of importance:

— The isotopic composition of the plant's feeding water, partly related to the water cycle, i.e., to the mean annual temperatures (cf. Dansgaard, 1964; Fig. 12-10).

— The isotopic fractionation (H_2 $^{18}O/H_2$ ^{16}O) of the sap with respect to the feeding water (influenced by evapotranspiration, climate, season and the type of hydrological budget of the plants; Fig. 12-11).

— The isotopic exchanges CO_2-H_2O-O_2 within the cell, their link with temperature; is any equilibrium achieved?

— The role of the initial isotopic condition of the dissolved CO_2, if the CO_2-H_2O equilibrium is not achieved? The relation with the amount of dissolved CO_2 and the assimilation rate? The possible influence of the atmospheric $C^{18}O^{16}O/C^{16}O_2$ ratio fluctuations.

— The contribution of expired CO_2 in the CO_2-H_2O equilibrium.

— The fractionation during the photosynthetic cycle; the three types of photosynthetic cycles (C_4, C_3, CAM).

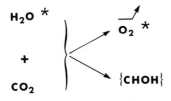

Fig. 12-9. Origin of the oxygen of organic matter (from Ruben et al., 1941).

— The exchanges between organic compounds and cell water.

Of course, these variables do not all have the same importance. The principal factors (cf. Fehri and Létolle, 1976, 1977; Hillaire-Marcel et al., 1976; Bricout, 1977; Epstein et al., 1977) remain the following (see Figs. 12-12 to 12-14):

— The isotopic composition of the feeding water.

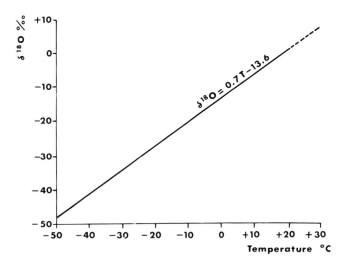

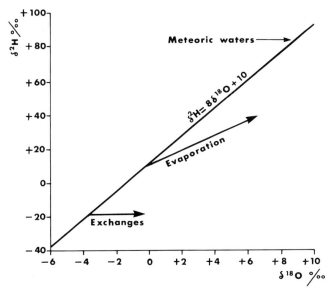

Fig. 12-10. Hydrogen and oxygen isotopes in precipitations, relationship with mean annual temperature and evaporation effect (from Dansgaard, 1964 and Fontes, 1976).

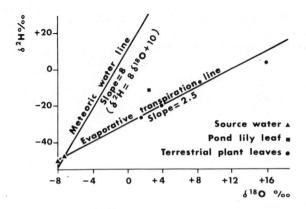

Fig. 12-11. Influence of evapotranspiration on the hydrogen and oxygen isotopic ratios in plants (from Epstein et al., 1977).

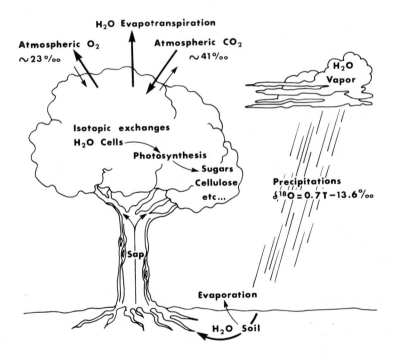

Fig. 12-12. Sketch of the various factors influencing the oxygen isotope composition of organic matter.

— The $^{18}O/^{16}O$ isotopic enrichment of the sap by evapotranspiration.
— The cellular CO_2-H_2O equilibrium.
— The relative contributions of CO_2 and H_2O to the organic oxygen in the cell.
— The fractionation during photosynthesis; the $^{18}O/^{16}O$ variations of the different cellular constituents; the isotopic exchanges with the cell water.
— The influence of temperature and/or climate in each of the preceding variables.

A detailed discussion of each of these variables would be beyond the framework of this study. Nevertheless, it does seem necessary to point out their areas of influence.

The isotopic composition of the feeding water. The aquifers of a specific region should theoretically possess a quite homogeneous isotopic composition related to the mean temperature of the precipitations and, in general, to the

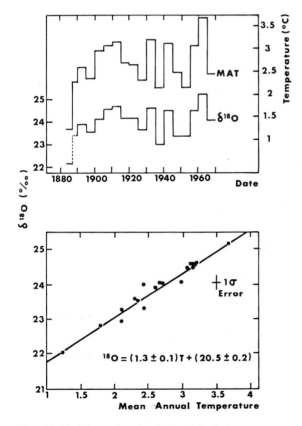

Fig. 12-13. Example of relationship between mean annual temperature and the oxygen isotope composition of cellulosis, in tree rings (from Gray and Thompson, 1976).

climate (evaporation). In continental regions, however, the Dansgaard (1964) relation between the isotopic composition of precipitations and the mean annual temperature is not valid anymore. Furthermore, an altitudinal effect (Dansgaard, 1964) can be added to this factor of continentality. Finally, the isotopic composition of groundwater can vary by a few permil, depending on (a) depth of the water table (a more intense evaporation occurs near the surface inducing a ^{18}O enrichment of the water); and (b) the season of the year (sudden replenishment of the aquifers during the spring by the ^{18}O-depleted snow-melt; intense evaporation during the summer, leading to an ^{18}O enrichment). One can understand that this variable will be particularly important in plants with shallow roots.

Isotopic enrichment of the sap by evapotranspiration. The water absorbed by the roots will be isotopically enriched in the leaves by evapotranspiration (cf. Gonfiantini et al., 1965).

The enrichment of the residual water, generally on the order of 10—15‰ (for ^{18}O), can vary greatly and depends on the plant's type of hydrological budget (e.g. the water stocking *Crassulaceae*) and on the relative humidity of

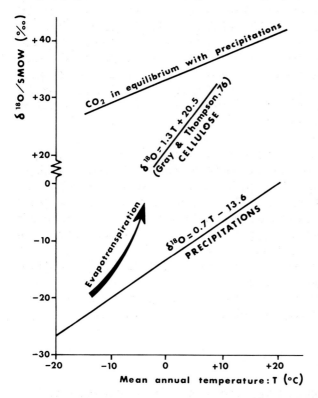

Fig. 12-14. Sketch showing the enrichment in ^{18}O induced by evapotranspiration.

the ambient air. Therefore, there is an indirect correlation between the ^{18}O enrichment of the leaf water and the climate.

CO_2-H_2O equilibrium in the chlorophyll-producing parenchimal cells. According to Craig (1954a), an isotopic enrichment of about $41^0/_{00}$ will occur in the CO_2 if it equilibrates with water at $25°C$. It is not known, however, if equilibrium is achieved during the periods of intense assimilation of CO_2. Furthermore, the amount of dissolved CO_2 is not negligible and the initial isotopic composition of the atmospheric CO_2 ($+41^0/_{00}$ vs. SMOW) might be a factor, especially since its composition is much higher than that of water. Let us finally note that the atmospheric CO_2 composition varies by a few permil, depending on the region and on the season (cf. Bottinga and Craig, 1969). In any case, temperature is a factor in the CO_2-H_2O equilibrium. A climatic dependence, therefore, comes in once more.

Fractionation during photosynthesis and between the cellular compounds. This aspect is very important because indeed, a carbon fractionation distinct from one cycle to another has been observed. One would have difficulty to imagine a different case for oxygen. However, even though the experimental data of Fehri and Létolle (1976, 1977) and Bricout (1977) suggest a fractionation on the order of $20^0/_{00}$, no systematic differences seem to exist between the three cycles, isotopic exchanges with cellular water might actually erase other effects. On the other hand, the chains of reactions, leading for example from carbohydrates to lipids, are accompanied by an ^{18}O depletion on the order of $14-20^0/_{00}$ (cf. Bricout, 1977; Carro et al., 1979). This is comparable to what was observed for carbon.

The influence of climate and temperature. We have seen that these variables could act at several levels. An example can be given for a better comprehension: A temperature drop of $10°C$ will be accompanied by a change of the $H_2^{18}O/H_2^{16}O$ ratio, as high as $-7^0/_{00}$ in extreme conditions; on the plant level, evapotranspiration will decrease due to the more rigorous climate; the cell water will therefore show a lesser enrichment in ^{18}O. Finally, during the H_2O-CO_2 equilibrium, a $-2^0/_{00}$ inverse effect will take place, almost negligible in comparison with the preceding ones. The total budget will result in a $^{18}O/^{16}O$ lowering of the organic matter.

With this background, we can examine some examples of the application of oxygen and deuterium isotope chemistry in the control of certain natural food products.

Examples of applications

Two distinct types of applications can be noted. The first pertains to fruit and vegetable juices and their derivatives (wines, ciders, etc.). There exists a

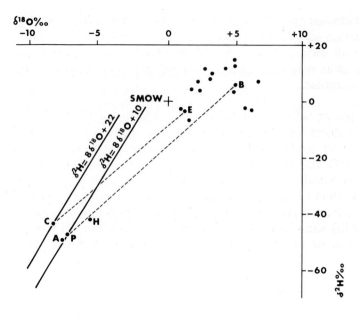

Fig. 12-15. Hydrogen and oxygen isotopes in precipitations and in orange juices (from Bricout et al., 1972). Natural waters: C = Corsica, P = Paris, A = Brasil; Orange juices: E = Corsica, B = Brasil; (•) = other production areas. H = orange juice diluted with water from Paris.

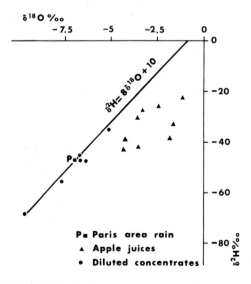

Fig. 12-16. Hydrogen and oxygen isotopes in natural and diluted apple juices (from Bricout et al., 1973).

direct relationship between the isotopic composition of their water and the climate of the production area which is linked to the variations of the isotopic composition of natural waters (precipitations, aquifer budgets). The second type of application deals with the isotopic composition of the organic substances and is more complex.

Fruit juices. The work of Bricout et al. (1973) on orange juices is one of the first studies on this subject. Indeed, the orange culture regions are associated with mediterranean and subtropical climates. Strong evaporation in these regions causes high ^{18}O and ^{2}H concentrations not only in natural waters (Fontes, 1976) but also and most importantly in the cellular water of plants. Bricout et al. (1973) thus demonstrated that it was possible to differentiate a natural orange juice from a concentrate diluted afterwards (Fig. 12-15).

Bricout et al. (1973) showed later on that the same principle could apply to the characterization of natural apple juices versus diluted concentrates

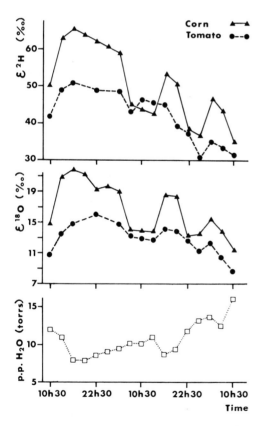

Fig. 12-17. Daily variations of the isotopic composition of leaf water (corn and tomato) in relation to relative humidity (from Lesaint et al., 1974).

(Fig. 12-16). In this case, however, the isotopic enrichment of the fruit juice is minimal, since evapotranspiration decreases toward the more northerly latitudes where the apple production regions are located. Similar studies have been accomplished on tomato and corn by Lesaint et al. (1974). Their merit lies in their demonstration of the relation between the partial pressure of the atmospheric water vapour and the isotopic enrichment of the cellular water, thereby showing clearly the determining influence of evapotranspiration (Fig. 12-17). Furthermore, Lesaint et al. (1974) observe a smaller enrichment in the fruit water than in the leaves. This suggests that unmetabolized sap which is isotopically close to the plants alimentation water, contributes in a great part to the fruit water (Fig. 12-18). Finally, the application of this method to the isotopic characterization of grape juices (Bricout et al., 1974) and wines (Hillaire-Marcel et al., unpublished) has important economic consequences. Even though systematic differences in isotopic composition are observed from one region to another, the local and seasonal climatic variations affect the isotopic composition range on the same crop (Fig. 12-19). A wine from Spain can nevertheless be distinguished from a Burgundy (Table 12-5).

Organic compounds. Due to the number of variables involved, the interpretation of ^{18}O and ^{2}H concentrations in organic compounds is difficult. The ^{18}O and ^{2}H concentrations can nevertheless be used to characterize certain food

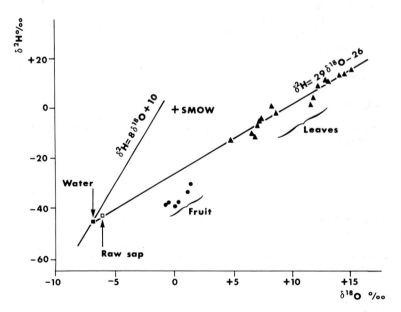

Fig. 12-18. Relationship between hydrogen and oxygen isotope ratios in sap, cellular water of leaves and fruit of corn (from Lesaint et al., 1974).

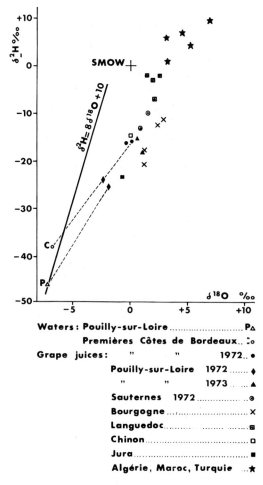

Fig. 12-19. Hydrogen and oxygen isotopes in grape juices from various production areas (from Bricout et al., 1974b).

TABLE 12-5

Example of oxygen isotope compositions of wine from different origins

Wines	$\delta^{18}O$ (‰ vs. SMOW)
Coronas (Spain)	+ 2.4
Merlot (Italy)	+ 1.6
Côtes de Provence (France)	+ 1.6
Bordeaux rouge (France)	+ 0.8
Bourgogne rouge (France)	− 1.1

products. Natural tracing can only be effected on non-exchangeable oxygen and hydrogen (cf. Epstein et al., 1977).

A few tests have already been done (Hillaire-Marcel et al., unpublished), for example on oils and fats (Table 12-6). An olive oil coming from a climatically well-defined production region can thus be clearly differentiated from a colza or soya oil. Similarly, peanut and palm oils have high concentrations in ^{18}O. However, the possibilities of complex mixtures are such that the isotopic characterization of these products cannot be recommended as a means of authenticity control. There are, however, a few specific cases where the method may be successful. For example, the admixture of corn alcohol to a rum cannot be detected by the ^{13}C content of the alcohol, since sugar

TABLE 12-6

Example of oxygen isotope compositions of oils from various origins

Type of oil	Origin	$\delta^{18}O$ (°/₀₀ vs. SMOW)
Soya	Iowa	16.6
Cotton seeds	Arkansas	14.6
Peanut	North Carolina	10.0
Cambra	Saskatchewan	15.8
Corn	unknown	13.7
Olive	Italy	29.4

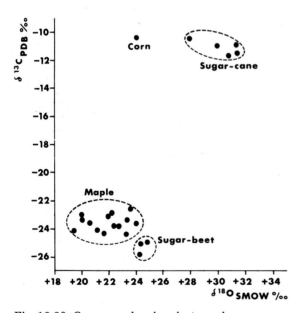

Fig. 12-20. Oxygen and carbon isotopes in some sugars.

cane and corn have the same photosynthetic cycle. Sugar cane, however, does show an enrichment in ^{18}O (due to its narrow geographic dispersion), which enables it to be isotopically differentiated from corn (Fig. 12-20). The variations in the isotopic composition of sugars are translated, with a fractionation factor, to their derived alcohols. Thus, it is possible to characterize rum alcohol by its ^{18}O content. Similarly, an orange blossom honey should show a much higher ^{18}O concentration than a regular one. This problem, however, does not present a great economic interest. One could, in the same manner suggest other examples of application (cf. Ziegler et al., 1979). But, as a general rule, the isotopic characterization by ^{18}O or by ^{2}H isotopes has less systematic applications than by ^{13}C.

THE RADIOACTIVE ISOTOPES ^{3}H AND ^{14}C

^{14}C, sometimes in conjunction with tritium, has been successfully used to characterize the addition of synthetic substances (derived from coal or petroleum) to natural biosynthesized products. Both of these isotopes have also been used to control the age of certain wines and alcohols. Before giving a few examples, we will briefly recapitulate the properties of these isotopes from our point of interest, i.e. the natural tagging of biosynthesized products.

^{14}C

^{14}C is produced in the upper atmosphere by the cosmic ray neutronic bombardment of ^{14}N (cf. Libby, 1955). It is quickly oxidized to $^{14}CO_2$.

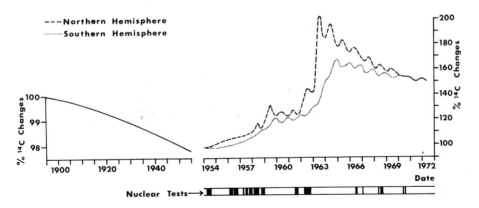

Fig. 12-21. Variations of the ^{14}C activity in the atmosphere in relation to the Suess effect (1890—1950: release of industrial CO_2) and to the nuclear test effect (since 1964). Data from Vogel (1972) and Suess (1970).

The $^{14}CO_2$, in turn, is homogeneously mixed in the atmosphere and equilibrates with the marine carbon reservoir (HCO_3^-, CO_3^{2-}, $MgHCO_3^+$, etc.). The thermonuclear tests at the end of the 50's and the start of the 60's have considerably increased the amount of atmospheric ^{14}C (Fig. 12-21). Due to the relative isolation of the two hemispheres, significant differences in ^{14}C content appear between the two. In any case, the ^{14}C is taken up in organic matter, at a fractionation factor close ($\epsilon^{14}C \simeq 2\epsilon^{13}C$; cf. Broecker and Olson, 1959, 1961), first by chlorophyll synthesis and then by the food chain.

The biosynthesized products thus have a ^{14}C activity, which reflects approximately that of the atmosphere during their synthesis. Indeed, the radioactive decay period of ^{14}C is sufficiently long (half-life: 5730 years) for this factor to be neglected.

The possible applications for the control of the authenticity of certain food products are of two types. The first type of application deals with the substitution or the adulteration of natural products by synthetic products, derived from petroleum and coal. These generally have a geological age of more than tens of millions of years and do not ordinarily show any measurable ^{14}C activity. In contrast, biosynthesized products are characterized by ^{14}C activities of the order of 13.6 (base level in 1950) to 25 (maximum level, 1964) dpm/g (disintegrations per minute per gram) of carbon. Thus, in the light of these facts, Francesco et al. (1970) have suggested the differentiation between industrial acetic acid and the natural wine vinegars. Guerain and Tourlière (1975), Resmini and Volonterio (1973), Eskola (1977), to mention only a few, have shown in the same manner that industrial alcohol could be differentiated from alcohol derived from the fermentation of fruit or vegetable juices. More recently, Hoffman and Sabb (in preparation) have suggested the differentiation between synthetic cinnamic aldehyde and the natural aldehyde derived from cinnamon. To generalize the above examples, one can propose this method to distinguish any biosynthesized product from a chemically similar product industrially synthesized from fuels and coals. This method can apply in part to the distinction between natural and artificial flavors and essential oils.

The second type of application of the ^{14}C method especially concerns the wines and alcohols produced since the start of the nuclear tests in the atmosphere. It is possible to control with an acceptable precision the aging of wines (Sousa-Lopes et al., 1975; Tarantola, 1976) or alcohols derived from grain (Reinhard, 1977), potatoes (Rauschenbach and Simon, 1975) or any other vegetable. This control is possible in so far as the atmospheric ^{14}C variations (due to the nuclear tests) are found correspondingly in natural alcohols (Fig. 12-22), as demonstrated by Zimen (1972). One will notice, however, that for any ^{14}C activity (measured on an alcohol) greater than a threshold on the order of 14 dpm/g of carbon, two approximate ages can be suggested, due to the behaviour of $^{14}CO_2$ in the atmosphere (cf. Fig. 12-21).

In this case, we will see that the ^{3}H content of the aqueous phase may settle the question (Guerain and Tourlière, 1975).

Tritium

Also produced in the upper atmosphere, tritium is injected into the water cycle by precipitation (cf. Fontes, 1976). Introduced in the feeding water of plants, it finally appears in their cellular water and organic matter. Due to its short disintegration period (half-life = 12.16 years), tritium has the advantage (over ^{14}C) of permitting the dating of, for example, wines produced before 1955 (Fig. 12-23). However, similarly to atmospheric ^{14}C, its concentration in precipitations increased considerably in the Northern Hemisphere (Fig. 12-24) during the nuclear tests. From less than 100 TU (1 TU = 1 ^{3}H atom/10^{18} ^{1}H atoms) in 1955 to more than 4000 TU in 1963, tritium was injected in great quantities in the water cycle. This increase, comparable to an artificial tracing signal, in conjunction with the ^{14}C activities of the corresponding alcohols, makes the more precise dating of wines and alcohols produced since 1955 possible (Guerain and Tourlière, 1975).

OTHER POSSIBLE APPLICATIONS

Some other applications of isotopic geochemistry in the control of food products could be mentioned. However, they usually involve specific cases. Let us mention, for example, the isotopic characterization of bottled spring waters or mineral waters. Even though a chemical analysis could

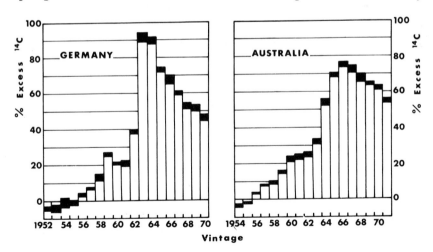

Fig. 12-22. ^{14}C activity in wines from Germany and Australia since beginning of nuclear tests (from Zimen, 1972).

suffice to characterize these waters, an isotopic analysis (^{18}O, ^3H, ^2H, ^{14}C, ^{13}C) could be done if needed. This is the case for carbonated waters notably, when one wants to verify if the CO_2 is of natural origin. Indeed, industrial CO_2 produced by combustion is characterized by a low ^{13}C concentration, which distinguishes it from hydrothermal CO_2 (Table 12-7).

TABLE 12-7

Distinction between natural CO_2 of mineral waters and industrial CO_2

Mineral water	$\delta^{13}CO_2$ (°/oo vs. PDB)
Saratoga spring	− 5.2
Perrier	− 4.4
Peter Val	− 4.1
Industrial CO_2 (average)	− 47.0

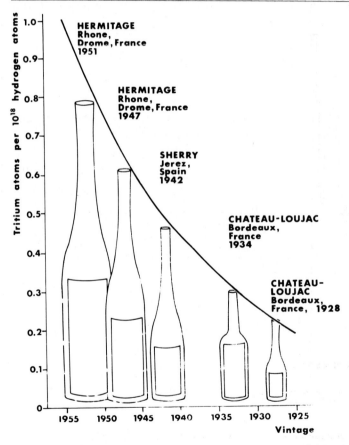

Fig. 12-23. Tritium dating of wines (Guerain and Tourlière, 1975).

These are minor problems, however. There is, nevertheless, one area where isotopic geochemistry could have interesting applications. It concerns the natural tracing of metabolic processes. The tagging of molecules by radioactive isotopes was traditionally used in biochemistry. In medicine, however,

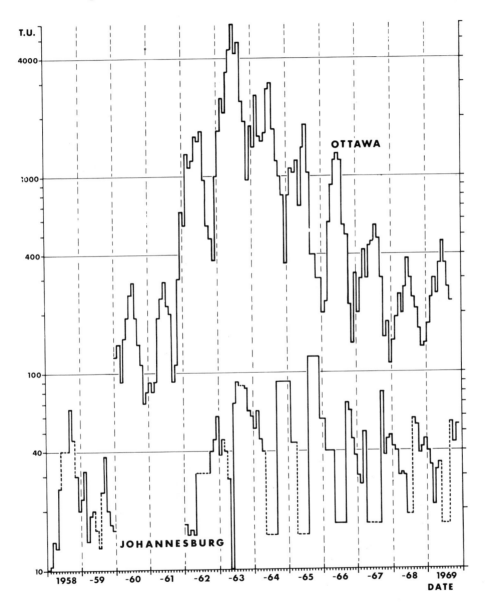

Fig. 12-24. Tritium in precipitations since the beginning of nuclear tests (from Brown, 1970, and Verhagen et al., 1970).

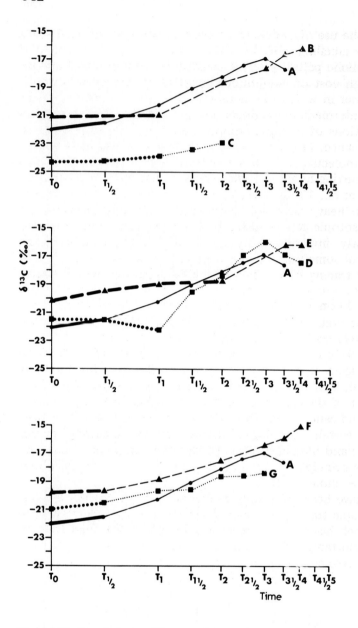

Fig. 12-25. Breath tests with natural isotopic tracing (C_4-type sugars) (from Dever et al., unpublished). Logarithmic scale for time (in hours). Curve A control subjects with 1/2-hour log time (thick line). Curves B-G: various pathologies which can be characterized either by the duration and the slope of the log time, or by the subsequent slope of the curve.

the interdiction of the use of radioactive tracers in some countries, has led to the stable isotope labelling (^{15}N, ^{13}C, ^{2}H) of compounds, in the hope of following their metabolic pathways. This method has several disadvantages, for example, the high cost of compounds enriched in rare stable isotopes. Furthermore, the order in which the metabolic reactions occurs can not be resolved, since the intermediate products are all so enriched in heavy isotopes that the variations of isotopic composition from one to another do not have any significance. Moreover, enzymatic reactions can be disrupted by the presence of molecules, which are quite different due to their high mass from those occurring naturally. These inconveniences have led certain researchers (Lacroix et al., 1973; Dever et al., in preparation) to use tracers naturally enriched in heavy isotopes. These studies, directly derived from the applications of isotopic geochemistry to the characterization of natural products, have mostly investigated by means of ^{13}C the metabolism of sugars, and cholesterol sometimes on the side. The principle at the origin of this natural isotopic tracing is that heterotrophic organisms have the same isotopic composition, within a few permil, as that of their diet (De Niro and Epstein, 1976; Lyon and Baxter, 1978). For a typical North American food regime for example, the mean isotopic composition of the human tissues is approximately $-23^o/_{oo}$ and is translated into that of the expired CO_2 (Lyon and Baxter, 1978; Dever et al., in preparation). The investigation on the metabolism of sugars, with $^{13}CO_2$ breath tests, is thus possible. One may use as a tracing signal, a C_6 or C_{12} sugar with an isotopic composition differing slightly from that of the food regime. This is the case (as we previously discussed) for corn and cane sugars. Thus, Dever et al. (Fig. 12-25) have been able to characterize certain pathological diseases by such a natural tracing. Experiments on hens and human subjects (Dever et al., in preparation) have shown that the same principle could apply to the study of other metabolic reaction chains (fats → cholesterol).

These few cases have been given only as example of the possible applications of natural isotopic tracing. The next few years will see a generalization of this method, which has the advantage of being less costly than artificial tracing and most importantly, permits normal enzymatic reactions since the compounds have isotopic compositions in the natural range of variations.

CONCLUSIONS

There are numerous examples of the application of isotopic geochemistry in the control of food products. Our perusal over some of these examples enables us to make some general remarks. The $^{13}C/^{12}C$ ratios, controlled by the photosynthetic cycles, have the advantage of distinguishing between products chemically identical but biosynthesized through a different cycle (case of sugars). This property is valid for all derived products (alcohols, vinegars) and is translated to the food chain (honey, grain-fed poultry) and into all metabolic processes.

The atoms present in the water molecule can often permit the characterization of the climates of production regions. The natural variability of the oxygen and hydrogen isotopic compositions is such that it is sometimes difficult to firmly conclude that a product has been adulterated.

^{14}C and ^3H methods are more conclusive but have a more restricted field of application. Let us retain that ^{14}C is an ideal tool for the differentiation between biosynthesized products and compounds synthesized from petroleum and coal, and that ^3H will give us the approximate age of a wine or an alcohol (even so it will not guarantee its quality).

Finally, even though one may expect a few new applications here and there, it is quite certain that most of them will be of limited range. We have seen that, in the best of cases, the adulteration detection thresholds were on the order of 5%. This method is thus far from being absolute. Nevertheless, the studies that we have briefly brought to light, have the merit of being at the origin of the natural tracing of metabolic processes. Stable isotope chemistry seems to have the more promising future in this field.

REFERENCES

Abelson, P.H. and Hoering, T.C., 1961. Carbon isotope fractionation in formation of aminoacids by photosynthetic organisms. *Biochemistry*, 47: 623—632.

Bender, M.M., 1968. Mass spectrometric studies of carbon-13 variations in corn and other grasses. *Radiocarbon*, 10: 468—472.

Bender, M.M., 1971. Variations in the $^{13}C/^{12}C$ ratios of plants in relation to the pathway of photosynthetic carbon dioxide fixation. *Phytochemistry*, 10: 1239—1244.

Berry, J., Troughton, J.H. and Björkman, O., 1972. Effect of oxygen concentration during growth on carbon isotope discrimination in C_3 and C_4. *Carnegie Inst. Washington Year.*

Bommer, P., Moser, H., Stichler, W., Trimborn, P. and Vetter, W., 1976. Herkunftsbestimmung von Arzneimitteln durch Messung von natürlichen Isotopen-verhäetnissen: D/H und $^{13}C/^{12}C$ Verhältnisse einiger Proben von Diazepam. *Z. Naturforsch., Teil C*, 31: 111—114.

Bottinga, Y. and Craig, H., 1969. Oxygen isotope fractionation between CO_2 and water and the isotopic composition of marine atmospheric CO_2. *Earth. Planet. Sci. Lett.*, 20: 250—265.

Bricout, J., 1977. Sur la teneur en oxygène 18 de la matière organique végétale. *C.R. Acad. Sci., Paris*, 284: 1891—1893.

Bricout, J. and Fontes, J.Ch., 1974. Distinction analytique entre sucre de canne et sucre de betterave. *Ann. Fals. Exp. Chim.*, 716: 211—215.

Bricout, J., Merlivat, L. and Fontes, J.Ch., 1973. Sur la composition en isotopes stables de l'eau des jus d'orange. *C.R. Acad. Sci., Paris*, 274: 1803—1806.

Bricout, J., Fontes, J.Ch. and Merlivat, L., 1974a. Detection of synthetic vanillin in vanilla extracts by isotopic analysis. *J. Assoc. Off. Anal. Chem.*, 57: 713—715.

Bricout, J., Fontes, J.Ch. and Merlivat, L., 1974b. Sur la composition en isotope stables de l'eau des jus de raisin. *Connaiss. Vigne Vin*, 2.

Broecker, W.S. and Olson, E.A., 1959. Lamont radiocarbon measurement. *Am. J. Sci., Radiocarbon Suppl.*, 1: 111—132.

Broecker, W.S. and Olson, E.A., 1961. Lamont radiocarbon measurement, VIII. *Radiocarbon*, 3: 176—204.
Brown, R.M., 1970. Distribution of hydrogen isotopes in canadian waters. In: *Isotope Hydrology 1970*, IAEA, Vienna, pp. 3—21.
Calvin, M. and Bassham, J.A., 1962. *The Photosynthesis of Carbon Compounds*. Benjamin, New York, N.Y.
Carro, O., Hillaire-Marcel, C. and Chevalier, R., 1979. Caractérisation isotopique des corps gras d'origine végétale et animale. *Minist. Agric., Québec, Rep.*, UQM-77684: 30 pp.
Carro, O., Hillaire-Marcel, C. and Gagnon, M., 1980. Detection of adulterated maple products by stable carbon isotope ratio. *J. Assoc. Off. Anal. Chem.*, 63: 840—844.
Cenci, P. and Cremonini, B., 1973. Addition of sugar to wine, III. Modification of a method for detecting the addition of beet sugar. *Riv. Vitic. Enol.*, 26: 194—198.
Cohn, M., 1964. Application of ^{18}O to biochemical studies. In: H. Craig et al. (Editors), *Isotopic and Cosmic Chemistry*. North-Holland, Amsterdam, pp. 45—59.
Cooper, T.G. and Wood, H.G., 1971. The carboxylation of phosphoenolpyruvate and pyruvate. *J. Biol. Chem.*, 246: 5488—5490.
Craig, H., 1953. The geochemistry of the stable carbon isotope. *Geochim. Cosmochim. Acta*, 3: 53—92.
Craig, H., 1954a. Carbon-13 in plants and the relationship between carbon-13 and carbon-14 variations in nature. *J. Geol.*, 62: 115—149.
Craig, H., 1954b. Geochemical implications of the isotopic composition of carbon in ancient rocks. *Geochim. Cosmochim. Acta*, 6: 186—196.
Craig, H. and Keeling, C.D., 1963. The effects of N_2O on the measured isotopic composition of atmospheric CO_2. *Geochim. Cosmochim. Acta*, 27: 549—551.
Dansgaard, W., 1964. Stable isotopes in precipitation. *Tellus*, 16: 436—467.
Degens, E.T., Guillard, R.R.L., Sackett, W.M. and Hellebust, J.A., 1968. Metabolic fractionation of carbon isotopes in marine plankton. *Deep-Sea Res.*, 15: 1—9.
Deines, P., 1980. The isotopic composition of reduced organic carbon. In: P. Fritz and J.Ch. Fontes (Editors), *Handbook of Environmental Isotope Geochemistry, 1. The Terrestrial Environment, A*. Elsevier, Amsterdam, pp. 329—406.
DeNiro, M.J. and Epstein, S., 1976. You are what you eat (plus a few $^0/_{00}$): the carbon isotope cycle in food chains. *Geol. Soc. Am. Abstr.*, 8(6): 834—835.
DeNiro, M.J. and Epstein, S., 1978. Carbon isotopic evidence for different feeding patterns in two hyrax species occupying the same habitat. *Science*, 201: 906—908.
Doner, L. and White, J.W., Jr., 1977. Carbon-13/carbon-12 ratio is relatively uniform among honeys. *Science*, 197: 891—892.
Epstein, S., Thompson, P. and Crayton, J., 1977. Oxygen and hydrogen isotopic ratios in plant cellulose. *Science*, 198: 1209—1215.
Eskola, A., 1977. Detection of synthesized mineral alcohol using a liquid scintillation counter. *LKB Appl. Note*, 500: 4 pp.
Evans, L.T. and Dunstow, R.L., 1970. Some physiological aspects of evolution in wheat. *Aust. J. Biol. Sci.*, 23: 725—741.
Ferhi, A. and Létolle, R., 1976. Les paramètres de variation de la composition isotopique de l'oxygène organique dans les parenchymes foliaires des plantes. (unpublished manuscript).
Ferhi, A. and Létolle, R., 1977. Variation de la composition isotopique de l'oxygène organique de quelques plantes en fonction de leur milieu de vie. *C.R. Acad. Sci., Paris*, 248: 1887—1889.
Ferhi, A.M., Létolle, R.R. and Lerman, J.C., 1975. Oxygen isotope ratios of natural compositions. *Int. Conf., Stable Isotopes, Oak Brook, Ill.*
Fontes, J.Ch., 1976. Les isotopes du milieu dans les eaux naturelles. *Houille Blanche*, 314: 205—221.

Francesco, F., Paris, A., Stefani, R. and Gianotti, L., 1970. Measurements of ^{14}C in some foods, I. Wine Vinegar. Sci. Aliment., 16: 379—388.
Galimov, E.M., 1979. $^{13}C/^{12}C$ in kerogen. In: B. Durand (Editor), Insoluble Organic Matter from Sedimentary Rocks. Technip, Paris, pp. 271—299.
Gonfiantini, R., Gratziu, S. and Tongiorgi, E., 1965. Oxygen isotope compositions of water in leaves. In: Isotopes and Radiation in Soil Plant Nutrition Studies. IAEA, Vienna, pp. 405—410.
Gray, J. and Thompson, P., 1976. Climatic information from $^{18}O/^{16}O$ ratios of cellulose in tree rings. Nature, 262: 481—482.
Guerain, J. and Tourlière, S., 1975. Carbon and tritium radioactivity in alcohol. Ind. Aliment. Agric., 92: 811—822.
Hardcastle, K.G. and Friedman, I., 1974. A method for oxygen isotope analysis of organic material. Geophys. Res. Lett., 1: 165—167.
Hatch, M.D., Slack, R.C. and Johnson, H.S., 1967. Further studies on a new pathway of photosynthetic carbon dioxide fixation in sugarcane and its occurrence in other plant species. Biochem. J., 10: 417—422.
Hillaire-Marcel, C. et al., 1976. Les isotopes du carbone et de l'oxygène dans Acer saccharum. Minist. Agric., Québec, Rep., UQM-76-662, 170 pp.
Hoffman, P.G. and Salb, M., 1979. Isolation and stable isotope ratio analysis of vanillin. J. Agric. Food Chem., 27: 352—355.
Kamen, M.D., 1946. Survey of contemporary knowledge of biochemistry. Bull. Am. Mus. Nat. Hist., 87: 103—138.
Keeling, C.D., 1961. The concentration and isotopic abundances of carbon dioxide in rural and marine air. Geochim. Cosmochim. Acta, 24: 277—298.
Lacroix, M., Moroso, F., Pontus, M., Lefebvre, P., Luyckx, A. and Lopes-Habib, G., 1973. Glucose naturally labeled with carbon 13. Use for metabolic studies in man. Science, 181: 445—446.
Lerman, J.C., 1972. ^{14}C dating: origin and correction of isotope fractionation errors in terrestrial living matter. Proc. 8th Conf., Radiocarbon Dating, Wellington 1972, 2: H-16, H-18.
Lerman, J.C. and Raynal, J., 1972. La teneur en isotopes stables du carbone chez les Cypéracées: sa valeur taxonomique. C.R. Acad. Sci., Paris, Sér. D, 275: 1391—1394.
Lesaint, C., Merlivat, L., Bricout, J., Fontes, J.Ch. and Gautheret, R., 1974. Sur la composition en isotopes stables de l'eau de la tomate et du maïs. C.R. Acad. Sci., Paris, 278.
Libby, W.F., 1955. Radiocarbon Dating. Unversity of Chicago Press, Chicago, Ill.
Libby, L.M., Pandolfi, L.J., Payton, P.H., Marshall III, J., Becker, B. and Giertz-Sienbenlist, 1976. Isotopic tree thermometers. Nature, 261: 284—289.
Lopes, J.S., Pinto, R.E., Almandra, M.E. and Maxhado, J.A., 1977. Variation of ^{14}C activity in Portugese wines from 1940 to 1970. In: Low Radioactivity Measurements and Applications. Slovenske Pedagogické Nabladatelstvo, Bratislava, pp. 265—268.
Lowdon, J.A. and Dyck, W., 1974. Seasonal variations in the isotope ratios of carbon in maple leaves and other plants. Can. J. Earth Sci., 11: 79—88.
Lyon, T.D.B. and Baxter, M.S., 1978. Stable carbon isotopes in human tissues. Nature, 273: 750—751.
Minson, D.J. and Ludlow, M.M., 1975. Differences in natural carbon isotope ratio of milk and hair from cattle grazing tropical and temperature pastures. Nature, 256: 602.
Morselli, M., 1975. Chemical composition of maple syrup (bibliographical review). University of Vermont, Burlington, Vt., 2 pp. (unpublished manuscript).
Nissenbaum, A., Lifshitz, A. and Stepek, Y., 1974. Detection of citrus fruit adulteration using the distribution of natural stable isotopes. Lebensm.-Wiss. Technol., 7: 152—154.
Park, R. and Epstein, S., 1960. Carbon isotope fractionation during photosynthesis. Geochim. Cosmochim. Acta, 21: 110—126.

Rauschenbach, P. and Simon, H., 1975. Further studies of the ^{14}C content of fermentation alcohol in relation to period of growth and locality of the fermented material. *Z. Lebensm.-Unters. Forsch.*, 157: 143—146.

Rausenbach, P., Simon, H., Stichler, W. and Moser, H., 1979. Vergleich der Deuterium- und Kohlenstoff-13-gehalte in Fermentations- und Synthese-ethanol. *Z. Naturforsh., Teil C*, 34: 141.

Reinhard, C., 1977. Analysis and evaluation of whisky. *Dent. Lebensm.-Runds*, 73: 124—129.

Resmini, P. and Volonterio, G., 1973. Liquid scintillation counting of low radiocarbon levels, II. Application to sparkling wines and food. *Sci. Technol. Aliment.*, 3: 71—79.

Rittenberg, D. and Ponticorvo, L., 1956. A method for the determination of the ^{18}O concentration of the oxygen of organic compounds. *Int. J. Appl. Radiat. Isot.*, 1: 208—214.

Ruben, S., Randall, M., Kamen, M.D. and Hyde, J.L., 1941. Heavy oxygen (^{18}O) as a tracer in the study of photosynthesis. *J. Am. Chem. Soc.*, 63: 877—879.

Schmid, E.R., Fogy, I. and Schwarz, P., 1978. Differentiation of vinegar produced by fermentation and vinegar made from synthetic acetic acid based on determination of the $^{13}C/^{12}C$ isotope ratio by mass spectrometry. *Z. Lebensm.-Unters. Forsch.*, 166: 89—92.

Silverman, S.R., 1964. Investigations of petroleum origin and evolution mechanism by carbon isotope studies. In: H. Craig et al. (Editors), *Isotopic and Cosmic Chemistry*. North-Holland, Amsterdam, pp. 92—102.

Silverman, S.R. and Epstein, S., 1958. Carbon isotopic compositions of petroleum and other sedimentary organic materials. *Bull. Am. Assoc. Pet. Geol.*, 42: 998—1012.

Smith, B.N. and Epstein, S., 1971. Two categories of $^{13}C/^{12}C$ ratios for higher plants. *Plant Physiol.*, 47: 380—384.

Sousa-Lopes, J., Pinto, R.E. and Almendra, M.E., 1975. Variation between 1950 and 1974 in the ^{14}C content of wines from the Douro region of Portugal. *Agron. Lusit.*, 36: 223—234.

Suess, H.E., 1970. Bristlecone-pine calibration of the radiocarbon time-scale 5200 B.P. to present. In: *12th Nobel Symposium, Radiocarbon Variations and Absolute Chronology*, pp. 303—312.

Tarantola, C., 1976. Possibility of characterization of wine according to origin and year of production. *Vini Ital.*, 18(104): 323—328.

Troughton, J.H., 1972. Carbon isotope fractionation in plants. *Proc. 8th Conf., Radiocarbon Dating, Wellington 1972*, 2: 39—57.

Verhagen, B.T., Sellschop, J.P.F. and Jennings, C.M.H., 1970. Contribution of environmental tritium measurements to some geohydrological problems in Southern Africa. In: *Isotope Hydrology 1970*. IAEA, Vienna, pp. 289—312.

Vogel, J.C., 1972. Radiocarbon in the surface waters of the Atlantic Ocean. *Proc. 8th Conf., Radiocarbon Dating, Wellington 1972*, 2: C43, C55.

Vogel, J.C., Grootes, M.M. and Mook, W.G., 1970. Isotopic fractionation between gaseous and dissolved carbon dioxide. *Z. Phys.*, 230: 225.

Whelan, T., Sackett, W.M. and Benedict, C.R., 1970. Carbon isotope discrimination in a plant possessing the C_4 dicarboxylic acid pathway. *Biochem. Biophys. Res. Comm.*, 41: 1205—1210.

Wickman, F.E., 1952. Variations in the relative abundance of the carbon isotopes in plants. *Geochim. Cosmochim. Acta*, 2: 243—000.

Ziegler, H., Stichler, W., Maurizio, A. and Vorwohl, G., 1977. Die Verwendung stabiler Isotope zur Charakterisierung von Honigen, ihrer Herkunft und ihrer Verfalschung. *Apidologie*, 8: 337—347.

Ziegler, H., Maurizio, A. and Stichler, W., 1979. Die Charakterisierung von Honigen nach ihrem Gehalt an Pollen und an stabilen Isotopen. *Apidologie*, 10: 301—311.

Zimen, K.E., 1972. The future CO_2 burden of the atmosphere and ^{14}C in the ethanol from wines. In: *Proc. 8th Int. Conf., Radiocarbon Dating, Wellington 1972*, pp. A-85, A-91.

SUBJECT INDEX

acetic acid, 523
activity of water, isotope effects, 125
adsorption, cesium-137, 173
adulteration of food, 516, 519, 536, 538
age
   distribution of water ages, 9
   sediments, 177, 180, 186
alcohol, 520, 538
aliphatic hydrocarbons in sediments, 193
amount effect, 88
analyses — *see* measurements
animal fat, carbon-13, 523
aqueous carbonate, 216
aquifers
   Carizzo aquifer, U.S.A., 453
   Foxhill aquifer, U.S.A., 450
   Great Artesian Basin, Australia, 444
   Milk River aquifer, Canada, 445
   stagnant zones, 28
aragonite, 208, 217, 234, 273
argon isotopes, 481
   ice cores, 496
   production, 495
archaeometry, 284
aromatic hydrocarbon in sediments, 193
atmospheric carbon dioxide, carbon-13, 211
atmospheric chlorine-36, 428
atmospheric nitrogen, 364
atmospheric noble gases, 481
atmospheric vapour
   over lakes, 132
   over oceans, 80
   sampling, 85

bacteria, nitrogen fixation, 385, 390
basalt, chlorine-36, 438
boundary layer, ocean surface, 80
brine
   chlorine-36, 457
   deuterium and oxygen-18, 314
   formation fluids, 306
burial diagenesis, 350

calcite
   -dolomite solid solution, 209
   precipitation, 208, 251, 287
   saturation, 210, 233, 247
calcrete, 241, 261
caliche, 436
carbon dioxide
   atmospheric, 211, 242, 540
   breath, 542
   decarbonation of carbonate, 213
   igneous, 213
   soil, 241
carbon-13
   alcohol, 538
   food, 516
   lake sediments, 188, 191
   plants, 510
   speleothems, 297
   soil carbonate, 257, 258
   travertines, 212
carbon-14
   alcohol, 538
   carbon-13 correction, 45, 275
   exchange with aquifer matrix, 31
   lake sediments, 186, 188, 191
   soil carbonate, 257, 259
   speleothems, 275
   thermonuclear, 191
   tracer in hydrology, 29, 43
   travertines, 225
   wine, 540
carbonate
   biogenic, 240
   cement, 335
   closed system dissolution, 247, 249
   diagenesis, 239
   geochemistry, 241, 245, 272, 286
   incongruent dissolution, 250
   marine, 312

open system dissolution, 246
soil carbonate, 243, 256, 259
speleothems, 207, 271
travertine, 207
weathering, 239
Carboniferous coal, nitrogen-15, 389
caves, 271
cesium-137, 172
chert, 312, 335, 341
chlorine-36
dating, 440
global inventory, 432
groundwaters, 431
lithosphere, 434
meteoric, 428
production, 428, 432, 438
chlorophyll, oxygen-18, 526
climate, krypton-85, 491
clouds
cloud types, isotopic composition, 71, 90
physics, 61, 68
water content, 101
coal
carbon-13, 522
nitrogen-15, 389
condensation of moisture, 66
connate waters, 305, 319, 328
corn oil, 524
Cretaceous
coal, nitrogen-15, 389
connate water, 313
crystallization, water of, 130

dating
groundwater, argon-39, 498
chlorine-36, 440
carbon-14, 45, 275
krypton isotopes, 492
lake sediments, 186, 440
speleothems, 276
travertines, 225
decay constant, *see* half-life
Deep Sea Drilling Program, 340
denitrification, 401
deuterium
fluid-inclusion in speleothems, 292
food, 525
hail, 95
precipitations, 79, 87
see also brines
deuterium excess, 82, 308

diagenesis
burial, 350
travertines, 222, 233
diagenetic water, 307
diffusion
coefficients, 62, 65, 147
eddy diffusion in lakes, 146, 156
gaseous, 52
loss of tracer from water, 51
krypton-85 in phreatic aquifers, 492
mixing of isotopes in sediment, 196
molecular, 5, 22, 65, 69, 84
radiocarbon in water, 2
soil carbon dioxide, 243
dispersion, 5, 17, 22, 46
dispersivity, 22, 473
dolomite, solid solution with calcite, 209

ecosystem, 171
electron spin resonance, 281
erosion rate, 436, 498
evaporation
in caves, 287
geothermal fluids, 344
hail stones, 96
isotope enrichment, 114
meteoric water, 527
nitrogen compounds, 405
oceans, 80, 83
pans, water budgets, 138
precipitations, 66, 84, 87
evapotranspiration, 34, 83, 246, 251, 529, 534
exchange, isotope
deuterium, clay-water, 339, 340
granodiorite-water, 336
H_2S-water, 348
H_2-water, 349
rock-water, 333
oxygen-18, clay-water, 324
carbon dioxide-water, 162, 348
granodiorite-water, 336
rock-water, 328, 333, 336
exchange of tracer with matrix, 30
exponential model, 5, 38, 42

fallout
carbon-14, 191
cesium-137, 173
chlorine-36, 428, 460, 474
lead-210, 175
krypton-85, 483

nitrogen-15, 369
fat, animal, 523
fluid-inclusions, speleothems, 291
flowstone, 273
foraminifera, 297
forest, carbon-13, 513
forest soil, nitrogen-15, 366
formation waters, 306, 315
fractionation factors, equilibrium
  deuterium, ice-vapour, 65, 69
    ice-water, 65
    water-biotite, 341
    water-chert, 341
    water-gibbsite, 339, 341
    water-glauconite, 339
    water-hornblende, 341
    water-illite, 339
    water-kaolinite, 339, 341
    water-montmorillonite, 341
    water-muscovite, 341
    water-smectite, 339
  carbon-13, CO_2-$CO_{2\ aq}$, 215, 248, 509
    CO_2-H_2CO_3, 248
    CO_2-HCO_3, 215, 248, 509
    CO_2-CO_3, 215
    CO_2-calcite, 215
    HCO_3-CO_3, 248
    HCO_3-calcite, 248
  nitrogen-15, NH_3-NH_4, 404
  oxygen-18, calcite-phosphate, 338
    calcite-CO_2, 217
    feldspar-plagioclase, 338
    vapour-ice, 65, 69
    water-anhydrite, 338
    water-calcite, 217, 248, 286, 337
    water-gibbsite, 339
    water-glauconite, 339
    water-ice, 65
    water-illite, 338, 339
    water-kaolinite, 339
    water-quartz, 338
    water-smectite, 338, 339
    water-sulphate, 338
fractionation factor, kinetic
  calcite-water, 217
  CO_2-$CO_{2\ aq}$, carbon-13, oxygen-18, 218
  CO_2 diffusion, 243
  denitrification, nitrogen-15, 395
  nitrification, nitrogen-15, 395
  vapour-water, deuterium and oxygen-18, 122

fruit juices, 520, 532
fumaroles, 314

gas exchange
  atmosphere-lakes, 161
geothermal waters, 314
  boiling and condensation, 344
  evaporation, 344
geothermometry, 337, 440
groundwater dating
  argon-39, 498
  chlorine-36, 440
  krypton isotopes, 492
groundwater recharge, 24
  chlorine-36, 460, 465, 473
groundwater, saline, 467
graupels, 77
growth rate, speleothems, 274

hail
  clouds, 73, 76
  stones, 71, 74, 77, 89
    growth temperature, 92
    stable isotopes, 95
    trajectories, 97
  storms, 78
  suppression, 102
half-life
  argon-37, argon-39, 481
  beryllium-7, 200
  bismuth-210, 178
  carbon-14, 31, 54
  cesium-137, 171
  chlorine-36, 54, 427
  krypton-81, 54, 436
  krypton-85, 481
  lead-210, 172
  polonium-210, 178
  radium-226, 172
  radon-222, 172
  silicon-32, 31
  tritium, 31
  xenon-133, 481
helium-3, 40
Holocene
  climate, 310
  meteoric water, 314, 317
honey, 519
hot springs, 215, 279, 305
hurricane, 86
hydration of ions, 127
hydrogen isotope shift, 315

ice crystals, 69
ice core, 83, 462, 496, 498
infiltration
    groundwater, 24, 52
    loss from lakes, 134, 140
input function
    carbon-14, 43, 45
    krypton-85, 42
    tritium, 35, 41
interconnections, lake-groundwater, 161
interstitial water, 311
inverse problem, 2, 19, 26, 35
ion exchange
    nitrogen-15, 381
    tracer experiments, 4
ion filtration, chlorine-36, 446, 452, 454
isotope fractionation, *see* fractionation factors
isotope effects
    activity of water, 125
    biologocal, nitrogen-15, 383, 396, 401
    carbon dioxide diffusion, 243
    ion exchange, nitrogen-15, 381
    membrane filtration, 346

juvenile water, 314

karst, 24, 51, 271
krypton isotopes, 476
krypton-81, 493
krypton-85
    production, 482
    tracer, 42

lakes, 114
    constant volumes, 133
    gas exchange with atmosphere, 162
    mixing, 146
    radon-222, 156
    saline, 144
    salt balance, 134
    sediments, 170
    terminal, 132
lead-210, 172
    decay scheme, 178
    detrital particles, 177
lipids, in sediments, 192
loess, 258

magmatic water, 314
maple syrup, 515

measurement
    argon isotopes, 499
    bismuth-214, 173
    cesium-137, 173
    chlorine-36, 427
    helium-3, 40
    krypton-85, 43, 484
    lead-210, 173, 179
    organic carbon in sediments, 191
    polonium-208, 173
    radium-226, 173
    thorium isotopes, 280
    uranium isotopes, 280
membrane filtration, 346
metamorphic water, 307
meteoric water line, 80, 87, 95, 308
methane, atmospheric, 79
mixing
    groundwaters, 456
    lakes, 146
    meteoric-connate waters, 328, 331
    sediments, 194
models
    mathematical, in hydrology, 2—46
    sediment deposition, 180
    sediment mixing, 195
molecular diffusion, 5, 22, 65, 69, 84

nitrification, 395
nitrogen-15
    fertilizer, 373, 394
    precipitations, 371
    soils, 365, 407
    see also fractionation factor
nitrogen assimilation, 383
Northern Hemisphere
    krypton-85, 487
    tritium, 541

ocean water
    isotope variations in time, 310
    Pacific, 319
    Pleistocene, 289
    Precambrian, 313
oceanic crust, 312
oceanic ridges, 313
oil field water, 306
oil
    edible, 524, 536
    petroleum, 522, 518
organic carbon
    food, 522

lake sediments, 191
plants, 528
output concentrations (tracer)
   carbon-14, 47
   krypton-85, 44
   tritium, 36
oxygen-18
   carbon dioxide, 213
   cherts, 312
   fluid inclusions in speleothems, 295
   hail, 95
   quartz, 335
   see also carbonate; isotope fractionation
oxygen isotope shift, 315

paleoclimate, 84, 289
paleomagnetic stratigraphy, 282
paleo-sea level, 282
paleo-temperatures, speleothems, 288
paleo-waters, 82, 289, 310
Paleozoicum, formation waters, 328
pedosphere, nitrogen-15, 365
particulate matter, lead-210, 177
petroleum, 522
photosynthesis, 242, 261, 298, 508, 526, 531
plant components
   carbon-13, 511, 520, 523, 530
   deuterium, 525
   nitrogen-15, 386, 393
   oxygen-18, 525
Pleistocene
   epoch, 284, 289, 296
   meteoric water, 317
plutonium-239, -240, 190, 195
Precambrium, ocean water, 313
precipitation
   amount effect, 89
   deuterium-oxygen-18, 80, 87, 95, 308, 310
   lead-210, 176
   nitrogen-15, 369
   tritium, 43, 541
   see also hail
protein, 523

radio-carbon, see carbon-14
radium-226, 172
radon-222, 156, 172
rain
   drops, 66, 77, 87

storms, 88
see also precipitation
Rayleigh distillation
   carbonate precipitation, 251
   in cloud processes, 72, 75, 79, 83, 86
recrystallization
   speleothems, 274
   travertines, 222, 233
residence times
   tracer in lake water, 197
   water age, distribution, 9, 19

salt
   chlorine-36, 427, 460
   in groundwater, 467
   salt balance of lakes, 134
   see also brine
sampling
   hail stones, 91
   precipitations, 85
   sediments, 172, 186
   water vapour, 85
saturation index, calcite, 210, 233, 247
saturated zone, 27
sebkhas, 144
secular equilibrium
   radon-radium, 174
   chlorine-36, 442
sediment flux, 199
sedimentation rate, 178, 194
sinter, siliceous, 228
soils
   carbonate, 243, 256
   carbon dioxide, 241
   lead-210, 176
   nitrogen-15, 365, 407
Southern Hemisphere
   chlorine-36, 474
   krypton-85, 487
   tritium, 541
spallation, chlorine-36, 428, 434
speleothems, 207, 271
stratosphere
   argon-37, argon-39, 494
   chlorine-36, 429, 461
   krypton-85, 487, 489
   tritium, 75
strontium
   isotopes, in travertines, 227
   trace element, 232
sugar, 514, 528, 536
sulphur-34, in travertines, 226

thermoluminescence, 281
thorium isotopes, 276
tortuosity, 22
trace elements
   in lake sediments, 189
   in carbonates, 208, 222
tracer for groundwater flow
   adsorption, 30
   chlorine-36, 460
   general considerations, 3
tracer for sedimentation, 173, 174
transit time
   distribution, 19
   mean transit time, 9, 12
transpiration, from lakes, 134, 142
travertine
   carbon-13, oxygen-18, 211
   deposition, 210
   diagenesis, 222, 233
   lead isotopes, 228
   strontium isotopes, 227
   sulphur-34, 226
tritium
   clouds, 75
   exchange, lakes-atmosphere, 162
   diffusion coefficients, 147
   hail, 93
   helium-3 method, 40, 154
   precipitation, 34, 541
   tracer, 4, 24, 27, 33
   wine, 540

tropopause, krypton-85, 487
troposphere
   argon isotopes, 494
   deuterium, oxygen-18, 75, 85
   lead-210, 175
   krypton isotopes, 86, 487
tufa, 209
turnover time, 8, 39

unsaturated zone, 24, 27, 29
unsupported lead, 172, 177
uranium
   activity ratio, 279
   decay series, 175, 277
   groundwater, 277
   isotopes, 276

vanilla, 520
velocity
   fracture, 29
   groundwater, mean transit, 6
   tracer, 29
vinegar, 523

water budget of lakes, 130
weathering, carbonates, 239, 256
weighting function, 13
Wiesenkalk, 261
Wisconsin glacial stage, 294, 297

xenon-133, 501

INDEX OF GEOGRAPHICAL NAMES

Afar, 225
Africa, 138, 147, 152, 161, 296
Alaska, 310, 317
Algeria, 137, 262
Antarctica, 83, 160, 389
Atlantic Ocean, 486, 492
Australia, 260, 444
Austria, 38, 136, 160, 258
Bahamas, 280
Belgium, 186
    Ardennes, 366
Bermuda, 260, 290, 293
Bolivia, 138
Botswana, 142
Brazil, 138, 468, 472

Canada, 173, 243, 328, 347, 367, 464
    Alberta, 97, 102, 290, 347, 431, 445
    British Columbia, 296
    Ontario, 435, 464
    Saskatchewan, 406
Chad, 147
Cyprus, 262
Czechoslovakia, 213, 215, 258

Egypt, Wadi El-Natrum, 414
England, 279, 283, 295, 298
Europe, 83, 211, 259, 296, 369, 430

Finland, 190
France, 96, 103, 256, 259, 291, 418, 488
    Brittany, 371
    Massive Central, 96, 103
    Melarchez Basin, 417
    Yerres River, 417

Germany, Democratic Republic, 371
Germany, Federal Republic, 211, 212, 215, 242, 259, 368, 369, 484, 501
Greart Britain, 405
Greece, 262
Greenland, 462, 497

Gulf Coast, 313, 320, 342, 347
Gulf of California, 197
Gulf of Mexico, 305, 313, 324, 350, 376

Hungary, 484

India, 262
Iran, 213
Italy, 212, 215, 262
    Latium, 213, 219, 228
    Tivoli, 207
    Tuscany, 220

Japan, 371

lakes
    Alkali Lake, U.S.A., 144
    Asal, Lake, Djibouti, 145
    Beyschir, Lake, Turkey, 137
    Biwa, Lake, Japan, 186
    Bracciano, Lake, Italy, 135, 161
    Burdur, Lake, Turkey, 137
    Canadurago, U.S.A., 190
    Chad, Lake, Africa, 133, 147, 161
    Chala, Lake, Africa, 144, 161
    Champlaon, U.S.A., 190
    Constance, Lake, Europe, 157, 162
    Crater Lake, Oregon, U.S.A., 151, 162
    Egridir, Lake, Turkey, 137, 161
    Erie, Lake, North America, 154, 194
    Fjndim, Sebka, Israel, 145
    Galilea, Sea of, Middle East, 123, 138, 143, 160, 162
    Gara Dibah, Guelta, Algeria, 139
    Geneva, Lake, Europe, 132, 150
    Great Bear Lake, Canada, 439
    Great Lakes, North America, 472
    Great Salt Lake, Utah, U.S.A., 435
    Green Lake, New York, U.S.A., 155
    Greifensee, Switzerland, 156
    G'vaoth, Sebkha, Israel, 145
    Hakone Caldera, Japan, 143

Huron, Lake, North America, 154, 197
Kainji, Lake, Nigeria, 149
Kinneret, Lake, Israel, 123, 138, 143, 160, 162
Lenore, Lake, Washington, U.S.A., 144
Lower Grand Coulee, Lakes, U.S.A., 144
Malawi, Lake, Africa, 152, 162
Melah, Sebkha el, Algeria, 139
Michigan, Lake, North America, 173, 185, 186
Mono Lake, U.S.A., 162, 201, 435, 440
Neusiedl, Lake, Europe, 136, 160
Okawango swamps, Botswana, 142
Ontario, Lake, North America, 154, 194, 472
Owens Lake, California, U.S.A., 145
Perch Lake, Ontario, Canada, 133
Piccolo, Lake, Italy, 160
Pitt Lake, Canada, 174
Pyramid Lake, U.S.A., 162, 435
Quebra Unhas, Brasil, 138
Ram Lake, Israel, 161
Sammamish, Lake, U.S.A., 198
Searles Lake, Washington, U.S.A., 440
Schwerin, Lake, G.D.R., 143
Shinji, Lake, Japan, 186
Sylvan, Lake, U.S.A., 190
Tahoe, Lake, U.S.A., 133, 151, 162, 185, 366
Tanganyika, Lake, Africa, 152, 162
Tiberias, Lake, Israel, 123, 138, 143, 160, 162
Titicaca, Lake, South America, 143, 185.
Vanda, Lake, Antarctica, 160
Waidsee, F.R.G., 140
Washington, Lake, U.S.A., 181, 186, 195
Wiesensee, F.R.G., 140
Windermere, Lake, U.K., 174
Lybia, 262

Mediterranean, 259, 308
Mexico, 288, 290
 Baja California, 185
Middle East, 296

Namibia, 414

Netherlands, The, 256, 258
New Zealand, 291
North America, 89, 308, 369
 Rocky Mountains, 283

Oman, 313

Pacific Ocean, 86, 317
Poland, 331

Rivers,
 Chari River, Chad, 147
 Cripple Creek, Colorado, U.S.A., 435, 440
 Dranse River, France, 151
 Koprucay River, Turkey, 161
 Milk River, Canada, 445
 Mississippi River, 343
 Niger River, Africa, 149
 Rhine, Europe, 158
 Rhône, Europe, 151
 Yerres River, France, 417

Sahara, 262
South Africa, 262, 291
 Kalahari, 262
Spain, 262, 431, 472
Switzerland, 78, 102, 496
Syria, 256

Tunesia, 262
Turkey, 435

U.S.S.R., 331
United States of America, 86, 89, 94, 101, 484
 Arkansas, 439, 435
 Arizona, 431, 472
 California, 185, 209, 211, 305, 329, 347, 435, 472
  Central Valley, 340
  Imperial Valley, 350
  Sacramento Valley, 340
  see also lakes
 Colorado, 101, 104, 435
 Florida, 243
 Illinois, 347
  Illinois Basin, 317
 Iowa, 290, 293, 365
 Kansas, 101
 Kentucky, 290, 293
 Louisiana, 320, 457
 Maine, 366

Michigan, 347
    Michigan Basin, 317
Missouri, 295
Montana, 445
    Yellowstone Park, 215, 220, 228
Nebraska, 413
Nevada, 431, 472
New Mexico, 431, 465
New York State, 190, 295, 501
    Long Island, 413
North Dakota, 450
Ohio, 102
Oklahoma, 94

Texas, 259, 262, 320, 334, 352, 431, 453, 472
Utah, 435
Washington, 187, 189
Wyoming, 220
Vermont, 472
Virginia, 288, 290, 293

Yugoslavia, 213
    Plitivice National Park, 225

Zaire, 439